# A BRIEF COURSE IN MATHEMATICAL STATISTICS

Elliot A. Tanis
*Hope College*

Robert V. Hogg
*University of Iowa*

PEARSON
Prentice
Hall

Upper Saddle River, NJ 07458

**Library of Congress Cataloging-in-Publication Data**

Tanis, Elliot A.
  A brief course in mathematical statistics / Elliot A. Tanis, Robert V.
Hogg.
     p. cm.
  Includes bibliographical references and index.
  ISBN 0-13-175139-5
  1. Mathematical statistics—Textbooks. I. Hogg, Robert V. II. Title.
  QA276.12.T359 2008
  519.5—dc22

                                                                2006031389

Editor-in-Chief: *Sally Yagan*
Executive Acquisition Editor: *Petra Recter*
Project Manager: *Michael Bell*
Production Management/Composition: *Laserwords Private Limited*
Senior Managing Editor: *Linda Mihatov Behrens*
Assistant Managing Editor: *Bayani Mendoza de Leon*
Executive Managing Editor: *Kathleen Schiaparelli*
Manufacturing Manager: *Alexis Heydt-Long*
Manufacturing Buyer: *Maura Zaldivar*
Marketing Manager: *Wayne Parkins*
Marketing Assistant: *Jennifer Leeuwerk*
Director of Creative Services: *Paul Belfanti*
Creative Director: *Juan R. López*
Art Director: *Maureen Eide*
Interior/Cover Designer: *Kristine Carney*
Art Editor: *Thomas Benfatti*
Editorial Assistant/Supplements Editor: *Joanne Wendelken*
Art Studio: *Laserwords Private Limited*
Cover Image: *Painting, "Toss-Up." ©Joel Schoon Tanis*

© 2008 by Prentice Hall, Inc.
Pearson Prentice Hall
Pearson Education, Inc.
Upper Saddle River, NJ 07458

Pearson Prentice Hall™ is a trademark of Pearson Education, Inc.

Printed in the United States of America

ISBN: 0-13-175139-5

Pearson Education, Ltd., *London*
Pearson Education Australia PTY. Limited, *Sydney*
Pearson Education Singapore, Pte., Ltd
Pearson Education North Asia Ltd, *Hong Kong*
Pearson Education Canada, Ltd., *Toronto*
Pearson Education de Mexico, S.A. de C.V.
Pearson Education—Japan, *Tokyo*
Pearson Education Malaysia, Pte. Ltd

# CONTENTS

# PREFACE

Many statisticians have mentioned the need of having a less expensive textbook for a one-semester course which, without so much probability, introduces important statistical concepts early. With a little algebra of sets and a standard course in calculus as the mathematical background, we do provide some basic probability in Chapter 1 that has been shortened somewhat from the usual presentation. We then study certain discrete distributions in Chapter 2. Included are the topics of expectations, maximum likelihood estimation, as well as expectations and variances of linear functions, in particular those of the sample mean; and this results in the introduction of confidence intervals for means of distributions. The introduction of estimation this early is really the main innovative feature of this textbook as it is not done this soon in most other books of this level. Also included in Chapter 2 is a section on multivariate discrete distributions. The third chapter concerns the continuous case and corresponding estimation problems. We summarize many of the important discrete and continuous distributions with their characteristics in the end pages at the front of this book.

Chapter 4 includes some standard statistical inferences. Tests of statistical hypotheses and confidence intervals are tied together throughout. Material on linear regression is included. The section on distribution-free confidence intervals for percentiles also includes the topic of tolerance intervals. The last two sections concern chi-square tests. We summarize confidence intervals and tests of hypotheses in the end pages at the back of this book.

Most instructors agree that almost all of the material in Chapters 1-3 and Sections 4.1–4.6 should be included in any one-semester course in elementary mathematical statistics. To complete the course, selection of topics can be made from Sections 4.7–4.11 and Chapters 5 and 6.

One of the authors (Tanis) has used the computer successfully with *MINITAB* for data analysis and *Maple* for certain mathematical operations and simulations. Clearly other computer packages and other computer algebra systems (CAS) could be used. We have devoted Chapter 5 to a discussion of some uses of the computer both for data analysis and also for theoretical solutions such as simulation and bootstrapping. This material can be used earlier in the text and references are incorporated throughout the first chapters.

The other author (Hogg) tends to like a more theoretical approach and introduces the moment-generating function from Chapter 6. This allows the student to see how certain important results in the theory are proved. It is noted by Remarks in the text that certain sections of Chapter 6 can be studied as early as Section 2.2.

Clearly the authors appreciate any feedback from instructors and students; in particular we like to know about typos and similar errors. While we believe the text to be almost error free, some errors always slip in.

All of the data in this text are available in a form that allows you to use them for data analysis with a variety of statistical packages. The data are stored by sections as ASCII files and as Minitab worksheets. The data are also stored in one folder named "R data" as ASCII files. These are in a format that was recommended by a statistician who uses R. The data are also stored by Chapter in a format used by *Maple*. To access these data go to www.prenhall.com/statistics and locate *A Brief Course in Mathematical Statistics*.

Several *Maple* applications use procedures that were written by Zaven Karian. These are stored as `stat.m` along with some other supplementary *Maple* procedures in a folder called "Maple Examples" at www.prenhall.com/statistics (*A Brief Course in Mathematical Statistics*). Short descriptions of these procedures are on the `Maple Card`, a pdf file. More extensive descriptions of the procedures are given in the laboratory manual, *Probability and Statistics: Explorations with MAPLE*, by Zaven Karian and Elliot Tanis.

To read the supplementary procedures, you must specify where they are. Copy the "Maple Examples" folder from the web site into a new folder, named "Tanis-Hogg" on your C drive. *Maple* should first be loaded. Then open the file named "Menu" in "Maple Examples." From this file, you can access the supplementary procedures and many of the examples and exercises that use *Maple*.

One of the authors (Hogg) has been involved in three books on mathematical statistics published by Prentice Hall. Let us make a few remarks to help the instructor decide which is best for the students. The present book is the most elementary and is intended for a one-semester course which needs only one year of calculus as a background (even though partial derivatives appear two times). It is a sophomore level course. The 7th edition of *Probability and Statistical Inference* (2006) by Hogg and Tanis is for a good two-semester course in probability and mathematical statistics and really requires multivariate calculus and hence is about a junior level course. The 6th edition of *Introduction to Mathematical Statistics* (2005) by Hogg, McKean, and Craig is somewhat more advanced, primarily for first-year graduate students (or the better seniors). It is helpful if the students have a good strong background in calculus and even some mathematical analysis (theoretical advanced calculus) plus some linear algebra.

Once again we want to thank our colleagues and reviewers who have made valuable suggestions. In particular, Cathy Mader, Chairperson of the Hope College Physics Department, has been extremely generous of her time solving LaTeX problems. Our very good friend, James Broffitt, Professor of Statistics and Actuarial Science at the University of Iowa, used a preliminary manuscript of this text and made excellent suggestions for improvement. We thank our accuracy checker, Joe Kupresanin. Comments and suggestions for improvement from the following reviewers were very helpful: Javier Rogo, Qin Shao, M. L. Aggarwal, Charles Sommer, and Ishwar Basawa. We also thank the University of Iowa and Hope College for providing office space and encouragement. Of course, the support and love of our wives, Elaine and Ann, has been absolutely necessary and truly appreciated.

E.A.T.
tanis@hope.edu
R.V.H.

# CHAPTER

# 1

# PROBABILITY

 **BASIC CONCEPTS**

It is usually difficult to explain to the general public what statisticians do. Many think of us as "math nerds" who seem to enjoy dealing with numbers. And there is some truth to that concept. But if we consider the bigger picture, many recognize that statisticians can be extremely helpful in many investigations.

Consider the following:

1. There is some problem or situation that needs to be considered; so statisticians are often asked to work with investigators or research scientists.
2. Suppose that some measure (or measures) are needed to help us understand the situation better. The measurement problem is often extremely difficult, and creating good measures is a valuable skill. For illustration, in higher education, how do we measure good teaching? This is a question to which we have not found a satisfactory answer, although several measures, such as student evaluations, have been used in the past.
3. After the measuring instrument has been developed, we must collect data through observation, possibly the results of a survey or an experiment.
4. Using these data, statisticians summarize the results, often using descriptive statistics and graphical methods.
5. These summaries are then used to analyze the situation. Here it is possible that statisticians make what are called *statistical inferences*.
6. Finally a report is presented, along with some recommendations that are based upon the data and the analysis of them. Frequently such a recommendation might be to perform the survey or experiment again, possibly changing some of the questions or factors involved. This is how statistics is used in what is referred to as the scientific method, because often the analysis of

the data suggests other experiments. Accordingly the scientist must consider different possibilities in his or her search for an answer and thus performs similar experiments over and over again.

The discipline of statistics deals with the *collection* and *analysis of data*. When measurements are taken, even seemingly under the same conditions, the results usually vary. Despite this variability, a statistician tries to find a pattern; yet due to the "noise," not all of the data lie on the pattern. In the face of this variability, he or she must still determine the best way to describe the pattern. Accordingly, statisticians know that mistakes will be made in data analysis, and they try to minimize those errors as much as possible and then give bounds on the possible errors. By considering these bounds, decision makers can decide how much confidence they want to place on these data and the analysis of them. If the bounds are wide, perhaps more data should be collected. If they are small, however, the person involved in the study might want to make a decision and proceed accordingly.

Variability is a fact of life, and proper statistical methods can help us understand data collected under inherent variability. Because of this variability, many decisions have to be made that involve uncertainties. In medical research, interest may center on the effectiveness of a new vaccine for mumps; an agronomist must decide if an increase in yield can be attributed to a new strain of wheat; a meteorologist is interested in predicting the probability of rain; the state legislature must decide whether increasing speed limits will result in more accidents; the admissions officer of a college must predict the college performance of an incoming freshman; a biologist is interested in estimating the clutch size for a particular type of bird; an economist desires to estimate the unemployment rate; an environmentalist tests whether new controls have resulted in a reduction in pollution.

In reviewing the preceding (relatively short) list of possible areas of applications of statistics, the reader should recognize that good statistics is closely associated with careful thinking in many investigations. For illustration, students should appreciate how statistics is used in the endless cycle of the scientific method. We observe nature and ask questions, we run experiments and collect data that shed light on these questions, we analyze the data and compare the results of the analysis to what we previously thought, we raise new questions, and on and on. Or if you like, statistics is clearly part of the important "plan–do–study–act" cycle: Questions are raised and investigations planned and carried out. The resulting data are studied and analyzed and then acted upon, often raising new questions.

There are many aspects of statistics. Some people get interested in the subject by collecting data and trying to make sense out of these observations. In some cases the answers are obvious and little training in statistical methods is necessary. But if a person goes very far in many investigations, he or she soon realizes that there is a need for some theory to help describe the error structure associated with the various estimates of the patterns. That is, at some point appropriate probability and mathematical models are required to make sense out of complicated data sets. Statistics and the probabilistic foundation on which statistical methods are based can provide the models to help people do this. So in this book, we are more concerned with the mathematical, rather than the applied, aspects of statistics. Still we give enough real examples so that the reader can get a good sense of a number of important applications of statistical methods.

In the study of statistics, we consider experiments for which the outcome cannot be predicted with certainty. Such experiments are called **random experiments**. Each experiment ends in an outcome that cannot be determined with certainty before the experiment is performed. However, the experiment is such that the collection of every possible outcome can be described and perhaps listed. This collection of all outcomes is called the **outcome space**, and is denoted by $O$. The following examples will help illustrate what we mean by random experiments, outcomes, and their associated outcome spaces.

**EXAMPLE 1.1-1**   Roll one die (one of a pair of dice). Let the outcome be the number of spots on the side that is "up." The outcome space is $O = \{1, 2, 3, 4, 5, 6\}$.

**EXAMPLE 1.1-2**   Flip a coin at random and let the outcome space be $O = \{\text{heads, tails}\}$.

**EXAMPLE 1.1-3**   Draw at random a card from an ordinary deck of cards. The outcome space is the set of 52 possible resulting cards, that is, $O = \{\text{Ace of Spades, King of Spades, ...,} \text{Two of Clubs}\}$.

**EXAMPLE 1.1-4**   A box of breakfast cereal contains one of four different prizes. The purchase of one box of cereal yields one of the prizes as the outcome, and the outcome space is the set of four different prizes.

**EXAMPLE 1.1-5**   To determine the percentage of body fat for a person, one measurement that is made is a person's weight under water. If $w$ denotes this weight in kilograms, then the outcome space could be $O = \{w: 0 < w < 7\}$, as we know from past experience that this weight does not exceed 7 kilograms.

**EXAMPLE 1.1-6**   An ornithologist is interested in the clutch size (number of eggs in a nest) for gallinules, a species of bird that lives in a marsh. If we let $c$ equal the clutch size then a possible outcome space would be $O = \{c: c = 0, 1, 2, \ldots, 15\}$, as 15 is the largest known clutch size.

So the collection of all possible outcomes (the *universal set*) of a random experiment is denoted by $O$ and is called the *outcome space*. Given an outcome space $O$, let $A$ be a part of the collection of outcomes in $O$, that is, $A \subset O$. Then $A$ is called an **event**. When the random experiment is performed and the outcome of the experiment is in $A$, we say that **event $A$ has occurred**.

Since, in studying probability, the words *set* and *event* are interchangeable, the reader might want to review **algebra of sets**. For convenience, however, here we remind the reader of some terminology:

- $\emptyset$ denotes the **null** or **empty** set;
- $A \subset B$ means $A$ is a **subset** of $B$;
- $A \cup B$ is the **union** of $A$ and $B$;
- $A \cap B$ is the **intersection** of $A$ and $B$;
- $A'$ is the **complement** of $A$ (i.e., all elements in $O$ that are not in $A$).

Some of these sets are depicted by the shaded regions in Figure 1.1-1, in which $O$ is the interior of each of the rectangles. Such figures are called Venn diagrams.

Special terminology associated with events that is often used by statisticians includes the following:

1. $A_1, A_2, \ldots, A_k$ are **mutually exclusive events** means that $A_i \cap A_j = \emptyset, i \neq j$, that is, $A_1, A_2, \ldots, A_k$ are disjoint sets;
2. $A_1, A_2, \ldots, A_k$ are **exhaustive events** means $A_1 \cup A_2 \cup \cdots \cup A_k = O$.

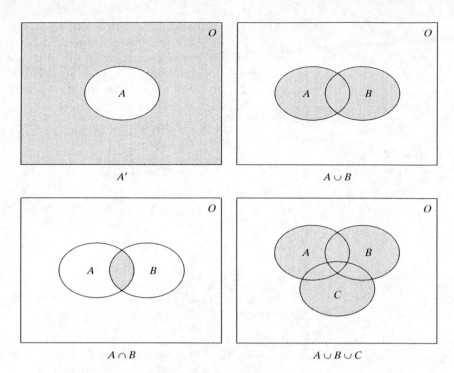

**Figure 1.1-1** Algebra of sets

So if $A_1, A_2, \ldots, A_k$ are **mutually exclusive and exhaustive** events, we know that $A_i \cap A_j = \emptyset, i \neq j$, and $A_1 \cup A_2 \cup \cdots \cup A_k = O$.

We are interested in defining what is meant by the probability of $A$, denoted by $P(A)$, and often called the chance of $A$ occurring. To help us understand what is meant by the probability of $A$, consider repeating the experiment a number of times, say $n$ times. Count the number of times that event $A$ actually occurred throughout these $n$ performances; this number is called the frequency of event $A$ and is denoted by $\mathcal{N}(A)$. The ratio $\mathcal{N}(A)/n$ is called the **relative frequency** of event $A$ in these $n$ repetitions of the experiment. A relative frequency is usually very unstable for small values of $n$, but it tends to stabilize as $n$ increases. This suggests that we associate with event $A$ a number, say $p$, that is equal to or approximately equal to the number about which the relative frequency tends to stabilize. This number $p$ can then be taken as the number that the relative frequency of event $A$ will be near in future performances of the experiment. Thus, although we cannot predict the outcome of a random experiment with certainty, we can, for a large value of $n$, predict fairly accurately the relative frequency associated with event $A$. The number $p$ assigned to event $A$ is called the **probability** of event $A$ and is denoted by $P(A)$. That is, $P(A)$ represents the proportion of outcomes of a random experiment that terminate in the event $A$ as the number of trials of that experiment increases without bound.

The following example will help to illustrate some of the ideas just presented.

EXAMPLE 1.1-7    A fair six-sided die is rolled six times. If the face numbered $k$ is the outcome on roll $k$ for $k = 1, 2, \ldots, 6$, we say that a match has occurred. The experiment is called a success if at least one match occurs during the six trials. Otherwise, the experiment is called a failure. The outcome space is $O = \{\text{success, failure}\}$. Let $A = \{\text{success}\}$. We would like to assign a value to $P(A)$. Accordingly, this

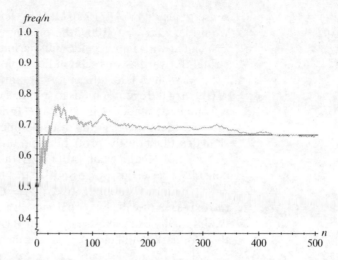

**Figure 1.1-2** Fraction of experiments having at least one match

experiment was simulated 500 times on a computer. Figure 1.1-2 depicts the results of this simulation, and the following table summarizes a few of the results:

| $n$ | $\mathcal{N}(A)$ | $\mathcal{N}(A)/n$ |
|---|---|---|
| 50 | 37 | 0.740 |
| 100 | 69 | 0.690 |
| 250 | 172 | 0.688 |
| 500 | 330 | 0.660 |

The probability of event $A$ is not intuitively obvious, but it will be shown in Example 1.4-6 that $P(A) = 1 - (1 - 1/6)^6 = 0.665$. This assignment is certainly supported by the simulation (although not proved by it).

Example 1.1-7 shows that at times intuition cannot be used to assign probabilities, although simulation can perhaps help to assign a probability empirically. The following example gives an illustration where intuition can help in assigning a probability to an event.

**EXAMPLE 1.1-8** A disk 2 inches in diameter is thrown at random on a tiled floor, where each tile is a square with sides 4 inches in length. Let $C$ be the event that the disk will land entirely on one tile. In order to assign a value to $P(C)$, consider the center of the disk. In what region must the center lie to assure that the disk lies entirely on one tile? If you draw a picture, it should be clear that the center must lie within a square having sides of length 2 and lying in the center of a tile. Since the area of this square is 4 and the area of a tile is 16, it makes sense to let $P(C) = 4/16$.

Sometimes the nature of an experiment is such that the probability of $A$ can be assigned easily. For example, when a state lottery randomly selects a three digit integer with outcome space $O = \{000, 001, 002, \ldots, 999\}$, we would expect each of the 1000 possible three-digit numbers to have the same chance of being selected, namely 1/1000. If we let $A = \{233, 323, 332\}$, then it makes sense to let $P(A) = 3/1000$. Or if we let $B = \{234, 243, 324, 342, 423, 432\}$, then we would let $P(B) = 6/1000$, the probability of the event $B$. Probabilities of events associated with many random experiments are perhaps not quite as obvious and straightforward as was seen in Example 1.1-7.

So we wish to associate with $A$ a number $P(A)$ about which the relative frequency $\mathcal{N}(A)/n$ of the event $A$ tends to stabilize with large $n$. A function such as $P(A)$, which is evaluated for a set A, is called a **set function**. In this section we consider the probability set function $P(A)$ and discuss some of its properties. In succeeding sections we shall describe how the probability set function is defined for particular experiments.

To help decide what properties the probability set function should satisfy, consider properties possessed by the relative frequency $\mathcal{N}(A)/n$. For example, $\mathcal{N}(A)/n$ is always nonnegative. If $A = O$, the outcome space, then the outcome of the experiment will always belong to $O$, and thus $\mathcal{N}(O)/n = 1$. Also, if $A$ and $B$ are two mutually exclusive events, then $\mathcal{N}(A \cup B)/n = \mathcal{N}(A)/n + \mathcal{N}(B)/n$. Hopefully, these remarks will help to motivate the following definition.

---

**Definition 1.1-1**

*Probability* is a real-valued set function $P$ that assigns to each event $A$ in the outcome space $O$ a number $P(A)$, called the probability of the event $A$, such that the following properties are satisfied:

**(a)** $P(A) \geq 0$,
**(b)** $P(O) = 1$,
**(c)** If $A_1, A_2, A_3, \ldots$ are events and $A_i \cap A_j = \emptyset, i \neq j$, then

$$P(A_1 \cup A_2 \cup \cdots \cup A_k) = P(A_1) + P(A_2) + \cdots + P(A_k)$$

for each positive integer $k$, and

$$P(A_1 \cup A_2 \cup A_3 \cup \cdots) = P(A_1) + P(A_2) + P(A_3) + \cdots$$

for an infinite, but countable, number of events.

---

The following theorems give some other important properties of the probability set function. When one considers these theorems, it is important to understand the theoretical concepts and proofs. However, if the reader keeps the relative frequency concept in mind, the theorems should also have some intuitive appeal.

---

**Theorem 1.1-1**

For each event $A$,

$$P(A) = 1 - P(A').$$

**Proof**   We have

$$O = A \cup A' \qquad \text{and} \qquad A \cap A' = \emptyset.$$

Thus, from properties **(b)** and **(c)**, it follows that

$$1 = P(A) + P(A').$$

Hence

$$P(A) = 1 - P(A').$$

EXAMPLE 1.1-9   A fair coin is flipped successively until the same face is observed on successive flips. If $x$ equals the number of flips needed, let $A = \{x : x = 3, 4, 5, \ldots\}$; that is, $A$ is the event that it will take three or more flips of the coin to observe the same face on two consecutive flips. To find $P(A)$, we first find the probability of $A' = \{x : x = 2\}$, the complement of $A$. In two flips of a coin, the possible outcomes are $\{HH, HT, TH, TT\}$, and we assume that each of these four outcomes has the same chance of being observed. Thus

$$P(A') = P(\{HH, TT\}) = \frac{2}{4}.$$

It follows from Theorem 1.1-1 that

$$P(A) = 1 - P(A') = 1 - \frac{2}{4} = \frac{2}{4}.$$

---

**Theorem 1.1-2**   $P(\emptyset) = 0.$

**Proof**   In Theorem 1.1-1, take $A = \emptyset$ so that $A' = O$. Thus

$$P(\emptyset) = 1 - P(O) = 1 - 1 = 0.$$

---

**Theorem 1.1-3**   If events $A$ and $B$ are such that $A \subset B$, then $P(A) \leq P(B)$.

**Proof**   Now

$$B = A \cup (B \cap A') \qquad \text{and} \qquad A \cap (B \cap A') = \emptyset.$$

Hence, from property (**c**),

$$P(B) = P(A) + P(B \cap A') \geq P(A)$$

because from property (**a**),

$$P(B \cap A') \geq 0.$$

---

**Theorem 1.1-4**   For each event $A$, $P(A) \leq 1$.

**Proof**   Since $A \subset O$, we have by Theorem 1.1-3 and property (**b**) that

$$P(A) \leq P(O) = 1,$$

which gives the desired result.

Property (**a**) along with Theorem 1.1-4 shows that, for each event $A$,

$$0 \leq P(A) \leq 1.$$

**Theorem 1.1-5**    If $A$ and $B$ are any two events, then

$$P(A \cup B) = P(A) + P(B) - P(A \cap B).$$

**Proof**    The event $A \cup B$ can be represented as a union of mutually exclusive events, namely,

$$A \cup B = A \cup (A' \cap B).$$

Hence, by property (**c**),

$$P(A \cup B) = P(A) + P(A' \cap B). \qquad (1.1\text{-}1)$$

However,

$$B = (A \cap B) \cup (A' \cap B),$$

which is a union of mutually exclusive events. Thus

$$P(B) = P(A \cap B) + P(A' \cap B)$$

and

$$P(A' \cap B) = P(B) - P(A \cap B).$$

If this result is substituted in Equation 1.1-1 we obtain

$$P(A \cup B) = P(A) + P(B) - P(A \cap B),$$

which is the desired result.

**EXAMPLE 1.1-10**    A faculty leader was meeting two students in Paris, one arriving by train from Amsterdam and the other arriving by train from Brussels at approximately the same time. Let $A$ and $B$ be the events that the trains are on time, respectively. Suppose from past experience we know that $P(A) = 0.93$, $P(B) = 0.89$, and $P(A \cap B) = 0.87$. Then

$$P(A \cup B) = P(A) + P(B) - P(A \cap B)$$

$$= 0.93 + 0.89 - 0.87 = 0.95$$

is the probability that at least one train is on time.

**Theorem 1.1-6**    If $A$, $B$, and $C$ are any three events, then

$$P(A \cup B \cup C) = P(A) + P(B) + P(C) - P(A \cap B)$$
$$- P(A \cap C) - P(B \cap C) + P(A \cap B \cap C).$$

**Proof**    Write

$$A \cup B \cup C = A \cup (B \cup C)$$

and then apply Theorem 1.1-5. The details are left as Exercise 1.1-12.

EXAMPLE 1.1-11 Continuing with Example 1.1-10, say that a third student is arriving from Cologne. Let $C$ be the event that this train is on time with $P(C) = 0.91$, $P(B \cap C) = 0.83$, $P(A \cap C) = 0.86$, and $P(A \cap B \cap C) = 0.81$. Then

$$P(A \cup B \cup C) = P(A) + P(B) + P(C) - P(A \cap B) - P(A \cap C)$$

$$- P(B \cap C) + P(A \cap B \cap C)$$

$$= 0.93 + 0.89 + 0.91 - 0.87 - 0.83 - 0.86 + 0.81$$

$$= 0.98$$

is the probability that at least one of these trains is on time.

Let a probability set function be defined on an outcome space $O$. Let $O = \{e_1, e_2, \ldots, e_m\}$, where each $e_i$ is a possible outcome of the experiment. The integer $m$ is called the total number of ways in which the random experiment can terminate. If each of these outcomes has the same probability of occurring, we say that the $m$ outcomes are **equally likely**. That is,

$$P(\{e_i\}) = \frac{1}{m}, \qquad i = 1, 2, \ldots, m.$$

If the number of outcomes in an event $A$ is $h$, the integer $h$ is called the number of ways that are favorable to the event $A$. In this case $P(A)$ is equal to the number of ways favorable to the event $A$ divided by the total number of ways in which the experiment can terminate. That is, under this assumption of equally likely outcomes, we have that

$$P(A) = \frac{h}{m} = \frac{N(A)}{N(O)},$$

where $h = N(A)$ is the number of ways $A$ can occur and $m = N(O)$ is the number of ways $O$ can occur. Exercise 1.1-11 considers this assignment of probability in a more theoretical manner.

It should be emphasized that in order to assign the probability $h/m$ to the event $A$, we must assume that each of the outcomes $e_1, e_2, \ldots, e_m$ has the same probability $1/m$. This assumption is then an important part of our probability model; if it is not realistic in an application, the probability of the event $A$ cannot be computed this way. Actually we have used this result in the simple case given in Example 1.1-9 because it seemed realistic to assume that each of the four possible outcomes in $\{HH, HT, TH, TT\}$ had the same chance of being observed.

EXAMPLE 1.1-12 Let a card be drawn at random from an ordinary deck of 52 playing cards. The outcome space $O$ is the set of $m = 52$ different cards, and it is reasonable to assume that each of these cards has the same probability for selection, 1/52. Accordingly, if $A$ is the set of outcomes that are kings, $P(A) = 4/52 = 1/13$ because there are $h = 4$ kings in the deck. That is, 1/13 is the probability of drawing a card that is a king provided that each of the 52 cards has the same probability of being drawn.

In Example 1.1-12, the computations are very easy because there is no difficulty in the determination of the appropriate values of $h$ and $m$. However, instead of drawing only one card, suppose that 13 are taken at random and without replacement. We can think of each possible 13-card hand as being an outcome in an outcome space, and it is reasonable to assume that each of these outcomes has the same probability. To use the above method to assign the probability of a hand, consisting of seven spades and six hearts, for illustration, we must be able to count the number $h$ of all such hands as well as the number $m$ of possible 13-card hands. In these more complicated situations, we need better methods of determining $h$ and $m$. We discuss some of these counting techniques in Section 1.2.

## EXERCISES 1.1

**1.1-1** Describe the outcome space for each of the following experiments:

**(a)** A student is selected at random from a statistics class, and the student's ACT score in mathematics is determined. HINT: ACT test scores in mathematics are integers between 1 and 36, inclusive.

**(b)** A candy bar with a 20.4-gram label weight is selected at random from a production line and is weighed.

**(c)** A coin is tossed three times, and the sequence of heads and tails is observed.

**1.1-2** A coin is tossed four times, and the sequence of heads and tails is observed.

**(a)** List each of the 16 sequences in the outcome space $O$.

**(b)** Let events $A$, $B$, $C$, and $D$ be given by $A = \{\text{at least 3 heads}\}$, $B = \{\text{at most 2 heads}\}$, $C = \{\text{heads on the third toss}\}$, and $D = \{\text{1 head and 3 tails}\}$. If the probability set function assigns 1/16 to each outcome in the outcome space, find **(i)** $P(A)$, **(ii)** $P(A \cap B)$, **(iii)** $P(B)$, **(iv)** $P(A \cap C)$, **(v)** $P(D)$, **(vi)** $P(A \cup C)$, and **(vii)** $P(B \cap D)$.

**1.1-3** Draw one card at random from a standard deck of cards. The outcome space $O$ is the collection of the 52 cards. Assume that the probability set function assigns 1/52 to each of these 52 outcomes. Let

$$A = \{x : x \text{ is a jack, queen, or king}\},$$

$$B = \{x : x \text{ is a 9, 10, or jack and } x \text{ is red}\},$$

$$C = \{x : x \text{ is a club}\},$$

$$D = \{x : x \text{ is a diamond, a heart, or a spade}\}.$$

Find **(a)** $P(A)$, **(b)** $P(A \cap B)$, **(c)** $P(A \cup B)$, **(d)** $P(C \cup D)$, **(e)** $P(C \cap D)$.

**1.1-4** If $P(A) = 0.4$, $P(B) = 0.5$, and $P(A \cap B) = 0.3$, find **(a)** $P(A \cup B)$, **(b)** $P(A \cap B')$, and **(c)** $P(A' \cup B')$.

**1.1-5** If $O = A \cup B$, $P(A) = 0.7$, and $P(B) = 0.9$, find $P(A \cap B)$.

**1.1-6** If $P(A) = 0.4$, $P(B) = 0.5$, and $P(A \cup B) = 0.7$, find

**(a)** $P(A \cap B)$.

**(b)** $P(A' \cup B')$.

**1.1-7** A typical roulette wheel used in a casino has 38 slots that are numbered $1, 2, 3, \ldots, 36, 0, 00$, respectively. The 0 and 00 slots are colored green. Half of the remaining slots are red and half are black. Also half of the integers between 1 and 36 inclusive are odd, half are even, and 0 and 00 are defined to be neither odd nor even. A ball is rolled around the wheel and ends up in one of the slots; we assume each slot has equal probability of 1/38 and we are interested in the number of the slot in which the ball falls.

**(a)** Define the outcome space $O$.

**(b)** Let $A = \{0, 00\}$. Give the value of $P(A)$.

**(c)** Let $B = \{14, 15, 17, 18\}$. Give the value of $P(B)$.

**(d)** Let $D = \{x : x \text{ is odd}\}$. Give the value of $P(D)$.

**1.1-8** The five numbers 1, 2, 3, 4, and 5 are written respectively on five disks of the same size and placed in a hat. Two disks are drawn without replacement from the hat, and the numbers written on them are observed.

    **(a)** List the 10 possible outcomes for this experiment as unordered pairs of numbers.

    **(b)** If each of the 10 outcomes has probability 1/10, assign a value to the probability that the sum of the two numbers drawn is **(i)** 3; **(ii)** between 6 and 8 inclusive.

**1.1-9** Divide a line segment into two parts by selecting a point at random. Use your intuition to assign a probability to the event that the larger segment is at least two times longer than the shorter segment.

**1.1-10** Let the interval $[-r, r]$ be the base of a semicircle. If a point is selected at random from this interval, assign a probability to the event that the length of the perpendicular segment from this point to the semicircle is less than $r/2$.

**1.1-11** Let $O = A_1 \cup A_2 \cup \cdots \cup A_m$, where events $A_1, A_2, \ldots, A_m$ are mutually exclusive and exhaustive.

    **(a)** If $P(A_1) = P(A_2) = \cdots = P(A_m)$, show that $P(A_i) = 1/m$, $i = 1, 2, \ldots, m$.

    **(b)** If $A = A_1 \cup A_2 \cup \cdots \cup A_h$, where $h < m$, and (a) holds, prove that $P(A) = h/m$.

**1.1-12** Prove Theorem 1.1-6.

# 1.2 METHODS OF ENUMERATION

In this section we develop counting techniques that are useful in determining the number of outcomes associated with the events of certain random experiments. We begin with a consideration of the multiplication principle.

    **Multiplication Principle:** Suppose that an experiment (or procedure) $E_1$ has $n_1$ outcomes and for each of these possible outcomes an experiment (procedure) $E_2$ has $n_2$ possible outcomes. The composite experiment (procedure) $E_1 E_2$ that consists of performing first $E_1$ and then $E_2$ has $n_1 n_2$ possible outcomes.

**EXAMPLE 1.2-1** Let $E_1$ denote the selection of a rat from a cage containing one female (F) rat and one male (M) rat. Let $E_2$ denote the administering of either drug A (A), drug B (B), or a placebo (P) to the selected rat. The outcome for the composite experiment can be denoted by an ordered pair, such as (F, P). In fact, the set of all possible outcomes, namely $(2)(3) = 6$ of them, can be denoted by the following rectangular array:

$$(F, A) \quad (F, B) \quad (F, P)$$
$$(M, A) \quad (M, B) \quad (M, P)$$

    Another way of illustrating the multiplication principle is with a tree diagram like that in Figure 1.2-1. The diagram shows that there are $n_1 = 2$ possibilities (branches) for the sex of the rat and that for each of these outcomes there are $n_2 = 3$ possibilities (branches) for the drug.

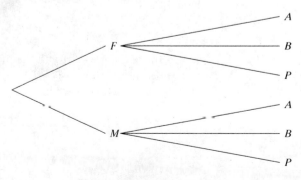

Figure 1.2-1 Tree diagram

Clearly, the multiplication principle can be extended to a sequence of more than two experiments or procedures. Suppose that the experiment $E_i$ has $n_i$ ($i = 1, 2, \ldots, m$) possible outcomes after previous experiments have been performed. The composite experiment $E_1 E_2 \cdots E_m$, which consists of performing $E_1$, then $E_2$, $\ldots$, and finally $E_m$, has $n_1 n_2 \cdots n_m$ possible outcomes.

**EXAMPLE 1.2-2** A certain food service gives the following choices for dinner: $E_1$, soup or tomato juice; $E_2$, steak or shrimp; $E_3$, french fried potatoes, mashed potatoes, or a baked potato; $E_4$, corn or peas; $E_5$, Jello, tossed salad, cottage cheese, or cole slaw; $E_6$, cake, cookies, pudding, brownie, vanilla ice cream, chocolate ice cream, or orange sherbet; $E_7$, coffee, tea, milk, or punch. How many different dinner selections are possible if one of the listed choices is made for each of $E_1$, $E_2$, $\ldots$, and $E_7$? By the multiplication principle there are

$$(2)(2)(3)(2)(4)(7)(4) = 2688$$

different combinations.

Although the multiplication principle is fairly simple and easy to understand, it will be extremely useful as we now develop various counting techniques.

Suppose that $n$ positions are to be filled with $n$ different objects. There are $n$ choices for filling the first position, $n - 1$ for the second, $\ldots$, and 1 choice for the last position. So by the multiplication principle there are

$$n(n - 1) \cdots (2)(1) = n!$$

possible arrangements. The symbol $n!$ is read "$n$ factorial." We define $0! = 1$; that is, we say that zero positions can be filled with zero objects in one way.

**Definition 1.2-1** Each of the $n!$ arrangements (in a row) of $n$ different objects is called a *permutation* of the $n$ objects.

**EXAMPLE 1.2-3** The number of permutations of the four letters a, b, c, and d is clearly $4! = 24$. However, the number of possible four-letter code words using the four letters a, b, c, and d if letters may be repeated is $4^4 = 256$, because in this case each selection can be performed in four ways.

If only $r$ positions are to be filled with objects selected from $n$ different objects, $r \leq n$, the number of possible ordered arrangements is

$$_nP_r = n(n - 1)(n - 2) \cdots (n - r + 1).$$

That is, there are $n$ ways to fill the first position, $(n - 1)$ ways to fill the second, and so on until there are $[n - (r - 1)] = (n - r + 1)$ ways to fill the $r$th position.

In terms of factorials we have that

$$_nP_r = \frac{n(n - 1) \cdots (n - r + 1)(n - r) \cdots (3)(2)(1)}{(n - r) \cdots (3)(2)(1)} = \frac{n!}{(n - r)!}.$$

**Definition 1.2-2** Each of the $_nP_r$ arrangements is called a *permutation of $n$ objects taken $r$ at a time.*

EXAMPLE 1.2-4    The number of possible four-letter code words, selecting from the 26 letters in the alphabet, in which all four letters are different is

$$_{26}P_4 = (26)(25)(24)(23) = \frac{26!}{22!} = 358,800.$$

EXAMPLE 1.2-5    The number of ways of selecting a president, a vice president, a secretary, and a treasurer in a club consisting of 10 persons is

$$_{10}P_4 = 10 \cdot 9 \cdot 8 \cdot 7 = \frac{10!}{6!} = 5,040.$$

Suppose that a set contains $n$ objects. Consider the problem of drawing $r$ objects from this set. The order in which the objects are drawn may or may not be important. In addition, it is possible that a drawn object is replaced before the next object is drawn. Accordingly, we give some definitions and show how the multiplication principle can be used to count the number of possibilities.

Definition  1.2-3    If $r$ objects are selected from a set of $n$ objects, and if the order of selection is noted, the selected set of $r$ objects is called an *ordered sample of size r*.

Definition  1.2-4    *Sampling with replacement* occurs when an object is selected and then replaced before the next object is selected.

By the multiplication principle, the number of possible ordered samples of size $r$ taken from a set of $n$ objects is $n^r$ when sampling with replacement.

EXAMPLE 1.2-6    A die is rolled five times. The number of possible ordered samples is $6^5 = 7776$. Note that rolling a die is equivalent to sampling with replacement.

EXAMPLE 1.2-7    An urn contains 10 balls numbered 0, 1, 2, ... , 9. If four balls are selected, one at a time and with replacement, the number of possible ordered samples is $10^4 = 10,000$. Note that this is the number of four-digit integers between 0000 and 9999, inclusive.

Definition  1.2-5    *Sampling without replacement* occurs when an object is not replaced after it has been selected.

By the multiplication principle, the number of possible ordered samples of size $r$ taken from a set of $n$ objects, sampling without replacement, is

$$n(n-1)\cdots(n-r+1) = \frac{n!}{(n-r)!}$$

which is equivalent to $_nP_r$, the number of permutations of $n$ objects taken $r$ at a time.

EXAMPLE 1.2-8   The number of ordered samples of five cards that can be drawn without replacement from a standard deck of 52 playing cards is

$$(52)(51)(50)(49)(48) = \frac{52!}{47!} = 311{,}875{,}200.$$

Often the order of selection is not important and interest centers only on the selected set of $r$ objects. That is, we are interested in the number of subsets of size $r$ that can be selected from a set of $n$ different objects. In order to find the number of (unordered) subsets of size $r$, we count, in two different ways, the number of ordered subsets of size $r$ that can be taken from the $n$ distinguishable objects. By equating these two answers, we are able to count the number of (unordered) subsets of size $r$.

Let $C$ denote the number of (unordered) subsets of size $r$ that can be selected from $n$ different objects. We can obtain each of the $_nP_r$ ordered subsets by first selecting one of the $C$ unordered subsets of $r$ objects and then ordering these $r$ objects. Since the latter can be carried out in $r!$ ways, the multiplication principle yields $(C)(r!)$ ordered subsets; so $(C)(r!)$ must equal $_nP_r$. Thus we have

$$(C)(r!) = \frac{n!}{(n-r)!}$$

or

$$C = \frac{n!}{r!\,(n-r)!}.$$

We denote this answer by either $_nC_r$ or $\binom{n}{r}$; that is,

$$_nC_r = \binom{n}{r} = \frac{n!}{r!\,(n-r)!}.$$

Accordingly, a set of $n$ different objects possesses

$$\binom{n}{r} = \frac{n!}{r!\,(n-r)!}$$

unordered subsets of size $r \leq n$.

We could also say that the number of ways in which $r$ objects can be selected without replacement from $n$ objects, when the order of selection is disregarded, is $\binom{n}{r} = {_nC_r}$, and this can be read as "$n$ choose $r$." This motivates the following definition.

**Definition 1.2-6**   Each of the $_nC_r$ unordered subsets is called a *combination of $n$ objects taken $r$ at a time*, where

$$_nC_r = \binom{n}{r} = \frac{n!}{r!\,(n-r)!}.$$

EXAMPLE 1.2-9   The number of possible five-card hands (hands in five-card poker) drawn from a deck of 52 playing cards is

$$_{52}C_5 = \binom{52}{5} = \frac{52!}{5!\,47!} = 2{,}598{,}960.$$

EXAMPLE 1.2-10   The number of possible 13-card hands (hands in bridge) that can be selected from a deck of 52 playing cards is

$$_{52}C_{13} = \binom{52}{13} = \frac{52!}{13!\,39!} = 635{,}013{,}559{,}600.$$

The numbers $\binom{n}{r}$ are frequently called **binomial coefficients**, since they arise in the expansion of a binomial. We illustrate this by giving a justification of the binomial expansion

$$(a+b)^n = \sum_{r=0}^{n} \binom{n}{r} b^r a^{n-r}. \qquad (1.2\text{-}1)$$

In the expansion of

$$(a+b)^n = (a+b)(a+b)\cdots(a+b),$$

either an $a$ or a $b$ is selected from each of the $n$ factors. One possible product is then $b^r a^{n-r}$; this occurs when $b$ is selected from each of $r$ factors and $a$ from each of the remaining $n-r$ factors. But this latter operation can be completed in $\binom{n}{r}$ ways, which then must be the coefficient of $b^r a^{n-r}$, as shown in Equation 1.2-1.

The binomial coefficients are given in Table I in the Appendix for selected values of $n$ and $r$. Note that for some combinations of $n$ and $r$, the table uses the fact that

$$\binom{n}{r} = \frac{n!}{r!\,(n-r)!} = \frac{n!}{(n-r)!\,r!} = \binom{n}{n-r}.$$

That is, the number of ways in which $r$ objects can be selected out of $n$ objects is equal to the number of ways in which $n-r$ objects can be selected out of $n$ objects.

EXAMPLE 1.2-11   Assume that each of the $\binom{52}{5} = 2{,}598{,}960$ five-card hands drawn from a deck of 52 playing cards has the same probability for being selected. The number of possible five-card hands that are all spades (event $A$) is

$$N(A) = \binom{13}{5}\binom{39}{0}$$

because the 5 spades can be selected from the 13 spades in $\binom{13}{5}$ ways after which zero nonspades can be selected in $\binom{39}{0} = 1$ way. We have

$$\binom{13}{5} = \frac{13!}{5!8!} = 1287$$

from Table I in the Appendix. Thus the probability of an all-spade five-card hand is

$$P(A) = \frac{N(A)}{N(O)} = \frac{1{,}287}{2{,}598{,}960} = 0.000495.$$

Suppose now that the event $B$ is the set of outcomes in which exactly three cards are kings and exactly two cards are queens. We can select the three kings in any one of $\binom{4}{3}$ ways and the two queens in any one of $\binom{4}{2}$ ways. By the multiplication principle, the number of outcomes in $B$ is

$$N(B) = \binom{4}{3}\binom{4}{2}\binom{44}{0},$$

where $\binom{44}{0}$ gives the number of ways in which 0 cards are selected out of the non-kings and non-queens and of course is equal to one. Thus

$$P(B) = \frac{N(B)}{N(O)} = \frac{\binom{4}{3}\binom{4}{2}\binom{44}{0}}{\binom{52}{5}} = \frac{24}{2{,}598{,}960} = 0.0000092.$$

Finally, let $C$ be the set of outcomes in which there are exactly two kings, two queens, and one jack. Then

$$P(C) = \frac{N(C)}{N(O)} = \frac{\binom{4}{2}\binom{4}{2}\binom{4}{1}\binom{40}{0}}{\binom{52}{5}} = \frac{144}{2{,}598{,}960} = 0.000055$$

because the numerator of this fraction is the number of outcomes in $C$.

Now suppose that a set contains $n$ objects of two types, $r$ of one type and $n - r$ of the other type. The number of permutations of $n$ different objects is $n!$. However, in this case, the objects are not all distinguishable. To count the number of distinguishable arrangements, first select $r$ out of the $n$ positions for the objects of the first type. This can be done in $\binom{n}{r}$ ways. Then fill in the remaining positions with the objects of the second type. Thus the number of distinguishable arrangements is

$$_nC_r = \binom{n}{r} = \frac{n!}{r!\,(n-r)!}.$$

**Definition 1.2-7**    Each of the $_nC_r$ permutations of $n$ objects, $r$ of one type and $n - r$ of another type, is called a *distinguishable permutation*.

**EXAMPLE 1.2-12**    A coin is flipped 10 times and the sequence of heads and tails is observed. The number of possible 10-tuplets that result in four heads and six tails is

$$\binom{10}{4} = \frac{10!}{4!\,6!} = \frac{10!}{6!\,4!} = \binom{10}{6} = 210.$$

EXAMPLE 1.2-13   Students on a boat send signals back to shore by arranging seven colored flags on a vertical flagpole. If they have four orange and three blue flags, they can send

$$\binom{7}{4} = \frac{7!}{4!\,3!} = 35$$

different signals. Note that if they had seven flags of different colors, they could send $7! = 5{,}040$ different signals.

## EXERCISES 1.2

**1.2-1** A boy found a bicycle lock for which the combination was unknown. The correct combination is a four-digit number, $d_1d_2d_3d_4$ where $d_i$, $i = 1, 2, 3, 4$, is selected from $1, 2, 3, 4, 5, 6, 7, 8$. How many different lock combinations are possible with such a lock?

**1.2-2** How many different signals can be made using four flags of different colors on a vertical flagpole if exactly three flags are used for each signal?

**1.2-3** How many different license plates are possible if a state uses

(a) Two letters followed by a four-digit integer (leading zeros permissible and the letters and digits can be repeated)?

(b) Three letters followed by a three-digit integer?

**1.2-4** In designing an experiment, the researcher can often choose many different levels of the various factors in order to try to find the best combination at which to operate. For illustration, suppose the research is studying a certain chemical reaction and can choose four levels of temperature, five different pressures, and two different catalysts.

(a) To consider all possible combinations, how many experiments would need to be conducted?

(b) Often in preliminary experimentation, each factor is restricted to two levels. With the three factors noted, how many experiments would need to be run to cover all possible combinations with each of the three factors at two levels? (NOTE: This is often called a $2^3$ design.)

**1.2-5** Some albatrosses return to the world's only mainland colony of royal albatrosses on Otago Peninsula near Dunedin, New Zealand, every two years to nest and raise their young. In order to learn more about the albatross, colored plastic bands are placed on their legs so that they can be identified from a distance. Suppose that three bands are placed on one leg, selecting from the colors red, yellow, green, white, and blue. Find the number of different color codes that are possible for banding an albatross if

(a) The three bands are different colors.

(b) Repeated colors are permissible.

**1.2-6** Suppose that Andy Roddick and Roger Federer are playing a tennis match in which the first player to win three sets wins the match. Considering the possible orderings for the winning player, in how many ways could this tennis match end?

**1.2-7** The World Series in baseball continues until either the American League team or the National League team wins four games. How many different orders are possible (e.g., *ANNAAA* means the American League team wins in six games) if the series goes

(a) Four games?

(b) Five games?

(c) Six games?

(d) Seven games?

**1.2-8** Pascal's triangle gives a method for calculating the binomial coefficients; it begins as follows:

$$
\begin{array}{ccccccccc}
 & & & & 1 & & & & \\
 & & & 1 & & 1 & & & \\
 & & 1 & & 2 & & 1 & & \\
 & 1 & & 3 & & 3 & & 1 & \\
1 & & 4 & & 6 & & 4 & & 1 \\
\end{array}
$$

1  5  10  10  5  1

$$
\vdots \quad \vdots \quad \vdots \quad \vdots \quad \vdots
$$

The $n$th row of this triangle gives the coefficients for $(a + b)^{n-1}$. To find an entry in the table other than a 1 on the boundary, add the two nearest numbers in the row directly above. The equation

$$
\binom{n}{r} = \binom{n-1}{r} + \binom{n-1}{r-1},
$$

called **Pascal's equation**, explains why Pascal's triangle works. Prove that this equation is correct.

**1.2-9** Among nine presidential candidates at a debate, three are Republicans and six are Democrats.

**(a)** In how many different ways can the nine candidates be lined up?

**(b)** How many lineups by party are possible if each candidate is labeled either $R$ or $D$?

**(c)** For each of the nine candidates, you are to decide whether the candidate did a good job or a poor job; that is, give each of the nine candidates a grade of $G$ or $P$. How many different "score cards" are possible?

**1.2-10** Prove:

$$
\sum_{r=0}^{n}(-1)^r\binom{n}{r} = 0 \qquad \text{and} \qquad \sum_{r=0}^{n}\binom{n}{r} = 2^n.
$$

HINT: Consider $(1 - 1)^n$ and $(1 + 1)^n$ or use Pascal's equation and proof by induction.

**1.2-11** A poker hand is defined as drawing five cards at random without replacement from a deck of 52 playing cards. Find the probability of each of the following poker hands:

**(a)** Four of a kind (four cards of equal face value and one card of a different value).

**(b)** Full house (one pair and one triple of cards with equal face value).

**(c)** Three of a kind (three equal face values plus two cards of different values).

**(d)** Two pairs (two pairs of equal face value plus one card of a different value).

**(e)** One pair (one pair of equal face value plus three cards of different values).

## 1.3 CONDITIONAL PROBABILITY

We introduce the idea of conditional probability by means of an example.

EXAMPLE 1.3-1    Suppose that we are given 20 tulip bulbs that are very similar in appearance and told that 8 tulips will bloom early, 12 will bloom late, 13 will be red, and 7 will be yellow, in accordance with the various combinations of Table 1.3-1. If one bulb is selected at random, the probability that it will produce a red tulip ($R$) is given by $P(R) = 13/20$, under the assumption that each bulb is "equally likely." Suppose,

| Table 1.3-1 Tulip Combinations | | | |
|---|---|---|---|
| | **Early ($E$)** | **Late ($L$)** | **Totals** |
| Red ($R$) | 5 | 8 | 13 |
| Yellow($Y$) | 3 | 4 | 7 |
| Totals | 8 | 12 | 20 |

however, that close examination of the bulb will reveal whether it will bloom early ($E$) or late ($L$). If we consider an outcome only if it results in a tulip bulb that will bloom early, only eight outcomes in the outcome space are now of interest. Thus it is natural to assign, under this limitation, the probability of 5/8 to $R$; that is, $P(R\,|\,E) = 5/8$, where $P(R\,|\,E)$ is read as the probability of $R$ given that $E$ has occurred. Note that

$$P(R\,|\,E) = \frac{5}{8} = \frac{N(R \cap E)}{N(E)} = \frac{N(R \cap E)/20}{N(E)/20} = \frac{P(R \cap E)}{P(E)},$$

where $N(R \cap E)$ and $N(E)$ are the numbers of outcomes in events $R \cap E$ and $E$, respectively.

This example is illustrative of a number of common situations. That is, in some random experiments, we are interested only in those outcomes that are elements of a subset $B$ of the outcome space $O$. This means, for our purposes, that the outcome space is effectively the subset $B$. We are now confronted with the problem of defining a probability set function with $B$ as the "new" outcome space. That is, for a given event $A$ we want to define $P(A\,|\,B)$, the probability of $A$ considering only those outcomes of the random experiment that are elements of $B$. The previous example gives us the clue to that definition. That is, for experiments in which each outcome is equally likely, it makes sense to define $P(A\,|\,B)$ by

$$P(A\,|\,B) = \frac{N(A \cap B)}{N(B)},$$

where $N(A \cap B)$ and $N(B)$ are the numbers of outcomes in $A \cap B$ and $B$, respectively. If we divide the numerator and the denominator of this fraction by $N(O)$, the number of outcomes in the outcome space, we have

$$P(A\,|\,B) = \frac{N(A \cap B)/N(O)}{N(B)/N(O)} = \frac{P(A \cap B)}{P(B)}.$$

We are thus led to the following definition.

**Definition 1.3-1**

The *conditional probability* of an event $A$ given that event $B$ has occurred, is defined by

$$P(A\,|\,B) = \frac{P(A \cap B)}{P(B)}$$

provided that $P(B) > 0$.

A formal use of the definition is given in the following example.

EXAMPLE 1.3-2   If $P(A) = 0.4$, $P(B) = 0.5$, and $P(A \cap B) = 0.3$, then $P(A \mid B) = 0.3/0.5 = 0.6$; $P(B \mid A) = P(A \cap B)/P(A) = 0.3/0.4 = 0.75$.

We can think of the "given $B$" as specifying the new outcome space for which we now want to calculate the probability of that part of $A$ that is contained in $B$ to determine $P(A \mid B)$. The following two examples illustrate this idea.

EXAMPLE 1.3-3   Suppose that $P(A) = 0.7$, $P(B) = 0.3$, and $P(A \cap B) = 0.2$. These probabilities are listed on the Venn diagram in Figure 1.3-1. Given that the outcome of the experiment belongs to $B$, what then is the probability of $A$? We are effectively restricting the outcome space to $B$; of the probability $P(B) = 0.3$, $0.2$ corresponds to $P(A \cap B)$ and hence to $A$. That is, $0.2/0.3 = 2/3$ of the probability of $B$ corresponds to $A$. Of course, formally by definition, we also obtain

$$P(A \mid B) = \frac{P(A \cap B)}{P(B)} = \frac{0.2}{0.3} = \frac{2}{3}.$$

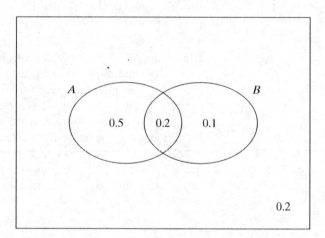

**Figure 1.3-1** Conditional probability illustration

EXAMPLE 1.3-4   A pair of four-sided dice is rolled and the sum is determined. Let $A$ be the event that a sum of 3 is rolled and let $B$ be the event that a sum of 3 or a sum of 5 is rolled. In a sequence of rolls the probability that a sum of 3 is rolled before a sum of 5 is rolled can be thought of as the conditional probability of a sum of 3 given that a sum of 3 or 5 has occurred; that is, the conditional probability of $A$ given $B$

$$P(A \mid B) = \frac{P(A \cap B)}{P(B)} = \frac{P(A)}{P(B)} = \frac{2/16}{6/16} = \frac{2}{6}.$$

Note that for this example, the only outcomes of interest are those having a sum of 3 or a sum of 5, and of these six equally likely outcomes, two have a sum of 3. See Figure 1.3-2. (See also Exercise 1.3-13.)

than $P(A \cap B)$. Then $P(A \cap B)$ can be computed with these assignments. This will be illustrated in Examples 1.3-6 and 1.3-7.

**EXAMPLE 1.3-6**  A bowl contains seven blue chips and three red chips. Two chips are to be drawn successively at random and without replacement. We want to compute the probability that the first draw results in a red chip ($A$) and the second draw results in a blue chip ($B$). It is reasonable to assign the following probabilities:

$$P(A) = \frac{3}{10} \quad \text{and} \quad P(B \,|\, A) = \frac{7}{9}.$$

The probability of red on the first draw and blue on the second draw is

$$P(A \cap B) = \frac{3}{10} \cdot \frac{7}{9} = \frac{7}{30}.$$

It should be noted that in many instances, it is possible to compute a probability by two seemingly different methods. For illustration, consider Example 1.3-6 but find the probability of drawing a red chip on each of the two draws. Following that example, it is

$$\frac{3}{10} \cdot \frac{2}{9} = \frac{1}{15}.$$

However, we can also find this probability using combinations as follows:

$$\frac{\binom{3}{2}\binom{7}{0}}{\binom{10}{2}} = \frac{\frac{(3)(2)}{(1)(2)}}{\frac{(10)(9)}{(1)(2)}} = \frac{1}{15}.$$

Thus we obtain the same answer, as we should, provided that our reasoning is consistent with the underlying assumptions.

**EXAMPLE 1.3-7**  From an ordinary deck of playing cards, cards are to be drawn successively at random and without replacement. The probability that the third spade appears on the sixth draw is computed as follows: Let $A$ be the event of two spades in the first five cards drawn, and let $B$ be the event of a spade on the sixth draw. Thus the probability that we wish to compute is $P(A \cap B)$. It is reasonable to take

$$P(A) = \frac{\binom{13}{2}\binom{39}{3}}{\binom{52}{5}} = 0.274 \quad \text{and} \quad P(B \,|\, A) = \frac{11}{47} = 0.234.$$

The desired probability $P(A \cap B)$ is the product of these numbers,

$$P(A \cap B) = (0.274)(0.234) = 0.064.$$

**EXAMPLE 1.3-8**  Continuing with Example 1.3-4, in which a pair of four-sided dice is rolled, the probability for rolling a sum of 3 on the first roll and then, continuing the sequence of rolls, rolling a sum of 3 before rolling a sum of 5 is

$$\frac{2}{16} \cdot \frac{2}{6} = \frac{4}{96} = \frac{1}{24}.$$

The multiplication rule can be extended to three or more events. In the case of three events we have, by using the multiplication rule for two events,

$$P(A \cap B \cap C) = P[(A \cap B) \cap C]$$
$$= P(A \cap B)P(C \mid A \cap B).$$

But

$$P(A \cap B) = P(A)P(B \mid A).$$

Hence

$$P(A \cap B \cap C) = P(A)P(B \mid A)P(C \mid A \cap B).$$

This type of argument can be used to extend the multiplication rule to more than three events, and the general formula for $k$ events can be officially proved by mathematical induction.

**EXAMPLE 1.3-9**   Four cards are to be dealt successively at random and without replacement from an ordinary deck of playing cards. The probability of receiving in order a spade, a heart, a diamond, and a club is

$$\frac{13}{52} \cdot \frac{13}{51} \cdot \frac{13}{50} \cdot \frac{13}{49},$$

a result that follows from the extension of the multiplication rule and reasonable assignments to the probabilities involved.

We close this section with three different types of example.

**EXAMPLE 1.3-10**   A grade school boy has five blue and four white marbles in his left pocket and four blue and five white marbles in his right pocket. If he transfers one marble at random from his left to his right pocket, what is the probability of his then drawing a blue marble from his right pocket? For notation let $BL$, $BR$, and $WL$ denote drawing blue from left pocket, blue from right pocket, and white from left pocket, respectively. Then

$$P(BR) = P(BL \cap BR) + P(WL \cap BR)$$
$$= P(BL)P(BR \mid BL) + P(WL)P(BR \mid WL)$$
$$= \frac{5}{9} \cdot \frac{5}{10} + \frac{4}{9} \cdot \frac{4}{10} = \frac{41}{90}$$

is the desired probability.

**EXAMPLE 1.3-11**   An insurance company sells a number of different policies; among these are 60% for autos, 40% for homeowners, and 20% are for both of these two. Let $A_1$ be people with only an auto policy, $A_2$ with only homeowners, $A_3$ with both, and $A_4$ are those with only other types of policies. If a person is selected at random from the policyholders, then $P(A_1) = 0.4$, $P(A_2) = 0.2$, $P(A_3) = 0.2$, and $P(A_4) = 0.2$, as these four events are mutually exclusive and exhaustive. Further let $B$ be the event that a policyholder will renew at least one of the auto or homeowner's policies. Say from past experience we can assign the conditional probabilities $P(B \mid A_1) = 0.6$, $P(B \mid A_2) = 0.7$, and $P(B \mid A_3) = 0.8$. Given that the person selected at random has an auto or homeowner policy, what is the conditional probability that the person will renew at least one of these policies? That is,

$$P(B \mid A_1 \cup A_2 \cup A_3) = \frac{P(A_1 \cap B) + P(A_2 \cap B) + P(A_3 \cap B)}{P(A_1) + P(A_2) + P(A_3)}$$

$$= \frac{(0.4)(0.6) + (0.2)(0.7) + (0.2)(0.8)}{0.4 + 0.2 + 0.2}$$

$$= \frac{0.54}{0.80} = \frac{27}{40}$$

$$= 0.675.$$

**EXAMPLE 1.3-12**

A device has two components, $C_1$ and $C_2$, but it will continue to operate with only one active component for a one-year period. The probability that each will fail when both are in operation is 0.01 in that one-year period. However, when one fails, the probability of the other failing is 0.03 in that period due to added strain. Thus the probability that the device fails in one year is

$$P(C_1 \text{ fails})P(C_2 \text{ fails} \mid C_1 \text{ fails}) + P(C_2 \text{ fails})P(C_1 \text{ fails} \mid C_2 \text{ fails}) =$$

$$(0.01)(0.03) + (0.01)(0.03) = 0.0006.$$

**EXERCISES 1.3**

**1.3-1** A common test for AIDS is called the ELISA (Enzyme-Linked Immunosorbent Assay) test. Among 1,000,000 people who are given the ELISA test, we can expect results similar to those given in the table.

| | $B_1$: Carry AIDS Virus | $B_2$: Do Not Carry AIDS Virus | Totals |
|---|---|---|---|
| $A_1$: Test positive | 4,885 | 73,630 | 78,515 |
| $A_2$: Test negative | 115 | 921,370 | 921,485 |
| Totals | 5,000 | 995,000 | 1,000,000 |

If one of these 1,000,000 people is selected randomly, find the following probabilities: **(a)** $P(B_1)$, **(b)** $P(A_1)$, **(c)** $P(A_1 \mid B_2)$, **(d)** $P(B_1 \mid A_1)$. **(e)** In words, what do parts (c) and (d) say?

**1.3-2** The following table classifies 1456 people by their gender and by whether or not they favor a gun law.

| | Male ($B_1$) | Female ($B_2$) | Totals |
|---|---|---|---|
| Favor ($A_1$) | 392 | 649 | 1041 |
| Oppose ($A_2$) | 241 | 174 | 415 |
| Totals | 633 | 823 | 1456 |

Compute the following probabilities if one of these 1456 persons is selected randomly: **(a)** $P(A_1)$, **(b)** $P(A_1 \mid B_1)$, **(c)** $P(A_1 \mid B_2)$. **(d)** Interpret your answers to parts (b) and (c).

**1.3-3** Let $A_1$ and $A_2$ be the events that a person is left eye dominant or right eye dominant, respectively. When a person folds their hands, let $B_1$ and $B_2$ be the events that their left thumb and right thumb, respectively, are on top. A survey in one statistics class yielded the following table:

|       | $B_1$ | $B_2$ | Totals |
|-------|-------|-------|--------|
| $A_1$ | 5     | 7     | 12     |
| $A_2$ | 14    | 9     | 23     |
| Totals| 19    | 16    | 35     |

If a student is selected randomly, find the following probabilities: **(a)** $P(A_1 \cap B_1)$, **(b)** $P(A_1 \cup B_1)$, **(c)** $P(A_1 \mid B_1)$, **(d)** $P(B_2 \mid A_2)$. **(e)** If the students had their hands folded and you hoped to select a right eye dominant student, would you select a "right thumb on top" or a "left thumb on top" student? Why?

**1.3-4** Two cards are drawn successively and without replacement from an ordinary deck of playing cards. Compute the probability of drawing

**(a)** Two hearts.

**(b)** A heart on the first draw, a club on the second draw.

**(c)** A heart on the first draw, an ace on the second draw.

HINT: In part (c), note that a heart can be drawn by getting the ace of hearts or one of the other 12 hearts.

**1.3-5** Suppose that the genes for eye color for a certain male fruit fly are $(R, W)$ and the genes for eye color for the mating female fruit fly are $(R, W)$, where $R$ and $W$ represent red and white, respectively. Their offspring receive one gene for eye color from each parent.

**(a)** Define the outcome space for the genes for eye color for the offspring.

**(b)** Assume that each of the four possible outcomes has equal probability. If an offspring ends up with either two red genes or one red and one white gene for eye color, its eyes will look red. Given that an offspring's eyes look red, what is the conditional probability that it has two red genes for eye color?

**1.3-6** Suppose that there are 14 songs on a compact disk (CD) and you like 8 of them. When using the random button selector on a CD player, each of the 14 songs is played once in a random order. Find the probability that among the first 2 songs that are played,

**(a)** You like both of them.

**(b)** You like neither of them.

**(c)** You like exactly one of them.

**1.3-7** An urn contains four colored balls: two orange and two blue. Two balls are selected at random without replacement, and you are told that at least one of them is orange. What is the probability that the other ball is also orange?

**1.3-8** A small grocery store had 10 cartons of milk, 2 of which were sour. If you are going to buy the sixth carton of milk sold that day at random, compute the probability of selecting a carton of sour milk.

**1.3-9** Consider the birthdays of the students in a class of size $r$. Assume that the year consists of 365 days.

**(a)** How many different ordered outcomes of birthdays are possible (for $r$ students) allowing repetitions (with replacement)?

**(b)** The same as part (a) except requiring that all the students have different birthdays (without replacement)?

**(c)** If we can assume that each ordered outcome in part (a) has the same probability, what is the probability that at least two students have the same birthday?

**(d)** For what value of $r$ is the probability in part (c) about equal to 1/2? Is this number surprisingly small? HINT: Use a calculator or computer to find $r$.

**1.3-10** You are a member of a class of 18 students. A bowl contains 18 chips, 1 blue and 17 red. Each student is to take 1 chip from the bowl without replacement. The student who draws the blue chip is guaranteed an A for the course.

   **(a)** If you have a choice of drawing first, fifth, or last, which position would you choose? Justify your choice using probability.

   **(b)** Suppose the bowl contains 2 blue and 16 red chips. What position would you now choose?

**1.3-11** A drawer contains four black, six brown, and eight olive socks. Two socks are selected at random from the drawer.

   **(a)** Compute the probability that both socks are the same color.

   **(b)** Compute the probability that both socks are olive if it is known that they are the same color.

**1.3-12** Bowl *A* contains three red and two white chips, and bowl *B* contains four red and three white chips. A chip is drawn at random from bowl *A* and transferred to bowl *B*. Compute the probability of then drawing a red chip from bowl *B*.

**1.3-13** In the gambling game "craps" a pair of dice is rolled and the outcome of the experiment is the sum of the points on the up-sides of the six-sided dice. The bettor wins on the first roll if the sum is 7 or 11. The bettor loses on the first roll if the sum is 2, 3, or 12. If the sum is 4, 5, 6, 8, 9, or 10, that number is called the bettor's "point." Once the point is established, the rule is, If the bettor rolls a 7 before the "point," the bettor loses; but if the "point" is rolled before a 7, the bettor wins. (See Examples 1.3-4 and 1.3-8 and Exercise 5.3-8.)

   **(a)** List the 36 outcomes in the outcome space for the roll of a pair of dice. Assume that each of them has a probability of 1/36.

   **(b)** Find the probability that the bettor wins on the first roll. That is, find the probability of rolling a 7 or 11, $P(7 \text{ or } 11)$.

   **(c)** Given that 8 is the outcome on the first roll, find the probability that the bettor now rolls the point 8 before rolling a 7 and thus wins. Note that at this stage in the game the only outcomes of interest are 7 and 8. Thus find $P(8 \mid 7 \text{ or } 8)$.

   **(d)** The probability that a bettor rolls an 8 on the first roll and then wins is given by $P(8)P(8 \mid 7 \text{ or } 8)$. Show that this probability is (5/36)(5/11).

   **(e)** Show that the total probability that a bettor wins in the game of craps is 0.49293. HINT: Note that the bettor can win in one of several mutually exclusive ways: by rolling a 7 or 11 on the first roll or by establishing one of the points 4, 5, 6, 8, 9, or 10 on the first roll and then obtaining that point before a 7 on successive rolls.

## 1.4 INDEPENDENT EVENTS

For certain pairs of events, the occurrence of one of them may or may not change the probability of the occurrence of the other. In the latter case they are said to be **independent events**. However, before giving the formal definition of independence, let us consider an example.

EXAMPLE 1.4-1

Flip a coin twice and observe the sequence of heads and tails. The outcome space is then

$$O = \{HH, HT, TH, TT\}.$$

It is reasonable to assign a probability of 1/4 to each of these four outcomes. Let

$$A = \{\text{heads on the first flip}\} = \{HH, HT\},$$

$$B = \{\text{tails on the second flip}\} = \{HT, TT\},$$

$$C = \{\text{tails on both flips}\} = \{TT\}.$$

Now $P(B) = 2/4 = 1/2$. However, if we are given that $C$ has occurred, then $P(B \mid C) = 1$ because $C \subset B$. That is, the knowledge of the occurrence of $C$ has changed the probability of $B$. On the other hand, if we are given that A has occurred,

$$P(B \mid A) = \frac{P(A \cap B)}{P(A)} = \frac{1/4}{2/4} = \frac{1}{2} = P(B).$$

So the occurrence of $A$ has not changed the probability of $B$. Hence the probability of $B$ does not depend upon knowledge about event $A$, so we say that $A$ and $B$ are independent events. That is, events $A$ and $B$ are independent if the occurrence of one of them does not affect the probability of the occurrence of the other. A more mathematical way of saying this is

$$P(B \mid A) = P(B) \qquad \text{or} \qquad P(A \mid B) = P(A),$$

provided that $P(A) > 0$ or, in the latter case, $P(B) > 0$. With the first of these equalities and the multiplication rule (Definition 1.3-2), we have

$$P(A \cap B) = P(A)P(B \mid A) = P(A)P(B).$$

The second of these equalities, namely $P(A \mid B) = P(A)$, gives us the same result

$$P(A \cap B) = P(B)P(A \mid B) = P(B)P(A).$$

This example motivates the following definition of independent events.

**Definition 1.4-1**

Events $A$ and $B$ are *independent* if and only if $P(A \cap B) = P(A)P(B)$. Otherwise $A$ and $B$ are called *dependent* events.

Events that are independent are sometimes called **statistically independent**, **stochastically independent**, or **independent in a probabilistic sense**, but in most instances we use independent without a modifier if there is no possibility of misunderstanding. It is interesting to note that the definition always holds if $P(A) = 0$ or $P(B) = 0$ because then $P(A \cap B) = 0$, since $(A \cap B) \subset A$ and $(A \cap B) \subset B$. Thus the left-hand and right-hand member of $P(A \cap B) = P(A)P(B)$ are both equal to zero and thus are equal to each other.

**EXAMPLE 1.4-2**

A red die and a white die are rolled. Let event $A = \{4$ on the red die$\}$ and event $B = \{$sum of dice is odd$\}$. Of the 36 equally likely outcomes, 6 are favorable to $A$, 18 are favorable to $B$, and 3 are favorable to $A \cap B$. Thus

$$P(A)P(B) = \frac{6}{36} \cdot \frac{18}{36} = \frac{3}{36} = P(A \cap B).$$

Hence $A$ and $B$ are independent by Definition 1.4-1.

**EXAMPLE 1.4-3**

A red die and a white die are rolled. Let event $C = \{5$ on red die$\}$ and event $D = \{$sum of dice is 11$\}$. Of the 36 equally likely outcomes, 6 are favorable to $C$, 2 are favorable to $D$, and 1 is favorable to $C \cap D$. Thus

$$P(C)P(D) = \frac{6}{36} \cdot \frac{2}{36} = \frac{1}{108} \neq \frac{1}{36} = P(C \cap D).$$

Hence $C$ and $D$ are dependent events by Definition 1.4-1.

**Theorem 1.4-1**

If $A$ and $B$ are independent events, then the following pairs of events are also independent:

    **(a)** $A$ and $B'$.
    **(b)** $A'$ and $B$.
    **(c)** $A'$ and $B'$.

*Proof*

We know that conditional probability satisfies the axioms for a probability function. Hence, if $P(A) > 0$, then $P(B'\,|\,A) = 1 - P(B\,|\,A)$. Thus

$$P(A \cap B') = P(A)P(B'\,|\,A) = P(A)[1 - P(B\,|\,A)]$$

$$= P(A)[1 - P(B)]$$

$$= P(A)P(B'),$$

since $P(B\,|\,A) = P(B)$ by hypothesis. Thus $A$ and $B'$ are independent events. The proofs for parts **(b)** and **(c)** are left as exercises.

Before extending the definition of independent events to more than two events, we present the following example.

**EXAMPLE 1.4-4**

An urn contains four balls numbered 1, 2, 3, and 4. One ball is to be drawn at random from the urn. Let the events $A$, $B$, and $C$ be defined by $A = \{1, 2\}$, $B = \{1, 3\}$, $C = \{1, 4\}$. Then $P(A) = P(B) = P(C) = 1/2$. Furthermore,

$$P(A \cap B) = \frac{1}{4} = P(A)P(B),$$

$$P(A \cap C) = \frac{1}{4} = P(A)P(C),$$

$$P(B \cap C) = \frac{1}{4} = P(B)P(C),$$

which implies that $A$, $B$, and $C$ are independent in pairs (called **pairwise independence**). However, since $A \cap B \cap C = \{1\}$, we have

$$P(A \cap B \cap C) = \frac{1}{4} \neq \frac{1}{8} = P(A)P(B)P(C).$$

That is, something seems to be lacking for the complete independence of $A$, $B$, and $C$.

This example illustrates the reason for the second condition in Definition 1.4-2.

**Definition 1.4-2**

Events $A$, $B$, and $C$ are *mutually independent* if and only if the following two conditions hold:

    **(a)** They are pairwise independent; that is,

$$P(A \cap B) = P(A)P(B), \qquad P(A \cap C) = P(A)P(C),$$

    and

$$P(B \cap C) = P(B)P(C).$$

    **(b)** $P(A \cap B \cap C) = P(A)P(B)P(C)$.

Definition 1.4-2 can be extended to mutual independence of four or more events. In this extension, each pair, triple, quartet, and so on, must satisfy this type of multiplication rule. If there is no possibility of misunderstanding, *independent* is often used without the modifier *mutually* when considering several events.

**EXAMPLE 1.4-5** A rocket has a built-in redundant system. In this system, if component $K_1$ fails, it is bypassed and component $K_2$ is used. If component $K_2$ fails, it is bypassed and component $K_3$ is used. (An example of such components is computer systems.) Suppose that the probability of failure of any one of these components is 0.15 and assume that the failures of these components are mutually independent events. Let $A_i$ denote the event that component $K_i$ fails for $i = 1, 2, 3$. Because the system fails if $K_1$ fails and $K_2$ fails and $K_3$ fails, the probability that the system does not fail is given by

$$P[(A_1 \cap A_2 \cap A_3)'] = 1 - P(A_1 \cap A_2 \cap A_3)$$
$$= 1 - P(A_1)P(A_2)P(A_3)$$
$$= 1 - (0.15)^3$$
$$= 0.9966.$$

One way to increase the reliability of such a system is to add more components (realizing that this also adds weight and takes up space). For example, if a fourth component $K_4$ were added to this system, the probability that the system does not fail is

$$P[(A_1 \cap A_2 \cap A_3 \cap A_4)'] = 1 - (0.15)^4 = 0.9995.$$

The proof and illustration of the following results are left as exercises. If $A$, $B$, and $C$ are mutually independent events, then the following events are also independent:

**(a)** $A$ and $(B \cap C)$;
**(b)** $A$ and $(B \cup C)$;
**(c)** $A'$ and $(B \cap C')$.

In addition, $A'$, $B'$, and $C'$ are mutually independent.

Many experiments consist of a sequence of $n$ trials that are mutually independent. If the outcomes of the trials, in fact, do not have anything to do with one another, then events, such that each is associated with a different trial, should be independent in the probability sense. That is, if the event $A_i$ is associated with the $i$th trial, $i = 1, 2, \ldots, n$, then

$$P(A_1 \cap A_2 \cap \cdots \cap A_n) = P(A_1)P(A_2) \cdots P(A_n).$$

**EXAMPLE 1.4-6** A fair six-sided die is rolled six independent times. Let $A_i$ be the event that side $i$ is observed on the $i$th roll, called a match on the $i$th trial, $i = 1, 2, \ldots, 6$. Thus $P(A_i) = 1/6$, and $P(A_i') = 1 - 1/6 = 5/6$. If we let $B$ denote the event that at least one match occurs, then $B'$ is the event that no matches occur. Thus

$$P(B) = 1 - P(B') = 1 - P(A_1' \cap A_2' \cap \cdots \cap A_6')$$

$$= 1 - \frac{5}{6} \cdot \frac{5}{6} \cdot \frac{5}{6} \cdot \frac{5}{6} \cdot \frac{5}{6} \cdot \frac{5}{6} = 1 - \left(\frac{5}{6}\right)^6$$

is the probability of $B$.

The outcome space for an experiment of $n$ trials is a set of $n$-tuples, where the $i$th component denotes the outcome on the $i$th trial. For example, if a six-sided die is rolled five times,

$$O = \{(O_1, O_2, O_3, O_4, O_5) : O_i = 1, 2, 3, 4, 5, \text{ or } 6, \text{ for } i = 1, 2, 3, 4, 5\}.$$

That is, $O$ is a set of five-tuples, where each component is one of the first six positive integers.

If a coin is tossed two times,

$$O = \{(O_1, O_2) : O_i = H \text{ or } T, i = 1, 2\}.$$

We often drop the commas and parentheses and let, for example, $(H, T) = HT$, as in Example 1.4.1

**EXAMPLE 1.4-7**    The probability that a company's work force has at least one accident in a month is $(0.01)k$, where $k$ is the number of days in that month (say February has 28 days). Assume the numbers of accidents is independent from month to month. If the company's year starts with January, the probability that the first accident is in April is

$P(\text{none in Jan., none in Feb., none in March, at least one in April}) =$
$$(1 - 0.31)(1 - 0.28)(1 - 0.31)(0.30) = (0.69)(0.72)(0.69)(0.30) = 0.103.$$

**EXAMPLE 1.4-8**    Three inspectors look at a critical component of a product. Their probabilities of detecting a defect are different, namely 0.99, 0.98, and 0.96, respectively. If we assume independence the probability of at least one detecting the defect is

$$1 - (0.01)(0.02)(0.04) = 0.999992.$$

The probability of only one finding the defect is

$$(0.99)(0.02)(0.04) + (0.01)(0.98)(0.04) + (0.01)(0.02)(0.96) = 0.001576.$$

As an exercise, compute the probability that: **(a)** exactly two find the defect, **(b)** all three find the defect.

**EXAMPLE 1.4-9**    Suppose that on five consecutive days an "instant winner" lottery ticket is purchased and the probability of winning is 1/5 on each day. Assuming independent trials,

$$P(WWLLL) = \left(\frac{1}{5}\right)^2 \left(\frac{4}{5}\right)^3,$$

$$P(LWLWL) = \frac{4}{5} \cdot \frac{1}{5} \cdot \frac{4}{5} \cdot \frac{1}{5} \cdot \frac{4}{5} = \left(\frac{1}{5}\right)^2 \left(\frac{4}{5}\right)^3.$$

In general, the probability of purchasing two winning tickets and three losing tickets is

$$\binom{5}{2}\left(\frac{1}{5}\right)^2\left(\frac{4}{5}\right)^3 = \frac{5!}{2!3!}\left(\frac{1}{5}\right)^2\left(\frac{4}{5}\right)^3 = 0.2048$$

because there are $\binom{5}{2}$ ways to select the positions (or the days) for the winning tickets, and each of these $\binom{5}{2}$ ways has the probability $(1/5)^2(4/5)^3$.

**1.4-1** Let $A$ and $B$ be independent events with $P(A) = 0.7$ and $P(B) = 0.2$. Compute **(a)** $P(A \cap B)$, **(b)** $P(A \cup B)$, and **(c)** $P(A' \cup B')$.

**1.4-2** Let $P(A) = 0.3$ and $P(B) = 0.6$.

   **(a)** Find $P(A \cup B)$ when $A$ and $B$ are independent.

   **(b)** Find $P(A \mid B)$ when $A$ and $B$ are mutually exclusive.

**1.4-3** Let $A$ and $B$ be independent events with $P(A) = 1/4$ and $P(B) = 2/3$. Compute: **(a)** $P(A \cap B)$, **(b)** $P(A \cap B')$, **(c)** $P(A' \cap B')$, **(d)** $P[(A \cup B)']$, and **(e)** $P(A' \cap B)$.

**1.4-4** If $A$, $B$, and $C$ are mutually independent, show that the following pairs of events are independent: $A$ and $(B \cap C)$, $A$ and $(B \cup C)$, $A'$ and $(B \cap C')$. Also show that $A'$, $B'$, and $C'$ are mutually independent.

**1.4-5** Each of three football players will attempt to kick a field goal from the 25-yard line. Let $A_i$ denote the event that the field goal is made by player $i$, $i = 1, 2, 3$. Assume that $A_1$, $A_2$, $A_3$ are mutually independent and that $P(A_1) = 0.5$, $P(A_2) = 0.7$, $P(A_3) = 0.6$.

   **(a)** Compute the probability that exactly one player is successful.

   **(b)** Compute the probability that exactly two players make a field goal (i.e., one misses).

**1.4-6** Die A has orange on one face and blue on five faces, Die B has orange on two faces and blue on four faces, Die C has orange on three faces and blue on three faces. These are unbiased dice. If the three dice are rolled, find the probability that exactly two of the three dice come up orange.

**1.4-7** Suppose that $A$, $B$, and $C$ are mutually independent events and that $P(A) = 0.5$, $P(B) = 0.8$, and $P(C) = 0.9$. Find the probabilities that **(a)** all three events occur, **(b)** exactly two of the three events occur, and **(c)** none of the events occur.

**1.4-8** Let $D_1, D_2, D_3$ be three four-sided dice whose sides have been labeled as follows:

$$D_1 : 0\,3\,3\,3 \qquad D_2 : 2\,2\,2\,5 \qquad D_3 : 1\,1\,4\,6$$

These dice are rolled at random. Let $A$, $B$, and $C$ be the events that the outcome on die $D_1$ is larger than the outcome on $D_2$, the outcome on $D_2$ is larger than the outcome on $D_3$, and the outcome on $D_3$ is larger than the outcome on $D_1$, respectively. Show that **(a)** $P(A) = 9/16$, **(b)** $P(B) = 9/16$, **(c)** $P(C) = 10/16$. Do you find it interesting that each of the probabilities that $D_1$ "beats" $D_2$, $D_2$ "beats" $D_3$, and $D_3$ "beats" $D_1$ is greater than 1/2? Thus it is difficult to determine the "best" die.

**1.4-9** An urn contains two red balls and four white balls. Sample successively five times at random and with replacement so that the trials are independent. Compute the probability of each of the two sequences $WWRWR$ and $RWWWR$.

**1.4-10** In Example 1.4-5, suppose that the probability of failure of a component is $p = 0.4$. Find the probability that the system does not fail if the number of redundant components is

   **(a)** 3.

   **(b)** 8.

**1.4-11** An urn contains 10 red and 10 white balls. The balls are drawn from the urn at random, one at a time. Find the probability that the fourth white ball is the sixth ball drawn if the sampling is done

   **(a)** With replacement.

   **(b)** Without replacement.

   **(c)** In the World Series the American League (red) and National League (white) teams play until one team wins four games. Do you think that this urn model could be used to describe the probabilities of a 4-, 5-, 6-, or 7-game series? If your answer is yes, would you choose sampling with or without replacement in your model? (For your information, the numbers of 4-, 5-, 6-, and 7-game series, up to and including 2004, were 18, 21, 21, 36. The World Series was cancelled in

1994. In 1903 and 1919–1921, winners had to take 5 out of 9 games. Three of those series went eight, one went seven.)

**1.4-12** An eight-team single elimination tournament is set up as follows:

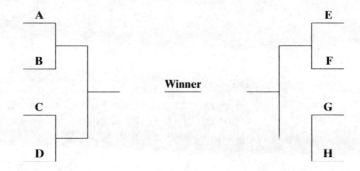

For example, eight students (called A–H) set up a tournament among themselves. The top listed student in each bracket calls Heads or Tails when their opponent flips a coin. If the call is correct, the student moves on to the next bracket.

**(a)** How many coin flips are required to determine the tournament winner?

**(b)** What is the probability that you can predict all of the winners?

**(c)** In NCAA Division I basketball, after a play-in game between the 64th and 65th seeds, 64 teams participate in a single elimination tournament to determine the national champion. Considering only the remaining 64 teams, how many games are required to determine the national champion?

**(d)** Assume that for any given game, either team has an equal chance of winning. (That is probably not true.) *Time*, on page 43 of the March 22, 1999 issue, claimed that the "mathematical odds of predicting all 63 NCAA games correctly is 1 in 75 million." Do you agree with this statement? If not, why not?

**1.4-13** Extend Example 1.4-6 to an $n$-sided die. That is, suppose that a fair $n$-sided die is rolled $n$ independent times. A match occurs if side $i$ is observed on the $i$th trial, $i = 1, 2, \ldots, n$.

**(a)** Show that the probability of at least one match is

$$1 - \left(\frac{n-1}{n}\right)^n = 1 - \left(1 - \frac{1}{n}\right)^n.$$

**(b)** Find the limit of this probability as $n$ increases without bound.

# 1.5 BAYES'S THEOREM

We consider Bayes's theorem in this section, beginning with an example.

Bowl $B_1$ contains two red and four white chips; bowl $B_2$ contains one red and two white chips; and bowl $B_3$ contains five red and four white chips. Say that the probabilities for selecting the bowls are not the same but are given by $P(B_1) = 1/3$, $P(B_2) = 1/6$, and $P(B_3) = 1/2$, where $B_1$, $B_2$, and $B_3$ are the events that bowls $B_1$, $B_2$, and $B_3$ are chosen, respectively. The experiment consists of selecting a bowl with these probabilities and then drawing a chip at random from that bowl. Let us compute the probability of event $R$, drawing a red chip, say $P(R)$. Note that $P(R)$ is dependent first of all on which bowl is selected and then on the probability of drawing a red chip from the selected bowl. That is, the event

$R$ is the union of the mutually exclusive events $B_1 \cap R$, $B_2 \cap R$, and $B_3 \cap R$. Thus

$$P(R) = P(B_1 \cap R) + P(B_2 \cap R) + P(B_3 \cap R)$$

$$= P(B_1)P(R \mid B_1) + P(B_2)P(R \mid B_2) + P(B_3)P(R \mid B_3)$$

$$= \frac{1}{3} \cdot \frac{2}{6} + \frac{1}{6} \cdot \frac{1}{3} + \frac{1}{2} \cdot \frac{5}{9} = \frac{4}{9}.$$

Suppose now that the outcome of the experiment is a red chip, but we do not know from which bowl it was drawn. Accordingly, we compute the conditional probability that the chip was drawn from bowl $B_1$, namely, $P(B_1 \mid R)$. From the definition of conditional probability and the result above, we have that

$$P(B_1 \mid R) = \frac{P(B_1 \cap R)}{.P(R)}$$

$$= \frac{P(B_1)P(R \mid B_1)}{P(B_1)P(R \mid B_1) + P(B_2)P(R \mid B_2) + P(B_3)P(R \mid B_3)}$$

$$= \frac{(1/3)(2/6)}{(1/3)(2/6) + (1/6)(1/3) + (1/2)(5/9)} = \frac{2}{8}.$$

Similarly, we have that

$$P(B_2 \mid R) = \frac{P(B_2 \cap R)}{P(R)} = \frac{(1/6)(1/3)}{4/9} = \frac{1}{8}$$

and

$$P(B_3 \mid R) = \frac{P(B_3 \cap R)}{P(R)} = \frac{(1/2)(5/9)}{4/9} = \frac{5}{8}.$$

Note that the conditional probabilities $P(B_1 \mid R)$, $P(B_2 \mid R)$, and $P(B_3 \mid R)$ have changed from the original probabilities $P(B_1)$, $P(B_2)$, and $P(B_3)$ in a way that agrees with your intuition. Namely, once the red chip has been observed, the probability concerning $B_3$ seems more favorable than originally because $B_3$ has a larger percentage of red chips than do $B_1$ and $B_2$. The conditional probabilities of $B_1$ and $B_2$ decrease from their original ones once the red chip is observed. Frequently, the original probabilities are called *prior probabilities*, and the conditional probabilities are the *posterior probabilities*.

We generalize the result of Example 1.5-1. Let $B_1$, $B_2$, ..., $B_m$ constitute a *partition* of the outcome space $O$. That is,

$$O = B_1 \cup B_2 \cup \cdots \cup B_m \text{ and } B_i \cap B_j = \emptyset, i \neq j.$$

Of course, the events $B_1$, $B_2$, ..., $B_m$ are mutually exclusive and exhaustive (since the union of the disjoint sets equals the outcome space $O$). Furthermore, suppose the **prior probability** of the event $B_i$ is positive; that is $P(B_i) > 0$, $i = 1, \ldots, m$. If $A$ is an event, then $A$ is the union of $m$ mutually exclusive events, namely,

$$A = (B_1 \cap A) \cup (B_2 \cap A) \cup \cdots \cup (B_m \cap A).$$

Thus

$$P(A) = \sum_{i=1}^{m} P(B_i \cap A)$$

$$= \sum_{i=1}^{m} P(B_i)P(A \mid B_i). \tag{1.5-1}$$

If $P(A) > 0$, we have that

$$P(B_k \mid A) = \frac{P(B_k \cap A)}{P(A)}, \qquad k = 1, 2, \ldots, m. \tag{1.5-2}$$

Using Equation 1.5-1 and replacing $P(A)$ in Equation 1.5-2, we have **Bayes's theorem**:

$$P(B_k \mid A) = \frac{P(B_k)P(A \mid B_k)}{\sum_{i=1}^{m} P(B_i)P(A \mid B_i)}, \qquad k = 1, 2, \ldots, m.$$

The conditional probability $P(B_k \mid A)$ is often called the **posterior probability** of $B_k$. The following example illustrates one application of Bayes's result.

**EXAMPLE 1.5-2** In a certain factory, machines I, II, and III are all producing springs of the same length. Of their production, machines I, II, and III produce 2%, 1%, and 3% defective springs, respectively. Of the total production of springs in the factory, machine I produces 35%, machine II produces 25%, and machine III produces 40%. If one spring is selected at random from the total springs produced in a day, the probability that it is defective, in an obvious notation, equals

$$P(D) = P(I)P(D \mid I) + P(II)P(D \mid II) + P(III)P(D \mid III)$$

$$= \left(\frac{35}{100}\right)\left(\frac{2}{100}\right) + \left(\frac{25}{100}\right)\left(\frac{1}{100}\right) + \left(\frac{40}{100}\right)\left(\frac{3}{100}\right) = \frac{215}{10,000}.$$

If the selected spring is defective, the conditional probability that it was produced by machine III is, by Bayes's formula,

$$P(III \mid D) = \frac{P(III)P(D \mid III)}{P(D)} = \frac{(40/100)(3/100)}{215/10,000} = \frac{120}{215}.$$

Note how the posterior probability of III increased from the prior probability of III after the defective spring was observed because III produces a larger percentage of defectives than do I and II.

**EXAMPLE 1.5-3** A Pap smear is a screening procedure used to detect cervical cancer. For women with this cancer, there are about 16% *false negatives*; that is,

$$P(T^- = \text{test negative} \mid C = \text{cancer}) = 0.16.$$

Thus

$$P(T^+ = \text{test positive} \mid C = \text{cancer}) = 0.84.$$

For women without cancer, there are about 19% false positives; that is,

$$P(T^+ \mid C' = \text{not cancer}) = 0.19.$$

Hence,

$$P(T^- \mid C' = \text{not cancer}) = 0.81.$$

In the United States, there are about 8 women in 100,000 who have this cancer; that is,

$$P(C) = 0.00008; \quad \text{so} \quad P(C') = 0.99992.$$

By Bayes's theorem,

$$P(C \mid T^+) = \frac{P(C \text{ and } T^+)}{P(T^+)}$$

$$= \frac{(0.00008)(0.84)}{(0.00008)(0.84) + (0.99992)(0.19)}$$

$$= \frac{672}{672 + 1899848} = 0.000354.$$

What this means is that for every million positive Pap smears, only 354 represent true cases of cervical cancer and makes one question the value of the procedure. The reason for this ineffective procedure is the percentage of women having that cancer is so small and the error rates of the procedure, namely 0.16 and 0.19, are so high. (See Yobs, A. R., Swanson, R. A., and Lamotte, L. C. "Laboratory Reliability of the Papanicolaou Smear." *Obstetrics and Gynecology*, Volume 65, February 1985, pp. 235–244.)

## EXERCISES 1.5

**1.5-1** Bowl $B_1$ contains 2 white chips, bowl $B_2$ contains 2 red chips, bowl $B_3$ contains 2 white and 2 red chips, and bowl $B_4$ contains 3 white chips and 1 red chip. The probabilities of selecting bowl $B_1, B_2, B_3,$ or $B_4$ are 1/2, 1/4, 1/8, and 1/8, respectively. A bowl is selected using these probabilities, and a chip is then drawn at random. Find

**(a)** $P(W)$, the probability of drawing a white chip.

**(b)** $P(B_1 \mid W)$, the conditional probability that bowl $B_1$ had been selected, given that a white chip was drawn.

**1.5-2** Bean seeds from supplier $A$ have an 85% germination rate and those from supplier $B$ have a 75% germination rate. A seed packaging company purchases 40% of their bean seeds from supplier $A$ and 60% from supplier $B$ and mixes these seeds together.

**(a)** Find the probability $P(G)$ that a seed selected at random from the mixed seeds will germinate.

**(b)** Given that a seed germinates, find the probability that the seed was purchased from supplier $A$.

**1.5-3** The Belgium 20-franc coin ($B20$), the Italian 500-lire coin ($I500$), and the Hong Kong 5-dollar coin ($HK5$) are approximately the same size. Coin purse 1 ($C1$) contains 6 of each of these coins. Coin purse 2 ($C2$) contains 9 $B20$'s, 6 $I500$'s, and 3 $HK5$'S. A fair four-sided die is rolled. If the outcome is $\{1\}$, a coin is selected randomly from $C1$. If the outcome belongs to $\{2, 3, 4\}$, a coin is selected randomly from $C2$. Find

**(a)** $P(B20)$, the probability of selecting a Belgian coin.

**(b)** $P(C1 \mid B20)$, the probability that the coin was selected from $C1$, given that it was a Belgian coin.

**1.5-4** Assume that an insurance company knows the following probabilities relating to automobile accidents:

| Age of Driver | Probability of Accident | Fraction of Company's Insured Drivers |
|---|---|---|
| 16–25 | 0.05 | 0.10 |
| 26–50 | 0.02 | 0.55 |
| 51–65 | 0.03 | 0.20 |
| 66–90 | 0.04 | 0.15 |

A randomly selected driver from the company's insured drivers has an accident. What is the conditional probability that the driver is in the 16–25 age group?

**1.5-5** At a hospital's emergency room, patients are classified and 20% of them are critical, 30% are serious, and 50% are stable. Of the critical ones, 30% die; of the serious, 10% die; and of the stable, 1% die. Given that a patient dies, what is the conditional probability that the patient was classified as critical?

**1.5-6** A life insurance company issues standard, preferred, and ultra-preferred policies. Of the company's policyholders of a certain age, 60% are standard with a probability of 0.01 of dying in the next year, 30% preferred with a probability of 0.008 of dying in the next year, and 10% are ultra-preferred with a probability of 0.007 of dying in the next year. A policyholder of that age dies in the next year. What are the conditional probabilities of the deceased being standard, preferred, and ultra-preferred?

**1.5-7** Among 60-year-old college professors, 10% are smokers and 90% are nonsmokers. The probability of a nonsmoker dying in the next year is 0.005 and the probability for smokers is 0.05. Given that one of this group of college professors dies in the next year, what is the conditional probability that the professor is a smoker?

**1.5-8** A store sells four brands of VCRs. The least expensive brand $B_1$ accounts for 40% of the sales. The other brands (in order of their price) have the following percentages of sales: $B_2, 30\%$; $B_3, 20\%$; and $B_4, 10\%$. The respective probabilities of needing repair during warranty are 0.10 for $B_1$, 0.05 for $B_2$, 0.03 for $B_3$, and 0.02 for $B_4$. A randomly selected purchaser has a VCR that needed repair under warranty. What are the four conditional probabilities of being brand $B_i$, $i = 1, 2, 3, 4$?

**1.5-9** There is a new diagnostic test for a disease that occurs in about 0.05% of the population. The test is not perfect but will detect a person with the disease 99% of the time. It will, however, say that a person without the disease has the disease about 3% of the time. A person is selected at random from the population and the test indicates that this person has the disease. What are the conditional probabilities that

**(a)** The person has the disease?

**(b)** The person does not have the disease?

Discuss. HINT: Note the fraction 0.0005 of diseased persons in the population is much smaller than the error probabilities of 0.01 and 0.03.

**1.5-10** Suppose we want to investigate the percentage of abused children in a certain population. To do this, doctors examine some of these children taken at random from this population. However, doctors are not perfect: They sometimes classify an abused child ($A$) as one not abused ($ND$) or they classify a nonabused child ($N$) as one that is abused ($AD$). Suppose these error rates are $P(ND \mid A) = 0.08$ and $P(AD \mid N) = 0.05$, respectively; thus $P(AD \mid A) = 0.92$ and $P(ND \mid N) = 0.95$ are probabilities of the correct decisions. Let us pretend that only two percent of all children are abused; that is, $P(A) = 0.02$ and $P(N) = 0.98$.

**(a)** Select a child at random. What is the probability that the doctor classifies this child as abused? That is, compute

$$P(AD) = P(A)P(AD \mid A) + P(N)P(AD \mid N).$$

**(b)** Given the child is classified by the doctor as abused, compute $P(N \mid AD)$ and $P(A \mid AD)$.

**(c)** Also compute $P(N \mid ND)$ and $P(A \mid ND)$.

**(d)** Are these probabilities in (b) and (c) alarming? This happens because the error rates of 0.08 and 0.05 are high relative to the fraction 0.02 of abused children in the population.

# CHAPTER ONE COMMENTS

With only these few sections on probability, the reader is certainly not an expert in that subject. However, it is interesting to note that we have enough background to solve certain problems that experts in that area consider the beginning of probability. The solutions are relatively easy today, but they were extremely difficult in earlier days.

Most probabilists would say that the mathematics of probability began when in 1654, Chevalier de Méré, a French nobleman who liked to gamble, challenged Blaise Pascal to explain a puzzle and a problem created from his observations concerning rolls of dice. Of course, there was gambling well before this; and actually almost 200 years before this challenge, a Franciscan monk, Luca Paccioli, proposed that puzzle.

"*A* and *B* are playing a fair game of balla. (*A* and *B* have an equal chance of winning.) They agree to continue until one has won six rounds. However, the game actually stops when *A* has won five and *B* three. How should the stakes be divided?"

And over 100 years before de Méré's challenge, a sixteenth-century doctor, Girolamo Cardano, who was also a gambler, had figured out many dice problems, but not the one that de Méré proposed. Chevalier de Méré had observed this. If a single die is tossed 4 times, the probability of obtaining at least one "six" was slightly greater than 1/2. However, keeping the same proportions, if a pair of dice is tossed 24 times, the probability of obtaining at least one double-six seemed to be slightly less than 1/2; at least he was losing money betting on it. This is when he approached Blaise Pascal with the challenge. Not wanting to work on the problems alone, Pascal formed a partnership with Pierre de Fermat, a great selection by Pascal. It was this 1654 correspondence between Pascal and Fermat that started the theory of probability.

Today an average student in probability could solve both easily. For the puzzle, note that *B* could only win six rounds by winning the next three rounds, which has probability of $(1/2)^3 = 1/8$ because it was a fair game of balla. Thus *A*'s probability of winning six rounds is $1 - 1/8 = 7/8$. So the stakes should be divided seven units to one.

For the dice problem, the probability of at least one six in four rolls of a die is

$$1 - \left(\frac{5}{6}\right)^4 = 0.518, \qquad \text{approximately,}$$

while the probability of rolling at least one double-six in 24 rolls of a pair of dice is

$$1 - \left(\frac{35}{36}\right)^{24} = 0.491, \qquad \text{approximately.}$$

It seems amazing to us that de Méré could have observed enough trials of those events to detect the slight difference in those probabilities. However, he won betting on the first but lost by betting on the second.

# CHAPTER

# 2

# DISCRETE DISTRIBUTIONS

**DISCRETE PROBABILITY DISTRIBU-TIONS**

Let us consider the following situation. Place ten chips of the same size in a bowl: one of which has the number one on it, two have the number two, three have the number three, and four have the number four. Then a chip is selected at random, and we observe the number on that chip. Of course, the answer must be one of the numbers from the set $S = \{1, 2, 3, 4\}$. If, before the random experiment is performed, we denote the future observed value by $X$, we say that $X$ is a **random variable** with **space** or **support** $S$. If $P(X = x)$ represents the probability $X = x$, where $x \, \varepsilon \, S$, we could assign the probabilities

$$P(X = 1) = \frac{1}{10}, \quad P(X = 2) = \frac{2}{10}, \quad P(X = 3) = \frac{3}{10}, \quad P(X = 4) = \frac{4}{10}$$

to obtain a distribution of probability on the support $S$. More simply, we could call $P(X = x) = f(x)$ and write

$$f(x) = \frac{x}{10}, \qquad x = 1, 2, 3, 4.$$

Such a function is called a **probability mass function** and abbreviated **p.m.f.** (As we will see later, the function corresponding to $f(x)$ in the continuous case is called the probability density function, abbreviated p.d.f.; and sometimes, particularly in *Maple*, the discrete p.m.f. is referred to as the PDF. Note the capital letters and no periods.) We could depict these probabilities with a bar graph called a **probability histogram** consisting of rectangles with heights $f(x)$, $x = 1, 2, 3, 4$, and bases of length one, each centered over its respective $x$ value. See the rectangles with the dashed tops in Figure 2.1-1.

**Remark**   Often in a formal definition of a random variable, we first consider an outcome space of a random experiment. A **random variable** is a real valued function defined on that outcome space. The range of this function is called the *space* or *support* of the random variable. In our first illustration, we could consider each of the 10 chips as one of the outcomes of the random experiment of selecting a chip. A reasonable assignment of probability to each of these outcomes is 1/10. In our illustration, the function (the random variable, say $X$) is the number on the selected chip and $X$ has the space $S = \{1, 2, 3, 4\}$. The probabilities *induced* on $S$ are:

$$P(X = 1) = \frac{1}{10}, \qquad\qquad P(X = 2) = \frac{1}{10} + \frac{1}{10} = \frac{2}{10}$$

$$P(X = 3) = \frac{1}{10} + \frac{1}{10} + \frac{1}{10} = \frac{3}{10}, \quad P(X = 4) = \frac{1}{10} + \frac{1}{10} + \frac{1}{10} + \frac{1}{10} = \frac{4}{10}.$$

These define the probability mass function (p.m.f.) $f(x) = P(X = x)$, $x \, \varepsilon \, S$, which in our illustration can be written $f(x) = x/10$, $x = 1, 2, 3, 4$.

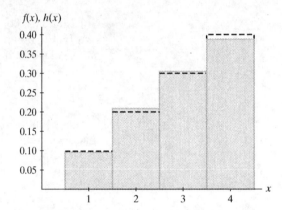

**Figure 2.1-1** Relative frequency histogram $h(x)$ (shaded), the probability histogram $f(x)$ (dashed) superimposed

Of course, using this p.m.f. $f(x)$, we can compute probabilities of events defined by expressions like $\{X = 3 \text{ or } 4\}$ and $\{X \leq 3\}$ by adding probabilities of mutually exclusive events together. In particular, for those respective probabilities, we have

$$P(X = 3, 4) = f(3) + f(4) = \frac{3}{10} + \frac{4}{10} = \frac{7}{10}$$

and

$$P(X \leq 3) = f(1) + f(2) + f(3) = \frac{1}{10} + \frac{2}{10} + \frac{3}{10} = \frac{6}{10}.$$

In general, if $A \subset S$,

$$P(A) = \sum_{x \, \varepsilon \, A} f(x).$$

We actually performed this experiment $n = 1000$ times. We could have done it by selecting a chip 1000 times, but we simulated the experiment using a computer obtaining the respective frequencies 98, 209, 305, 388 for $x = 1, 2, 3, 4$. The respective relative frequencies are

$$h(1) = \frac{98}{1000} = 0.098, \qquad h(2) = \frac{209}{1000} = 0.209,$$

$$h(3) = \frac{305}{1000} = 0.305, \qquad h(4) = \frac{388}{1000} = 0.388,$$

and these approximate the corresponding probabilities, as they should. The corresponding function $h(x)$ is called the **relative frequency histogram**, and it is plotted as the shaded part of Figure 2.1-1. The collection of $n = 1000$ observations is sometimes referred to as a **sample** from this distribution of probability.

There are some interesting characteristics associated with these two functions which we now explore. First both are nonnegative functions such that

$$\sum_{x=1}^{4} f(x) = 1 \quad \text{and} \quad \sum_{x=1}^{4} h(x) = 1.$$

That is, the probabilities and relative frequencies sum to one when the summations are taken over all possible $x$ values.

Second, we can find a weighted average of the numbers $1, 2, 3, 4$ for each situation in which the weights are $f(x)$ and $h(x)$, $x = 1, 2, 3, 4$, respectively. The weighted average associated with $f(x)$ is

$$(1)(0.1) + (2)(0.2) + (3)(0.3) + (4)(0.4) = \sum_{x=1}^{4} xf(x),$$

which equals 3.0. We call this the **mean** of the (probability) distribution or, more simply, the **mean** of $X$; and we denote it by the Greek letter $\mu$ (the Greek mu for mean). That is, here

$$\mu = \sum_{x=1}^{4} xf(x) = 3.0.$$

In this situation, $\mu = 3.0$ is one of the numbers in the support $S = \{1, 2, 3, 4\}$, but this need not be the case; that is, the mean $\mu$ could equal a number not in $S$, like 2.6 or 3.15, in other situations.

The weighted average associated with $h(x)$ is

$$(1)(0.098) + (2)(0.209) + (3)(0.305) + (4)(0.388) = \sum_{x=1}^{4} xh(x),$$

which equals 2.983. We call this the **mean** of the sample of size $n = 1000$ and denote it by $\bar{x}$. That is, $\bar{x}$ is the mean of the 1000 observed values of $x$. If we think of placing the weight $1/n = 1/1000$ on each observed value of $x$, then of course the weight on $x = 1$ would be $98/1000 = 0.098$ since there were 98 values of one in the sample of $n = 1000$. Likewise the weights for $x = 2, 3, 4$ are as given earlier. Note these weights sum to one. Of course, if $x_1, x_2, \ldots, x_{1000}$ represent the $n = 1000$ individual observed values, then

$$\bar{x} = \frac{1}{n} \sum_{i=1}^{n} x_i = \sum_{x=1}^{4} xh(x) = \frac{1}{n} \sum_{j=1}^{4} f_j u_j,$$

where $f_j$ is the frequency of $u_j$; here $u_j = 1, 2, 3, 4$. Placing weight $1/n$ on each observed $x_i$ creates an **empirical distribution**, and $\bar{x}$ is sometimes referred to as the **mean of the empirical distribution** or as the **sample mean**.

The next characteristic of a distribution is the weighted average of $(x - \mu)^2$, where here $x = 1, 2, 3, 4$ and $\mu = 3$. It is

$$(1 - 3)^2(0.1) + (2 - 3)^2(0.2) + (3 - 3)^2(0.3) + (4 - 3)^2(0.4) = \sum_{x=1}^{4} (x - \mu)^2 f(x),$$

which equals 1.0. This is called the **variance** of the distribution or, more simply, the **variance** of $X$. It is denoted by $\sigma^2$ or Var($X$). The Greek s is $\sigma$, pronounced sigma, and the positive square root of the variance, $\sigma = \sqrt{\sigma^2} = 1$ is called the **standard deviation** of the distribution or, more simply, of $X$.

At this point of the study of statistics, it is difficult to see why the standard deviation is an important characteristic of a distribution. However, we can note that it is some measure of spread; for if we had the same probabilities $0.1, 0.2, 0.3, 0.4$ on $x = 1, 3, 5, 7$, respectively, the new mean is $\mu = 5$ and the new standard deviation is $\sigma = 2$. Of course, this new distribution of probability is spread out twice as much as the old; the standard deviation is twice as large.

There is another observation that should be made about $\sigma^2$, namely in general, summing for all $x$,

$$\sigma^2 = \sum_x (x - \mu)^2 f(x) = \sum_x (x^2 - 2x\mu + \mu^2) f(x)$$

$$= \sum_x x^2 f(x) - 2\mu \sum_x x f(x) + \mu^2 \sum_x f(x),$$

because $\mu$ can be taken outside the summation as it is a constant. However,

$$\mu = \sum_x x f(x) \qquad \text{and} \qquad \sum_x f(x) = 1;$$

thus

$$\sigma^2 = \sum_x x^2 f(x) - (2\mu)(\mu) + \mu^2 = \sum_x x^2 f(x) - \mu^2.$$

In our illustration, we see

$$\sigma^2 = \sum_{x=1}^{4} x^2 f(x) - \mu^2 = (1^2)(0.1) + (2^2)(0.2) + (3^2)(0.3) + (4^2)(0.4) - 3^2$$

$$= 10.0 - 9.0 = 1.0,$$

which is the same result as before, as it must be.

> **Remark**  In the physical sciences, a summation like $\sum_x x^r f(x)$ is often referred to as the **$r$th moment about the origin**. Hence $\mu$ is the first moment about the origin and $\sum_x x^2 f(x)$ is the second moment about the origin.

If we return to the sample of size $n = 1000$, the corresponding weighted average of $(x - \bar{x})^2$ is

$$\sum_{x=1}^{4} (x - 2.983)^2 h(x) = \sum_{x=1}^{4} x^2 h(x) - 2.983^2$$

$$= 1^2(0.098) + 2^2(0.209) + 3^2(0.305) + 4^2(0.388) - 2.983^2$$

$$= 9.887 - 8.898 = 0.989.$$

We use the letter $v$ to denote this **variance of the empirical distribution**. However we do <u>not</u> call $v$ the variance of the sample, but rather we use

$$s^2 = \frac{n}{n-1} v = \left( \frac{1000}{999} \right)(0.989) = 0.990$$

as the **variance of the sample**. We do this because later we discover that, in some sense, $s^2$ is a better estimate of $\sigma^2$ than is $v$. Of course, here $s = \sqrt{s^2} = 0.995$, which is an excellent estimate of $\sigma = 1$ and $\bar{x} = 2.983$ is an excellent estimate of $\mu = 3$. These should be rather close since $h(x)$ approximates $f(x)$ quite well in this case, although here the results of our simulation were somewhat closer than might be the usual case. We were lucky!

**EXAMPLE 2.1-1**   Let $X$ equal the number of spots on the side that is up after a six-sided die (one of a pair of dice) is cast at random. If everything is fair about this experiment, a reasonable probability model is given by the p.m.f.

$$f(x) = P(X = x) = \frac{1}{6}, \qquad x = 1, 2, 3, 4, 5, 6.$$

The mean of $X$ is

$$\mu = \sum_{x=1}^{6} x\left(\frac{1}{6}\right) = \frac{1 + 2 + 3 + 4 + 5 + 6}{6} = \frac{7}{2}.$$

The second moment about the origin is

$$\sum_{x=1}^{6} x^2\left(\frac{1}{6}\right) = \frac{1^2 + 2^2 + 3^2 + 4^2 + 5^2 + 6^2}{6} = \frac{91}{6}.$$

Thus the variance equals

$$\sigma^2 = \frac{91}{6} - \left(\frac{7}{2}\right)^2 = \frac{182 - 147}{12} = \frac{35}{12}.$$

The standard deviation is $\sigma = \sqrt{35/12} = 1.708$.

**EXAMPLE 2.1-2**   Let the p.m.f. of $X$ be defined by $f(x) = x/6, x = 1, 2, 3$. The mean of $X$ is

$$\mu = 1\left(\frac{1}{6}\right) + 2\left(\frac{2}{6}\right) + 3\left(\frac{3}{6}\right) = \frac{7}{3}.$$

To find the variance and standard deviation of $X$ we first find the second moment about the origin

$$1^2\left(\frac{1}{6}\right) + 2^2\left(\frac{2}{6}\right) + 3^2\left(\frac{3}{6}\right) = \frac{36}{6} = 6.$$

Thus the variance of $X$ is

$$\sigma^2 = 6 - \left(\frac{7}{3}\right)^2 = \frac{5}{9}$$

and the standard deviation of $X$ is

$$\sigma = \sqrt{5/9} = 0.745.$$

Although most students understand that $\mu$ is, in some sense, a measure of the middle of the distribution of $X$, it is difficult to get much of a feeling for the variance and the standard deviation. The following example illustrates that the standard deviation is a measure of dispersion or spread of the points belonging to the space $S$.

**EXAMPLE 2.1-3**   Let $X$ have the p.m.f. $f(x) = 1/3$, $x = -1, 0, 1$. Here the mean is

$$\mu = \sum_{x=-1}^{1} xf(x) = (-1)\left(\frac{1}{3}\right) + (0)\left(\frac{1}{3}\right) + (1)\left(\frac{1}{3}\right) = 0.$$

Accordingly, the variance, denoted by $\sigma_X^2$, is

$$\sigma_X^2 = \sum_{x=-1}^{1} x^2 f(x) - 0^2$$

$$= (-1)^2\left(\frac{1}{3}\right) + (0)^2\left(\frac{1}{3}\right) + (1)^2\left(\frac{1}{3}\right)$$

$$= \frac{2}{3},$$

so the standard deviation is $\sigma_X = \sqrt{2/3}$. Next let another random variable $Y$ have the p.m.f. $g(y) = 1/3$, $y = -2, 0, 2$. Its mean is also zero, and it is easy to show that $\text{Var}(Y) = 8/3$, so the standard deviation of $Y$ is $\sigma_Y = 2\sqrt{2/3}$. Here the standard deviation of $Y$ is twice that of the standard deviation of $X$, reflecting the fact that the probability of $Y$ is spread out twice as much as that of $X$.

**EXAMPLE 2.1-4**   Rolling a fair six-sided die five independent times could result in the following sample of $n = 5$ observations:

$$x_1 = 3, \quad x_2 = 1, \quad x_3 = 2, \quad x_4 = 6, \quad x_5 = 3.$$

In this case

$$\bar{x} = \frac{3 + 1 + 2 + 6 + 3}{5} = 3$$

and

$$s^2 = \frac{(3-3)^2 + (1-3)^2 + (2-3)^2 + (6-3)^2 + (3-3)^2}{4} = \frac{14}{4} = 3.5.$$

It follows that $s = \sqrt{14/4} = 1.87$. Roughly, $s$ can be thought of as the average distance that the $x$-values are away from the sample mean $\bar{x}$. Clearly this is not exact because in this example the distances from $\bar{x} = 3$ are 0, 2, 1, 3, 0, with an average of 1.2. In general $s$ will be somewhat larger than this average distance.

**EXAMPLE 2.1-5**   Let us cast an unbiased die a number of independent times until the first "five" or "six" appears on the upside of the die. Let the random variable $X$ be the number of times needed to obtain the first five or six. Let us call a five or a six "success." Of course, the support of $X$ is $S = \{x: x = 1, 2, 3, 4, \ldots\}$, the set of all positive integers. The probability of casting a five or six (success) is $2/6 = 1/3$. Thus the p.m.f. of $X$ is

$$f(x) = P(X = x) = \left(\frac{2}{3}\right)^{x-1}\left(\frac{1}{3}\right), \qquad x = 1, 2, 3, 4, \ldots,$$

because if $X = x$ we need $x - 1$ failures, followed by a success on the $x^{th}$ trial.

The mean of $X$ is

$$\mu = \sum_{x=1}^{\infty} xf(x) = 1\left(\frac{1}{3}\right) + 2\left(\frac{2}{3}\right)\left(\frac{1}{3}\right) + 3\left(\frac{2}{3}\right)^2\left(\frac{1}{3}\right) + \cdots.$$

If we compare this series to the Taylor's expansion

$$(1 - y)^{-2} = 1 + 2y + 3y^2 + \cdots, \qquad |y| < 1,$$

we see that

$$\mu = \left(\frac{1}{3}\right)\left(1 - \frac{2}{3}\right)^{-2} = \left(\frac{1}{3}\right)\left(\frac{1}{3}\right)^{-2} = 3.$$

It can also be shown that $\sigma^2 = 6$ and thus the standard deviation is $\sigma = \sqrt{6} = 2.449$. A *Maple* solution is given in Example 5.2-1.

We simulated this experiment, using the computer, $n = 200$ times resulting in the respective frequencies for $x = 1, 2, 3, \ldots$ given by

| $x$ | 1 | 2 | 3 | 4 | 5 | 6 | 7 | 8 | 9 | 10 | 11 | 12 |
|-----|-----|-----|-----|-----|-----|-----|-----|-----|-----|-----|-----|-----|
| $f$ | 61 | 48 | 26 | 26 | 12 | 12 | 4 | 5 | 2 | 2 | 0 | 2 |

The sample mean is

$$\bar{x} = \frac{1}{200}[(61)(1) + (48)(2) + (26)(3) + \cdots + (2)(12)] = \frac{601}{200} = 3.005$$

and

$$v = \frac{1}{200}[(61)(1)^2 + (48)(2)^2 + (26)(3)^2 + \cdots + (2)(12)^2] - 3.005^2$$

$$= \frac{198,999}{40,000} = 4.974975.$$

Accordingly the sample variance is

$$s^2 = \frac{200}{199} \cdot v = 4.999975$$

and the sample standard deviation is

$$s = 2.236.$$

The probability model and the simulated data are compared in Figure 2.1-2.

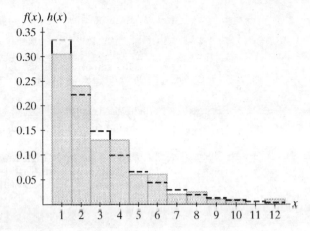

$f(x), h(x)$

**Figure 2.1-2** Relative frequency histogram $h(x)$ (shaded), the probability histogram $f(x)$ (dashed) superimposed

## EXERCISES 2.1

**2.1-1** Find the mean and variance for the following discrete distributions:

**(a)** $f(x) = \dfrac{1}{5}$,    $x = 5, 10, 15, 20, 25$.

**(b)** $f(x) = 1$,    $x = 5$.

**(c)** $f(x) = \dfrac{4 - x}{6}$,    $x = 1, 2, 3$.

**2.1-2** Let us draw a card at random from an ordinary deck of cards. Define the random variable $X$ in the following way: $X = 4$ if an ace is drawn, $X = 3$ if a king is drawn, $X = 2$ if a queen is drawn, $X = 1$ if a jack is drawn, and $X = 0$ for any other card.

**(a)** Give the p.m.f. $f(x)$ of $X$.

**(b)** Compute $P(X \geq 1)$.

**(c)** Evaluate the mean $\mu$ and the variance $\sigma^2$ of $X$.

**(d)** Draw a probability histogram for this p.m.f.

**2.1-3** For each of the following, determine the constant $c$ so that $f(x)$ satisfies the conditions of being a p.m.f.

**(a)** $f(x) = cx$,    $x = 1, 2, 3, 4, 5, 6$.

**(b)** $f(x) = c\left(\dfrac{2}{3}\right)^x$,    $x = 1, 2, 3, 4, \ldots$.

**(c)** $f(x) = c(x + 1)^2$,    $x = 0, 1, 2$.

**(d)** $f(x) = c(x^2 + 1)$,    $x = -1, 0, 1$.

Depict each $f(x)$ as a probability histogram.

**2.1-4** For the p.m.f. in parts **(a)**, **(c)**, **(d)** of Exercise 2.1-3, compute the mean $\mu$ and the variance $\sigma^2$ of the distribution. If available, use a CAS to compute the mean and variance for part **(b)**.

**2.1-5** The State of Arizona generates a three-digit number at random six days a week for their daily lottery. The numbers are generated one digit at a time. Consider the following set of 50 three-digit numbers as 150 one-digit integers that were generated at random from May 1, 2000 to July 3, 2000:

| | | | | | | | | | |
|---|---|---|---|---|---|---|---|---|---|
| 9 5 1 | 7 2 8 | 8 1 8 | 9 2 2 | 8 5 0 | 8 3 5 | 0 0 3 | 4 0 6 | 2 0 3 | 6 0 3 |
| 0 1 1 | 2 1 7 | 8 0 3 | 7 7 6 | 3 9 7 | 0 1 9 | 7 8 5 | 1 8 5 | 6 3 2 | 2 4 5 |
| 9 4 5 | 9 2 9 | 5 0 8 | 8 4 9 | 5 1 6 | 7 2 9 | 3 0 6 | 3 0 5 | 2 7 8 | 1 0 0 |
| 0 8 9 | 8 6 0 | 9 1 8 | 1 2 4 | 6 7 5 | 2 2 0 | 7 2 8 | 7 5 1 | 7 8 6 | 6 0 9 |
| 0 7 6 | 3 2 0 | 7 3 2 | 9 1 1 | 9 1 3 | 5 5 6 | 3 6 7 | 8 9 7 | 5 4 0 | 9 7 9 |

Let $X$ denote the outcome when a single digit is generated.

(a) With true random numbers, what is the p.m.f. of $X$? Draw the probability histogram.

(b) For the 150 observations, determine the relative frequencies of 0, 1, 2, 3, 4, 5, 6, 7, 8, and 9, respectively.

(c) Construct a probability histogram and a relative frequency histogram like Figures 2.1-1 and 2.1-2.

(d) Compute the means, $\mu$ and $\bar{x}$, and the variances, $\sigma^2$ and $s^2$, of the distribution and of the sample, respectively.

(e) The following three-digit numbers were generated from May 4, 1998–June 9, 1998. Answer **(b)–(d)** for these data. (For a more complete discussion of these numbers, see "The Case of the Missing Lottery Number," by W. D. Kaigh in *The College Mathematics Journal*, January, 2001, pages 15–19.)

```
287   215   846   873   485   812   432   415   348   383
334   075   655   871   824   704   533   014   822   477
880   316   012   010   557   881   222   256   282   704
111   124
```

**2.1-6** The p.m.f. of $X$ is $f(x) = (5-x)/10$, $x = 1, 2, 3, 4$.

(a) For the following 100 independent observations of $X$, simulated on a computer, determine the relative frequencies of 1, 2, 3, and 4.

```
3   1   2   2   3   2   2   2   1   3   3   2   3   2   4   4   2   1   1   3
3   1   2   2   1   1   4   2   3   1   1   1   2   1   3   1   1   3   3   1
1   1   1   1   1   4   1   3   1   2   4   1   1   2   3   4   3   1   4   2
2   1   3   2   1   4   1   1   1   2   1   3   4   3   2   1   4   4   1   3
2   2   2   1   2   3   1   1   4   2   1   4   2   1   2   3   1   4   2   3
```

(b) Construct a probability histogram and a relative frequency histogram like Figures 2.1-1 and 2.1-2.

(c) Compute the means, $\mu$ and $\bar{x}$, and the variances, $\sigma^2$ and $s^2$, of the distribution and of the sample, respectively.

**2.1-7** Let a random experiment be the cast of a pair of unbiased dice, each having six faces, and let the random variable $X$ denote the sum of the dice.

(a) With reasonable assumptions, determine the p.m.f. $f(x)$ of $X$. HINT: Picture the outcome space consisting of the 36 points (result on first die, result on second die), and assume that each has probability 1/36. Find the probability for each possible value of $X$, namely, $x = 2, 3, 4, \ldots, 12$.

(b) Draw the probability histogram for $f(x)$.

(c) Compute the mean $\mu$ and the variance $\sigma^2$ of the distribution.

**2.1-8** Let a random experiment be the cast of a pair of unbiased six-sided dice and let $X$ equal the smaller of the outcomes if they are different and the common value if they are equal.

(a) With reasonable assumptions, find the p.m.f. of $X$.

(b) Draw a probability histogram.

(c) Let $Y$ equal the range of the two outcomes, i.e., the absolute value of the difference of the largest and the smallest outcomes. Determine the p.m.f. $g(y)$ of $Y$ for $y = 0, 1, 2, 3, 4, 5$.

(d) Draw the probability histogram for $g(y)$.

**2.1-9** In a lot of 100 light bulbs, there are 5 bad bulbs. An inspector inspects 10 bulbs selected at random and without replacement. Find the probability of finding at least one defective bulb. HINT: First compute the probability of finding no defectives in the sample.

**2.1-10** (Birthday Problem) (Also see Exercise 5.3-7.) Consider the birthdays of the students in a class of size $r$. Assume that the year consists of 365 days and that birthdays are distributed uniformly on the integers from 1 to 365, inclusive. That is, each day has probability 1/365.

**(a)** Show that the probability that all of the students have different birthdays is given by

$$\frac{365!}{(365-r)!\,365^r}.$$

**(b)** Define the probability as a function of $r$, say $P(r)$, that at least two students have the same birthday.

**(c)** Graph the set of points, $\{[r, P(r)], r = 1, 2, \ldots, 80\}$.

**(d)** Find the smallest value of $r$ for which $P(r) > 1/2$. Either use your graph in part (c) or use your calculator or use a Computer Algebra System, if it is available. For example, in *Maple* you could use:

Maple:
```
>   fsolve({1-365!/(365-r)!/365^r=1/2}, {r}, {r=1..60});
>   assign(%);
>   r := r;
>   r := ceil(r);
```
★

**(e)** Find the smallest value of $r$ for which **(i)** $P(r) > 3/4$, **(ii)** $P(r) > 9/10$.

**(f)** For $p = 0.05, 0.10, \ldots, 0.95$, find the smallest integer $r$ for which $P(r) > p$. Plot the 19 pairs of points, $\{[p, r]\}$.

**2.1-11** Continuing with the birthday problem, given a large class of students, let the students in turn give their birthday. Suppose the students are numbered $1, 2, \ldots$. Let the random variable $Y$ equal the number of the first student whose birthday matches one of the previous students. Renumber the remaining students, beginning with 1, and repeat the experiment.

**(a)** Define the p.m.f. of $Y$.

**(b)** Find $\mu = E(Y)$ and $\sigma^2 = \text{Var}(Y)$.

**(c)** Illustrate these results empirically. Note that this experiment is equivalent to rolling a 365-sided die and letting $Y$ equal the roll on which the first match is observed.

**2.2   EXPECTATIONS**   If $X$ is a random variable with p.m.f. $f(x)$ with space (support) $S$, we have defined the mean of the distribution (or, more simply, of $X$) to be

$$\mu = \sum_{x \varepsilon S} x f(x).$$

In some books, it is pointed out that we really should have a certain absolute convergence, namely,

$$\sum_{x \varepsilon S} |x| f(x) \quad \text{should exist,}$$

before we can say that the **expectation of a random variable**, given by

$$E(X) = \sum_{x \varepsilon S} x f(x),$$

exists and then, of course, is equal to $\mu$. In most cases in this book, we do not concern ourselves with this absolute convergence, for we do not deal much with these unusual distributions.

Remark    It is possible to find a constant $c$ so that

$$f(x) = \frac{c}{x^2}, \qquad x = \pm 1, \pm 2, \cdots$$

is a p.m.f. The support $S$ of $X$ is all positive and negative integers. Since this p.m.f. is symmetric about the point zero, we might guess that the mean is zero with the positive products $(x)\frac{c}{x^2}$ cancelling the negative products $(-x)\frac{c}{x^2}$, $x = 1, 2, 3, \ldots$. However,

$$\sum_{x \varepsilon S} |x| \frac{c}{x^2} = \sum_{x \varepsilon S} \frac{c}{|x|}$$

clearly does not exist because we are dealing with twice the sum of the harmonic series

$$1 + \frac{1}{2} + \frac{1}{3} + \cdots,$$

which does not exist. So the mean $\mu$ does not exist.

To find the value of $c$, we use the following known result:

$$\sum_{x=1}^{\infty} \frac{1}{x^2} = \frac{\pi^2}{6}.$$

Thus

$$\sum_{x \varepsilon S} \frac{c}{x^2} = \frac{2c\pi^2}{6}.$$

It follows that

$$c = \frac{3}{\pi^2}.$$

You could also find the value of $c$ to make $f(x)$ a p.m.f. with the following *Maple* command:

Maple:
```
> solve({sum(c/x^2 + c/(-x)^2, x = 1 .. infinity) = 1}, {c});
```
$$\{c = 3\frac{1}{\pi^2}\} \qquad \qquad \bigstar$$

That is,

$$c = \frac{3}{\pi^2}.$$

We gave several examples of the computation of $\mu$ in Section 2.1. Let us now suppose that $X$ has the p.m.f.

$$f(x) = \frac{1}{3}, \qquad x = -1, 0, 1.$$

It is a simple exercise to show that $\mu = E(X) = 0$. We define $Y = X^2$, which is a function of the random variable $X$ and is itself a random variable. Clearly the support of $Y$ is $y = 0, 1$ and the probabilities concerning $Y$ are *induced* from those of $X$, namely

$$P(Y = 0) = P(X^2 = 0) = P(X = 0) = \frac{1}{3}$$

and
$$P(Y = 1) = P(X^2 = 1) = P(X = -1) + P(X = 1) = \frac{1}{3} + \frac{1}{3} = \frac{2}{3}.$$

That is, $Y$ is a random variable with p.m.f.

$$g(y) = \begin{cases} \dfrac{1}{3}, & y = 0, \\[2mm] \dfrac{2}{3}, & y = 1, \end{cases}$$

or, more simply, $g(y) = (y + 1)/3$, $y = 0, 1$. Of course, the mean $\mu_Y$ is

$$E(Y) = \sum_{y=0}^{1} y\, g(y) = 0 \cdot \frac{1}{3} + 1 \cdot \frac{2}{3} = \frac{2}{3}.$$

This last summation could be rewritten as follows:

$$(-1)^2 \cdot \frac{1}{3} + 0^2 \cdot \frac{1}{3} + 1^2 \cdot \frac{1}{3} = \sum_{x=-1}^{1} x^2 f(x).$$

That is, in this case,

$$\sum_{y=0}^{1} y\, g(y) = \sum_{x=-1}^{1} x^2 f(x).$$

We denote the last summation by $E(X^2)$ and call it the expectation of $X^2$. So $E(Y) = E(X^2)$.

This is true in general. If $X$ has a p.m.f. $f(x)$ on space $S_X$, then the random variable defined by $Y = u(X)$ has a p.m.f. $g(y)$ on a space $S_Y = \{y: y = u(x),\ x \,\varepsilon\, S_X\}$. If

$$\mu_Y = E(Y) = \sum_{y \varepsilon S_Y} y\, g(y)$$

exists, it can be computed by finding the expectation of $u(X)$, which is denoted by

$$E[u(X)] = \sum_{x \varepsilon S_X} u(x) f(x).$$

Let $X$ have p.m.f.
$$f(x) = \frac{x}{6}, \qquad x = 1, 2, 3.$$

Let $Y = X^3$. Then $E(Y)$ can be computed by

$$E(X^3) = \sum_{x=1}^{3} (x^3) \frac{x}{6} = \frac{1^4}{6} + \frac{2^4}{6} + \frac{3^4}{6} = \frac{98}{6}.$$

More difficult summations can be computed by *Maple* or *Mathematica*.

EXAMPLE 2.2-2    Let $X$ be the number of trials needed to cast the first "six" in a sequence of independent casts of an unbiased die. Here the support of $X$ is $x = 1, 2, 3, \ldots$ and the p.m.f. is

$$f(x) = \left(\frac{5}{6}\right)^{x-1}\left(\frac{1}{6}\right), \qquad x = 1, 2, 3, \ldots.$$

Let $Y = X^2$ so that $E(Y)$ is equal to

$$E(X^2) = \sum_{x=1}^{\infty} x^2 \left(\frac{5}{6}\right)^{x-1}\left(\frac{1}{6}\right) = 66$$

which is found using *Maple*:

```
>  sum(x^2*(5/6)^(x-1)*(1/6), x = 1 .. infinity);
                           66
```

There are a few observations we must make about expectations when they exist.

**1.** If $k$ is a constant, then $E(k) = k$ because

$$E(k) = \sum_{x \varepsilon S} k f(x) = k \sum_{x \varepsilon S} f(x) = k \cdot 1 = k.$$

That is, the expectation of a constant equals that constant.

**2.** If $k$ is a constant and $v$ is a function, then, with $u(x) = kv(x)$, we have

$$E[kv(X)] = \sum_{x \varepsilon S} kv(x)f(x) = k \sum_{x \varepsilon S} v(x)f(x) = kE[v(X)].$$

**3.** If $k_1$ and $k_2$ are constants and $v_1$ and $v_2$ are functions, then, with $u = k_1 v_1 + k_2 v_2$, we have

$$E[k_1 v_1(X) + k_2 v_2(X)] = \sum_{x \varepsilon S} [k_1 v_1(x) + k_2 v_2(x)]f(x)$$

$$= k_1 \sum_{x \varepsilon S} v_1(x)f(x) + k_2 \sum_{x \varepsilon S} v_2(x)f(x)$$

$$= k_1 E[v_1(X)] + k_2 E[v_2(X)].$$

This can be extended to $k_1 v_1 + k_2 v_2 + \cdots + k_m v_m$, and thus we call the expectation $E$ a **linear** or **distributive operator**; that is

$$E\left[\sum_{i=1}^{m} k_i v_i(X)\right] = \sum_{i=1}^{m} k_i E[v_i(X)].$$

With this expectation notation, we see that the variance is

$$\sigma^2 = E[(X - \mu)^2],$$

which, because $E$ is a linear operator, is equal to

$$E(X^2 - 2\mu X + \mu^2) = E(X^2) - 2\mu E(X) + \mu^2$$

$$= E(X^2) - 2\mu^2 + \mu^2 = E(X^2) - \mu^2.$$

That is, $\sigma^2 = E(X^2) - \mu^2$, a result that was established and used in Section 2.1.

**EXAMPLE 2.2-3**   Let the random variable $X$ have p.m.f.

$$f(x) = \left(\frac{1}{2}\right)^x, \qquad x = 1, 2, 3, \ldots.$$

We have

$$\mu = E(X) = \sum_{x=1}^{\infty} x\left(\frac{1}{2}\right)^x = 1 \cdot \left(\frac{1}{2}\right) + 2 \cdot \left(\frac{1}{2}\right)^2 + 3 \cdot \left(\frac{1}{2}\right)^3 + \cdots$$

$$= \frac{1}{2}\left[1 + 2 \cdot \left(\frac{1}{2}\right) + 3 \cdot \left(\frac{1}{2}\right)^2 + \cdots\right]$$

$$= \frac{1}{2}\left(1 - \frac{1}{2}\right)^{-2} = 2,$$

because $(1 - z)^{-2} = 1 + 2z + 3z^2 + \cdots$, $|z| < 1$. Also

$$E[X(X+1)] = \sum_{x=1}^{\infty} x(x+1)\left(\frac{1}{2}\right)^x = 1 \cdot 2\left(\frac{1}{2}\right)^1 + 2 \cdot 3\left(\frac{1}{2}\right)^2 + 3 \cdot 4\left(\frac{1}{2}\right)^3 + \cdots$$

$$= 1 \cdot 2 \cdot \frac{1}{2}\left[1 + \frac{3 \cdot 2}{2 \cdot 1}\left(\frac{1}{2}\right) + \frac{4 \cdot 3}{2 \cdot 1}\left(\frac{1}{2}\right)^2 + \cdots\right]$$

$$= \left(1 - \frac{1}{2}\right)^{-3} = 8.$$

However, $E[X(X+1)] = E(X^2) + E(X)$; so

$$E(X^2) = E[X(X+1)] - E(X) = 8 - 2 = 6$$

and

$$\sigma^2 = E(X^2) - \mu^2 = 6 - 4 = 2.$$

See Example 5.2-9 for a *Maple* solution for this example.

If $X$ is a random variable with mean $\mu_X$ and variance $\sigma_X^2$, consider the random variable $Y = aX + b$, where $a$ and $b$ are constants. Then

$$\mu_Y = E(Y) = E(aX + b) = aE(X) + b = a\mu_X + b$$

and

$$\sigma_Y^2 = E[(Y - \mu_Y)^2] = E[(aX + b - a\mu_X - b)^2]$$

$$= E[a^2(X - \mu_X)^2] = a^2\sigma_X^2.$$

In particular, if $Y = X - b$ so that $a = 1$ and the distribution of $Y$ is just a translation of the distribution of $X$, then $\sigma_Y^2 = \sigma_X^2$.

We now model a situation which leads to an important discrete probability distribution, the binomial. We repeat a certain experiment, which has one of two possible outcomes, say "success" ($S$) or "failure" ($F$), a number of times. Assume the probability of success on each trial is the constant $p$, and the outcomes from different trials are mutually independent. These are called **Bernoulli** trials. In a sequence of $n$ Bernoulli trials, say we are interested in the number of successes, and we denote this random variable by $X$. In finding the p.m.f. of $X$, we recognize that the possible values of $X$ are $0, 1, 2, \ldots, n$. If $x$ successes occur, where $x = 0, 1, 2, \ldots, n$, then $n - x$ failures occur. The number of ways of selecting $x$ positions for the $x$ successes in the $n$ trials is

$$\binom{n}{x} = \frac{n!}{x!(n-x)!}.$$

Since the trials are independent and since the probabilities of success and failure on each trial are, respectively, $p$ and $1 - p$, the probability of each of the $n$ trials having $x$ successes and $n - x$ failures is $p^x(1-p)^{n-x}$. Thus the p.m.f. of $X$, say $f(x)$, is the sum of the probabilities of these $\binom{n}{x}$ mutually exclusive events; that is,

$$f(x) = \binom{n}{x}p^x(1-p)^{n-x} = \frac{n!}{x!(n-x)!}p^x(1-p)^{n-x}, \qquad x = 0, 1, 2, \ldots, n,$$

where sometimes $1 - p$ is replaced by $q$; that is, $q = 1 - p$. These probabilities are called binomial probabilities, and the random variable $X$ is said to have a **binomial distribution**. For notation we say that $X$ is $b(n, p)$.

In Equation 1.2-1 we gave the binomial expansion

$$(a + b)^n = a^n + \binom{n}{1}a^{n-1}b + \binom{n}{2}a^{n-2}b^2 + \cdots + \binom{n}{n-1}ab^{n-1} + b^n.$$

Letting $a = 1 - p$ and $b = p$, we see that

$$(1 - p + p)^n = \sum_{x=0}^{n}\binom{n}{x}p^x(1-p)^{n-x} = \sum_{x=0}^{n}f(x).$$

That is,

$$\sum_{x=0}^{n}f(x) = 1.$$

To find $\mu = E(X)$, we must find the value of

$$E(X) = \sum_{x=0}^{n}x\frac{n!}{x!(n-x)!}p^x(1-p)^{n-x} = \sum_{x=1}^{n}\frac{n!}{(x-1)!(n-x)!}p^x(1-p)^{n-x}.$$

Let $k = x - 1$ or $x = k + 1$. Then

$$E(X) = \sum_{k=0}^{n-1}\frac{n!}{k!(n-1-k)!}p^{k+1}(1-p)^{n-1-k}$$

$$= np\sum_{k=0}^{n-1}\frac{(n-1)!}{k!(n-1-k)!}p^k(1-p)^{n-1-k}$$

$$= np(1-p+p)^{n-1}$$

$$= np.$$

That is, if the distribution of $X$ is $b(n,p)$, then $\mu = E(X) = np$. To find $\sigma^2 = \text{Var}(X)$, we first find the value of $E[X(X-1)] = E(X^2) - E(X)$. Then $\text{Var}(X) = E[X(X-1)] + E(X) - \mu^2$. Now

$$E[X(X-1)] = \sum_{x=0}^{n} x(x-1)\frac{n!}{x!(n-x)!}p^x(1-p)^{n-x}$$

$$= \sum_{x=2}^{n} \frac{n!}{(x-2)!(n-x)!}p^x(1-p)^{n-x}.$$

Let $k = x - 2$ or $x = k + 2$. Then

$$E[X(X-1)] = \sum_{k=0}^{n-2} \frac{n!}{k!(n-2-k)!}p^{k+2}(1-p)^{n-2-k}$$

$$= n(n-1)p^2 \sum_{k=0}^{n-2} \frac{(n-2)!}{k!(n-2-k)!}p^k(1-p)^{n-2-k}$$

$$= n(n-1)p^2(1-p+p)^{n-2}$$

$$= n(n-1)p^2.$$

Thus the variance of $X$ is

$$\sigma^2 = n(n-1)p^2 + np - (np)^2 = np(1-p).$$

Is is also possible to use *Maple* to find the mean and variance. See Example 5.2-2 for a *Maple* solution.

**EXAMPLE 2.2-4** In an instant lottery with 20% winning tickets, if $X$ is equal to the number of winning tickets among $n = 8$ that are purchased, the probability of purchasing 2 winning tickets is

$$f(2) = P(X = 2) = \binom{8}{2}(0.2)^2(0.8)^6 = 0.2936.$$

The distribution of the random variable $X$ is $b(8, 0.2)$.

**EXAMPLE 2.2-5** Suppose that the probability of germination of a beet seed is 0.8 and the germination of a seed is called a success. The number $X$ of seeds that germinate in $n = 10$ independent trials is $b(10, 0.8)$: that is,

$$f(x) = \binom{10}{x}(0.8)^x(0.2)^{10-x}, \qquad x = 0, 1, 2, \ldots, 10.$$

In particular,

$$P(X \le 8) = 1 - P(X = 9) - P(X = 10)$$

$$= 1 - 10(0.8)^9(0.2) - (0.8)^{10} = 0.6242.$$

Also, we could compute, with a little more work,

$$P(X \le 6) = \sum_{x=0}^{6} \binom{10}{x}(0.8)^x(0.2)^{10-x}.$$

**EXAMPLE 2.2-6**    Let $X$ be $b(n,p)$. Then $Y = X/n$ has mean and variance given by

$$\mu_Y = \frac{np}{n} = p \quad \text{and} \quad \sigma_Y^2 = \frac{np(1-p)}{n^2} = \frac{p(1-p)}{n},$$

since here $a = 1/n$ and $b = 0$ in $aX + b$. We say $Y = X/n$ is an **unbiased estimator** of $p$ because $E(Y) = p$; its variance is $p(1-p)/n$.

**EXAMPLE 2.2-7**    Let $u(x) = (x-b)^2$, where $b$ is not a function of $X$, and suppose $E[(X-b)^2]$ exists. To find that value of $b$ for which $E[(X-b)^2]$ is a minimum, we write

$$g(b) = E[(X-b)^2] = E[X^2 - 2bX + b^2]$$
$$= E(X^2) - 2bE(X) + b^2$$

because $E(b^2) = b^2$. To find the minimum, we differentiate $g(b)$ with respect to $b$, set $g'(b) = 0$ and solve for $b$ as follows:

$$g'(b) = -2E(X) + 2b = 0,$$
$$b = E(X).$$

Since $g''(b) = 2 > 0$, $E(X)$ is the value of $b$ that minimizes $E[(X-b)^2]$. (See Exercise 2.2-11.)

There is an interesting inequality that gives us a little more feeling for the meaning of the standard deviation. Note that

$$\sigma^2 = E[(X-\mu)^2] = \sum_{x \varepsilon S}(x-\mu)^2 f(x) \ge \sum_{x \varepsilon A}(x-\mu)^2 f(x),$$

where $A = \{x: |x-\mu| \ge k\sigma\}$ and $k$ is a positive constant. In $A$ we know that $|x-\mu| \ge k\sigma$; so if we replace in the last summation $(x-\mu)^2$ by $k^2\sigma^2$, we obtain even a smaller value so

$$\sigma^2 \ge \sum_{x \varepsilon A} k^2\sigma^2 f(x) = k^2\sigma^2 \sum_{x \varepsilon A} f(x) = k^2\sigma^2 P(X \varepsilon A).$$

That is

$$\frac{1}{k^2} \ge P(|X-\mu| \ge k\sigma),$$

which is called **Chebyshev's Inequality**. For illustration, we note with $k = 2$ that

$$\frac{1}{4} \ge P(|X-\mu| \ge 2\sigma).$$

This is equivalent to

$$\frac{3}{4} \leq 1 - P(\,|X - \mu| \geq 2\sigma) = P(\,|X - \mu| < 2\sigma).$$

That is, the probability that $X$ is inside the $2\sigma$ distance from its mean $\mu$ is greater than or equal to 3/4. This is true of every random variable having a mean and standard deviation. Clearly if $3\sigma$ replaces $2\sigma$, the probability is greater than or equal to 8/9.

If we now reconsider $Y = X/n$ of Example 2.2-6, we see that, for every $\epsilon > 0$,

$$P\left(\left|\frac{X}{n} - p\right| \geq \epsilon\right) = P\left[\left|\frac{X}{n} - p\right| \geq \left(\frac{\epsilon\sqrt{n}}{\sqrt{p(1-p)}}\right)\sqrt{\frac{p(1-p)}{n}}\right]$$

$$\leq \frac{1}{\left(\dfrac{\epsilon\sqrt{n}}{\sqrt{p(1-p)}}\right)^2} = \frac{p(1-p)}{\epsilon^2 n}$$

since the standard deviation of $Y = X/n$ is $\sqrt{p(1-p)/n}$; thus $k = \epsilon\sqrt{n}/\sqrt{p(1-p)}$. Note no matter how small the given $\epsilon$ is, it is true that

$$\lim_{n \to \infty} P\left(\left|\frac{X}{n} - p\right| \geq \epsilon\right) \leq \lim_{n \to \infty} \frac{p(1-p)}{n\epsilon^2} = 0.$$

That is, the probability that the estimator $Y = X/n$ is more than $\epsilon$ away from $p$ goes to zero as $n$ increases without bound. This is one form of the **weak law of large numbers**, and we say that $Y = X/n$ **converges in probability** to $p$. Intuitively, this means that the relative frequency $X/n$ will be close to $p$ as $n$ becomes large, and we used this in first discussing probability in Section 1.1.

**Remark**   If the instructor so chooses, the first part of Section 6.1 on the moment-generating function (m.g.f.) of a discrete random variable can be studied at this time.

## EXERCISES 2.2

**2.2-1**  Suppose $X$ has p.m.f. $f(x) = 1/5$, $x = 1, 2, 3, 4, 5$. Compute $E(X)$ and $E(X^2)$, and use those two results to find $E[(X + 2)^2]$. HINT: $(X + 2)^2 = X^2 + 4X + 4$.

**2.2-2**  Let $X$ be a number selected at random from the first 10 positive integers. Assuming equal probabilities on these 10 integers, compute $E[X(11 - X)]$.

**2.2-3**  Let the random variable $X$ have the p.m.f.

$$f(x) = \frac{(|x| + 1)^2}{9}, \qquad x = -1, 0, 1.$$

Compute $E(X)$, $E(X^2)$, and $E(3X^2 - 2X + 4)$.

**2.2-4**  An insurance company sells an automobile policy with a deductible of one unit. Let $X$ be the amount of the loss having p.m.f.

$$f(x) = \begin{cases} 0.9, & x = 0, \\ \dfrac{c}{x}, & x = 1, 2, 3, 4, 5, 6, \end{cases}$$

where $c$ is a constant. Determine $c$ and the expected value of the amount the insurance company must pay.

**2.2-5** In a state lottery a non-negative three-digit integer is selected at random (this includes 000). If a player bets $1 on a particular number and if that number is selected, the payoff is $500 minus the $1 paid for the ticket. Let $X$ equal the payoff to the bettor, namely $-$1$ or $499, and find $E(X)$.

**2.2-6** Let $X$ be a number selected at random from the first $n$ positive integers. Assign a reasonable p.m.f. to $X$.

**(a)** Compute $E(1/X)$ when $n = 5$.

**(b)** Approximate $E(1/X)$ when $n = 100$. HINT: Find reasonable upper and lower bounds for $E(1/X)$ using two integrals bounding it.

**(c)** If available, use a CAS to find the exact value of $E(1/X)$.

**2.2-7** Suppose that a school has 20 classes: 16 with 25 students in each, three with 100 students in each, and one with 300 students for a total of 1000 students.

**(a)** What is the average class size?

**(b)** Select a student randomly out of the 1000 students. Let the random variable $X$ equal the size of the class to which this student belongs and define the p.m.f. of $X$.

**(c)** Find the mean of $X$. Does this answer surprise you?

**(d)** To help understand the last answer, you could simulate 15 observations of $X$ and find the value of $\bar{x}$.

**2.2-8** Let $X$ have mean $\mu$ and variance $\sigma^2$. Find the mean and the variance of $Y = (X - \mu)/\sigma$.

**2.2-9** A measure of **skewness** is defined by

$$\frac{E[(X - \mu)^3]}{\{E[(X - \mu)^2]\}^{3/2}} = \frac{E[(X - \mu)^3]}{(\sigma^2)^{3/2}} = \frac{E[(X - \mu)^3]}{\sigma^3}.$$

When a distribution is symmetrical about the mean, the skewness is equal to zero. If the probability histogram has a longer "tail" to the right than to the left, the measure of skewness is positive, and we say that the distribution is skewed positively or to the right. If the probability histogram has a longer tail to the left than to the right, the measure of skewness is negative, and we say that the distribution is skewed negatively or to the left. If the p.m.f. of $X$ is given by $f(x)$, **(i)** depict the p.m.f. as a probability histogram and find the values of **(ii)** the mean, **(iii)** the standard deviation and **(iv)** skewness.

**(a)**

$$f(x) = \begin{cases} \dfrac{2^{6-x}}{64}, & x = 1, 2, 3, 4, 5, 6, \\ \dfrac{1}{64}, & x = 7. \end{cases}$$

**(b)**

$$f(x) = \begin{cases} \dfrac{1}{64}, & x = 1, \\ \dfrac{2^{x-2}}{64}, & x = 2, 3, 4, 5, 6, 7. \end{cases}$$

**2.2-10** A measure of **kurtosis** is defined by

$$E\left[\left(\frac{X - \mu}{\sigma}\right)^4\right] = \frac{E[(X - \mu)^4]}{\sigma^4}.$$

Let $X$ have p.m.f.

$$f(x) = \begin{cases} p, & x = -1, 1, \\ 1 - 2p, & x = 0, \end{cases}$$

where $0 < p < 1/2$.

(a) Compute the measure of kurtosis as a function of $p$.

(b) Determine its value when $p = 1/3, p = 1/5, p = 1/10$, and $p = 1/100$. Note that the kurtosis increases as $p$ becomes smaller.

(c) Graph kurtosis as a function of $p$.

**2.2-11** Let $X$ be a random variable with support $\{1, 2, 3, 5, 15, 25, 50\}$, each point of which has the same probability $1/7$. Argue that $c = 5$ is the value that minimizes $h(c) = E(|X - c|)$. Compare this to the value of $b$ that minimizes $g(b) = E[(X - b)^2]$. (See Example 2.2-7.)

**2.2-12** Let $X$ be $b(2, p)$ and let $Y$ be $b(4, p)$. If $P(X \geq 1) = 5/9$, find $P(Y \geq 1)$.

**2.2-13** On a six-question multiple-choice test there are five possible answers for each question, of which one is correct (C) and four are incorrect (I). If a student guesses randomly and independently, find the probability of

(a) Being correct only on questions 1 and 4 (i.e., scoring C, I, I, C, I, I).

(b) Being correct on exactly two questions.

(c) If $X$ equals the number of correct answers in the six guesses, what is the p.m.f.; the mean; and the variance of $X$?

## 2.3 SPECIAL DISCRETE DISTRIBU-TIONS

We now consider the binomial distribution more and model three other situations which lead to three other important distributions: the Poisson, the negative binomial, and the hypergeometric.

**Binomial Distribution**: Cumulative probabilities like those in Example 2.2-5 are often of interest. We call the function defined by

$$F(x) = P(X \leq x)$$

the **cumulative distribution function**, or more simply, the **distribution function** of the random variable $X$. Values of the distribution function of a random variable $X$ that is $b(n, p)$ are given in Table II in the Appendix for selected values of $n$ and $p$.

For the binomial distribution given in Example 2.2-5, namely the $b(10, 0.8)$ distribution, the distribution function is defined by

$$F(x) = P(X \leq x) = \sum_{y=0}^{\lfloor x \rfloor} \binom{10}{y}(0.8)^y(0.2)^{10-y},$$

where $\lfloor x \rfloor$ is the floor or greatest integer in $x$. A figure of this distribution function is shown in Figure 2.3-1. Note very carefully that the jumps at the integers in this step function are equal to the probabilities associated with those respective integers.

Figure 2.3-1 Distribution function for the $b(10, 0.8)$ distribution

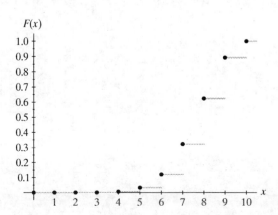

EXAMPLE 2.3-1 Leghorn chickens are raised for laying eggs. Let $p = 0.5$ be the probability of a female chick hatching. Assuming independence, let $X$ equal the number of female chicks out of 10 newly hatched chicks selected at random. Then the distribution of $X$ is $b(10, 0.5)$. The probability of 5 or fewer female chicks is, using Table II in the Appendix (see Example 5.1-1 for a Minitab solution),

$$P(X \le 5) = 0.6230.$$

The probability of exactly 6 female chicks is

$$P(X = 6) = \binom{10}{6}\left(\frac{1}{2}\right)^6\left(\frac{1}{2}\right)^4$$
$$= P(X \le 6) - P(X \le 5)$$
$$= 0.8281 - 0.6230 = 0.2051,$$

since $P(X \le 6) = 0.8281$. The probability of at least 6 female chicks is

$$P(X \ge 6) = 1 - P(X \le 5) = 1 - 0.6230 = 0.3770.$$

Although probabilities for the binomial distribution $b(n, p)$ are given in Table II in the Appendix for selected values of $p$ that are less than or equal to 0.5, the next example demonstrates that this table can also be used for values of $p$ that are greater than 0.5. (In later sections we learn how to approximate certain binomial probabilities with those of other distributions.)

EXAMPLE 2.3-2 Suppose that we are in one of those rare times when 65% of the American public approve of the way the President of the United States is handling his job. Select randomly eight ($n = 8$) Americans and let $Y$ equal the number who give approval. Then the distribution of $Y$ is $b(8, 0.65)$. To find $P(Y \ge 6)$, note that

$$P(Y \ge 6) = P(8 - Y \le 8 - 6) = P(X \le 2),$$

where $X = 8 - Y$ counts the number who disapprove. Since $q = 1 - p = 0.35$ equals the probability of disapproval by each person selected, the distribution of $X$ is $b(8, 0.35)$ (see Figure 2.3-2). From Table II in the Appendix, since $P(X \le 2) = 0.4278$, it follows that $P(Y \ge 6) = 0.4278$.

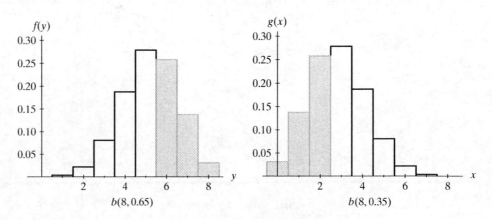

**Figure 2.3-2** Presidential approval histogram

Similarly,

$$P(Y \leq 5) = P(8 - Y > 8 - 5)$$
$$= P(X \geq 3) = 1 - P(X \leq 2)$$
$$= 1 - 0.4278 = 0.5722$$

and

$$P(Y = 5) = P(8 - Y = 8 - 5)$$
$$= P(X = 3) = P(X \leq 3) - P(X \leq 2)$$
$$= 0.7064 - 0.4278 = 0.2786.$$

**EXAMPLE 2.3-1** Say a lot (collection) of items (maybe fuses) is large, possibly $N = 10,000$ or more. We take a sample of size $n = 100$ from the lot, and accept the lot if the number of defectives in this sample of 100 is zero or one. We assume that the lot is so large that the probability, $p$, of a defective essentially remains the same as we select items at random. For the same reason, the trials seem independent. Hence $X$, the number of defectives in the sample, has an approximate binomial distribution with p.m.f.

$$f(x) = \binom{100}{x} p^x (1 - p)^{100-x}, \qquad x = 0, 1, 2, \ldots, 100.$$

The probability of accepting the lot for various $p$ is called the **operating characteristic curve** and for this testing procedure is given by

$$OC(p) = f(0) + f(1) = (1 - p)^{100} + 100p(1 - p)^{99}.$$

Of course, if $p = 0.01$,

$$OC(0.01) = 0.366 + 0.370 = 0.736;$$

while if $p = 0.05$,

$$OC(0.05) = 0.006 + 0.031 = 0.037.$$

If the lot is good with a small fraction of defectives, we want the $OC$ curve to be high. If the lot is bad with a large fraction of defectives, like 0.05, then we want the probability of accepting the lot to be small; the small probability of 0.037 seems appropriate.

**Poisson Distribution:** Suppose we have a binomial distribution with large $n$ and small $p$. Let us model this by saying that $p = \lambda/n$, where $\lambda$ is some constant. Then

$$P(X = x) = \frac{n!}{x!(n-x)!} \left(\frac{\lambda}{n}\right)^x \left(1 - \frac{\lambda}{n}\right)^{n-x}.$$

If $n$ increases without bound, we have that

$$\lim_{n \to \infty} \frac{n!}{x!(n-x)!} \left(\frac{\lambda}{n}\right)^x \left(1 - \frac{\lambda}{n}\right)^{n-x}$$

$$= \lim_{n \to \infty} \frac{n(n-1)\cdots(n-x+1)}{n^x} \frac{\lambda^x}{x!} \left(1 - \frac{\lambda}{n}\right)^n \left(1 - \frac{\lambda}{n}\right)^{-x}$$

Now, for fixed $x$, we have

$$\lim_{n\to\infty} \frac{n(n-1)\cdots(n-x+1)}{n^x} = \lim_{n\to\infty} \left[ (1)\left(1-\frac{1}{n}\right)\cdots\left(1-\frac{x-1}{n}\right)\right] = 1,$$

$$\lim_{n\to\infty} \left(1-\frac{\lambda}{n}\right)^n = e^{-\lambda},$$

$$\lim_{n\to\infty} \left(1-\frac{\lambda}{n}\right)^{-x} = 1.$$

Thus

$$\lim_{n\to\infty} \frac{n!}{x!\,(n-x)!}\left(\frac{\lambda}{n}\right)^x\left(1-\frac{\lambda}{n}\right)^{n-x} = \frac{\lambda^x e^{-\lambda}}{x!}.$$

The distribution of probability associated with this model has a special name. We say that the random variable $X$ has a **Poisson distribution** if its p.m.f. is of the form

$$f(x) = \frac{\lambda^x e^{-\lambda}}{x!}, \qquad x = 0, 1, 2, \ldots,$$

where $\lambda > 0$.

We can use the Maclaurin's expansion of

$$e^x = \sum_{k=0}^{\infty} \frac{x^k}{k!}$$

to find the mean of $X$. We have

$$\mu = \sum_{x=0}^{\infty} x\frac{\lambda^x e^{-\lambda}}{x!} = \sum_{x=1}^{\infty} \frac{\lambda^x e^{-\lambda}}{(x-1)!}.$$

Let $k = x - 1$ or $x = k + 1$. Then

$$\mu = \sum_{k=0}^{\infty} \frac{\lambda^{k+1} e^{-\lambda}}{k!} = \lambda \sum_{k=0}^{\infty} \frac{\lambda^k e^{-\lambda}}{k!}$$

$$= \lambda e^{-\lambda} e^{\lambda} = \lambda.$$

It can also be shown that $\sigma^2 = \text{Var}(X) = \lambda$. (See Example 5.2-3 for a *Maple* solution.) That is, for the Poisson distribution, $\mu = \sigma^2 = \lambda$.

Table III in the Appendix gives values of the distribution function of a Poisson random variable for selected values of $\lambda$. This table is illustrated in the next example.

**EXAMPLE 2.3-4**    Let $X$ have a Poisson distribution with a mean of $\lambda = 5$. Then using Table III in the Appendix (see Example 5.1-2 for a Minitab solution),

$$P(X \leq 6) = \sum_{x=0}^{6} \frac{5^x e^{-5}}{x!} = 0.762,$$

$$P(X > 5) = 1 - P(X \leq 5) = 1 - 0.616 = 0.384,$$

and

$$P(X = 6) = P(X < 6) - P(X < 5) = 0.762 - 0.616 = 0.146.$$

**EXAMPLE 2.3-5** Let us refer to Example 2.3-3 and consider the case in which $p = 0.05$ and $n = 100$. Since $n$ is large and $p$ is small, we have that, with $np = \lambda = 5$,

$$OC(0.05) = f(0) + f(1) \approx e^{-5} + \frac{5e^{-5}}{1!} = 0.0067 + 0.0337 = 0.0404,$$

which is a reasonable approximation to the probability 0.037. (Note that Table III in the Appendix can be used.) The approximation is even better when $p = 0.01$ for then, with $\lambda = 100(0.01) = 1$,

$$OC(0.01) \approx e^{-1} + \frac{e^{-1}}{1!} = 0.368 + 0.368 = 0.736,$$

which is equal to that probability in Example 2.3-3 to three decimal places.

Not only is the Poisson distribution a good approximation to the binomial distribution of probabilities when $n$ is large and $p$ is small, but it is often a good model in the following situation.

Some experiments result in counting the number of times particular events occur in given times or on given physical objects. For example, we could count the number of phone calls arriving at a switchboard between 9 and 10 A.M., the number of flaws in 100 feet of wire, the number of customers that arrive at a ticket window between 12 noon and 2 P.M., or the number of defects in a 100-foot roll of aluminum screen that is 2 feet wide. Often these counts can be looked upon as a random variable having a Poisson distribution, at least as an approximation.

**EXAMPLE 2.3-6** Flaws (bad records) on a used computer tape occur on the average of one flaw per 1200 feet. Let $X$ be the number of flaws on a 1200-foot roll, and assume a Poisson distribution with $\mu = 1$. Then, with $\mu = 1$,

$$P(X \leq 2) = \sum_{x=0}^{2} \frac{\mu^x e^{-\mu}}{x!} = \frac{e^{-1}}{0!} + \frac{e^{-1}}{1!} + \frac{e^{-1}}{2!} = (2.5)e^{-1} = 0.920$$

is the probability of having two or fewer flaws on that roll.

If we had a 4800-foot roll, it would be reasonable to assume that if $Y$ equals the number of flaws on it, then $Y$ has a Poisson distribution with $\mu = 4$, because the roll is four times as long as the first. Then

$$P(Y = 0) = \frac{4^0 e^{-4}}{0!} = 0.018.$$

Other probabilities could be computed. These probabilities can also be found using Table III in the Appendix or a computer program.

**Negative Binomial Distribution:** Let us observe a sequence of Bernoulli trials until exactly $r$ successes occur, where $r$ is a fixed positive integer. Let the random variable $X$ denote the number of trials needed to observe the $r$th success. That is, $X$ is the trial number on which the $r$th success is observed. By the multiplication rule of probabilities, the p.m.f. of $X$, say $g(x)$, equals the product of the probability

$$\binom{x-1}{r-1} p^{r-1}(1-p)^{x-r}$$

of obtaining exactly $r-1$ successes in the first $x-1$ trials and the probability $p$ of a success on the $r$th trial. Thus the p.m.f. of $X$ is

$$g(x) = \binom{x-1}{r-1} p^r(1-p)^{x-r} = \binom{x-1}{r-1} p^r q^{x-r}, \qquad x = r, r+1, \ldots,$$

where $q = 1 - p$. We say that $X$ has a **negative binomial distribution**. If $r = 1$, we say that $X$ has a **geometric distribution** since the p.m.f. consists of terms of a geometric series, namely

$$g(x) = p(1-p)^{x-1}, \qquad x = 1, 2, 3, \cdots.$$

Using *Maple* it is easy to show that the sum of the terms of the negative binomial p.m.f. is equal to one and to find the mean and variance. (See Example 5.2-4.)
  They are

$$\mu = \frac{r}{p} \qquad \text{and} \qquad \sigma^2 = \frac{r(1-p)}{p^2},$$

which in the special case of $r = 1$ (the geometric distribution) become

$$\mu = \frac{1}{p} \qquad \text{and} \qquad \sigma^2 = \frac{1-p}{p^2}.$$

Sometimes we use the **translated negative binomial distribution** by letting the random variable be $Y = X - r$ so that its p.m.f. is

$$h(y) = \binom{y+r-1}{r-1} p^r(1-p)^y, \qquad y = 0, 1, 2, \ldots.$$

That is, $Y$ equals the number of failures before achieving the $r$th success. Its mean and variance are

$$\mu_Y = \frac{r}{p} - r = \frac{r(1-p)}{p} \qquad \text{and} \qquad \sigma_Y^2 = \frac{r(1-p)}{p^2}.$$

Of course, since this is just a translation, your intuition should suggest that the variance does not change. (See Section 2.2 for a proof.)

**EXAMPLE 2.3-7**    A certain basketball player is excellent at free-throw shooting and makes $p = 0.90 = 90\%$ of her free throws. Assume that we can treat these as Bernoulli trials and let $X$ equal the number of attempts needed to make $r = 10$ free throws. Then

$$g(x) = \binom{x-1}{9} (0.9)^{10}(0.1)^{x-10}, \qquad x = 10, 11, 12, \ldots.$$

Then
$$g(10) = P(X = 10) = (0.9)^{10} = 0.349$$

and
$$g(11) = 10(0.9)^{10}(0.1) = 0.349.$$

Hence the probability that the sequence ends at or before the $11^{th}$ trial is 0.698. The mean and variance are

$$\mu = \frac{10}{0.9} = 11.11 \quad \text{and} \quad \sigma^2 = \frac{10(0.1)}{0.9^2} = 1.235.$$

**Hypergeometric Distribution** Consider a collection of $N = N_1 + N_2$ similar objects, $N_1$ of them belonging to one of two dichotomous classes (orange chips, say) and $N_2$ of them belonging to the second class (blue chips, say). A collection of $n$ objects is selected from these $N$ objects at random and without replacement. Find the probability that exactly $x$ of these $n$ objects are orange (i.e., $x$ belong to the first class and $n - x$ belong to the second and where the non-negative integer $x$ satisfies $x \leq n, x \leq N_1$, and $n - x \leq N_2$). Of course, we can select $x$ orange chips in any one of $\binom{N_1}{x}$ ways and $n - x$ blue chips in any one of $\binom{N_2}{n - x}$ ways. By a multiplication principle, the product $\binom{N_1}{x}\binom{N_2}{n - x}$ equals the number of ways the joint operation can be performed. If we assume that each of the $\binom{N}{n}$ ways of selecting $n$ objects from $N = N_1 + N_2$ objects has the same probability, we have that the probability of selecting exactly $x$ orange chips is

$$f(x) = P(X = x) = \frac{\binom{N_1}{x}\binom{N_2}{n - x}}{\binom{N}{n}},$$

where the space $S$ is the collection of nonnegative integers $x$ that satisfies the inequalities $x \leq n, x \leq N_1$, and $n - x \leq N_2$. We say that the random variable $X$ has a **hypergeometric distribution**.

The mean and variance of this distribution are not easy to find. However, they can be found using *Maple* or *Mathematica*. We shall be using *Maple*. (See Example 5.2-5.)

The mean and the variance of the hypergeometric distribution are

$$\mu = n\left(\frac{N_1}{N}\right) = np$$

and

$$\sigma^2 = n\left(\frac{N_1}{N}\right)\left(\frac{N_2}{N}\right)\left(\frac{N - n}{N - 1}\right) = np(1 - p)\left(\frac{N - n}{N - 1}\right),$$

where $p = N_1/N$ is the fraction of orange chips and $1 - p = N_2/N$ is the fraction of blue chips in the bowl.

**EXAMPLE 2.3-8** A lot, consisting of 50 fuses, is inspected by the following procedure. Four fuses are selected at random and without replacement from the $N = 50$ and tested; if all

$n = 4$ blow at the correct amperage, the lot is accepted. Suppose the lot contains $N_1 = 5$ defective fuses. Let the random variable $X$ equal the number of defective fuses in the sample of $n = 4$. Then the p.m.f. of $X$ is

$$f(x) = \frac{\binom{5}{x}\binom{45}{4-x}}{\binom{50}{4}}, \qquad x = 0, 1, 2, 3, 4.$$

The mean equals $(4)(5/50) = 0.4$ and the variance is $(4)(0.1)(0.9)(46)/49 = 0.338$; so the standard deviation is $0.581$. In particular, the probability $P(X = 0)$ is

$$f(0) = \frac{\binom{5}{0}\binom{45}{4}}{\binom{50}{4}} = \frac{4257}{6580} = 0.647.$$

That is, the probability of accepting a lot having 5 defective fuses in the $N = 50$ fuses is about 0.65. (See Example 5.1-3 for a Minitab solution.) This is quite high; but our sampling procedure is rather limited as the lot size is small and hence the sample size must be small in this destructive testing.

**Remark**   For the convenience of the reader, all of the characteristics of these discrete distributions are summarized in the end pages at the front of the book. Moreover, the important continuous distributions presented in Chapter 3 and the moment generating functions (m.g.f.s) of all of these distributions are also summarized in the end pages. The m.g.f.s are not studied until Chapter 6, but it is convenient to have all of these characteristics in this one summary.

## EXERCISES 2.3

**2.3-1** It is claimed that 15% of the ducks in a particular region have patent schistosome infection. Suppose that seven ducks are selected at random. Let $X$ equal the number of ducks that are infected.

(a) Assuming independence, how is $X$ distributed?

(b) Find (i) $P(X \geq 2)$, (ii) $P(X = 1)$, and (iii) $P(X \leq 3)$.

**2.3-2** According to a *CNN/USA Today* poll, approximately 70% of Americans believe the IRS abuses its power. Let $X$ equal the number of people who believe the IRS abuses its power in a sample of $n = 25$ Americans. Assuming that the poll results are still valid, find the probability that

(a) $X$ is at least 13.

(b) $X$ is at most 11.

(c) $X$ is equal to 12.

(d) Give the mean, variance, and standard deviation of $X$.

**2.3-3** A national study showed that approximately 45% of college students binge drink. Let $X$ equal the number of students in a random sample of size $n = 12$ who binge drink. Find the probability that

(a) $X$ is at most 5.

(b) $X$ is at least 6.

(c) $X$ is equal to 7.

(d) Give the mean, variance, and standard deviation of $X$.

**2.3-4** It is believed that 20% of Americans do not have any health insurance. Suppose this is true and let $X$ equal the number with no health insurance in a sample of $n = 15$ Americans.

(a) How is $X$ distributed?

(b) Give the mean, variance, and standard deviation of $X$.

(c) Find $P(X \geq 5)$.

**2.3-5** Let $X$ be the number of successes throughout $n$ independent repetitions of a random experiment having probability of success $p = 1/4$. Determine the smallest value of $n$ so that $P(1 \leq X) \geq 0.70$.

**2.3-6** Let $X$ equal the number of "ones" if a fair die is tossed two independent times by someone out in the hall. If $X = x$, that person gives the player a bowl consisting of $10 - 3x$ red chips and $3x$ white chips. The player then selects one chip at random from the bowl. What is the conditional probability that $X = 0$ given that the chip is red? How does this compare to the prior probability $P(X = 0)$?

**2.3-7** If a fair coin is tossed at random five independent times, find the conditional probability of five heads, given that there are at least four heads.

**2.3-8** For each question on a multiple-choice test, there are five possible answers of which exactly one is correct.

(a) If a student selects answers at random, give the probability that the first question answered correctly is question 4.

(b) If $X$ equals the first time the student is correct, give the p.m.f., the mean, and the variance of $X$. Assume that there is a large number of questions on the test so that we may use the geometric distribution as an approximating distribution.

**2.3-9** Let $p = 0.05$ that a man in a certain age group dies in the next four years. Say we observe 60 such men and we assume independence.

(a) Find the probability that fewer than five of them die in the next four years.

(b) Find an approximation to (a) using the Poisson distribution.

**2.3-10** Suppose that the probability of suffering a side effect from a certain flu vaccine is 0.005. If 1000 persons are inoculated, find approximately the probability that

(a) At most 1 person suffers a side effect.

(b) 4, 5, or 6 persons suffer side effects.

**2.3-11** Customers arrive at a travel agency at a mean rate of 11 per hour. Assuming that the number of arrivals per hour has a Poisson distribution, give the probability that more than 10 customers arrive in a given hour.

**2.3-12** Flaws in a certain type of drapery material appear on the average of one in 150 square feet. If we assume the Poisson distribution, find the probability of at most one flaw in 225 square feet.

**2.3-13** Let $X$ have a Poisson distribution with mean $\mu = 4$. Show that

$$P(\mu - 2\sigma < X < \mu + 2\sigma) = 0.931.$$

**2.3-14** A grocer stocks a certain article for which the average demand is three per week. How many of these should be in stock so that the probability of the grocer running out within a week will be less than 0.01? Assume a Poisson distribution.

**2.3-15** A lot (collection) of 10,000 items is accepted if a sample of $n = 400$ has no more than $Ac = 3$ defective items. $Ac$ is called the acceptance number. Determine the approximate values of $OC(0.002)$, $OC(0.004)$, $OC(0.006)$, $OC(0.01)$, and $OC(0.02)$ and plot the operating characteristic curve.

**2.3-16** Let $X$ have a geometric distribution. Show that

$$P(X > k + j \mid X > k) = P(X > j),$$

where $k$ and $j$ are nonnegative integers. NOTE: We sometimes say that in this situation, $X$ is memoryless.

**2.3-17** Suppose that a basketball player can make a free throw 60% of the time. Let $X$ equal the number of free throws that this player must attempt to make a total of 10 shots.

(a) Give the mean, variance, and standard deviation of $X$.

(b) Find $P(X = 16)$.

**2.3-18** One of four different prizes was randomly put into each box of a cereal. If a family decided to buy this cereal until it obtained at least one of each of the four different prizes, use your intuition to determine the expected number of boxes of cereal that must be purchased.

**2.3-19** A lot (collection) consisting of 200 fuses is inspected by taking 10 fuses at random and without replacement. If all 10 "blow" at the correct amperage, the lot is accepted. If there are 20 defective fuses in the lot and the random variable $X$ equals the number of defectives among the 10 selected, find the p.m.f., the mean, and the variance of $X$. What is the probability of accepting the lot? Find the probability using the hypergeometric distribution, then the binomial approximation, and finally the Poisson approximation.

**2.3-20** A bowl contains five chips of the same size: three are marked $1 and the remaining two are marked $4. A player draws two chips at random and without replacement. If $X$ equals the number of chips marked $1 in the two selected, give the p.m.f., the mean, and the variance of $X$. Using your intuition, if the player is to receive the sum of the values on the two chips, would the player want to play the game for a protracted period of time if the charge to play is $4.75? HINT: Note that the player receives $X + 4(2 - X)$ each play. Use the mean of $X$ to see if the charge is reasonable.

**2.3-21** In LOTTO 49, Michigan's lottery game, a player selects 6 integers out of the first 49 positive integers. The state then randomly selects 6 out of the first 49 integers. Cash prizes are given to a player who matches 4, 5, or 6 integers. Let $X$ equal the number of integers selected by a player that match integers selected by the state.

(a) State the p.m.f. of $X$.

(b) Calculate the mean, variance, and standard deviation of $X$.

(c) What value of $X$ is most likely to occur?

(d) On February 25, 1995, the jackpot was worth $45,000,000. When the prize is this large, many bets are placed. Out of the 25,000,000 bets that were placed, 3 people matched all 6 numbers with each winning $15,000,000 (most of this paid by losers during preceding games), 390 matched 5 numbers to win $2500, and 22,187 matched 4 numbers to win $100. Are these numbers of winners consistent with the probability model?

## 2.4 ESTIMATION

In earlier sections we have alluded to estimating characteristics of the distribution from the corresponding ones of the sample, hoping that the latter would be reasonably close to the former. For example, the sample mean $\bar{x}$ can be thought of as an estimate of the distribution mean $\mu$, and the sample variance $s^2$ can be used as an estimate of the distribution variance $\sigma^2$. Even the relative frequency histogram associated with a sample can be taken as an estimate of the probability histogram associated with the p.m.f. of the underlying distribution. But how good are these estimates? What makes an estimate good? Can we say anything about the closeness of an estimate to an unknown **parameter**?

In this section we consider random variables for which the functional form of the p.m.f. is known, but the distribution depends on an unknown parameter, say $\theta$, that may have any value in a set, say $\Omega$, which is called the **parameter space**. For example, perhaps it is known that $f(x; \theta) = \theta^x e^{-\theta}/x!$, $x = 0, 1, 2, \ldots$, and that $\theta \in \Omega = \{\theta : 0 < \theta < \infty\}$. In certain instances, it might be necessary for the experimenter to select precisely one member of the family $\{f(x, \theta), \theta \in \Omega\}$ as the most likely p.m.f. of the random variable. That is, the experimenter needs a point estimate of the parameter $\theta$, namely the value of the parameter that corresponds to the selected p.m.f.

In estimation we take a **random sample** from the distribution to elicit some information about the unknown parameter $\theta$. That is, we repeat the same experiment $n$ independent times, observe the sample, $X_1, X_2, \ldots, X_n$, in which each $X_i$ has the same distribution, and try to estimate the value of $\theta$ using the observations $x_1, x_2, \ldots, x_n$. The function of $X_1, X_2, \ldots, X_n$ used to estimate $\theta$, say the **statistic** $u(X_1, X_2, \ldots, X_n)$, is called an **estimator** of $\theta$. We want it to be such that the computed **estimate** $u(x_1, x_2, \ldots, x_n)$ is usually close to $\theta$, where $x_1, x_2, \ldots, x_n$ are the observed values of $X_1, X_2, \ldots, X_n$. Since the estimate is a single number or single point (one member of $\theta \in \Omega$), the estimate is called a **point estimate**. The corresponding estimator is often called a **point estimator**.

The random variables $X_1, X_2, \ldots, X_n$ associated with a random sample from a distribution with p.m.f. $f(x; \theta)$ are said to be **independent and identically distributed**. This means that the events $\{X_1 = x_1\}, \{X_2 = x_2\}, \ldots, \{X_n = x_n\}$ are mutually independent, where $x_1, x_2, \ldots, x_n$ are possible values of these random variables, respectively. Moreover, from the independence, the probability of the joint event $\{X_1 = x_1, X_2 = x_2, \ldots, X_n = x_n\}$ is

$$P(X_1 = x_1, X_2 = x_2, \ldots, X_n = x_n) = f(x_1; \theta)f(x_2; \theta) \cdots f(x_n; \theta).$$

The latter product, namely $f(x_1; \theta)f(x_2; \theta) \cdots f(x_n; \theta)$ is called the **joint p.m.f.** of the random variables $X_1, X_2, \ldots, X_n$.

The following illustration should help motivate one principle that is often used in finding point estimates. Suppose that $X$ is $b(1, p)$ so that the p.m.f. of $X$ is

$$f(x; p) = p^x(1 - p)^{1-x}, \qquad x = 0, 1, \qquad 0 < p < 1.$$

We note that $p \in \Omega = \{p : 0 < p < 1\}$ where $\Omega$ represents the parameter space, that is, the space of all possible values of the parameter, $p$. Given a random sample $X_1, X_2, \ldots, X_n$, the problem is to find an estimator $u(X_1, X_2, \ldots, X_n)$ such that $u(x_1, x_2, \ldots, x_n)$ is a good point estimate of $p$, where $x_1, x_2, \ldots, x_n$ are the observed values of the random sample. Because $X_1, X_2, \ldots, X_n$ are independent and have the same p.m.f. $f(x)$, the probability that $X_1, X_2, \ldots, X_n$ takes these particular values is (with $\Sigma x_i$ denoting $\sum_{i=1}^{n} x_i$)

$$P(X_1 = x_1, \ldots, X_n = x_n) = \prod_{i=1}^{n} p^{x_i}(1 - p)^{1-x_i} = p^{\Sigma x_i}(1 - p)^{n - \Sigma x_i},$$

which is the joint p.m.f. of $X_1, X_2, \ldots, X_n$ evaluated at the observed values. One reasonable way to proceed toward finding a good estimate of $p$ is to regard this probability (or joint p.m.f.) as a function of $p$ and find the value of $p$ that maximizes it. That is, find the value of $p$ that is most likely to have produced the sample values $x_1, x_2, \ldots, x_n$. The joint p.m.f., when regarded as a function of $p$, is frequently called the **likelihood function**. Here the likelihood function is

$$L(p) = L(p; x_1, x_2, \ldots, x_n)$$
$$= f(x_1; p)f(x_2; p) \cdots f(x_n; p)$$
$$= p^{\Sigma x_i}(1 - p)^{n - \Sigma x_i}, \qquad 0 < p < 1.$$

To find the value of $p$ that maximizes $L(p)$, we first take its derivative:

$$\frac{dL(p)}{dp} = (\Sigma x_i)p^{\Sigma x_i - 1}(1 - p)^{n - \Sigma x_i} - (n - \Sigma x_i)p^{\Sigma x_i}(1 - p)^{n - \Sigma x_i - 1}.$$

Setting this first derivative equal to zero gives us,

$$p^{\Sigma x_i}(1-p)^{n-\Sigma x_i}\left[\frac{\Sigma x_i}{p} - \frac{n - \Sigma x_i}{1-p}\right] = 0.$$

Since $0 < p < 1$, this equals zero when

$$\frac{\Sigma x_i}{p} - \frac{n - \Sigma x_i}{1-p} = 0. \tag{2.4-1}$$

Multiplying each member of Equation 2.4-1 by $p(1-p)$ and simplifying, we obtain

$$\sum_{i=1}^{n} x_i - np = 0$$

or, equivalently,

$$p = \frac{\sum_{i=1}^{n} x_i}{n} = \bar{x}.$$

The corresponding statistic, namely $(\sum_{i=1}^{n} X_i)/n = \bar{X}$, is called the **maximum likelihood estimator** and is denoted by $\hat{p}$; that is,

$$\hat{p} = \frac{1}{n}\sum_{i=1}^{n} X_i = \bar{X}. \tag{2.4-2}$$

When finding a maximum likelihood estimator, it is often easier to find the value of the parameter that maximizes the natural logarithm of the likelihood function rather than the value of the parameter that maximizes the likelihood function itself. Because the natural logarithm function is an increasing function, the solutions will be the same. To see this, the example we have been considering gives us

$$\ln L(p) = \left(\sum_{i=1}^{n} x_i\right)\ln p + \left(n - \sum_{i=1}^{n} x_i\right)\ln(1-p).$$

To find the maximum, we set the first derivative equal to zero to obtain

$$\frac{d\,[\ln L(p)]}{dp} = \left(\sum_{i=1}^{n} x_i\right)\left(\frac{1}{p}\right) + \left(n - \sum_{i=1}^{n} x_i\right)\left(\frac{-1}{1-p}\right) = 0,$$

which is the same as Equation 2.4-1. Thus the solution is $p = \bar{x}$ and the maximum likelihood estimator for $p$ is $\hat{p} = \bar{X}$.

**Remark**  It would be possible for $\hat{p}$ to equal zero or one by having each $x$ value equal zero or one. Yet we have the restriction $0 < p < 1$ so that the student would not need to worry about dividing by $p$ or $1 - p$ in our development. In these trivial cases in which $p = 0$ and $p = 1$, $\hat{p} = \bar{X}$ is still the maximum likelihood estimator.

Motivated by the preceding illustration, we present the formal definition of a maximum likelihood estimator. This definition is used in both the discrete and continuous cases, where the latter is considered in Chapter 3.

Let $X_1, X_2, \ldots, X_n$ be a random sample from a distribution that depends on an unknown parameter $\theta$ with p.m.f. denoted by $f(x; \theta)$. Suppose that $\theta$ is restricted to a given parameter space $\Omega$. Then the joint p.m.f. of $X_1, X_2, \ldots, X_n$, namely

$$L(\theta) = \prod_{i=1}^{n} f(x_i; \theta), \qquad \theta \in \Omega,$$

when regarded as a function of $\theta$, is called the **likelihood function**. If the function of the sample values that maximizes $L(\theta)$ in $\Omega$ is $u(x_1, x_2, \ldots, x_n)$, then $\widehat{\theta} = u(X_1, X_2 \ldots, X_n)$ is a **maximum likelihood estimator** of $\theta$. The corresponding observed value of this statistic, namely $u(x_1, \ldots, x_n)$, is called a **maximum likelihood estimate**. In many practical cases, estimators (and estimates) are unique.

Some additional examples will help clarify these definitions.

**EXAMPLE 2.4-1** Let $X$ have a Poisson distribution with parameter $\theta$. We take a random sample $X_1, X_2, \ldots, X_n$ from this distribution, which means that each $X_i$ has the same Poisson distribution and the events associated with the random variables are mutually independent. So the likelihood function is

$$L(\theta) = \left(\frac{\theta^{x_1} e^{-\theta}}{x_1!}\right)\left(\frac{\theta^{x_2} e^{-\theta}}{x_2!}\right) \cdots \left(\frac{\theta^{x_n} e^{-\theta}}{x_n!}\right) = \frac{\theta^{\Sigma x_i} e^{-n\theta}}{x_1! x_2! \cdots x_n!}, \qquad \theta > 0.$$

To find that value of $\theta$ that maximizes $L(\theta)$, let us consider the logarithm of $L(\theta)$, $\ln L(\theta)$, because this function and $L(\theta)$ are maximized with the same value of $\theta$; and, in most cases, $\ln L(\theta)$ is easier to maximize. Here

$$\ln L(\theta) = \left(\sum_{i=1}^{n} x_i\right) \ln \theta - n\theta - \ln(x_1! x_2! \cdots x_n!)$$

and the derivative with respect to $\theta$ is

$$\frac{d[\ln L(\theta)]}{d\theta} = \frac{\sum_{i=1}^{n} x_i}{\theta} - n.$$

Setting this equal to zero, we see that

$$\theta = \frac{\sum_{i=1}^{n} x_i}{n} = \overline{x}.$$

Since the second derivative of $\ln L(\theta)$ is

$$-\frac{\sum_{i=1}^{n} x_i}{\theta^2} = -\frac{n\overline{x}}{\overline{x}^2} = \frac{-n}{\overline{x}} < 0$$

at $\theta = \overline{x}$, this value of $\theta$ does maximize $\ln L(\theta)$ and $L(\theta)$. We denote this maximizing value of $\theta$ by $\widehat{\theta} = \overline{X}$, and call it the maximum likelihood estimator of the parameter $\theta$.

**EXAMPLE 2.4-2** Let $X_1, X_2, \ldots, X_n$ be a random sample from the geometric distribution with p.m.f. $f(x, p) = (1-p)^{x-1} p, x = 1, 2, 3, \ldots$. The likelihood function is given by

$$L(p) = (1-p)^{x_1-1} p (1-p)^{x_2-1} p \cdots (1-p)^{x_n-1} p$$
$$= p^n (1-p)^{\Sigma x_i - n}, \qquad 0 < p < 1.$$

The natural logarithm of $L(p)$ is

$$\ln L(p) = n \ln p + \left( \sum_{i=1}^{n} x_i - n \right) \ln(1 - p), \qquad 0 < p < 1.$$

Thus, we have

$$\frac{d \ln L(p)}{dp} = \frac{n}{p} - \frac{\sum_{i=1}^{n} x_i - n}{1 - p} = 0.$$

Solving for $p$, we obtain

$$p = \frac{n}{\sum_{i=1}^{n} x_i} = \frac{1}{\bar{x}},$$

and this solution provides a maximum. So the maximum likelihood estimator of $p$ is

$$\widehat{p} = \frac{n}{\sum_{i=1}^{n} X_i} = \frac{1}{\bar{X}}.$$

This estimator agrees with our intuition because, in $n$ observations of a geometric random variable, there are $n$ successes in the $\sum_{i=1}^{n} x_i$ trials. Thus the estimate of $p$ is the number of successes divided by the total number of trials.

EXAMPLE 2.4-3 Let $X$ have the uniform distribution (constant p.m.f. on space $S$) with p.m.f. $f(x) = 1/\theta$, $x = 1, 2, \ldots, \theta$, where $\theta$ is a positive integer. If $X_1, X_2, \ldots, X_n$ is a random sample from this distribution, the likelihood function is

$$L(\theta) = \left( \frac{1}{\theta} \right)^n, \qquad x_i = 1, 2, \ldots, \theta, \quad i = 1, 2, \ldots, n.$$

We note that

$$\ln L(\theta) = -n \ln \theta$$

and

$$\frac{d \left[ \ln L(\theta) \right]}{d\theta} = -\frac{n}{\theta},$$

where here we take derivatives as if $\theta$ is any positive number (not just an integer). Clearly, $-n/\theta < 0$ for all positive $\theta$ so that we want to make $\theta$ as small as possible. However, $\theta \geq x_i$ for $i = 1, 2, \ldots, n$. Thus the smallest possible value of $\theta$ is

$$\widehat{\theta} = \max(X_1, X_2, \ldots, X_n),$$

and this is the maximum likelihood estimator of $\theta$. (See Exercise 5.3-9.)

EXAMPLE 2.4-4   Let us consider the practical problem of finding how many fish are in a lake. This could be extended to many different situations: finding the number of deer in a certain area, finding the number of birds in a certain region, or finding the number of whales in an ocean.

We catch a number, $N_1$, of fish in the lake and tag them, and then release them so that these $N_1$ fish are back in the lake. We do this in such a way that the tagged fish are distributed around the lake. We then catch $n$ fish from various parts of the lake and find that $x$ of them are tagged; these $n$ fish are released back to the lake. It seems reasonable that the number, $N$, of fish in the lake should be such that

$$\frac{x}{n} = \frac{N_1}{N}, \qquad \text{approximately.}$$

Thus an estimate of $N$ is

$$N = \frac{nN_1}{x}.$$

Is this a maximum likelihood estimate of $N$?

Before solving this mathematical problem, we might ask about the practical problem of tagging deer, birds, and whales. We have all heard about banding birds and tagging the deer does not seem like a major problem. However, how do you "tag" whales? A foot-long metal cylinder is fired into the thick blubber that lies just under the skin of each of $N_1$ whales. These cylinders do not hurt the whales and they seem to carry on normal activities even though they are carrying these cylinders. Later on the whaling industry captures $n$ whales, and they find out that $x$ of them are carrying the cylinders. From that they can estimate the number of whales in the ocean before the capture and hence how many remain after the capture.

Now to solve the mathematical problem of finding the maximum likelihood estimate of $N$. It seems reasonable to assume that the number $X$ of tagged fish in the $n$ that are in the catch after $N_1$ have been tagged has a hypergeometric distribution with p.m.f.

$$f(x) = \frac{\binom{N_1}{x}\binom{N - N_1}{n - x}}{\binom{N}{n}}, \qquad x \le n, \ x \le N_1, \ n - x \le N - N_1.$$

This is a function of $N$, and we wish to maximize it with respect to $N$. Clearly we know $N_1$ and $x$ and hence $\binom{N_1}{x}$ is known, and we need not consider it. What remains is

$$\frac{\binom{N - N_1}{n - x}}{\binom{N}{n}} = \frac{\dfrac{(N - N_1)!}{(n - x)!(N - N_1 - n + x)!}}{\dfrac{N!}{n!(N - n)!}}.$$

Of course, $(n - x)!$ and $n!$ are known and need not be considered. Hence we wish to maximize

$$h(N) = \frac{(N - N_1)!(N - n)!}{N!(N - N_1 - n + x)!}$$

However, due to the factorials and the fact $N$ must be an integer, we can not differentiate $h(N)$ to help us in search of the $N$ that maximizes $h(N)$.

We can consider

$$\frac{h(N+1)}{h(N)} = \frac{(N+1-N_1)!(N+1-n)!N!(N-N_1-n+x)!}{(N+1)!(N+1-N_1-n+x)!(N-N_1)!(N-n)!}$$

$$= \frac{(N+1-N_1)(N+1-n)}{(N+1)(N+1-N_1-n+x)} > 1.$$

For all $N$ values for which this ratio is greater than one, we know that we can increase $h(N)$ by making $N$ one larger. The inequality can be rewritten

$$[(N+1) - N_1][(N+1) - n] > (N+1)[(N+1) - n - N_1 + x].$$

Making some obvious cancellations, we have

$$-N_1[(N+1) - n] > (N+1)(-N_1 + x)$$

or, equivalently,

$$N_1 n > (N+1)x \qquad \text{and thus} \qquad \frac{N_1 n}{x} - 1 > N.$$

Hence if $N_1 n/x$ is an integer, we take $N$ equal to that integer to maximize $h(N)$ and, of course, $f(x)$. If $N_1 n/x$ is not an integer, we take $N$ to be the greatest integer in $N_1 n/x$, namely $\lfloor N_1 n/x \rfloor = \widehat{N}$, because then

$$\frac{h(\widehat{N})}{h(\widehat{N} - 1)} > 1 \qquad \text{and} \qquad \frac{h(\widehat{N} + 1)}{h(\widehat{N})} < 1.$$

Thus, our earlier estimate based on our intuition is equal to the maximum likelihood estimate

$$\widehat{N} = \left\lfloor \frac{N_1 n}{x} \right\rfloor. \tag{2.4-3}$$

So if we tag $N_1 = 1000$ whales in the ocean and later capture $n = 500$ whales and $x = 50$ are tagged, then we estimate that there were

$$\widehat{N} = \left\lfloor \frac{(1000)(500)}{50} \right\rfloor = 10{,}000$$

whales in the ocean before the capture. Of course, we estimate that there are $10{,}000 - 500 = 9{,}500$ after the capture. (See Exercise 5.3-10 for a simulation of this application.)

# APPLICATION

Suppose we generate some data from one of the three distributions: the binomial, the Poisson, or the translated negative binomial. Which one and which parameters are used are unknown to the reader at the moment and we want to walk through a possible data analysis procedure. All we provide are the frequencies of the non-negative integers $0, 1, 2, \ldots$ that are possible in each of these three distributions. If, in practice, we were faced with this situation, we would certainly question the investigator about how the experiment was performed. We can not do that here, but let us see if we can select a reasonable model to fit to our data.

To help us, we recall the mean and the variances of the three respective distributions, where here $m$ is the number of trials in the binomial distribution. Note the following:

binomial: $\qquad\qquad\qquad\qquad \mu = mp > \sigma^2 = mp(1-p)$

Poisson: $\qquad\qquad\qquad\qquad\quad \mu = \lambda = \sigma^2 = \lambda$

translated negative binomial: $\quad \mu = \dfrac{r(1-p)}{p} < \sigma^2 = \dfrac{r(1-p)}{p^2}$

Here we do not provide any of the parameters, like $m$ or $r$, because that would give away the answer. Clearly if it were the binomial model, $m$ (number of trials) would need to be at least as great as the largest observation. Let us hope that $\bar{x}$ and $s^2$ give us some clues as they are estimates of $\mu$ and $\sigma^2$. The data, with $n = 500$ simulations, are:

| $x$ | 0 | 1 | 2 | 3 | 4 |
|---|---|---|---|---|---|
| $f$ | 351 | 118 | 27 | 3 | 1 |

and, for these data, $\bar{x} = 0.370$ and $s^2 = 0.402$.

Here $\bar{x}$ and $s^2$ are about equal so we could consider a Poisson distribution with $\lambda$ equal to the maximum likelihood estimate, i.e., $\lambda = 0.37$. In this case the probabilities would be given by

$$f(x) = \frac{(0.37)^x e^{-0.37}}{x!}, \qquad x = 0, 1, 2, 3, 4, \ldots.$$

For example,
$$f(0) = 0.6907;$$

and if that experiment were run $n = 500$ times, with probability 0.6907, we would expect $500(0.6907) = 345.4$ zeros. That is in quite good agreement with 351. We give the other probabilities along with their expectations $nf(x) = 500(0.37)^x e^{-0.37}/x!$.

| $x$ | 0 | 1 | 2 | 3 | 4 |
|---|---|---|---|---|---|
| $f(x)$ | 0.6907 | 0.2556 | 0.0473 | 0.0058 | 0.0005 |
| $nf(x)$ | 345.4 | 127.8 | 23.6 | 2.9 | 0.3 |

These seem to be in quite good agreement with the frequencies $351, 118, 27, 3, 1$. So maybe we have found our model.

However, $\bar{x} < s^2$; thus maybe we should at least consider the translated negative binomial. Recall that $r$ must be an integer, but possibly we can get some idea about $p$ and $r$ by equating

$$\bar{x} = \frac{r(1-p)}{p} \quad \text{and} \quad s^2 = \frac{r(1-p)}{p^2}.$$

That is, let

$$0.370 = \frac{r(1-p)}{p} \quad \text{and} \quad 0.402 = \frac{r(1-p)}{p^2}.$$

Dividing the first of these equations by the second, we obtain

$$p = \frac{0.370}{0.402} = 0.9204.$$

Hence

$$r = (0.370)\frac{0.9204}{0.0796} = 4.278.$$

Since $r$ must be an integer we let $r = 4$.

With $r = 4$ and equating $\bar{x}$ and $\mu$, we get a solution for $p$ from

$$0.370 = \frac{(4)(1-p)}{p}.$$

That is, $p = 4/4.370 = 0.915$. Using $r = 4$ and $p = 0.915$, we obtain the probabilities

$$f(x) = \binom{x+r-1}{r-1}p^r(1-p)^x = \binom{x+4-1}{4-1}(0.915)^4(0.085)^x, \qquad x = 0, 1, 2, 3, \ldots$$

and $nf(x)$ is given by

| $x$ | 0 | 1 | 2 | 3 | 4 |
|---|---|---|---|---|---|
| $f(x)$ | 0.7009 | 0.2383 | 0.0506 | 0.0086 | 0.0013 |
| $nf(x)$ | 350.5 | 119.2 | 25.3 | 4.3 | 0.6 |

Note here the $nf(x)$ sum to only 499.9 instead of 500 due to some roundoff error. However it looks like an excellent fit, possibly better than the Poisson.

Maybe we should continue with our third model, the binomial, even though we are not encouraged by the fact that $\bar{x} < s^2$. Here we have equating $\bar{x} = mp$ and $s^2 = mp(1-p)$

$$0.370 = mp \quad \text{and} \quad 0.402 = mp(1-p),$$

remembering that $m$ must be an integer greater than or equal to 4. Dividing the second equation by the first, we obtain

$$1-p = \frac{0.402}{0.370} = 1.086.$$

However, $0 \leq 1 - p \leq 1$ and thus we can not obtain a sensible solution. Hence we must rule out the binomial model.

So we have two strong candidates for the model. How do we decide which is better? Actually by comparing the observed frequencies, say $O_i$, to the expected frequencies $nf(x)$, say $E_i$. With the Poisson fit using $\lambda = 0.37$ we have

| $O_i$ | 351 | 118 | 27 | 3 | 1 |
|---|---|---|---|---|---|
| $E_i$ | 345.4 | 127.8 | 23.6 | 2.9 | 0.3 |

On the other hand, the translated negative binomial with $r = 4$ and $p = 0.915$ gives

| $O_i$ | 351 | 118 | 27 | 3 | 1 |
|---|---|---|---|---|---|
| $E_i$ | 350.5 | 119.2 | 25.3 | 4.3 | 0.6 |

Only in the case where $O_4 = 3$ does the $E_4 = 2.9$ of the Poisson come closer than the $E_4 = 4.3$ of the translated negative binomial. So it seems as if the translated negative binomial is the best. The fit is so good it almost looks as if we cheated, but we must confess that we simulated from a translated negative binomial distribution with $r = 3$ and $p = 0.9$, not with $r = 4$. It is extremely difficult to select the right model with correct parameters. No model is ever exactly right; but a good fitting model is extremely useful in practice, and we are very satisfied with the translated negative binomial with $r = 4$ and $p = 0.915$ in this case.

A graphical comparison confirms what we just did numerically. In Figure 2.4-1, the shaded portions of the histograms shows the relative frequency histogram of the data. In (a) the comparison is with the Poisson probability histogram and in (b) the comparison is with the translated negative binomial probability histogram.

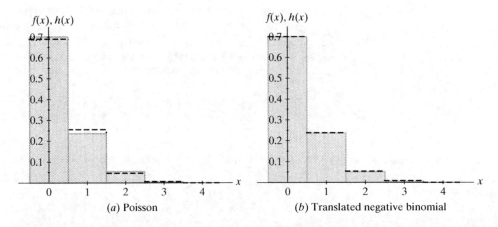

(a) Poisson          (b) Translated negative binomial

**Figure 2.4-1** Relative frequency histogram $h(x)$ (shaded), the probability histogram $f(x)$ (dashed) superimposed

**Remark** In the above illustration the translated negative binomial had a slight advantage over the Poisson, as it had two parameters to be selected and the Poisson only had one. We discuss this situation later in the book. However, right now the question might be why we used $\lambda = \bar{x} = 0.370$ instead of

$\lambda = s^2 = 0.402$, as $\lambda$ is really equal to the common value of $\mu$ and $\sigma^2$. As a matter of fact, why not use some value between 0.370 and 0.402? We have already addressed one excellent method of finding an estimate of a parameter in Example 2.4-1 and this indicates that the maximum likelihood estimate $\lambda = \theta = \bar{x} = 0.370$ was a good choice.

## EXERCISES 2.4

**2.4-1** Let $X$ equal the number of telephone calls per hour that are received by 911 between midnight and noon and reported in the *Holland Sentinel*. On October 29 and October 30, the following numbers of calls were reported:

$$
\begin{array}{cccccccccccc}
0 & 1 & 1 & 1 & 0 & 1 & 2 & 1 & 4 & 1 & 2 & 3 \\
0 & 3 & 0 & 1 & 0 & 1 & 1 & 2 & 3 & 0 & 2 & 2
\end{array}
$$

(a) Calculate the sample mean and sample variance for these data.

(b) Using the maximum likelihood estimate of $\lambda$, fit a Poisson distribution. Does this seem reasonable?

**2.4-2** Let $X_1, X_2, \ldots, X_n$ be a random sample from the negative binomial distribution with parameters where $r$ is known and $0 < p < 1$. Find the maximum likelihood estimator of $p$. Does this agree with your intuition?

**2.4-3** A Geiger counter was set up in the physics laboratory to record the number of alpha particle emissions of carbon-14 in 0.5 second. The following are 50 observations.

$$
\begin{array}{cccccccccc}
4 & 6 & 6 & 12 & 11 & 11 & 10 & 5 & 10 & 7 \\
9 & 6 & 9 & 11 & 9 & 6 & 4 & 9 & 11 & 8 \\
10 & 11 & 7 & 4 & 5 & 8 & 6 & 5 & 8 & 7 \\
6 & 3 & 4 & 6 & 4 & 12 & 7 & 14 & 5 & 9 \\
7 & 10 & 9 & 6 & 6 & 12 & 10 & 7 & 12 & 13
\end{array}
$$

(a) Calculate the sample mean and sample variance for these data.

(b) Of the three models considered in the application of this section, which seems to be the most reasonable considering the computed $\bar{x}$ and $s^2$?

**2.4-4** For determining the half-lives of radioactive isotopes, it is important to know what the background radiation is in a given detector over a period of time. Data taken in a $\gamma$-ray detection experiment over 300 one-second intervals yielded the following data:

$$
\begin{array}{cccccccccccccccccccc}
0 & 2 & 4 & 6 & 6 & 1 & 7 & 4 & 6 & 1 & 1 & 2 & 3 & 6 & 4 & 2 & 7 & 4 & 4 & 2 \\
2 & 5 & 4 & 4 & 4 & 1 & 2 & 4 & 3 & 2 & 2 & 5 & 0 & 3 & 1 & 1 & 0 & 0 & 5 & 2 \\
7 & 1 & 3 & 3 & 3 & 2 & 3 & 1 & 4 & 1 & 3 & 5 & 3 & 5 & 1 & 3 & 3 & 0 & 3 & 2 \\
6 & 1 & 1 & 4 & 6 & 3 & 6 & 4 & 4 & 2 & 2 & 4 & 3 & 3 & 6 & 1 & 6 & 2 & 5 & 0 \\
6 & 3 & 4 & 3 & 1 & 1 & 4 & 6 & 1 & 5 & 1 & 1 & 4 & 1 & 4 & 1 & 1 & 1 & 3 & 3 \\
4 & 3 & 3 & 2 & 5 & 2 & 1 & 3 & 5 & 3 & 2 & 7 & 0 & 4 & 2 & 3 & 3 & 5 & 6 & 1 \\
4 & 2 & 6 & 4 & 2 & 0 & 4 & 4 & 7 & 3 & 5 & 2 & 2 & 3 & 1 & 3 & 1 & 3 & 6 & 5 \\
4 & 8 & 2 & 2 & 4 & 2 & 2 & 1 & 4 & 7 & 5 & 2 & 1 & 1 & 4 & 1 & 4 & 3 & 6 & 2 \\
1 & 1 & 2 & 2 & 2 & 3 & 5 & 4 & 3 & 2 & 2 & 3 & 3 & 2 & 4 & 4 & 3 & 2 & 2 \\
3 & 6 & 1 & 1 & 3 & 3 & 2 & 1 & 4 & 5 & 5 & 1 & 2 & 3 & 3 & 1 & 3 & 7 & 2 & 5 \\
4 & 2 & 0 & 6 & 2 & 3 & 2 & 3 & 0 & 4 & 4 & 5 & 2 & 5 & 3 & 0 & 4 & 6 & 2 & 2 \\
2 & 2 & 2 & 5 & 2 & 2 & 3 & 4 & 2 & 3 & 7 & 1 & 1 & 7 & 1 & 3 & 6 & 0 & 5 & 3 \\
0 & 0 & 3 & 3 & 0 & 2 & 4 & 3 & 1 & 2 & 3 & 3 & 3 & 4 & 3 & 2 & 2 & 7 & 5 & 3 \\
5 & 1 & 1 & 2 & 2 & 6 & 1 & 3 & 1 & 4 & 4 & 2 & 3 & 4 & 5 & 1 & 3 & 4 & 3 & 1 \\
0 & 3 & 7 & 4 & 0 & 5 & 2 & 5 & 4 & 4 & 2 & 2 & 3 & 2 & 4 & 6 & 5 & 5 & 3 & 4
\end{array}
$$

Do these look like observations of a Poisson random variable with mean $\lambda = 3$? To help answer this question, do the following.

(a) Find the frequencies of $0, 1, 2, \ldots, 8$.

(b) Calculate the sample mean and sample variance.

**(c)** Of the three models considered in the application of this section, which seems to give the best fit?

**2.4-5** Let $X$ equal the number of green peanut m&m's in packages of size 22. Forty-five observations of $X$ yielded the following frequencies for the possible outcomes of $X$:

| Outcome ($x$): | 0 | 1 | 2 | 3 | 4 | 5 | 6 | 7 | 8 | 9 |
|---|---|---|---|---|---|---|---|---|---|---|
| Frequency: | 0 | 2 | 4 | 5 | 7 | 9 | 8 | 5 | 3 | 2 |

**(a)** Calculate the sample mean and sample variance for these data.

**(b)** Of the three models considered in the application in this section, which seems to give the best fit?

**2.4-6** Let $X$ equal the number of chocolate chips in a chocolate-chip cookie. Sixty-two observations of $X$ yielded the following frequencies for the possible outcomes of $X$:

| Outcome ($x$): | 0 | 1 | 2 | 3 | 4 | 5 | 6 | 7 | 8 | 9 | 10 |
|---|---|---|---|---|---|---|---|---|---|---|---|
| Frequency: | 0 | 0 | 2 | 8 | 7 | 13 | 13 | 10 | 4 | 4 | 1 |

**(a)** Calculate the sample mean and sample variance for these data.

**(b)** Of the three models considered in the application in this section, which seems to give the best fit?

**2.4-7** Let $X$ have the binomial p.m.f.

$$f(x) = \binom{n}{x} p^x (1-p)^{n-x}, \qquad x = 0, 1, 2, \ldots, n.$$

For what value of $x$ is $f(x)$ maximized? HINT: This is not a maximum likelihood problem but a search for the mode of the distribution. However, it is solved like Example 2.4-4 by considering $f(x+1)/f(x) > 1$.

## 2.5  LINEAR FUNCTIONS OF INDEPENDENT RANDOM VARIABLES

Let us return to a situation in which we have more than one random variable. In Section 2.4 we said that $X_1, X_2, \ldots, X_n$ were observations of a random sample from a distribution with p.m.f. $f(x)$ if they were independent and had the same distribution. That is,

$$P(X_1 = x_1, X_2 = x_2, \ldots, X_n = x_n) = f(x_1)f(x_2) \cdots f(x_n).$$

This can be generalized somewhat. Suppose $X_1, X_2, \ldots, X_n$ are independent but they might have different p.m.f.s, say $f_1(x_1), f_2(x_2), \ldots, f_n(x_n)$, respectively. If $x_1, x_2, \ldots, x_n$ are possible values of $X_1, X_2, \ldots, X_n$, respectively, then, from the independence,

$$P(X_1 = x_1, X_2 = x_2, \ldots, X_n = x_n) = P(X_1 = x_1)P(X_2 = x_2) \cdots P(X_n = x_n)$$

$$= f_1(x_1)f_2(x_2) \cdots f_n(x_n).$$

This product $f_1(x_1)f_2(x_2) \cdots f_n(x_n)$ is called the **joint p.m.f.** of the independent random variables $X_1, X_2, \ldots, X_n$.

**EXAMPLE 2.5-1** Let the independent random variables $X_1$ and $X_2$ have Poisson distributions with means $\mu_1 = 2$ and $\mu_2 = 3$, respectively. Then

$$P(X_1 = 3 \text{ and } X_2 = 4) = \left(\frac{2^3 e^{-2}}{3!}\right)\left(\frac{3^4 e^{-3}}{4!}\right) = \frac{9}{2} e^{-5}$$

and

$$P(X_1 + X_2 = 2) = P(X_1 = 0, X_2 = 2) + P(X_1 = 1, X_2 = 1) + P(X_1 = 2, X_2 = 0)$$

$$= \left(\frac{2^0 e^{-2}}{0!}\right)\left(\frac{3^2 e^{-3}}{2!}\right) + \left(\frac{2^1 e^{-2}}{1!}\right)\left(\frac{3^1 e^{-3}}{1!}\right) + \left(\frac{2^2 e^{-2}}{2!}\right)\left(\frac{3^0 e^{-3}}{0!}\right)$$

$$= \frac{(2+3)^2 e^{-5}}{2!} = \frac{5^2 e^{-5}}{2!},$$

which might suggest that the random variable $Y = X_1 + X_2$ has a Poisson distribution with mean $\mu_Y = 5$. This is true. It will be proved in Section 6.2.

To make the following discussion easier, let us consider only $n = 2$ independent random variables $X_1$ and $X_2$ with p.m.f.s $f_1(x_1)$, $x_1 \varepsilon S_1$, and $f_2(x_2)$, $x_2 \varepsilon S_2$, and a linear function of $X_1$ and $X_2$, namely $Y = a_1 X_1 + a_2 X_2$. Again $Y$ is a random variable and, by advanced methods, its p.m.f. $g(y)$ might be found and $\mu_Y$ and $\sigma_Y^2$ computed. However, the mean of $Y$ can be computed by

$$\mu_Y = E(a_1 X_1 + a_2 X_2),$$

where the expectation of $u(X_1, X_2)$ is given by the sum

$$E[u(X_1, X_2)] = \sum_{x_1 \varepsilon S_1, x_2 \varepsilon S_2} u(x_1, x_2) f_1(x_1) f_2(x_2).$$

Thus, since this $E$ is also a linear operator,

$$\mu_Y = a_1 E(X_1) + a_2 E(X_2) = a_1 \mu_1 + a_2 \mu_2,$$

where $\mu_1$ and $\mu_2$ are the respective means of $X_1$ and $X_2$. The variance of $Y$ is

$$\sigma_Y^2 = E[(a_1 X_1 + a_2 X_2 - a_1 \mu_1 - a_2 \mu_2)^2]$$

$$= E\{[a_1(X_1 - \mu_1) + a_2(X_2 - \mu_2)]^2\}$$

$$= E[a_1^2(X_1 - \mu_1)^2 + a_2^2(X_2 - \mu_2)^2 + 2a_1 a_2(X_1 - \mu_1)(X_2 - \mu_2)]$$

$$= a_1^2 E[(X_1 - \mu_1)^2] + a_2^2 E[(X_2 - \mu_2)^2] + 2a_1 a_2 E[(X_1 - \mu_2)(X_2 - \mu_2)]$$

$$= a_1^2 \sigma_1^2 + a_2^2 \sigma_2^2 + 2a_1 a_2 E[(X_1 - \mu_1)(X_2 - \mu_2)].$$

However

$$E[(X_1 - \mu_1)(X_2 - \mu_2)] = \sum_{x_1 \varepsilon S_1, x_2 \varepsilon S_2} (x_1 - \mu_1)(x_2 - \mu_2) f_1(x_1) f_2(x_2)$$

$$= \sum_{x_1 \varepsilon S_1} (x_1 - \mu_1) f_1(x_1) \sum_{x_2 \varepsilon S_2} (x_2 - \mu_2) f_2(x_2)$$

$$= E(X_1 - \mu_1) E(X_2 - \mu_2) = 0,$$

since $E(X_i - \mu_i) = E(X_i) - \mu_i = \mu_i - \mu_i$, $i = 1, 2$. Thus we have

$$\sigma_Y^2 = a_1^2 \sigma_1^2 + a_2^2 \sigma_2^2.$$

Remark   In this development, we have shown that

$$E[(X_1 - \mu_1)(X_2 - \mu_2)] = E(X_1 - \mu_1)E(X_2 - \mu_2)$$

because $X_1$ and $X_2$ are independent. However it is clear, using the same argument, that there is a more general result: If $u(X_1, X_2) = u_1(X_1)u_2(X_2)$, then

$$E[u_1(X_1)u_2(X_2)] = E[u_1(X_1)]E[u_2(X_2)].$$

That is, when $X_1$ and $X_2$ are independent, the expected value of the product $u_1(X_1)u_2(X_2)$ is the product of the expected values of $u_1(X_1)$ and $u_2(X_2)$.

**EXAMPLE 2.5-2**   Let $X_1$ and $X_2$ be independent binomial random variables with $n_1 = 100, p_1 = 1/2$ and $n_2 = 48, p_2 = 1/4$, respectively. Then $Y = X_1 - X_2$ has

$$\mu_Y = 100\left(\frac{1}{2}\right) - 48\left(\frac{1}{4}\right) = 38$$

and

$$\sigma_Y^2 = (1)^2(100)\left(\frac{1}{2}\right)\left(\frac{1}{2}\right) + (-1)^2(48)\left(\frac{1}{4}\right)\left(\frac{3}{4}\right)$$

$$= 25 + 9 = 34.$$

Note why the variances are added and not subtracted.

In the same way, these results can be extended to $n$ independent random variables $X_1, X_2, \ldots, X_n$ with respective means $\mu_1, \mu_2, \ldots, \mu_n$ and variances $\sigma_1^2, \sigma_2^2, \ldots, \sigma_n^2$. The linear function $Y = a_1 X_1 + a_2 X_2 + \cdots + a_n X_n$ is a random variable with mean and variance equal to

$$\mu_Y = a_1\mu_1 + a_2\mu_2 + \cdots + a_n\mu_n = \sum_{i=1}^{n} a_i\mu_i$$

and

$$\sigma_Y^2 = a_1^2\sigma_1^2 + a_2^2\sigma_2^2 + \cdots + a_n^2\sigma_n^2 = \sum_{i=1}^{n} a_i^2\sigma_i^2.$$

**EXAMPLE 2.5-3**   Let the independent random variables, $X_1, X_2, X_3$, measure the times needed in three steps to complete a certain project. Say, in hours, they have the respective means and variances $\mu_1 = 2.5, \mu_2 = 4.1, \mu_3 = 3.8$ and $\sigma_1^2 = 4.16, \sigma_2^2 = 7.18, \sigma_3^2 = 5.81$. The time needed to complete the three-step project is $Y = X_1 + X_2 + X_3$ which has mean

$$\mu_Y = 2.5 + 4.1 + 3.8 = 10.4 \text{ hours}$$

and variance

$$\sigma_Y^2 = 4.16 + 7.18 + 5.81 = 17.15 \text{ hours}^2.$$

Of course, $\sigma_Y = \sqrt{17.15} = 4.14$ hours.

Let us return to the case in which $X_1, X_2, \ldots, X_n$ are observations of a random sample from some distribution with p.m.f. $f(x)$, having mean $\mu$ and variance $\sigma^2$. The sample mean is the linear function

$$\overline{X} = \frac{X_1 + X_2 + \cdots + X_n}{n}$$

in which each $a_i = 1/n$. Then

$$\mu_{\overline{X}} = \sum_{i=1}^{n} \left(\frac{1}{n}\right)\mu = \mu$$

and

$$\sigma_{\overline{X}}^2 = \sum_{i=1}^{n} \left(\frac{1}{n}\right)^2 \sigma^2 = \frac{\sigma^2}{n}.$$

That is, $\overline{X}$ is an unbiased estimator of $\mu$ since $E(\overline{X}) = \mu$. Using Chebyshev's Inequality, with any given $\epsilon > 0$, we have that

$$P(|\overline{X} - \mu| \geq \epsilon) = P\left[|\overline{X} - \mu| \geq \left(\frac{\epsilon\sqrt{n}}{\sigma}\right)\left(\frac{\sigma}{\sqrt{n}}\right)\right] \leq \frac{\sigma^2}{n\epsilon^2},$$

because $\sigma_{\overline{X}} = \sigma/\sqrt{n}$ and $k = \epsilon\sqrt{n}/\sigma$. Thus

$$\lim_{n \to \infty} P(|\overline{X} - \mu| \geq \epsilon) \leq \lim_{n \to \infty} \frac{\sigma^2}{n\epsilon^2} = 0,$$

for every $\epsilon > 0$. Thus we note that $\overline{X}$ converges in probability to $\mu$, and this is another form of the **weak law of large numbers**.

We have noted in this chapter that $X/n$, where $X$ is $b(n, p)$, and the mean $\overline{X}$ of a random sample from a distribution with mean $\mu$ and variance $\sigma^2$ are, in some sense, very good estimators of the respective parameters $p$ and $\mu$. In particular, they are unbiased estimators which converge in probability to their respective parameters. Since, in most practical situations the parameters $p$, $\mu$, and $\sigma^2$ are unknown, it is reasonable to estimate the standard deviations of $X/n$ and $\overline{X}$ by

$$\sqrt{\frac{(X/n)(1 - X/n)}{n}} \quad \text{and} \quad \sqrt{\frac{S^2}{n}} = \frac{S}{\sqrt{n}},$$

respectively, where $S^2$ is the sample variance. Often these estimates of the standard deviations of $X/n$ and $\overline{X}$ are called **standard errors**. Moreover, it is shown later that the probability is about 0.95 that the true unknown $p$ is in the interval

$$\frac{X}{n} - 2\sqrt{\frac{(X/n)(1 - X/n)}{n}} \quad \text{to} \quad \frac{X}{n} + 2\sqrt{\frac{(X/n)(1 - X/n)}{n}}.$$

The same type of statement can be made about $\mu$, namely, the probability is about 0.95 that the true unknown $\mu$ is in the interval

$$\overline{X} - 2\frac{S}{\sqrt{n}} \quad \text{to} \quad \overline{X} + 2\frac{S}{\sqrt{n}}.$$

We call each of these intervals, once computed with observed values of $X$, $\overline{X}$, and $S$, an approximate 95% **confidence interval** for the respective parameters $p$ and $\mu$; and we feel quite certain that the true parameters are in the respective intervals.

> **Remark**   We are able to make these statements because both $X/n$ and $\overline{X}$ have approximate normal distributions which are studied in Chapter 3. The computed intervals,
>
> $$\frac{x}{n} \pm 2\sqrt{\frac{(x/n)(1-x/n)}{n}}$$
>
> and
>
> $$\overline{x} \pm 2\frac{s}{\sqrt{n}},$$
>
> are called approximate 95% confidence intervals for $p$ and $\mu$, respectively, where 95% refers back to the approximate probabilities obtained from those normal distributions.

EXAMPLE 2.5-4   Say $X$ is $b(n = 72, p)$. We observe $X$ to be $x = 28$. Then $\widehat{p} = 28/72 = 0.389$ is an estimate of the unknown $p$, and we are about 95% confident that $p$ is in the interval

$$0.389 \pm 2\sqrt{\frac{(0.389)(1 - 0.389)}{72}},$$

for simplicity. That is, the interval 0.274 to 0.504 serves as an approximate 95% confidence interval for $p$.

EXAMPLE 2.5-5   Say $\overline{X}$ is the mean of a random sample of size $n = 47$ from a distribution with unknown mean $\mu$ and unknown variance $\sigma^2$. The computed values of the sample characteristics are $\overline{x} = 70.4$ and $s = 9.4$. Then the interval

$$70.4 \pm (2)\frac{9.4}{\sqrt{47}}$$

or, equivalently, from 67.66 to 73.14 is an approximate 95% confidence interval for $\mu$.

## EXERCISES 2.5

**2.5-1**  Let the independent random variables $X_1$ and $X_2$ have parameters $\mu_1 = 1$, $\mu_2 = 2$, $\sigma_1^2 = 4$, and $\sigma_2^2 = 9$, respectively. Find the mean and the variance of $Y = 3X_1 - 2X_2$.

**2.5-2**  Let $X_1$ and $X_2$ be independent random variables with respective variances $\sigma_1^2 = k$ and $\sigma_2^2 = 2$. Given that the variance of $Y = 3X_2 - X_1$ is 25, find $k$.

**2.5-3**  Let $\overline{X}$ be the mean of a random sample of size 9 from a distribution with p.m.f. $f(x) = x/6$, $x = 1, 2, 3$. Find the mean and the variance of $\overline{X}$.

**2.5-4**  If the independent random variables $X_1$ and $X_2$ have means $\mu_1$, $\mu_2$ and variances $\sigma_1^2$, $\sigma_2^2$, respectively, show that the mean and the variance of the product $Y = X_1 X_2$ are $\mu_1 \mu_2$ and $\sigma_1^2 \sigma_2^2 + \mu_1^2 \sigma_2^2 + \mu_2^2 \sigma_1^2$, respectively.

**2.5-5**  Let $X_1, X_2, X_3$ be three independent random variables with binomial distributions $b(4, 1/2)$, $b(6, 1/3)$, and $b(12, 1/6)$ respectively. Find

(a)  $P(X_1 = 2, X_2 = 2, X_3 = 5)$.

(b)  $E(X_1 X_2 X_3)$.

(c)  The mean and the variance of $Y = X_1 + X_2 + X_3$.

**2.5-6** Flip $n = 8$ fair coins and remove all that came up heads. Flip the other (tails) coins and remove the heads. Continue flipping the remaining coins until each has come up heads. We shall find the p.m.f. of $Y$, the number of trials needed. Let $X_i$ equal the number of flips required to observe heads on coin $i, i = 1, 2, \ldots, 8$. Then $Y = \max(X_1, X_2, \ldots, X_8)$ has space given by $y = 1, 2, 3, \ldots$.

**(a)** Show that $P(Y \leq y) = [1 - (1/2)^y]^8$.

**(b)** Show that $P(Y = y) = [1 - (1/2)^y]^8 - [1 - (1/2)^{y-1}]^8$, $\qquad y = 1, 2, \ldots$.

**(c)** Use a CAS such as *Maple* or *Mathematica* to show that

$$E(Y) = \frac{13{,}315{,}424}{3{,}011{,}805} = 4.421.$$

**(d)** What happens to the expected value of $Y$ if the number of coins is doubled?

**(e)** Simulate this experiment either physically or on the computer.

**2.5-7** To determine the bacteria count in the west basin of Lake Macatawa, $n = 37$ samples of water were taken from the west basin, and the number of bacteria colonies in 100 milliliters of water was counted. The sample characteristics were $\bar{x} = 11.95$ and $s = 11.80$, measured in hundreds of colonies. Find the approximate 95% confidence interval for the mean number of colonies, say $\mu_W$, in 100 milliliters of water in the west basin.

**2.5-8** A manufacturer of soap powder packages the soap in "6-pound" boxes. To check the filling machine, they took a sample of $n = 1219$ boxes and weighed them. Given that $\bar{x} = 6.05$ pounds and $s = 0.02$ pounds, give the endpoints for a 95% confidence interval for $\mu$, the mean weight of the boxes of soap filled by this machine.

**2.5-9** Let $p$ equal the proportion of letters mailed in the Netherlands that are delivered the next day. If $y = 142$ out of a random sample of $n = 200$ letters were delivered the day after they were mailed, find an approximate 95% confidence interval for $p$.

**2.5-10** Let $p$ equal the proportion of Americans who favor the death penalty. If a random sample of $n = 1234$ Americans yielded $y = 864$ who favored the death penalty, find an approximate 95% confidence interval for $p$.

**2.5-11** Assume a binomial model for the random variable $X$. If we desire a 95% confidence interval of the form $(X/n) \pm 0.02$, find $n$. Hint: Note that

$$\sqrt{\left(\frac{x}{n}\right)\left(1 - \frac{x}{n}\right)} \leq \sqrt{\left(\frac{1}{2}\right)\left(\frac{1}{2}\right)}.$$

## 2.6 MULTIVARIATE DISCRETE DISTRIBUTIONS

We already know something about **independent** random variables denoted by $X_1, X_2, \ldots, X_n$: If their p.m.f.s are $f_1(x_1), f_2(x_2), \ldots, f_n(x_n)$, respectively, their joint p.m.f. is $f_1(x_1)f_2(x_2) \cdots f_n(x_n)$. However $X_1, X_2, \ldots, X_n$ do not need to be independent as illustrated in the next example with two random variables.

**EXAMPLE 2.6-1** Suppose that $X_1$ and $X_2$ have the joint p.m.f. $f(x_1, x_2)$ given by Figure 2.6-1. This means that $f(1, 1) = P(X_1 = 1, X_2 = 1) = 4/10$, $f(1, 2) = P(X_1 = 1, X_2 = 2) = 3/10$, and so on. The **marginal p.m.f.s** are given by $f_1(x_1)$ and $f_2(x_2)$ in the "margins" of the display meaning $f_1(1) = P(X_1 = 1) = 7/10$, $f_2(2) = P(X_2 = 2) = 4/10$, and so on. Here $X_1$ and $X_2$ are not independent because

$$f(1, 2) = P(X_1 = 1, X_2 = 2) = \frac{3}{10} \neq P(X_1 = 1)P(X_2 = 2) = f_1(1)f_2(2) = \frac{7}{10} \cdot \frac{4}{10}$$

and, in general,

$$f(x_1, x_2) \neq f_1(x_1)f_2(x_2), \qquad \text{when} \qquad x_1 - 1, 2; \quad x_2 = 1, 2.$$

We say that $X_1$ and $X_2$ are **dependent**.

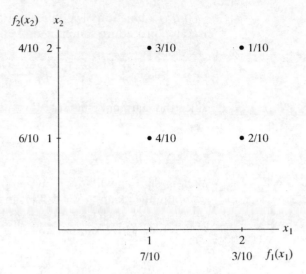

**Figure 2.6-1** Discrete joint p.m.f.

If $Y = u(X_1, X_2)$, then $Y$ is a random variable. Say its expected value, $E(Y)$, exists. Then it can be computed by

$$E[u(X_1, X_2)] = \sum_{x_1} \sum_{x_2} u(x_1, x_2) f(x_1, x_2).$$

While we do not prove this, we do illustrate it by using the joint p.m.f. of Example 2.6-1. Let $Y = X_1 + X_2$ so that the p.m.f. $g(y)$ of $Y$ is given by

$$g(2) = P(Y = 2) = P(X_1 = 1, X_2 = 1) = f(1, 1) = \frac{4}{10},$$

$$g(3) = P(Y = 3) = P(X_1 = 1, X_2 = 2) + P(X_1 = 2, X_2 = 1)$$
$$= \frac{3}{10} + \frac{2}{10} = \frac{5}{10},$$

$$g(4) = P(Y = 4) = P(X_1 = 2, X_2 = 2) = f(2, 2) = \frac{1}{10}.$$

So

$$E(Y) = 2 \cdot \frac{4}{10} + 3 \cdot \frac{5}{10} + 4 \cdot \frac{1}{10} = \frac{27}{10} = 2.7.$$

Moreover,

$$E(X_1 + X_2) = \sum_{x_2=1}^{2} \sum_{x_1=1}^{2} (x_1 + x_2) f(x_1, x_2)$$

$$= (1 + 1)\left(\frac{4}{10}\right) + (2 + 1)\left(\frac{2}{10}\right) + (1 + 2)\left(\frac{3}{10}\right) + (2 + 2)\left(\frac{1}{10}\right)$$

$$= \frac{27}{10} = 2.7,$$

which is the same as $E(Y) = 2.7$.

If $u$ is a function of $X_1$ alone, say $u(X_1)$, then computing the double sum by an iterated procedure summing on $x_2$ first, we have

$$E[u(X_1)] = \sum_{x_1} \sum_{x_2} u(x_1) f(x_1, x_2) = \sum_{x_1} u(x_1) f_1(x_1)$$

since the marginal p.m.f. of $X_1$ is

$$f_1(x_1) = \sum_{x_2} f(x_1, x_2).$$

That is, starting with the joint p.m.f. we can get the marginal p.m.f. of $X_1$ and then compute the mean, $\mu_1 = E(X_1)$, and the variance, $\sigma_1^2 = E[(X_1 - \mu_1)^2]$, of $X_1$.

EXAMPLE 2.6-2    Continuing with Example 2.6-1,

$$\mu_1 = E(X_1) = \sum_{x_1=1}^{2} x_1 f_1(x_1) = (1)\left(\frac{7}{10}\right) + (2)\left(\frac{3}{10}\right) = \frac{13}{10}$$

and

$$\sigma_1^2 = E[X_1^2] - \left(\frac{13}{10}\right)^2$$

$$= \sum_{x_1=1}^{2} x_1^2 f_1(x_1) - \left(\frac{13}{10}\right)^2$$

$$= (1)^2 \left(\frac{7}{10}\right) + (2)^2 \left(\frac{3}{10}\right) - \frac{169}{100}$$

$$= \frac{21}{100} = 0.21.$$

Similar statements can be made about $X_2$.

A joint p.m.f. $f(x_1, x_2)$ with space $(x_1, x_2) \in S$ has the properties that:

(a) $0 \le f(x_1, x_2) \le 1$,

(b) $\displaystyle\sum_{(x_1, x_2) \varepsilon S} \sum f(x_1, x_2) = 1$,

(c) $P[(X_1, X_2) \varepsilon A] = \displaystyle\sum_{(x_1, x_2) \varepsilon A} \sum f(x_1, x_2)$

where $A$ is a subset of the space $S$.

Moreover the expected value of $Y = u(X_1, X_2)$, if it exists, is

$$E[u(X_1, X_2)] = \sum_{(x_1, x_2) \varepsilon S} \sum u(x_1, x_2) f(x_1, x_2).$$

An important expectation that can not be computed from the marginal distributions is $E[(X_1 - \mu_1)(X_2 - \mu_2)]$ and is called the **covariance** of $X_1$ and $X_2$ and is denoted by $\text{Cov}(X_1, X_2) = \sigma_{12}$. If the standard deviations, $\sigma_1$ and $\sigma_2$, are positive then

$$\rho = \frac{\text{Cov}(X_1, X_2)}{\sigma_1 \sigma_2} = \frac{\sigma_{12}}{\sigma_1 \sigma_2}$$

is called the **correlation coefficient** of $X_1$ and $X_2$. (The reader is asked to show that $1 \le \rho \le 1$ in Exercise 2.6-4.) Moreover the covariance can be computed using the formula

$$\text{Cov}(X_1, X_2) = E(X_1 X_2) - \mu_1 \mu_2.$$

(See Exercise 2.6-2.)

**EXAMPLE 2.6-3**   Let $X_1$ and $X_2$ have the joint p.m.f.

$$f(x_1, x_2) = \frac{x_1 + 2x_2}{18}, \qquad x_1 = 1, 2, \quad x_2 = 1, 2.$$

It is easy to show that

$$\mu_1 = \frac{14}{9}, \qquad \sigma_1^2 = \frac{20}{81}, \qquad \mu_2 = \frac{29}{18}, \qquad \sigma_2^2 = \frac{77}{324}.$$

Since

$$E(X_1 X_2) = \sum_{x_2=1}^{2} \sum_{x_1=1}^{2} x_1 x_2 \frac{x_1 + 2x_2}{18}$$

$$= (1)(1)\left(\frac{3}{18}\right) + (2)(1)\left(\frac{4}{18}\right) + (1)(2)\left(\frac{5}{18}\right) + (2)(2)\left(\frac{6}{18}\right) = \frac{45}{18},$$

the covariance of $X_1$ and $X_2$ is

$$\text{Cov}(X_1, X_2) = \frac{45}{18} - \left(\frac{14}{9}\right)\left(\frac{29}{18}\right) = -\frac{1}{162}$$

and the correlation coefficient is

$$\rho = \frac{-1/162}{\sqrt{(20/81)(77/324)}} = -0.025.$$

**Remark**   If we have $n$ bivariate observations $(x_1, y_1), (x_2, y_2), \ldots, (x_n, y_n)$ and place the weight (probability) $1/n$ on each of them, we have created an empirical distribution. We not only can compute the means $\bar{x}, \bar{y}$ and the variances $v_x, v_y$ of the empirical distribution, but also the **sample correlation coefficient**, say

$$r = \frac{\sum_{i=1}^{n} (x_i - \bar{x})(y_i - \bar{y})\left(\frac{1}{n}\right)}{\sqrt{\sum_{i=1}^{n} (x_i - \bar{x})^2 \left(\frac{1}{n}\right)} \sqrt{\sum_{i=1}^{n} (y_i - \bar{y})^2 \left(\frac{1}{n}\right)}}$$

$$= \frac{\sum_{i=1}^{n} (x_i - \bar{x})(y_i - \bar{y})}{\sqrt{\sum_{i=1}^{n} (x_i - \bar{x})^2} \sqrt{\sum_{i=1}^{n} (y_i - \bar{y})^2}}$$

$$= \frac{\sum_{i=1}^{n} x_i y_i - n\bar{x}\,\bar{y}}{\sqrt{\sum_{i=1}^{n} x_i^2 - n\bar{x}^2}\,\sqrt{\sum_{i=1}^{n} y_i^2 - n\bar{y}^2}}.$$

In a sense, the correlation coefficient $r$ measures the linearity of the points $(x_1, y_1), (x_2, y_2), \ldots, (x_n, y_n)$. That is, note that if all the points are on a straight line with positive (negative) slope, then $r = 1$ $(r = -1)$, the largest (smallest) possible value of $r$.

While we have been considering only two random variables, these concepts can be extended to $n$ random variables. For illustration, we know that if $X_1, X_2, \ldots, X_n$ are independent, then the joint p.m.f. is equal to

$$f(x_1, x_2, \ldots, x_n) \equiv f_1(x_1) f_2(x_2) \cdots f_n(x_n),$$

where $f_i(x_i)$ is the marginal p.m.f. of $X_i$, $i = 1, 2, \ldots, n$. In this independent case, we know that $Y = \sum_{i=1}^{n} a_i X_i$ has mean and variance

$$\mu_Y = \sum_{i=1}^{n} a_i \mu_i \quad \text{and} \quad \sigma_Y^2 = \sum_{i=1}^{n} a_i^2 \sigma_i^2,$$

where $\mu_1, \mu_2, \ldots, \mu_n$ and $\sigma_1^2, \sigma_2^2, \ldots, \sigma_n^2$ are the respective means and variances of $X_1, X_2, \ldots, X_n$. If there is correlation among the random variables, it is not difficult to show that $\mu_Y$ is given by the same formula, but

$$\sigma_Y^2 = \sum_{i=1}^{n} a_i^2 \sigma_i^2 + 2 \sum \sum_{i<j} a_i a_j \sigma_{ij},$$

where $\sigma_{ij} = \text{Cov}(X_i, X_j)$, the covariance of $X_i$ and $X_j$, $i \neq j$. Of course, $\sigma_{ij}$ can be replaced in this formula by $\rho_{ij} \sigma_i \sigma_j$ because $\rho_{ij} = \sigma_{ij}/(\sigma_i \sigma_j)$, $i \neq j$.

**EXAMPLE 2.6-4**   Let $X_1, X_2, X_3$ have parameters $\mu_1 = 7$, $\mu_2 = 3$, $\mu_3 = -2$, $\sigma_1^2 = 4$, $\sigma_2^2 = 9$, $\sigma_3^2 = 25$, $\rho_{12} = 0.2$, $\rho_{13} = 0.3$, and $\rho_{23} = 0.1$, respectively. If $Y = 4X_1 - 2X_2 + 3X_3$, then

$$\mu_Y = (4)(7) + (-2)(3) + (3)(-2) = 16$$

and

$$\sigma_Y^2 = (4)^2(4) + (-2)^2(9) + (3)^2(25) + 2[(4)(-2)(0.2)(2)(3)$$
$$+ (4)(3)(0.3)(2)(5) + (-2)(3)(0.1)(3)(5)] = 359.8.$$

Let $X_1$ and $X_2$ have a joint discrete distribution with p.m.f. $f(x_1, x_2)$ on space $S$. Say the marginal probability mass functions are $f_1(x_1)$ and $f_2(x_2)$ with spaces $S_1$ and $S_2$, respectively. Let event $A = \{X_1 = x_1\}$ and event $B = \{X_2 = x_2\}$, $(x_1, x_2) \in S$. Thus $A \cap B = \{X_1 = x_1, X_2 = x_2\}$. Because

$$P(A \cap B) = P(X_1 = x_1, X_2 = x_2) = f(x_1, x_2)$$

Thus $h(x_2 \mid x_1)$ satisfies the conditions of a probability mass function, and so we can compute conditional probabilities such as

$$P(a < X_2 < b \mid X_1 = x_1) = \sum_{\{x_2 : a < x_2 < b\}} h(x_2 \mid x_1)$$

and conditional expectations such as

$$E[u(X_2) \mid X_1 = x_1] = \sum_{x_2} u(x_2) h(x_2 \mid x_1)$$

in a manner similar to those associated with probabilities and expectations.

Two special conditional expectations are the **conditional mean** of $X_2$, given $X_1 = x_1$, defined by

$$\mu_{X_2 \mid x_1} = E(X_2 \mid x_1) = \sum_{x_2} x_2 \, h(x_2 \mid x_1),$$

and the **conditional variance** of $X_2$, given $X_1 = x_1$, defined by

$$\sigma^2_{X_2 \mid x_1} = E\{[X_2 - E(X_2 \mid x_1)]^2 \mid x_1\} = \sum_{x_2} [x_2 - E(X_2 \mid x_1)]^2 \, h(x_2 \mid x_1),$$

which can be computed using

$$\sigma^2_{X_2 \mid x_1} = E(X_2^2 \mid x_1) - [E(X_2 \mid x_1)]^2.$$

The conditional mean $\mu_{X_1 \mid x_2}$ and the conditional variance $\sigma^2_{X_1 \mid x_2}$ are given by similar expressions.

**EXAMPLE 2.6-6**  We use the background of Example 2.6-5 and compute $\mu_{X_2 \mid x_1}$ and $\sigma^2_{X_2 \mid x_1}$, when $x_1 = 3$:

$$\mu_{X_2 \mid 3} = E(X_2 \mid X_1 = 3) = \sum_{x_2=1}^{2} x_2 \, h(x_2 \mid 3)$$

$$= \sum_{x_2=1}^{2} x_2 \left( \frac{3 + x_2}{9} \right) = 1\left( \frac{4}{9} \right) + 2\left( \frac{5}{9} \right) = \frac{14}{9},$$

and

$$\sigma^2_{X_2 \mid 3} = E\left[ \left( X_2 - \frac{14}{9} \right)^2 \,\middle|\, X_1 = 3 \right] = \sum_{x_2=1}^{2} \left( x_2 - \frac{14}{9} \right)^2 \left( \frac{3 + y}{9} \right)$$

$$= \frac{25}{81}\left( \frac{4}{9} \right) + \frac{16}{81}\left( \frac{5}{9} \right) = \frac{20}{81}.$$

The conditional mean of $X_1$, given $X_2 = x_2$, is a function of $x_2$ alone; the conditional mean of $X_2$, given $X_1 = x_1$, is a function of $x_1$ alone. Suppose that the latter conditional mean is a linear function of $x_1$; that is, $E(X_2 \mid x_1) = a + bx_1$. Let us find the constants $a$ and $b$ in terms of characteristics $\mu_{X_1}$, $\mu_{X_2}$, $\sigma^2_{X_1}$, $\sigma^2_{X_2}$, and $\rho$. This development will shed additional light on the correlation coefficient $\rho$; accordingly we assume that the respective standard deviations $\sigma_{X_1}$ and $\sigma_{X_2}$ are both positive so that the correlation coefficient will exist.

It is given that

$$\sum_{x_2} x_2 \, h(x_2 \mid x_1) = \sum_{x_2} x_2 \, \frac{f(x_1, x_2)}{f_1(x_1)} = a + bx_1, \qquad \text{for } x_1 \in S_1,$$

where $S_1$ is the space of $X_1$ and $S_2$ is the space of $X_2$. Hence

$$\sum_{x_2} x_2 f(x_1, x_2) = (a + bx_1)f_1(x_1), \qquad \text{for } x_1 \in S_1, \tag{2.6-1}$$

and

$$\sum_{x_1 \varepsilon S_1} \sum_{x_2} x_2 f(x_1, x_2) = \sum_{x_1 \varepsilon S_1} (a + bx_1)f_1(x_1).$$

That is, with $\mu_{X_1}$ and $\mu_{X_2}$ representing the respective means, we have

$$\mu_{X_2} = a + b\mu_{X_1}. \tag{2.6-2}$$

In addition, if we multiply both members of Equation 2.6-1 by $x_1$ and sum, we obtain

$$\sum_{x_1 \varepsilon S_1} \sum_{x_2} x_1 x_2 f(x_1, x_2) = \sum_{x_1 \varepsilon S_1} (ax_1 + bx_1^2)f_1(x_1).$$

That is,

$$E(X_1 X_2) = aE(X_1) + bE(X_1^2)$$

or, equivalently,

$$\mu_{X_1}\mu_{X_2} + \rho\sigma_{X_1}\sigma_{X_2} = a\mu_{X_1} + b(\mu_{X_1}^2 + \sigma_{X_1}^2). \tag{2.6-3}$$

The solution of Equations 2.6-2 and 2.6-3 is

$$a = \mu_{X_2} - \rho \, \frac{\sigma_{X_2}}{\sigma_{X_1}} \mu_{X_1} \qquad \text{and} \qquad b = \rho \, \frac{\sigma_{X_2}}{\sigma_{X_1}},$$

which implies that if $E(X_2 \mid x_1)$ is linear, it is given by

$$E(X_2 \mid x_1) = \mu_{X_2} + \rho \, \frac{\sigma_{X_2}}{\sigma_{X_1}} (x_1 - \mu_{X_1}).$$

So if the conditional mean of $X_2$ given $X_1 = x_1$ is linear, it is like the least squares regression line considered in Section 4.6.

By symmetry, if the conditional mean of $X_1$, given $X_2 = x_2$, is linear, it is given by

$$E(X_1 \mid x_2) = \mu_{X_1} + \rho \, \frac{\sigma_{X_1}}{\sigma_{X_2}} (x_2 - \mu_{X_2}).$$

We see that the point $[x_1 = \mu_{X_1}, E(X_2 \mid x_1) = \mu_{X_2}]$ satisfies the expression for $E(X_2 \mid x_1)$; and $[E(X_1 \mid x_2) = \mu_{X_1}, x_2 = \mu_{X_2}]$ satisfies the expression for $E(X_1 \mid x_2)$. That is, the point $(\mu_{X_1}, \mu_{X_2})$ is on each of the two lines. In addition, we note that the product of the coefficient of $x_1$ in $E(X_2 \mid x_1)$ and the coefficient of $x_2$ in $E(X_1 \mid x_2)$ equals $\rho^2$ and the ratio of these two coefficients equals $\sigma_{X_2}^2 / \sigma_{X_1}^2$. These observations sometimes prove useful in particular problems.

**Remark**   While Section 2.6 is about the multivariate discrete case, the same results hold in the multivariate continuous case with integrals replacing summations.

## EXERCISES 2.6

**2.6-1** Let $X_1$ and $X_2$ have the joint p.m.f. given by the following:

| $(x_1, x_2)$ | $f(x_1, x_2)$ |
|:---:|:---:|
| $(1, 1)$ | 3/8 |
| $(2, 1)$ | 1/8 |
| $(1, 2)$ | 1/8 |
| $(2, 2)$ | 3/8 |

Find

(a) the marginal p.m.f.s, $f_1(x_1)$ and $f_2(x_2)$;

(b) the means, $\mu_1$ and $\mu_2$;

(c) the variances, $\sigma_1^2$ and $\sigma_2^2$;

(d) the covariance, $\sigma_{12}$;

(e) the correlation coefficient, $\rho$.

(f) Are $X_1$ and $X_2$ independent?

**2.6-2** Use the fact that $E$ is a linear or distributive operator to show that

$$\sigma_{12} = E[(X_1 - \mu_1)(X_2 - \mu_2)] = E(X_1 X_2) - \mu_1 \mu_2.$$

**2.6-3** Let $X_1$ and $X_2$ be independent random variables with joint p.m.f.

$$f(x_1, x_2) = \frac{x_1 x_2}{18}, \qquad x_1 = 1, 2, 3, \qquad \text{and} \qquad x_2 = 1, 2.$$

(a) Determine the means and the variances of $X_1$ and $X_2$.

(b) Compute $P(X_1 = 3, X_2 = 2)$.

(c) Compute $P(X_1 + X_2 = 3)$.

(d) Find the p.m.f. of $Y = X_1 + X_2$, $y = 2, 3, 4, 5$.

(e) Determine the mean and the variance of $Y$.

(f) How do the answers in part (e) compare to $\mu_1 + \mu_2$ and $\sigma_1^2 + \sigma_2^2$?

**2.6-4** Show that the correlation coefficient $\rho$ of $X_1$ and $X_2$ is such that $-1 \leq \rho \leq 1$. HINT: Consider the discriminant of the quadratic function

$$h(v) = E\{[(X_1 - \mu_1) + v(X_2 - \mu_2)]^2\} \geq 0.$$

**2.6-5** Let $X_1, X_2$ be two random variables with means $\mu_1 = 6$, $\mu_2 = 4$; variances $\sigma_1^2 = \sigma_2^2 = 9$; and correlation coefficient $\rho$.

(a) Determine the mean and the variance of $Y = X_1 - X_2$.

**(b)** Find that value of $\rho$ that minimizes the variance of $Y$. Does this agree with your common sense? HINT: Use the result of Exercise 2.6-4.

**2.6-6** Let four random variables $X_1, X_2, X_3, X_4$ have common mean $\mu_1 = \mu_2 = \mu_3 = \mu_4 = 5$ and common variance $\sigma_1^2 = \sigma_2^2 = \sigma_3^2 = \sigma_4^2 = 6$ and common correlation coefficient $\rho_{ij} = 0.1$, $i \neq j$. Determine the mean and the variance of $Y = X_1 + X_2 + X_3 + X_4$.

**2.6-7** Let $X_1$ and $X_2$ have the joint p.m.f.

$$f(x_1, x_2) = \frac{x_1 + x_2}{32}, \qquad x_1 = 1, 2, \qquad x_2 = 1, 2, 3, 4.$$

**(a)** Display the joint p.m.f. and the marginal p.m.f.s on a graph like Figure 2.6-2($a$).
**(b)** Find $g(x_1 \mid x_2)$ and draw a figure like Figure 2.6-2($b$), depicting the conditional p.m.f.s for $x_2 = 1, 2, 3$, and 4.
**(c)** Find $h(x_2 \mid x_1)$ and draw a figure like Figure 2.6-2($c$), depicting the conditional p.m.f.s for $x_1 = 1$ and 2.
**(d)** Find **(i)** $P(1 \leq X_2 \leq 3 \mid X_1 = 1)$, **(ii)** $P(X_2 \leq 2 \mid X_1 = 2)$, **(iii)** $P(X_1 = 2 \mid X_2 = 3)$.
**(e)** Find $E(X_2 \mid X_1 = 1)$ and $\text{Var}(X_2 \mid X_1 = 1)$.

**2.6-8** Let the joint p.m.f. $f(x_1, x_2)$ of $X_1$ and $X_2$ be given by the following:

| $(x_1, x_2)$ | $f(x_1, x_2)$ |
|:---:|:---:|
| $(1, 1)$ | 3/8 |
| $(2, 1)$ | 1/8 |
| $(1, 2)$ | 1/8 |
| $(2, 2)$ | 3/8 |

Find the two conditional probability mass functions and the corresponding means and variances.

**2.6-9** The following data are the ACT composite score and the first year college Grade Point Average (GPA) for 18 students.

| ACT | GPA | ACT | GPA |
|:---:|:---:|:---:|:---:|
| 23 | 2.40 | 25 | 3.67 |
| 29 | 3.10 | 23 | 3.03 |
| 15 | 2.51 | 30 | 3.07 |
| 23 | 2.98 | 16 | 2.28 |
| 28 | 3.93 | 24 | 2.45 |
| 27 | 1.91 | 26 | 3.83 |
| 23 | 1.69 | 27 | 3.78 |
| 26 | 2.88 | 21 | 2.57 |
| 20 | 2.60 | 19 | 3.42 |

Compute the correlation coefficient.

**2.6-10** In a college health fitness program, let $X$ equal the weight in kilograms of a female freshman at the beginning of the program and let $Y$ equal her weight change during the semester. Compute the correlation coefficient for the following $n = 16$ observations of $(x, y)$.

| | | | |
|---|---|---|---|
| (61.4, −3.2) | (62.9, 1.4) | (58.7, 1.3) | (49.3, 0.6) |
| (71.3, 0.2) | (81.5, −2.2) | (60.8, 0.9) | (50.2, 0.2) |
| (60.3, 2.0) | (54.6, 0.3) | (51.1, 3.7) | (53.3, 0.2) |
| (81.0, −0.5) | (67.6, −0.8) | (71.4, −0.1) | (72.1, −0.1) |

# CHAPTER
# TWO COMMENTS

It was quite easy to explain with elementary probability the binomial, negative binomial, and hypergeometric models. Also we found the Poisson probabilities as the limit of binomial ones under certain restrictions. Moreover, we state that the Poisson model is very good in certain other situations. Let us consider the latter statement in a little more detail.

Suppose the number of *changes* that occur in some continuous interval are such that:

**(a)** The numbers of changes occurring in non-overlapping intervals are independent.

**(b)** The probability of exactly one change in a very short interval of length $h$ is approximately $\lambda h$, where the constant $\lambda > 0$.

**(c)** The probability of two or more changes in that short interval is essentially zero.

These three conditions are essentially the axioms of a **Poisson process**. If $X$ is the number of changes in a length $t$ (not necessarily short), then $X$ has a Poisson distribution with mean $\lambda t$. Clearly if $t = 1$, we have the Poisson distribution given in this chapter. For example, if phone calls arrive at a switchboard following a Poisson process at a mean rate of $\lambda = 3$ per minute, then the expected number of phone calls in $t = 5$ minutes is $(3)(5) = 15$. The number of phone calls in that 5-minute period has a Poisson distribution with mean 15.

The other comment we wish to make is about maximum likelihood estimation, which was initially proposed by Sir Ronald A. Fisher, certainly one of the greatest statisticians of all times. It is interesting to note here that an Englishman was knighted for his work in statistics and related areas. Fisher has been dead for a number of years, but Sir David Cox is an excellent statistician and is still alive. These two statisticians have ranks like a certain "Beatle." (Beatle McCartney was knighted Sir Paul by the Queen in 1997.) Maybe the monarch of England believes statisticians can do worthwhile things.

# CHAPTER 3

# CONTINUOUS DISTRIBUTIONS

## 3.1 DESCRIPTIVE STATISTICS AND EDA

Up to this point, we have considered only discrete probability distributions whose spaces contain a countable number of outcomes. The corresponding random variables are said to be of the discrete type. We have taken samples leading to data from these discrete probability distributions. However, many experiments or observations of random phenomena do not have integers as outcomes but instead are measurements selected from an interval of numbers. For example, you could find the length of time that it takes when waiting in line to buy frozen yogurt. Or, the weight of a "one-pound" package of hot dogs could perhaps be any number between 0.94 and 1.25 pounds. The weight of a miniature Baby Ruth candy bar could be any number between 20 and 27 grams. Even though such times and weights could be selected from an interval of values, times and weights are generally rounded off so that the data often look like discrete data. If conceptually the measurements could come from an interval of possible outcomes, we call it data from a continuous-type distribution or, more simply, **continuous-type data**.

Given a sample of size $n$ of continuous-type data, we shall group the data into classes and then construct a histogram of the grouped data. This will help us better visualize the data. The following guidelines and terminology will be used to group continuous-type data into classes of equal length. These guidelines can also be used for sets of discrete data that have a large range.

1. Determine the largest (maximum) and smallest (minimum) observations. The **range** is the difference, $R = maximum - minimum$.
2. Generally select from $k = 5$ to $k = 20$ classes which are non-overlapping intervals, often of equal length, so that these classes cover the interval from

the minimum to the maximum. Often $k$ is small when $n$ is small and gets bigger as $n$ increases. A very rough rule some statisticians use is $k \approx \sqrt{n}$, when $n \leq 400$.

3. Each interval begins and ends halfway between two possible values of the measurements which have been rounded off to a given number of decimal places.

4. The first interval should begin about as much below the smallest value as the last interval ends above the largest.

5. The intervals are called **class intervals** and the boundaries are called **class boundaries** or **cut points**. We shall denote these $k$ class intervals by

$$(c_0, c_1), (c_1, c_2), \ldots, (c_{k-1}, c_k).$$

6. The **class limits** are the smallest and the largest possible observed (recorded) values in a class.

7. The **class mark** is the midpoint of a class interval.

A frequency table is constructed that lists the class intervals, the class limits, a tabulation of the measurements in the various classes, the frequency $f_i$ of each class, the class marks and (optionally) a column that is used to construct a relative frequency (density) histogram. When the class intervals are of equal lengths, a frequency histogram is constructed by drawing a rectangle for each class having as its base the class interval and height equal to the frequency of the class. For the relative frequency histogram, each rectangle has an **area** equal to the relative frequency $f_i/n$ of the observations for the class. That is, the function defined by

$$h(x) = \frac{f_i}{(n)(c_i - c_{i-1})}, \qquad \text{for} \quad c_{i-1} < x \leq c_i, \quad i = 1, 2, \ldots, k,$$

is called a **relative frequency histogram** or **density histogram**, where $f_i$ is the frequency of the $i$th class and $n$ is the total number of observations. Clearly, if the class intervals are of equal length, the relative frequency histogram, $h(x)$, is proportional to the **frequency histogram** $f_i$, for $c_{i-1} < x \leq c_i, i = 1, 2, \ldots, k$. The frequency histogram should be used only in those situations in which the class intervals are of equal length.

**EXAMPLE 3.1-1**   The weights in grams of 40 miniature Baby Ruth candy bars, with the weights ordered, are given in Table 3.1-1.

| Table 3.1-1 Candy Bar Weights | | | | | | | | | |
|---|---|---|---|---|---|---|---|---|---|
| 20.5 | 20.7 | 20.8 | 21.0 | 21.0 | 21.4 | 21.5 | 22.0 | 22.1 | 22.5 |
| 22.6 | 22.6 | 22.7 | 22.7 | 22.9 | 22.9 | 23.1 | 23.3 | 23.4 | 23.5 |
| 23.6 | 23.6 | 23.6 | 23.9 | 24.1 | 24.3 | 24.5 | 24.5 | 24.8 | 24.8 |
| 24.9 | 24.9 | 25.1 | 25.1 | 25.2 | 25.6 | 25.8 | 25.9 | 26.1 | 26.7 |

The range of the data is

$$R = 26.7 - 20.5 = 6.2.$$

We shall group these data and then construct a histogram to visualize the distribution of weights. The interval, $[20.5, 26.7]$, could be covered with $k = 8$ classes of width $0.8$, $k = 9$ classes of width $0.7$, and there are other possibilities. We shall use $k = 7$ classes of width $0.9$. The first class interval will be $(20.45, 21.35)$ and the last class interval will be $(25.85, 26.75)$. The data are grouped in Table 3.1-2.

| Table 3.1-2 Frequency Table of Candy Bar Weights | | | | | |
|---|---|---|---|---|---|
| Class Interval | Class Limits | Tabulation | Frequency $(f_i)$ | $h(x)$ | Class Marks |
| $(20.45, 21.35)$ | 20.5-21.3 | ⦀ | 5 | 5/36 | 20.9 |
| $(21.35, 22.25)$ | 21.4-22.2 | ⦀⦀ | 4 | 4/36 | 21.8 |
| $(22.25, 23.15)$ | 22.3-23.1 | ⦀ ⦀ | 8 | 8/36 | 22.7 |
| $(23.15, 24.05)$ | 23.2-24.0 | ⦀ ⦀ | 7 | 7/36 | 23.6 |
| $(24.05, 24.95)$ | 24.1-24.9 | ⦀ ⦀ | 8 | 8/36 | 24.5 |
| $(24.95, 25.85)$ | 25.0-25.8 | ⦀ | 5 | 5/36 | 25.4 |
| $(25.85, 26.75)$ | 25.9-26.7 | ⦀ | 3 | 3/36 | 26.3 |

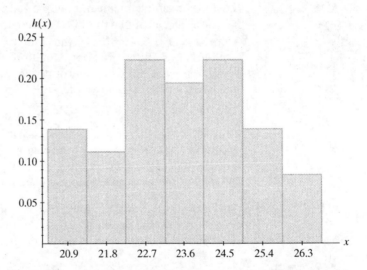

**Figure 3.1-1** Relative frequency histogram of weights of candy bars

A relative frequency histogram of these data is given in Figure 3.1-1. Note that the total area of this histogram is equal to one. We could also construct a frequency histogram in which the heights of the rectangles would be equal to the frequencies of the classes. The shapes of the two histograms are the same. Later we will see the reason for preferring the relative frequency histogram. In particular, we will be superimposing on the relative frequency histogram the graph of a continuous function, called a probability density function, that corresponds to a probability model. Both have total areas equal to one.

Now suppose that continuous-type data for an experiment have been grouped in a frequency table with $k$ classes. In such a case we could assume that the possible outcomes are approximately equal to the midpoints of the classes, say $u_1, u_2, \ldots, u_k$. Recall that these midpoints are called the **class marks**. Denote the respective frequencies as $f_1, f_2, \ldots, f_k$ and note that $n = \sum_{i=1}^{k} f_i$. In these cases we have $k$

points $u_1, u_2, \ldots, u_k$ with respective frequencies $f_1, f_2, \ldots, f_k$. The sample mean and the sample variance of the $u$ values of these grouped data are, respectively,

$$\bar{u} = \frac{1}{n} \sum_{i=1}^{k} f_i u_i,$$

$$s_u^2 = \frac{1}{n-1} \sum_{i=1}^{k} f_i (u_i - \bar{u})^2 = \frac{\sum_{i=1}^{k} f_i u_i^2 - \frac{1}{n} \left( \sum_{i=1}^{k} f_i u_i \right)^2}{n-1}$$

$$= \frac{\sum_{i=1}^{k} f_i u_i^2 - n \bar{u}^2}{n-1}.$$

For grouped "continuous data," these statistics can be used as respective approximations of $\bar{x}$ and $s_x^2$ which are computed before grouping. In the discrete case, we note that $\bar{x} = \bar{u}$ and $s_x^2 = s_u^2$.

EXAMPLE 3.1-2

To find the sample mean and sample variance of the grouped data in Example 3.1-1 we use a calculator like the TI-83 or a computer. They are $\bar{u} = 23.51$ and $s_u^2 = 2.5671$. Here $s_u = \sqrt{2.5671} = 1.60$ is the standard deviation of the grouped data.

Note that $\bar{u}$ and $s_u$ are approximations for $\bar{x}$ and $s_x$, the sample mean and sample standard deviation of the ungrouped data. Usually the approximation is quite good and in this particular example, $\bar{x} = 23.505$ and $s_x = 1.64$, so the approximations are very good.

In some situations it is not necessarily desirable to use class intervals of equal widths in the construction of the relative frequency histogram. This is particularly true if the data are skewed with a very long tail. We now present such an illustration in which it seems desirable to use class intervals of unequal lengths.

EXAMPLE 3.1-3

The following 40 losses, due to wind-related catastrophes, were recorded to the nearest one million dollars. These data include only those losses of $2 million or more; and, for convenience, they have been ordered and recorded in millions.

| | | | | | | | | | |
|---|---|---|---|---|---|---|---|---|---|
| 2 | 2 | 2 | 2 | 2 | 2 | 2 | 2 | 2 | 2 |
| 2 | 2 | 3 | 3 | 3 | 3 | 4 | 4 | 4 | 5 |
| 5 | 5 | 5 | 6 | 6 | 6 | 6 | 8 | 8 | 9 |
| 15 | 17 | 22 | 23 | 24 | 24 | 25 | 27 | 32 | 43 |

The selection of class boundaries is more subjective in this case. It makes sense to let $c_0 = 1.5$ and $c_1 = 2.5$ because only values of $2 million or more are recorded and there are 12 observations equal to 2. We could let $c_2 = 6.5$, $c_3 = 29.5$, and $c_4 = 49.5$. This would yield the following relative frequency histogram:

$$h(x) = \begin{cases} \dfrac{12}{40}, & 1.5 < x \le 2.5, \\[2mm] \dfrac{15}{(40)(4)}, & 2.5 < x \le 6.5, \\[2mm] \dfrac{11}{(40)(23)}, & 6.5 < x \le 29.5, \\[2mm] \dfrac{2}{(40)(20)}, & 29.5 < x \le 49.5. \end{cases}$$

It is displayed in Figure 3.1-2. It takes some experience before a person can display a relative frequency histogram that is most meaningful.

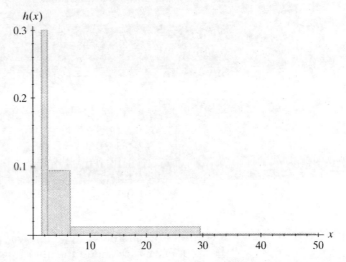

**Figure 3.1-2** Relative frequency histogram of losses

The areas of the four rectangles, 0.300, 0.375, 0.275, 0.050, are the respective relative frequencies. It is important to note in the case of unequal lengths among class intervals that the areas, not the heights, of the rectangles are proportional to the frequencies. In particular, the first and second classes have frequencies $f_1 = 12$ and $f_2 = 15$, and yet the height of the first is greater than the height of the second while here $f_1 < f_2$. If we have equal lengths among the class intervals, then the heights are proportional to the frequencies.

To explore the characteristics of an unknown distribution, we need to take a sample of $n$ observations, $x_1, x_2, \ldots, x_n$, from that distribution and often need to order them from the smallest to the largest. One convenient way of doing this is using a stem-and-leaf display which was devised by John W. Tukey. For more details of Tukey's **Exploratory Data Analysis (EDA)**, see the books by Tukey (1977) and Velleman and Hoaglin (1981).

Possibly the easiest way to begin is with an example to which all of us can relate. Say we have the following 50 test scores on a statistics examination:

| | | | | | | | | | |
|---|---|---|---|---|---|---|---|---|---|
| 93 | 77 | 67 | 72 | 52 | 83 | 66 | 84 | 59 | 63 |
| 75 | 97 | 84 | 73 | 81 | 42 | 61 | 51 | 91 | 87 |
| 34 | 54 | 71 | 47 | 79 | 70 | 65 | 57 | 90 | 83 |
| 58 | 69 | 82 | 76 | 71 | 60 | 38 | 81 | 74 | 69 |
| 68 | 76 | 85 | 58 | 45 | 73 | 75 | 42 | 93 | 65 |

We can do much the same thing as a frequency table and histogram, but keep the original values, through a **stem-and-leaf display**. For this particular data set, we could use the following procedure. The first number in the set, 93, is recorded as follows: the 9 (in the "tens" place) is treated as the stem, and the 3 (in the "units" place) is the corresponding leaf. Note that this leaf of 3 is the first digit after the stem of 9 in Table 3.1-3. The second number, 77, is given by the leaf of 7 after the stem of 7; the third number, 67, by the leaf of 7 after the stem of 6; the fourth number, 72, as the leaf of 2 after the stem of 7 (note that this is the second leaf on the 7 stem); and so on. Table 3.1-3 is called a stem-and-leaf display. If the leaves are carefully aligned vertically, this table has the same effect as a histogram, but the original numbers are not lost.

**Table 3.1-3  Stem-and-Leaf Display of Scores from 50 Statistics Examinations**

| Stems | Leaves | Frequency | Depths |
|-------|--------|-----------|--------|
| 3 | 4 8 | 2 | 2 |
| 4 | 2 7 5 2 | 4 | 6 |
| 5 | 2 9 1 4 7 8 8 | 7 | 13 |
| 6 | 7 6 3 1 5 9 0 9 8 5 | 10 | 23 |
| 7 | 7 2 5 3 1 9 0 6 1 4 6 3 5 | 13 | (13) |
| 8 | 3 4 4 1 7 3 2 1 5 | 9 | 14 |
| 9 | 3 7 1 0 3 | 5 | 5 |

Another term in these stem-and-leaf displays is **depths**. In the "depths" column the frequencies are added together (accumulated) from the low end and the high end until the row is reached that contains the middle value in the ordered display. That middle value is called the **median** of the sample, and the frequency of that row simply placed in parentheses. That is, the frequency of the row containing this median is not included in either the sum from the low end or the high end.

**Table 3.1-4  Ordered Stem-and-Leaf Display of Statistics Examinations**

| Stems | Leaves | Frequency | Depths |
|-------|--------|-----------|--------|
| 3 | 4 8 | 2 | 2 |
| 4 | 2 2 5 7 | 4 | 6 |
| 5 | 1 2 4 7 8 8 9 | 7 | 13 |
| 6 | 0 1 3 5 5 6 7 8 9 9 | 10 | 23 |
| 7 | 0 1 1 2 3 3 4 5 5 6 6 7 9 | 13 | (13) |
| 8 | 1 1 2 3 3 4 4 5 7 | 9 | 14 |
| 9 | 0 1 3 3 7 | 5 | 5 |

It is useful to modify the stem-and-leaf display by ordering the leaves in each row from smallest to largest. The resulting stem-and-leaf diagram is called an **ordered stem-and-leaf display**. Table 3.1-4 gives an ordered stem-and-leaf using the data in Table 3.1-3. Most computer programs provide ordered stem-and-leaf displays.

There is another modification that can also be helpful. Suppose that we want two rows of leaves with each original stem. We can do this by recording leaves 0, 1, 2, 3, and 4 with a stem adjoined with an asterisk (∗) and leaves 5, 6, 7, 8, and 9 with a stem adjoined with a dot (●). Of course, in our example, by going from seven original classes to 14 classes, we lose a certain amount of smoothness with this particular data set, as illustrated in Table 3.1-5, which is also ordered.

Tukey suggested another modification used in the next example.

**EXAMPLE 3.1-4**  The following numbers represent ACT composite scores for 60 entering freshmen at a certain college:

## Table 3.1-5 Ordered Stem-and-Leaf Display of Statistics Examinations

| Stems | Leaves | Frequency | Depths |
|---|---|---|---|
| 3∗ | 4 | 1 | 1 |
| 3● | 8 | 1 | 2 |
| 4∗ | 2 2 | 2 | 4 |
| 4● | 5 7 | 2 | 6 |
| 5∗ | 1 2 4 | 3 | 9 |
| 5● | 7 8 8 9 | 4 | 13 |
| 6∗ | 0 1 3 | 3 | 16 |
| 6● | 5 5 6 7 8 9 9 | 7 | 23 |
| 7∗ | 0 1 1 2 3 3 4 | 7 | (7) |
| 7● | 5 5 6 6 7 9 | 6 | 20 |
| 8∗ | 1 1 2 3 3 4 4 | 7 | 14 |
| 8● | 5 7 | 2 | 7 |
| 9∗ | 0 1 3 3 | 4 | 5 |
| 9● | 7 | 1 | 1 |

| | | | | | | | | | | |
|---|---|---|---|---|---|---|---|---|---|---|
| 26 | 19 | 22 | 28 | 31 | 29 | 25 | 23 | 20 | 33 | 23 | 26 |
| 30 | 27 | 26 | 29 | 20 | 23 | 18 | 24 | 29 | 27 | 32 | 24 |
| 25 | 26 | 22 | 29 | 21 | 24 | 20 | 28 | 23 | 26 | 30 | 19 |
| 27 | 21 | 32 | 28 | 29 | 23 | 25 | 21 | 28 | 22 | 25 | 24 |
| 19 | 24 | 35 | 26 | 25 | 20 | 31 | 27 | 23 | 26 | 30 | 29 |

An ordered stem-and-leaf display is given in Table 3.1-6. Here leaves are recorded as zeros and ones with a stem adjoined with an asterisk (∗), twos and threes with a stem adjoined with $t$, fours and fives with a stem adjoined with $f$, sixes and sevens with a stem adjoined with $s$, and eights and nines with a stem adjoined with a dot (●). (See Example 5.1-5 for a Minitab solution.)

## Table 3.1-6 Ordered Stem-and-Leaf Display of 60 ACT Scores

| Stems | Leaves | Frequency | Depths |
|---|---|---|---|
| 1● | 8 9 9 9 | 4 | 4 |
| 2∗ | 0 0 0 0 1 1 1 | 7 | 11 |
| 2t | 2 2 2 3 3 3 3 3 3 | 9 | 20 |
| 2f | 4 4 4 4 4 5 5 5 5 5 | 10 | 30 |
| 2s | 6 6 6 6 6 6 6 7 7 7 7 | 11 | 30 |
| 2● | 8 8 8 8 9 9 9 9 9 9 | 10 | 19 |
| 3∗ | 0 0 0 1 1 | 5 | 9 |
| 3t | 2 2 3 | 3 | 4 |
| 3f | 5 | 1 | 1 |

There is a reason for constructing ordered stem-and-leaf diagrams. For a sample of $n$ observations, $x_1, x_2, \ldots, x_n$, when the observations are ordered from small to large, the resulting ordered data are called the **order statistics** of the sample. For years statisticians have found that order statistics and certain functions of the order statistics are extremely valuable. It is very easy to determine the values of the sample in order from an ordered stem-and-leaf display. For illustration, consider the values in Table 3.1-4 or Table 3.1-5. The order statistics of the 50 test scores are given in Table 3.1-7. Sometimes we give ranks to these order statistics and use the rank as the subscript on $y$. The first order statistic $y_1 = 34$ has rank 1; the second order statistic $y_2 = 38$ has rank 2; the third order statistic $y_3 = 42$ has rank 3; the fourth order statistic $y_4 = 42$ has rank 4, ... ; and the 50th order statistic $y_{50} = 97$ has rank 50. It is also about as easy to determine these values from the ordered stem-and-leaf display. We see that $y_1 \leq y_2 \leq \cdots \leq y_{50}$.

| Table 3.1-7  Order Statistics of 50 Exam Scores | | | | | | | | | |
|----|----|----|----|----|----|----|----|----|----|
| 34 | 38 | 42 | 42 | 45 | 47 | 51 | 52 | 54 | 57 |
| 58 | 58 | 59 | 60 | 61 | 63 | 65 | 65 | 66 | 67 |
| 68 | 69 | 69 | 70 | 71 | 71 | 72 | 73 | 73 | 74 |
| 75 | 75 | 76 | 76 | 77 | 79 | 81 | 81 | 82 | 83 |
| 83 | 84 | 84 | 85 | 87 | 90 | 91 | 93 | 93 | 97 |

From either these order statistics or the corresponding ordered stem-and-leaf display, it is rather easy to find the **sample percentiles**. If $0 < p < 1$, the $(100p)$th sample percentile has *approximately np* sample observations less than it and also $n(1 - p)$ sample observations greater than it. One way of achieving this is to take the $(100p)$th sample percentile as the $(n + 1)p$th order statistic, provided that $(n + 1)p$ is an integer. If $(n + 1)p$ is not an integer but is equal to an integer $r$ plus some proper fraction, say $a/b$, use a weighted average of the $r$th and the $(r + 1)$st order statistics. That is, define the $(100p)$th sample percentile as

$$\tilde{\pi}_p = y_r + (a/b)(y_{r+1} - y_r) = (1 - a/b)y_r + (a/b)y_{r+1}.$$

Note that this is simply a linear interpolation between $y_r$ and $y_{r+1}$. (If $p < 1/(n+1)$ or $p > n/(n + 1)$, that sample percentile is not defined.)

For illustration, consider the 50 ordered test scores. With $p = 1/2$, we find the 50th percentile by averaging the 25th and 26th order statistics, since $(n + 1)p = (51)(1/2) = 25.5$. Thus the 50th percentile is

$$\tilde{\pi}_{0.50} = (1/2)y_{25} + (1/2)y_{26} = (71 + 71)/2 = 71.$$

With $p = 1/4$, we have $(n + 1)p = (51)(1/4) = 12.75$; and thus the 25th sample percentile is

$$\tilde{\pi}_{0.25} = (1 - 0.75)y_{12} + (0.75)y_{13} = (0.25)(58) + (0.75)(59) = 58.75.$$

With $p = 3/4$, so that $(n + 1)p = (51)(3/4) = 38.25$, the 75th sample percentile is

$$\tilde{\pi}_{0.75} = (1 - 0.25)y_{38} + (0.25)y_{39} = (0.75)(81) + (0.25)(82) = 81.25.$$

Note that *approximately* 50%, 25%, and 75% of the sample observations are less than 71, 58.75, and 81.25, respectively.

Special names are given to certain percentiles. The 50th percentile is called the **median** of the sample. The 25th, 50th, and 75th percentiles are the **first, second**, and

**third quartiles** of the sample. For notation we let $\tilde{q}_1 = \tilde{\pi}_{0.25}$, $\tilde{q}_2 = \tilde{m} = \tilde{\pi}_{0.50}$, and $\tilde{q}_3 = \tilde{\pi}_{0.75}$. The 10th, 20th, ... , and 90th percentiles are the **deciles** of the sample. So note that the 50th percentile is also the median, the second quartile, and the fifth decile. By using the set of 50 test scores, since $(51)(2/10) = 10.2$ and $(51)(9/10) = 45.9$, the second and ninth deciles are

$$\tilde{\pi}_{0.20} = (0.8)y_{10} + (0.2)y_{11} = (0.8)(57) + (0.2)(58) = 57.2$$

and

$$\tilde{\pi}_{0.90} = (0.1)y_{45} + (0.9)y_{46} = (0.1)(87) + (0.9)(90) = 89.7,$$

commonly called the 20th and 90th percentiles, respectively.

Let us consider, along with the three quartiles, the extremes of the sample, namely the first and $n$th order statistics. (In the test scores data, they are 34 and 97, respectively.) These provide a **five-number summary** of a set of data consisting of the **minimum**, the **first quartile**, the **median**, the **third quartile**, and the **maximum**, written in that order. Furthermore, the difference between the third and first quartiles is called the **interquartile range, IQR**. That is, $IQR = \tilde{q}_3 - \tilde{q}_1$, which equals $81.25 - 58.75 = 22.50$ with the test scores data.

There is a graphical means for displaying the five-number summary of a set of data that is called a **box-and-whisker diagram**. To construct a horizontal box-and-whisker diagram, or more simply, a **box plot**, draw a horizontal axis that is scaled to the data. Above the axis draw a rectangular box with the left and right sides drawn at $\tilde{q}_1$ and $\tilde{q}_3$ with a vertical line segment drawn at the median, $\tilde{q}_2 = \tilde{m}$. A left whisker is drawn as a horizontal line segment from the minimum to the midpoint of the left side of the box and a right whisker is drawn as a horizontal line segment from the midpoint of the right side of the box to the maximum. Note that the length of the box is equal to the $IQR$. The left and right whiskers represent the first and fourth quarters of the data while the two middle quarters of the data are represented, respectively, by the two sections of the box, one to the left and one to the right of the median line.

**EXAMPLE 3.1-5**   Using the test score data, we have that the five-number summary is given by

$$y_1 = 34, \quad \tilde{q}_1 = 58.75, \quad \tilde{q}_2 = \tilde{m} = 71, \quad \tilde{q}_3 = 81.25, \quad y_{50} = 97.$$

The box-and-whisker diagram or box plot of these data is given in Figure 3.1-3. The facts that the long whisker is to the left and the box from 58.75 to 71 is slightly longer than that from 71 to 81.25 lead us to say that these data are slightly *skewed to the left*.

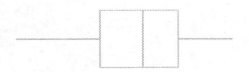

**Figure 3.1-3** Box plot of scores from 50 statistics examinations

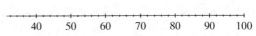

# APPLICATIONS

The following four illustrations of the use of stem-and-leaf displays and box plots have been adapted from data sets given in the book *A Data-Based Approach to Statistics* (1994) by Ronald L. Iman, Wadsworth Publishing Co, Belmont CA 94002.

**1.** In response to President George Bush's call in February, 1992, to phase out chlorofluorocarbons (CFCs), a major electronics manufacturer developed a soldering process which eliminated the use of CFCs. To show that this process produced good solder joints, a mechanical pull strength device was used to pull 98 soldered resistor leads until they broke or came out of the board. These 98 values are given to the nearest tenth of a pound. We record them in the ordered stem-and-leaf display, Table 3.1-8, in which the stems are pounds and the leaves are the tenths.

**Table 3.1-8 Ordered Stem-and-Leaf Display of 98 Pull Strengths of Solder Joints**

| Stems | Leaves | Frequency | Depths |
|---|---|---|---|
| 57 | 2 | 1 | 1 |
| 58 | | 0 | 1 |
| 59 | 2 4 9 | 3 | 4 |
| 60 | 0 1 2 7 9 | 5 | 9 |
| 61 | 0 0 5 7 | 4 | 13 |
| 62 | 1 2 4 5 5 6 6 6 | 8 | 21 |
| 63 | 0 1 2 2 4 5 5 5 8 8 9 9 | 12 | 33 |
| 64 | 1 2 2 2 2 3 3 3 3 3 3 3 3 4 4 4 4 4 5 5 6 6 7 8 8 8 9 | 27 | (27) |
| 65 | 0 1 1 1 1 1 1 2 3 3 3 3 5 5 5 6 6 7 8 8 9 | 21 | 38 |
| 66 | 1 2 3 3 4 4 5 5 6 6 7 8 8 | 13 | 17 |
| 67 | 1 4 7 | 3 | 4 |
| 68 | 9 | 1 | 1 |

From this ordered stem-and-leaf display, Table 3.1-8, we find that the median is the $(98 + 1)/2 = 49.5$ order statistic; that is, the average of the 49th and the 50th order statistics, namely

$$\widetilde{m} = \frac{64.4 + 64.4}{2} = 64.4.$$

The first quartile is found by considering $(98 + 1)/4 = 24.75$ and computing

$$\widetilde{q}_1 = \left(\frac{1}{4}\right)(63.2) + \left(\frac{3}{4}\right)(63.2) = 63.2$$

since $y_{24} = 63.2$ and $y_{25} = 63.2$. The third quartile is found by first considering $(98 + 1)(3/4) = 74.25$ and then computing

$$\tilde{q}_3 = \left(\frac{3}{4}\right)(65.5) + \left(\frac{1}{4}\right)(65.5) = 65.5$$

since $y_{74} = 65.5$ and $y_{75} = 65.5$. The five-number summary is

$$y_1 = 57.2, \quad \tilde{q}_1 = 63.2, \quad \tilde{m} = 64.4, \quad \tilde{q}_3 = 65.5, \quad y_{98} = 68.9.$$

The box plot is shown in Figure 3.1-4. The outlier $y_1 = 57.2$ has caused the long left whisker, and it seems that it should have been investigated. Persons in that industry can use these displays to help decide whether the new process is satisfactory.

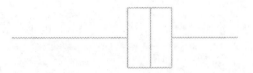

**Figure 3.1-4** Box plot of 98 pull strengths of solder joints

57  58  59  60  61  62  63  64  65  66  67  68  69

**2.** Edwin Moses in a period of a little over nine and one half years, September 2, 1977 to May 29, 1987, won 122 consecutive races in the 400-meter hurdles. Such a record in any sport is incredible. We record these 122 times in an ordered stem-and-leaf display, Table 3.1-9, using stems of $47*$, $47t$, $47f$, $47s$, $47\bullet$, $\cdots$, $49\bullet$, $50*$ seconds and leaves that represent tenths of a second.

| Table 3.1-9 Race Times for Edwin Moses | | | |
|---|---|---|---|
| **Stems** | **Leaves** | **Frequency** | **Depths** |
| $47*$ | 0 1 1 1 | 4 | 4 |
| $47t$ | 2 3 3 3 3 | 5 | 9 |
| $47f$ | 4 5 5 5 5 5 5 5 | 8 | 17 |
| $47s$ | 6 6 6 6 7 7 | 6 | 23 |
| $47\bullet$ | 8 8 8 9 9 9 9 9 9 9 | 10 | 33 |
| $48*$ | 0 1 1 | 3 | 36 |
| $48t$ | 2 2 2 2 2 2 2 2 3 3 3 | 11 | 47 |
| $48f$ | 4 4 4 4 4 5 5 5 5 5 5 5 5 5 5 5 5 | 17 | (17) |
| $48s$ | 6 6 6 6 6 6 6 6 7 7 7 7 7 7 7 7 7 | 17 | 58 |
| $48\bullet$ | 8 8 8 8 8 8 8 9 9 9 | 10 | 41 |
| $49*$ | 0 0 0 0 0 1 1 1 1 | 9 | 31 |
| $49t$ | 2 2 2 3 3 | 5 | 22 |
| $49f$ | 4 4 5 5 5 5 5 | 7 | 17 |
| $49s$ | 6 7 7 7 7 | 5 | 10 |
| $49\bullet$ | 8 9 | 2 | 5 |
| $50*$ | 1 1 1 | 3 | 3 |

By considering $(122 + 1)/2 = 61.5$, $123/4 = 30.75$, and $123(3/4) = 92.25$, the 50th, 25th, and 75th percentiles are given respectively by

$$\tilde{\pi}_{0.50} = \left(\frac{1}{2}\right)(48.5) + \left(\frac{1}{2}\right)(48.5) = 48.5,$$

$$\tilde{\pi}_{0.25} = \left(\frac{1}{4}\right)(47.9) + \left(\frac{3}{4}\right)(47.9) = 47.9,$$

$$\tilde{\pi}_{0.75} = \left(\frac{3}{4}\right)(49.0) + \left(\frac{1}{4}\right)(49.0) = 49.0.$$

These three quartiles along with the minimum, $y_1 = 47.0$, and the maximum, $y_{122} = 50.1$, yield the five-number summary

$$y_1 = 47.0, \quad \tilde{q}_1 = 47.9, \quad \tilde{m} = 48.5, \quad \tilde{q}_3 = 49.0, \quad y_{98} = 50.1$$

and the box plot shown in Figure 3.1-5.

**Figure 3.1-5** Box plot for times for Edwin Moses

Both displays indicate that the distribution of these times is quite symmetric. Moreover it is easy to see that $33/122 \approx 0.27 = 27\%$ of the times were less than 48.0 seconds and $31/122 \approx 0.254 = 25.4\%$ were greater than or equal to 49.0 seconds.

**3.** Often box plots are used to compare two or more processes. Here a pull strength device was used to test soldering at a low solder temperature and a high soldering temperature. Ten experimental values were found at each temperature, and we record the order statistics.

Low:  63.4  64.3  64.3  64.8  64.8  65.3  65.6  66.1  66.7  67.1
High:  61.2  62.0  63.5  64.5  64.6  65.1  66.4  67.3  67.5  69.7

By considering $(10 + 1)(0.5) = 5.5$, the two medians are respectively

$$\left(\frac{1}{2}\right)(64.8) + \left(\frac{1}{2}\right)(65.3) = 65.05 \quad \text{and} \quad \left(\frac{1}{2}\right)(64.6) + \left(\frac{1}{2}\right)(65.1) = 64.85.$$

By considering $(11)(0.25) = 2.75$ and $(11)(0.75) = 8.25$, the respective first and third quartiles are

$$\left(\frac{1}{4}\right)(64.3) + \left(\frac{3}{4}\right)(64.3) = 64.3, \quad \left(\frac{1}{4}\right)(62.0) + \left(\frac{3}{4}\right)(63.5) = 63.125$$

and

$$\left(\frac{3}{4}\right)(66.1) + \left(\frac{1}{4}\right)(66.7) = 66.25, \quad \left(\frac{3}{4}\right)(67.3) + \left(\frac{1}{4}\right)(67.5) = 67.35.$$

Using the five-number summary for each set of data,

Low:  63.4,  64.3,  65.05,  66.25,  67.1
High:  61.2,  63.125,  64.85,  67.35,  69.7,

the box plots are shown in Figure 3.1-6.

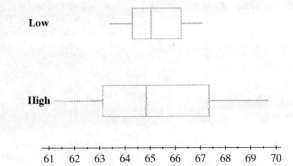

**Figure 3.1-6** Box plots comparing low and high soldering temperature

So while the middles of these results are fairly similar, the spread of the high soldering temperature is much greater than that of the low and that is undesirable. Hence the lower soldering temperature was used.

**4.** Pollution is a major concern in large cities and hence is closely monitored to see if the Air Pollution Index is within the Environmental Protection Agency requirements. Samples of size $n = 10$ were taken from each of five cities and ordered to give the following data set.

| City | Air Pollution Index |
|------|---------------------|
| L.A. | 55.9 56.3 56.8 57.2 61.2 61.9 62.5 63.8 64.4 68.2 |
| N.Y. | 55.7 55.8 57.0 57.4 59.0 59.5 59.9 60.4 64.2 67.7 |
| K.C. | 53.0 54.6 54.7 54.8 57.6 58.6 62.4 63.5 65.5 66.6 |
| D.C. | 57.3 58.1 58.6 58.7 59.0 61.9 62.6 64.4 64.9 66.7 |
| S.F. | 50.5 54.4 54.8 56.3 58.3 59.0 61.2 61.6 62.2 63.1 |

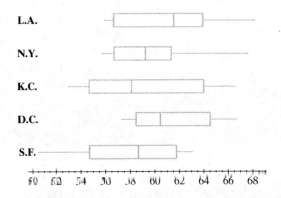

**Figure 3.1-7** Box Plots comparing the air pollution index for five cities

The five respective box plots are given in Figure 3.1-7. With these few data, this very rough comparison suggests that L.A. is the worst followed closely by D.C., then N.Y. is in third position, then K.C. is somewhat better than N.Y. but a little worse than S.F., which seems to be best as far as low pollution is concerned. Clearly we would need more data to rank these cities more carefully.

## EXERCISES 3.1

**3.1-1** Decks have become very popular items for American homeowners. Lumber yards sell many styles of decks during the summer months, but they all use the same type of nail. The question that always arises is how many nails to buy. If 800 nails are needed, how many grams of nails should be purchased? The weights (in grams) of the following 50 nails should help to answer the question.

| | | | | | | | | | |
|------|------|------|------|------|------|------|------|------|------|
| 9.42 | 8.69 | 8.93 | 8.27 | 8.82 | 8.66 | 8.90 | 8.31 | 9.15 | 9.63 |
| 9.41 | 8.56 | 8.82 | 8.58 | 8.43 | 8.05 | 8.56 | 8.55 | 8.88 | 8.73 |
| 8.29 | 8.79 | 8.51 | 8.85 | 9.34 | 9.21 | 8.38 | 8.51 | 8.41 | 8.98 |
| 8.58 | 9.21 | 8.27 | 8.76 | 9.26 | 8.59 | 8.36 | 8.71 | 8.51 | 8.88 |
| 9.20 | 8.24 | 8.57 | 8.85 | 8.69 | 8.85 | 9.08 | 9.40 | 9.25 | 8.79 |

**(a)** Group these data into seven classes using as class intervals (7.995, 8.245), (8.245, 8.495), and so on.

**(b)** Construct a histogram.

**(c)** Calculate the sample mean and sample standard deviation for either the grouped or the ungrouped data.

**(d)** How many grams of nails would you recommend?

**3.1-2** In the Old Kent River Bank Run (a 25K race) the times are listed separately for males and females and also by age groupings. A 53-year-old male runner was interested in the distribution of race times for males, 50-54 years old. There were 125 runners in this category and their race times are listed in order in minutes and seconds.

| | | | | | | | | |
|--------|--------|--------|--------|--------|--------|--------|--------|--------|
| 96:21  | 98:24  | 100:26 | 101:23 | 101:29 | 101:38 | 103:11 | 103:58 | 104:28 |
| 105:02 | 105:14 | 106:00 | 106:14 | 108:49 | 109:28 | 109:32 | 109:42 | 110:49 |
| 110:54 | 111:48 | 111:54 | 112:17 | 112:33 | 112:48 | 112:54 | 112:55 | 113:09 |
| 113:10 | 113:15 | 113:43 | 114:30 | 114:36 | 115:28 | 115:39 | 115:44 | 116:11 |
| 116:34 | 116:56 | 117:45 | 118:10 | 118:11 | 118:12 | 118:28 | 118:30 | 118:44 |
| 119:10 | 119:27 | 119:53 | 120:17 | 120:19 | 121:12 | 122:15 | 122:17 | 122:46 |
| 122:57 | 122:59 | 123:46 | 124:04 | 124:06 | 124:49 | 124:55 | 125:14 | 125:15 |
| 125:16 | 126:09 | 126:12 | 126:29 | 126:43 | 127:21 | 127:25 | 127:57 | 128:02 |
| 128:23 | 129:13 | 129:21 | 129:39 | 130:01 | 130:07 | 130:29 | 130:34 | 131:07 |
| 131:22 | 131:58 | 132:27 | 132:51 | 132:58 | 133:51 | 134:04 | 134:40 | 134:53 |
| 135:05 | 135:41 | 136:27 | 136:51 | 136:52 | 136:53 | 137:09 | 137:13 | 137:31 |
| 137:48 | 138:01 | 138:13 | 138:44 | 139:34 | 139:51 | 140:03 | 140:05 | 141:20 |
| 142:28 | 143:08 | 143:46 | 144:10 | 144:35 | 145:40 | 147:05 | 148:12 | 151:18 |
| 151:39 | 152:00 | 156:13 | 160:43 | 161:12 | 162:29 | 166:42 | 169:54 | |

**(a)** Group these data into 10 classes with a class width of eight minutes.

**(b)** Draw a histogram of the grouped data.

**(c)** Describe the shape of the histogram. Is it what you would have expected? (Similar data can be found following most road races.)

**3.1-3** The distances in feet of the 51 home runs hit by Andrew Jones in 2005 are the following:

| | | | | | | | | |
|-----|-----|-----|-----|-----|-----|-----|-----|-----|
| 382 | 383 | 395 | 380 | 416 | 390 | 420 | 400 | 360 |
| 340 | 360 | 370 | 374 | 408 | 359 | 419 | 411 | 423 |
| 418 | 415 | 432 | 420 | 375 | 432 | 373 | 350 | 410 |
| 390 | 340 | 380 | 370 | 390 | 434 | 395 | 379 | 407 |
| 393 | 385 | 379 | 416 | 420 | 440 | 354 | 405 | 452 |
| 380 | 380 | 390 | 390 | 425 | 427 | | | |

The minimum equals 340 and the maximum is 452; thus the range is $452 - 340 = 112$. There are a number of possibilities for $k$. So that each member of your class gets the same result, take $k = 12$ and class intervals each with length 10. That is, $(12)(10) = 120 > 112$. We start with 339.5 and accordingly end at 459.5.

(a) Construct a stem-and-leaf diagram with stems of $34, 35, \ldots, 45$.

(b) Construct a frequency table.

(c) Plot the relative frequency histogram.

(d) Draw a box-and-whisker diagram.

(e) Compute $\bar{x}$ and $s_x^2$.

(f) Compute $\bar{u}$ and $s_u^2$ of the grouped data.

**3.1-4** When researching ground water it is often important to know the characteristics of the soil at a certain site. Many of these characteristics, such as porosity, are at least partially dependent upon the grain size. The diameter of individual grains of soil can be measured. Here are the diameters (in mm) of 30 grains:

| 1.24 | 1.36 | 1.28 | 1.31 | 1.35 | 1.20 | 1.39 | 1.35 | 1.41 | 1.31 |
|------|------|------|------|------|------|------|------|------|------|
| 1.28 | 1.26 | 1.37 | 1.49 | 1.32 | 1.40 | 1.33 | 1.28 | 1.25 | 1.39 |
| 1.38 | 1.34 | 1.40 | 1.27 | 1.33 | 1.36 | 1.43 | 1.33 | 1.29 | 1.34 |

(a) Calculate the sample mean and the sample variance for these data.

(b) Construct a histogram of these data.

**3.1-5** Let $X$ denote the concentration of $CaCO_3$ in milligrams per liter. Twenty observations of $X$ are:

| 130.8 | 129.9 | 131.5 | 131.2 | 129.5 |
|-------|-------|-------|-------|-------|
| 132.7 | 131.5 | 127.8 | 133.7 | 132.2 |
| 134.8 | 131.7 | 133.9 | 129.8 | 131.4 |
| 128.8 | 132.7 | 132.8 | 131.4 | 131.3 |

(a) Construct an ordered stem-and-leaf display, using stems of $127, 128, \ldots, 134$.

(b) Find the range, interquartile range, median, sample mean, and sample variance.

(c) Draw a box-and-whisker diagram.

**3.1-6** A company manufactures mints that have a label weight of 20.4 grams. It regularly samples from the production line and weighs the selected mints. During two mornings of production it sampled 81 mints with the following weights:

| 21.8 | 21.7 | 21.7 | 21.6 | 21.3 | 21.6 | 21.5 | 21.3 | 21.2 |
|------|------|------|------|------|------|------|------|------|
| 21.0 | 21.6 | 21.6 | 21.6 | 21.5 | 21.4 | 21.8 | 21.7 | 21.6 |
| 21.6 | 21.3 | 21.9 | 21.9 | 21.6 | 21.0 | 20.7 | 21.8 | 21.7 |
| 21.7 | 21.4 | 20.9 | 22.0 | 21.3 | 21.2 | 21.0 | 21.0 | 21.9 |
| 21.7 | 21.5 | 21.5 | 21.1 | 21.3 | 21.3 | 21.2 | 21.0 | 20.8 |
| 21.6 | 21.6 | 21.5 | 21.5 | 21.2 | 21.5 | 21.4 | 21.4 | 21.3 |
| 21.2 | 21.8 | 21.7 | 21.7 | 21.6 | 20.5 | 21.8 | 21.7 | 21.5 |
| 21.4 | 21.4 | 21.9 | 21.8 | 21.7 | 21.4 | 21.3 | 20.9 | 21.9 |
| 20.7 | 21.1 | 20.8 | 20.6 | 20.6 | 22.0 | 22.0 | 21.7 | 21.6 |

(a) Construct an ordered stem-and-leaf display using stems of $20f$, $20s$, $20\bullet$, $21*, \ldots, 22*$.

(b) Find (i) the three quartiles, (ii) the 60th percentile, and (iii) the 15th percentile.

(c) Construct a box-and-whisker display.

**3.1-7** *PC Magazine* in February, 1994, includes "Going Mainstream" by John R. Quain in which they report tests of 26 CD-ROM drives. One of the tests was a random read of a 4K (4,000 byte) block of data. The read rates, recorded in kilobytes (K) per second, were:

| 29.21 | 14.50 | 25.10 | 22.90 | 30.80 | 30.62 | 27.88 | 20.65 | 30.90 |
|-------|-------|-------|-------|-------|-------|-------|-------|-------|
| 30.60 | 6.12 | 19.94 | 19.61 | 21.51 | 28.80 | 21.72 | 21.95 | 18.89 |
| 15.60 | 29.38 | 25.45 | 30.41 | 30.11 | 30.51 | 28.17 | 21.42 | |

(a) Construct an ordered stem-and-leaf display using three-digit leaves. That is, 29.21 would have the stem of 2 and leaf of (921). Would much be lost if, before constructing the stem-and-leaf display, each number was rounded to the nearest integer?

(b) Find the five-number summary of the data.

(c) Construct a box-and-whisker display.

(d) What type of skewness is indicated by your displays?

**3.1-8** In the casino game roulette, if a player bets $1 on red (or on black or on odd or on even), the probability of winning $1 is 18/38 and the probability of losing $1 is 20/38. Suppose that a player begins with $5 and makes successive $1 bets. Let $Y$ equal the player's maximum capital before losing the $5. One hundred observations of $Y$ were simulated on a computer, yielding the following data:

| | | | | | | | | | |
|---|---|---|---|---|---|---|---|---|---|
| 25 | 9 | 5 | 5 | 5 | 9 | 6 | 5 | 15 | 45 |
| 55 | 6 | 5 | 6 | 24 | 21 | 16 | 5 | 8 | 7 |
| 7 | 5 | 5 | 35 | 13 | 9 | 5 | 18 | 6 | 10 |
| 19 | 16 | 21 | 8 | 13 | 5 | 9 | 10 | 10 | 6 |
| 23 | 8 | 5 | 10 | 15 | 7 | 5 | 5 | 24 | 9 |
| 11 | 34 | 12 | 11 | 17 | 11 | 16 | 5 | 15 | 5 |
| 12 | 6 | 5 | 5 | 7 | 6 | 17 | 20 | 7 | 8 |
| 8 | 6 | 10 | 11 | 6 | 7 | 5 | 12 | 11 | 18 |
| 6 | 21 | 6 | 5 | 24 | 7 | 16 | 21 | 23 | 15 |
| 11 | 8 | 6 | 8 | 14 | 11 | 6 | 9 | 6 | 10 |

**(a)** Construct an ordered stem-and-leaf display.

**(b)** Find the five-number summary of the data and draw a box-and-whisker diagram.

**(c)** Find the 90th percentile.

**3.1-9** The weights (in grams) of 25 indicator housings used on gauges are as follows:

| | | | | |
|---|---|---|---|---|
| 102.0 | 106.3 | 106.6 | 108.8 | 107.7 |
| 106.1 | 105.9 | 106.7 | 106.8 | 110.2 |
| 101.7 | 106.6 | 106.3 | 110.2 | 109.9 |
| 102.0 | 105.8 | 109.1 | 106.7 | 107.3 |
| 102.0 | 106.8 | 110.0 | 107.9 | 109.3 |

**(a)** Construct an ordered stem-and-leaf display using integers as the stems and tenths as the leaves.

**(b)** Find the five-number summary of the data and draw a box plot.

## 3.2 CONTINUOUS PROBABILITY DISTRIBUTIONS

The relative frequency (or density) histogram $h(x)$ associated with $n$ observations of a random variable of the continuous type is a non-negative function defined so that the total area between its graph and the $x$ axis equals one. In addition, $h(x)$ is constructed so that the integral

$$\int_a^b h(x)\,dx$$

is approximately equal to the relative frequency of the interval $\{x: a < x < b\}$ and hence can be thought of as an estimate of the probability that the random variable $X$ falls in the interval $(a, b)$. This probability is often written $P(a < X < b)$.

Let us now consider what happens to the function $h(x)$ in the limit as $n$ increases without bound and as the lengths of the class intervals decrease to zero. It is to be hoped that $h(x)$ will become closer and closer to some function, say $f(x)$, that gives the true probabilities, such as $P(a < X < b)$, through the integral

$$P(a < X < b) = \int_a^b f(x)\,dx.$$

That is, $f(x)$ should be a non-negative function such that the total area between its graph and the $x$ axis equals one. Moreover, the probability $P(a < X < b)$ is the area bounded by the graph of $f(x)$, the $x$ axis, and the lines $x = a$ and $x = b$. Thus we say that the **probability density function (p.d.f.)** of a random variable $X$ of the **continuous type**, with space $S$ that is an interval or union of intervals, is an integrable function $f(x)$ satisfying the following conditions:

**(a)** $f(x) > 0$,    $x \varepsilon S$,

   $f(x) = 0$,    $x \notin S$.

**(b)** $\int_S f(x)\, dx = 1$.

**(c)** If $(a, b) \subset S$, the probability of the event $a < X < b$ is

$$P(a < X < b) = \int_a^b f(x)\, dx;$$

or, in general, if $A \subset S$, then

$$P(A) = \int_A f(x)\, dx.$$

The corresponding distribution of probability is said to be one of the continuous type.

EXAMPLE 3.2-1    Suppose a balanced spinner is constructed so that the result of a spin is a random variable $X$ whose space is $S = \{x: 0 \le x < 1\}$. Since the spinner is balanced, it is reasonable to say that $X$ has the p.d.f.

$$f(x) = 1, \qquad 0 \le x < 1.$$

Outside the space $S$, we always take the p.d.f. to be equal to zero. A p.d.f. that is constant on its space $S$ is said to be **uniform**. With this assignment, we have that

$$P(0.25 < X < 0.75) = \int_{0.25}^{0.75} 1\, dx = [x]_{0.25}^{0.75} = 0.5,$$

$$P(0.5 < X < 0.6) = \int_{0.5}^{0.6} 1\, dx = [x]_{0.5}^{0.6} = 0.1,$$

and

$$P(0.1 < X < 0.4) = \int_{0.1}^{0.4} 1\, dx = [x]_{0.1}^{0.4} = 0.3;$$

these probabilities certainly agree with our intuition. It is interesting to note that if we ask for the probability that $X$ equals one given value, like $X = 0.5$, which could be written $P(X = 0.5)$, we get the answer

$$P(X = 0.5) = \int_{0.5}^{0.5} 1\, dx = [x]_{0.5}^{0.5} = 0.$$

The fact that the probability that $X$ equals a given number is zero also agrees with our intuition once we recognize that there are an infinite (uncountable) number of possible outcomes for $X$ and we want the probability of just one of them.

EXAMPLE 3.2-2    Let the random variable $X$ be the lengths of time in minutes between calls to 911 in a small city that were reported in the newspaper on February 26 and 27. Suppose that a reasonable probability model for $X$ is given by the p.d.f.

$$f(x) = \frac{1}{20}\, e^{-x/20}, \qquad 0 \le x < \infty.$$

Note that $S = \{x : 0 \le x < \infty\}$ and $f(x) > 0$ for $x \varepsilon S$. Also,

$$\int_S f(x)\,dx = \int_0^\infty \frac{1}{20}\,e^{-x/20}\,dx$$

$$= \lim_{b \to \infty}\left[-e^{-x/20}\right]_0^b$$

$$= 1 - \lim_{b \to \infty} e^{-b/20} = 1.$$

The probability that the time between calls is greater than 20 minutes is

$$P(X > 20) = \int_{20}^\infty \frac{1}{20}\,e^{-x/20}\,dx = e^{-1} = 0.368.$$

The p.d.f. and the probability of interest are depicted in Figure 3.2-1.

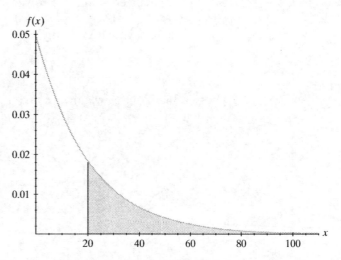

**Figure 3.2-1** Times between calls to 911

So that we can avoid repeated references to the space (or support) $S$ of the random variable $X$, we shall adopt the same convention when describing probability density functions of the continuous type as we did in the discrete case. We extend the definition of the p.d.f. $f(x)$ to the entire set of real numbers by letting it equal zero when $x \notin S$. For example,

$$f(x) = \begin{cases} \dfrac{1}{20}\,e^{-x/20}, & 0 \le x < \infty, \\ 0, & \text{elsewhere}, \end{cases}$$

has the properties of a p.d.f. of a continuous-type random variable $X$ having support $\{x : 0 \le x < \infty\}$. It will always be understood that $f(x) = 0$ when $x \notin S$, even when this is not explicitly written out.

The **distribution function** of a random variable $X$ of the continuous type, defined in terms of the p.d.f. of $X$, is given by

$$F(x) = P(X \le x) = \int_{-\infty}^x f(t)\,dt.$$

Here, again, $F(x)$ accumulates (or more simply, cumulates) all of the probability less than or equal to $x$. (This function is sometimes called a **cumulative distribution function** or **c.d.f.**) From the fundamental theorem of calculus we have, for $x$ values for which the derivative $F'(x)$ exists, that $F'(x) = f(x)$.

**EXAMPLE 3.2-3**   Continuing with Example 3.2-2, if the p.d.f. of $X$ is

$$f(x) = \begin{cases} 0, & -\infty < x < 0, \\ \dfrac{1}{20} e^{-x/20}, & 0 \leq x < \infty, \end{cases}$$

the distribution function of $X$ is $F(x) = 0$ for $x < 0$ and, for $x \geq 0$,

$$F(x) = \int_{-\infty}^{x} f(t)\, dt = \int_{0}^{x} \frac{1}{20} e^{-t/20}\, dt$$

$$= \left[ -e^{-t/20} \right]_{0}^{x} = 1 - e^{-x/20}.$$

Note that

$$F'(x) = \begin{cases} 0, & -\infty < x < 0 \\ \dfrac{1}{20} e^{-x/20}, & 0 < x < \infty. \end{cases}$$

Also $F'(0)$ does not exist. (Sketch a graph of $y = F(x)$ to see why this is true.)

For continuous-type random variables, the definitions associated with mathematical expectation are the same as those in the discrete case except that integrals replace summations. For illustration, let $X$ be a continuous random variable with a p.d.f. $f(x)$.

The **expected value of $X$** or **mean of $X$** is

$$\mu = E(X) = \int_{-\infty}^{\infty} x f(x)\, dx.$$

The **variance of $X$** is

$$\sigma^2 = \text{Var}(X) = E[(X - \mu)^2] = \int_{-\infty}^{\infty} (x - \mu)^2 f(x)\, dx.$$

Again it is true that
$$\sigma^2 = \text{Var}(X) = E(X^2) - \mu^2.$$

The **standard deviation of $X$** is

$$\sigma = \sqrt{\text{Var}(X)}.$$

**EXAMPLE 3.2-4**   Let the random variable $Y$ have p.d.f. $g(y) = 2y$, $0 < y < 1$. Then

$$\mu = E(Y) = \int_{0}^{1} y\,(2y)\, dy = \left[ \left( \frac{2}{3} \right) y^3 \right]_{0}^{1} = \frac{2}{3}$$

and

$$\sigma^2 = \text{Var}(Y) = E(Y^2) - \mu^2$$

$$= \int_0^1 y^2(2y)\,dy - \left(\frac{2}{3}\right)^2 = \left[\left(\frac{1}{2}\right)y^4\right]_0^1 - \frac{4}{9} = \frac{1}{18}$$

are the mean and variance, respectively, of $Y$.

The $(100p)$th **percentile** of a continuous-type distribution with p.d.f. $f(x)$ is a number $\pi_p$ such that the area under $f(x)$ to the left of $\pi_p$ is $p$. That is,

$$p = \int_{-\infty}^{\pi_p} f(x)\,dx = F(\pi_p).$$

The 50th percentile is called the **median**. We let $m = \pi_{0.50}$. The 25th and 75th percentiles are called the **first** and **third quartiles**, respectively, denoted by $q_1 = \pi_{0.25}$ and $q_3 = \pi_{0.75}$. Of course, the median $m = \pi_{0.50} = q_2$ is also called the second quartile.

**EXAMPLE 3.2-5**    The time $X$ in months until failure of a certain product has a p.d.f. of the Weibull type defined by

$$f(x) = \frac{3x^2}{4^3}\,e^{-(x/4)^3}, \qquad 0 < x < \infty.$$

Its distribution function is

$$F(x) = 1 - e^{-(x/4)^3}, \qquad 0 \le x < \infty.$$

Thus, for example, the 30th percentile, $\pi_{0.3}$, is given by

$$F(\pi_{0.3}) = 0.3$$

or, equivalently,

$$1 - e^{-(\pi_{0.3}/4)^3} = 0.3$$

$$\ln(0.7) = -(\pi_{0.3}/4)^3$$

$$\pi_{0.3} = -4\sqrt[3]{\ln(0.7)} = 2.84.$$

Likewise, $\pi_{0.9}$ is found by

$$F(\pi_{0.9}) = 0.9;$$

so

$$\pi_{0.9} = -4\sqrt[3]{\ln(0.1)} = 5.28.$$

Thus

$$P(2.84 < X < 5.28) = 0.6.$$

The 30th and 90th percentiles are shown in Figure 3.2-2. (See Example 5.2-6 for a *Maple* solution and Example 5.1-6 for a Minitab solution.)

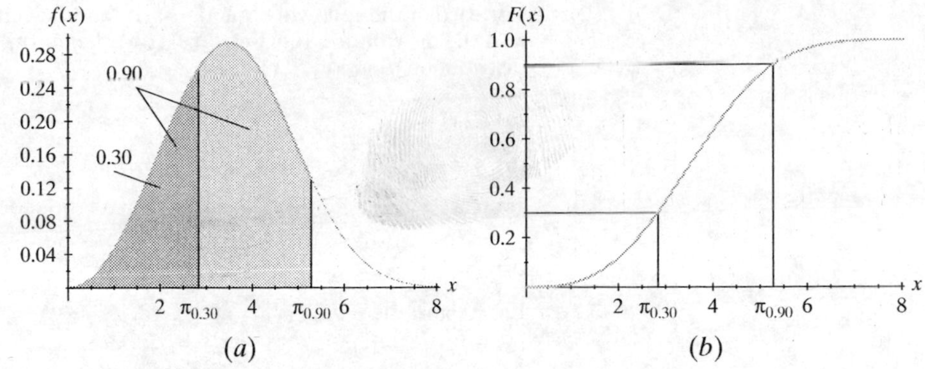

Figure 3.2-2 Illustration of percentiles $\pi_{0.30}$ and $\pi_{0.90}$

## EXERCISES 3.2

**3.2-1** Let $X$ have the p.d.f. $f(x) = x^3/4$, $0 < x < 2$. If $A_1 = \{x : 0 < x < 1\}$ and $A_2 = \{x : 1/2 < x < 3/2\}$, compute

(a) $P(A_1)$,

(b) $P(A_2)$,

(c) $P(A_1 \cup A_2)$.

**3.2-2** Let $X$ be a continuous-type random variable with space $S = \{x : 0 < x < 10\}$. If $P(A_1) = 1/4$, where $A_1 = \{x : 1 < x < 5\}$, show that $P(A_2) \leq 3/4$, where $A_2 = \{x : 5 < x < 10\}$.

**3.2-3** For each of the following p.d.f.s of $X$, compute

$$P(|X| < 1) \quad \text{and} \quad P(X^2 < 9/4).$$

(a) $f(x) = (x^3 + 8)/32$, $\quad -2 < x < 2$.

(b) $f(x) = (x + 2)/18$, $\quad -2 < x < 4$.

**3.2-4** A **mode** of a distribution of a random variable $X$ is a value of $x$ that maximizes $f(x)$. If there is only one such $x$, it is called the mode of the distribution. Find the mode of the distribution for each of the following and sketch their graphs.

(a) $f(x) = 20x^3(1 - x)$, $\quad 0 < x < 1$.

(b) $f(x) = (x^3 e^{-x})/6$, $\quad 0 < x < \infty$.

**3.2-5** Find the three quartiles of the following distributions given by their respective p.d.f.s $f(x)$.

(a) $f(x) = 4x^3$, $\quad 0 < x < 1$.

(b) $f(x) = \dfrac{1}{\pi(1 + x^2)}$, $\quad -\infty < x < \infty$.

**3.2-6** Compute the mean and the variance and find the distribution function for each of the p.d.f.s defined in Exercise 3.2-3. Also sketch the graphs of the distribution functions.

**3.2-7** Find the mean and the variance, if they exist, of each of the following distributions.

(a) $f(x) = 6x(1 - x)$, $\quad 0 < x < 1$.

(b) $f(x) = 2/x^3$, $\quad 1 < x < \infty$.

(c) $f(x) = 1/x^2$, $\quad 1 < x < \infty$.

**3.2-8** Compute the measures of skewness and kurtosis for each of the following distributions. Sketch their graphs.

(a) $f(x) = 1$, $\quad 0 < x < 1$.

(b) $f(x) = 2x$, $\quad 0 < x < 1$.

(c) $f(x) = 2(1 - x)$, $\quad 0 < x < 1$.

**3.2-9** For each of the following functions: **(i)** find the constant $c$ so that $f(x)$ is a p.d.f., **(ii)** find the distribution function $F(x)$, **(iii)** sketch the graphs of the p.d.f. $f(x)$ and the distribution function $F(x)$.

**(a)** $f(x) = \dfrac{x^3}{4}, \quad 0 < x < c.$

**(b)** $f(x) = c\sqrt{x}, \quad 0 < x < 4.$

**(c)** $f(x) = \dfrac{c}{\sqrt{x}}, \quad 0 < x < 1.$ Is this p.d.f. bounded?

**(d)** $f(x) = \dfrac{c}{x^2}, \quad 1 < x < \infty.$

**3.2-10** Let $X$ have the p.d.f. $f(x) = (x+1)/2, \ -1 < x < 1$. Find

**(a)** $\pi_{0.64}$,

**(b)** $q_1 = \pi_{0.25}$,

**(c)** $\pi_{0.81}$.

**3.2-11** The **logistic distribution** is associated with the distribution function $F(x) = (1 + e^{-x})^{-1}, \ -\infty < x < \infty$. Find the p.d.f. of the logistic distribution and show that its graph is symmetric about $x = 0$.

**3.2-12** One hundred observations of $X$ having a certain distribution were taken and then ordered, yielding

| | | | | | | | | | |
|------|------|------|------|------|------|------|------|------|------|
| 0.19 | 0.25 | 0.26 | 0.26 | 0.28 | 0.31 | 0.39 | 0.42 | 0.44 | 0.45 |
| 0.48 | 0.53 | 0.54 | 0.57 | 0.66 | 0.69 | 0.70 | 0.72 | 0.73 | 0.79 |
| 0.83 | 0.85 | 0.87 | 0.90 | 0.91 | 0.94 | 0.96 | 0.99 | 1.01 | 1.02 |
| 1.02 | 1.03 | 1.05 | 1.07 | 1.13 | 1.13 | 1.14 | 1.20 | 1.22 | 1.27 |
| 1.28 | 1.30 | 1.34 | 1.34 | 1.37 | 1.39 | 1.40 | 1.43 | 1.43 | 1.46 |
| 1.49 | 1.50 | 1.51 | 1.53 | 1.54 | 1.56 | 1.56 | 1.57 | 1.60 | 1.60 |
| 1.61 | 1.62 | 1.62 | 1.62 | 1.63 | 1.64 | 1.66 | 1.67 | 1.70 | 1.71 |
| 1.71 | 1.71 | 1.72 | 1.73 | 1.73 | 1.74 | 1.75 | 1.76 | 1.77 | 1.80 |
| 1.81 | 1.84 | 1.85 | 1.85 | 1.86 | 1.87 | 1.87 | 1.88 | 1.90 | 1.91 |
| 1.91 | 1.92 | 1.92 | 1.93 | 1.93 | 1.93 | 1.94 | 1.95 | 1.98 | 1.99 |

**(a)** Calculate the sample mean $\bar{x}$.

**(b)** Calculate the sample standard deviation $s$.

**(c)** Group these data into 10 classes with class boundaries 0.005, 0.205, 0.405, ... , 2.005.

**(d)** Construct a relative frequency histogram and then superimpose the p.d.f. $f(x) = x/2, \ 0 < x < 2$. Is there a good fit?

**(e)** Find $\mu$ and $\sigma$, assuming that the p.d.f. of $X$ is $f(x) = x/2, 0 < x \le 2$. Is $\bar{x}$ close to $\mu$? Is $s$ close to $\sigma$?

## 3.3 SPECIAL CONTINUOUS DISTRIBUTIONS

Let the random variable $X$ denote the outcome when a point is selected at random from an interval $[a, b], -\infty < a < b < \infty$. If the experiment is performed in a fair manner, it is reasonable to assume that the probability that the point is selected from the interval $[a, x], a \le x < b$ is $(x - a)/(b - a)$. That is, the probability is proportional to the length of the interval so that the distribution function of $X$ is

$$F(x) = \begin{cases} 0, & x < a, \\ \dfrac{x - a}{b - a}, & a \le x < b, \\ 1, & b \le x. \end{cases}$$

Because $X$ is a continuous-type random variable, $F'(x)$ is equal to the p.d.f. of $X$ whenever $F'(x)$ exists; thus when $a < x < b$, we have $f(x) = F'(x) = 1/(b - a)$.

The random variable $X$ has a **uniform distribution** if its p.d.f. is equal to a constant on its support. In particular, if the support is the interval $[a, b]$, then

$$f(x) = \frac{1}{b - a}, \qquad a \leq x \leq b.$$

Moreover, we shall say that $X$ is $U(a, b)$. This distribution is also referred to as **rectangular** because the graph of $f(x)$ suggests that name. See Figure 3.3-1 for the graph of $f(x)$ and the distribution function $F(x)$ when $a = 0.30$ and $b = 1.55$. Note that we could have taken $f(a) = 0$ or $f(b) = 0$ without altering the probabilities since this is a continuous-type distribution, and we shall do this in some cases.

The mean and variance of $X$ are not difficult to calculate (see Exercise 3.3-1). They are

$$\mu = \frac{a + b}{2}, \qquad \sigma^2 = \frac{(b - a)^2}{12}.$$

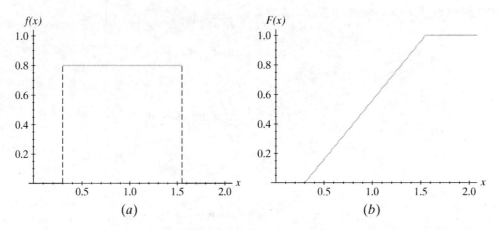

**Figure 3.3-1**  Uniform p.d.f. and distribution function

An important uniform distribution is that for which $a = 0$ and $b = 1$, namely $U(0, 1)$. If $X$ is $U(0, 1)$, approximate values of $X$ can be simulated on most computers using a **random number generator**. In fact, it should be called a **pseudo-random number generator** because the programs that produce the random numbers are usually such that if the starting number (the seed number) is known, all subsequent numbers in the sequence may be determined by simple arithmetical operations. Yet, despite their deterministic origin, these computer-produced numbers do behave as if they were truly randomly generated, and we shall not encumber our terminology by adding *pseudo*. Examples of computer-produced random numbers are given in the Appendix in Table VIII. Treat each of these numbers as a decimal; that is, divide each of them by $10^4$.

**EXAMPLE 3.3-1**   Let $X$ have the p.d.f.

$$f(x) = \frac{1}{100}, \qquad 0 < x < 100,$$

so that $X$ is $U(0, 100)$. The mean and the variance are

$$\mu = \frac{0 + 100}{2} = 50 \qquad \text{and} \qquad \sigma^2 = \frac{(100 - 0)^2}{12} = \frac{10{,}000}{12}.$$

The standard deviation is $\sigma = 100/\sqrt{12}$, which is 100 times that of the $U(0, 1)$ distribution. This agrees with our intuition since the standard deviation is a measure of spread and $U(0, 100)$ is clearly spread out 100 times more than $U(0, 1)$.

We say that the random variable $X$ has an **exponential distribution** if its p.d.f. is of the form

$$f(x) = \frac{1}{\theta} e^{-x/\theta}, \qquad 0 \le x < \infty,$$

where the parameter $\theta > 0$. The mean of $X$ is

$$\mu = E(X) = \int_0^\infty x \left(\frac{1}{\theta}\right) e^{-x/\theta}\, dx = \left[-x e^{-x/\theta} - \theta e^{-x/\theta}\right]_0^\infty = \theta$$

and

$$E(X^2) = \int_0^\infty x^2 \left(\frac{1}{\theta}\right) e^{-x/\theta}\, dx = \left[-x^2 e^{-x/\theta} - (2x)\theta e^{-x/\theta} - 2\theta^2 e^{-x/\theta}\right]_0^\infty = 2\theta^2.$$

Thus the variance of $X$ is

$$\sigma^2 = E(X^2) - \mu^2 = 2\theta^2 - \theta^2 = \theta^2.$$

The distribution function of $X$ is

$$F(x) = P(X \le x) = \int_{-\infty}^x f(w)\, dw = \begin{cases} 0, & -\infty < x < 0, \\ 1 - e^{-x/\theta}, & 0 \le x < \infty. \end{cases}$$

See Example 5.2-7 for calculations of these characteristics using *Maple*.

The p.d.f. and distribution function are graphed in Figure 3.3-2 for $\theta = 5$. The median, $m$, is found by solving $F(m) = 0.5$. That is,

$$1 - e^{-m/\theta} = 0.5.$$

Thus

$$m = -\theta \ln(0.5) = \theta \ln(2).$$

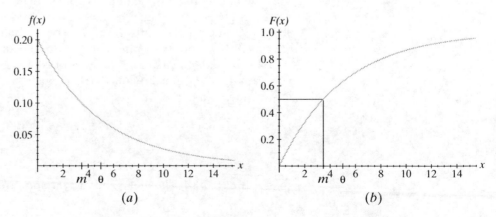

(a)                                        (b)

Figure 3.3-2 Exponential p.d.f. and distribution function

So with $\theta = 5$, the median is $m = -5\ln(0.5) = 5\ln(2) = 3.466$. Both the median and the mean $\theta = 5$ are indicated on the graphs.

It is useful to note that for an exponential random variable, $X$, we have that

$$P(X > x) = 1 - F(x) = 1 - (1 - e^{-x/\theta})$$

$$= e^{-x/\theta},$$

when $x > 0$.

**EXAMPLE 3.3-2** Customers arrive in a certain shop at a mean rate of 20 per hour. What is the probability that the shopkeeper will have to wait more than 5 minutes for the arrival of the first customer? Let $X$ denote the waiting time in minutes until the first customer arrives and assume that $X$ is exponentially distributed with the mean $\theta = 60/20 = 3$. Thus

$$f(x) = \frac{1}{3} e^{-(1/3)x}, \qquad 0 \le x < \infty.$$

Hence

$$P(X > 5) = \int_5^\infty \frac{1}{3} e^{-(1/3)x}\, dx = e^{-5/3} = 0.1889.$$

The median time until the first arrival is

$$m = -3\ln(0.5) = 3\ln(2) = 2.0794.$$

**EXAMPLE 3.3-3** Suppose that the life of a certain type of electronic component has an exponential distribution with a mean life of 500 hours. If $X$ denotes the life of this component (or the time to failure of this component), then

$$P(X > x) = \int_x^\infty \frac{1}{500} e^{-t/500}\, dt = e^{-x/500}.$$

Suppose that the component has been in operation for 300 hours. The conditional probability that it will last for more than 600 additional hours is

$$P(X > 900 \mid X > 300) = \frac{P(\{X > 900\} \cap \{X > 300\})}{P(X > 300)}$$

$$= \frac{P(X > 900)}{P(X > 300)}$$

$$= \frac{e^{-900/500}}{e^{-300/500}} = e^{-6/5}.$$

It is important to note that this conditional probability is exactly equal to $P(X > 600) = e^{-6/5}$. That is, the probability that it will last longer than an additional 600 hours, given that it has operated 300 hours, is the same as the probability that it would last longer than 600 hours when first put into operation. Thus, for such components, an old component is as good as a new one, and we say that the failure rate is constant. Certainly, with constant failure rate, there is no advantage in replacing components that are operating satisfactorily. Obviously, this is not true in practice because most would have an increasing failure rate with time; hence the exponential distribution is probably not the best model for the probability distribution of such a life.

Before determining the characteristics of the gamma distribution, let us consider the gamma function for which the distribution is named. The **gamma function** is defined by

$$\Gamma(t) = \int_0^\infty y^{t-1} e^{-y} \, dy, \qquad 0 < t.$$

This integral is positive for $0 < t$ because the integrand is positive. Values of it are often given in a table of integrals. If $t > 1$, integration of the gamma function of $t$ by parts yields

$$\Gamma(t) = \left[ -y^{t-1} e^{-y} \right]_0^\infty + \int_0^\infty (t-1) y^{t-2} e^{-y} \, dy$$

$$= (t-1) \int_0^\infty y^{t-2} e^{-y} \, dy = (t-1)\Gamma(t-1).$$

For example, $\Gamma(6) = 5\Gamma(5)$ and $\Gamma(3) = 2\Gamma(2) = (2)(1)\Gamma(1)$. Whenever $t = n$, a positive integer, we have, by repeated application of $\Gamma(t) = (t-1)\Gamma(t-1)$, that

$$\Gamma(n) = (n-1)\Gamma(n-1) = (n-1)(n-2)\cdots(2)(1)\Gamma(1).$$

However,

$$\Gamma(1) = \int_0^\infty e^{-y} \, dy = 1.$$

Thus, when $n$ is a positive integer, we have that

$$\Gamma(n) = (n-1)!$$

and, for this reason, the gamma function is called the generalized factorial. Incidentally, $\Gamma(1)$ corresponds to $0!$, and we have noted that $\Gamma(1) = 1$, which is consistent with earlier discussions.

The random variable $X$ has a **gamma distribution** if its p.d.f. is of the form

$$f(x) = \frac{1}{\Gamma(\alpha)\theta^\alpha} x^{\alpha-1} e^{-x/\theta}, \qquad 0 \le x < \infty,$$

where the parameters $\alpha$ and $\theta$ are positive. (See Exercise 3.3-12.) Note that the exponential distribution is a special case of the gamma distribution when $\alpha = 1$.

To determine the mean of the gamma distribution, integrate

$$\mu = E(X) = \int_0^\infty x \frac{1}{\Gamma(\alpha)\theta^\alpha} x^{\alpha-1} e^{-x/\theta} \, dx$$

by changing variables $y = x/\theta$ or, equivalently, $x = \theta y$ so $dx/dy = \theta$. We obtain

$$\mu = \int_0^\infty (\theta y) \frac{1}{\Gamma(\alpha)\theta^\alpha} (\theta y)^{\alpha-1} e^{-y} \theta \, dy$$

$$= \frac{\theta}{\Gamma(\alpha)} \int_0^\infty y^{(\alpha+1)-1} e^{-y} \, dy$$

$$= \frac{\theta \Gamma(\alpha+1)}{\Gamma(\alpha)} = \frac{\theta \alpha \Gamma(\alpha)}{\Gamma(\alpha)} = \alpha\theta,$$

using the property $\Gamma(\alpha+1) = \alpha\Gamma(\alpha)$ of the gamma function. Likewise $E(X^2)$, by the same change of variables, equals

$$E(X^2) = \theta^2 \frac{\Gamma(\alpha+2)}{\Gamma(\alpha)} = \theta^2 \frac{(\alpha+1)(\alpha)\Gamma(\alpha)}{\Gamma(\alpha)} = \alpha(\alpha+1)\theta^2.$$

Thus the variance of $X$ is

$$\sigma^2 = \alpha(\alpha+1)\theta^2 - (\alpha\theta)^2 = \alpha\theta^2.$$

These characteristics can also be calculated using *Maple*. (See Example 5.2-8 .)

**Remark**   If the instructor so chooses, the second part of Section 6.1 on the m.g.f. of a continuous random variable can now be studied.

In order to see the effect of the parameters on the shape of the p.d.f., several combinations of $\alpha$ and $\theta$ have been used for graphs that are displayed in Figure 3.3-3.

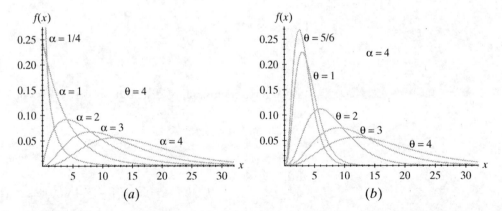

**Figure 3.3-3**  Gamma p.d.f.s: $\theta = 4$, $\alpha = 1/4, 1, 2, 3, 4$; $\alpha = 4$, $\theta = 5/6, 1, 2, 3, 4$

We now consider a special case of the gamma distribution that plays an important role in statistics. Let $X$ have a gamma distribution with $\theta = 2$ and $\alpha = r/2$, where $r$ is a positive integer. The p.d.f. of $X$ is

$$f(x) = \frac{1}{\Gamma(r/2)2^{r/2}} x^{r/2-1} e^{-x/2}, \qquad 0 \le x < \infty.$$

We say that $X$ has a **chi-square distribution with $r$ degrees of freedom**, which we abbreviate by saying $X$ is $\chi^2(r)$. The mean and the variance of this chi-square distribution are

$$\mu = \alpha\theta = \left(\frac{r}{2}\right)2 = r \qquad \text{and} \qquad \sigma^2 = \alpha\theta^2 = \left(\frac{r}{2}\right)2^2 = 2r.$$

That is, the mean equals the number of degrees of freedom, and the variance equals twice the number of degrees of freedom. An explanation of "number of degrees of freedom" is given later.

In Figure 3.3-4 the graphs of chi-square p.d.f.s for $r = 2, 3, 5, 8,$ and 14 are given. Note the relationship between the mean, $\mu = r$, and the point at which the p.d.f. obtains its maximum.

Because the chi-square distribution is so important in applications, tables have been prepared giving the values of the distribution function

$$F(x) = \int_0^x \frac{1}{\Gamma(r/2)2^{r/2}} w^{r/2-1} e^{-w/2} \, dw$$

for selected values of $r$ and $x$. For an example, see Table IV in the Appendix.

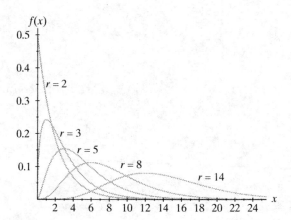

**Figure 3.3-4** Chi-square p.d.f.s with $r = 2, 3, 5, 8, 14$

EXAMPLE 3.3-4 Let $X$ have a chi-square distribution with $r = 5$ degrees of freedom. Then, using Table IV in the Appendix,

$$P(1.145 \le X \le 12.83) = F(12.83) - F(1.145) = 0.975 - 0.050 = 0.925$$

and

$$P(X > 15.09) = 1 - F(15.09) = 1 - 0.99 = 0.01.$$

EXAMPLE 3.3-5 If $X$ is $\chi^2(7)$, two constants, $a$ and $b$, such that

$$P(a < X < b) = 0.95$$

are $a = \chi^2_{0.975}(7) = 1.690$ and $b = \chi^2_{0.025}(7) = 16.01$, where, in general, $\chi^2_\alpha(r)$ is the point in a $\chi^2(r)$ distribution with $\alpha$ probability to the right of it. Other constants $a$

and $b$ can be found, and we are only limited in our choices by the table. It is very easy to find these other constants using Minitab, *Maple*, or some other statistical package.

**3.3-1** Show that the mean and the variance of the uniform distribution $U(a, b)$ are

$$\mu = \frac{a+b}{2} \quad \text{and} \quad \sigma^2 = \frac{(b-a)^2}{12}.$$

**3.3-2** Let $f(x) = 1/2$, $-1 \leq x \leq 1$, be the p.d.f. of $X$. Graph the p.d.f. and distribution function, and record the mean and variance of $X$.

**3.3-3** Let $Y$ have a uniform distribution $U(0, 1)$ and let

$$W = a + (b-a)Y, \quad a < b.$$

(a) Find the distribution function of $W$. HINT: Find $P[a + (b-a)Y \leq w]$.

(b) How is $W$ distributed?

**3.3-4** The *Holland Sentinel* reported the following numbers of calls **per hour** received by 911 between noon, February 26, and all day February 27.

| 3 | 0 | 3 | 4 | 9 | 1 | 6 | 2 | 2 | 5 | 7 | 6 | 4 | 2 | 2 | 4 | 1 | 0 |
|---|---|---|---|---|---|---|---|---|---|---|---|---|---|---|---|---|---|
| 3 | 1 | 3 | 3 | 4 | 2 | 1 | 3 | 3 | 2 | 3 | 2 | 5 | 2 | 1 | 0 | 2 | 4 |

They also reported the following lengths of time **per minute** between calls.

| 30 | 17 | 65 | 8 | 38 | 35 | 4 | 19 | 7 | 14 | 12 | 4 | 5 | 4 | 2 |
|----|----|----|---|----|----|---|----|---|----|----|---|---|---|---|
| 7 | 5 | 12 | 50 | 33 | 10 | 15 | 2 | 10 | 1 | 5 | 30 | 41 | 21 | 31 |
| 1 | 18 | 12 | 5 | 24 | 7 | 6 | 31 | 1 | 3 | 2 | 22 | 1 | 30 | 2 |
| 1 | 3 | 12 | 12 | 9 | 28 | 6 | 50 | 63 | 5 | 17 | 11 | 23 | 2 | 46 |
| 90 | 13 | 21 | 55 | 43 | 5 | 19 | 47 | 24 | 4 | 6 | 27 | 4 | 6 | 37 |
| 16 | 41 | 68 | 9 | 5 | 28 | 42 | 3 | 42 | 8 | 52 | 2 | 11 | 41 | 4 |
| 35 | 21 | 3 | 17 | 10 | 16 | 1 | 68 | 105 | 45 | 23 | 5 | 10 | 12 | 17 |

(a) Show graphically that the numbers of calls per hour have an approximate Poisson distribution with a mean of $\lambda = 3$.

(b) Show that the sample mean and the sample standard deviation of the times between calls are both approximately equal to 20.

(c) If $X$ is an exponential random variable with mean $\theta = 20$, compare $P(X > 15)$ with the proportion of times that are greater than 15.

(d) Compare $P(X > 45.5 \mid X > 30.5)$ with the proportion of observations that satisfy this condition.

**3.3-5** Let $X$ equal the time (in minutes) between calls that are made over the public safety radio. On four different days (February 14, 21, 28, and March 6) and during a period of one hour on each day, the following observations of $X$ were made:

| 5 | 7 | 8 | 20 | 17 | 2 | 24 | 8 | 8 | 6 | 4 |
|---|---|---|----|----|---|----|---|---|---|---|
| 3 | 42 | 10 | 18 | 5 | 7 | 8 | 4 | 5 | 10 | |

(a) Calculate the values of the sample mean and sample standard deviation. Are they close to each other in value?

(b) Construct a box-and-whisker diagram. Does it indicate that the data are skewed, as would be true for observations from an exponential distribution?

**3.3-6** Let $X$ have an exponential distribution with mean $\theta$,

(a) Find the first quartile, $q_1$.

(b) How far is the first quartile below the mean?

(c) Find the third quartile, $q_3$.

(d) How far is the third quartile above the mean?

**3.3-7** The waiting times in minutes until two calls to 911, as reported by the *Holland Sentinel* on November 13 between noon and midnight, were:

| | | | | | | | | | | |
|---|---|---|---|---|---|---|---|---|---|---|
| 20 | 28 | 81 | 4 | 9 | 41 | 9 | 11 | 10 | 24 | 20 |
| 44 | 18 | 30 | 16 | 53 | 15 | 38 | 50 | 84 | 44 | 69 |

Could these times represent a random sample from a gamma distribution with $\alpha = 2$ and $\theta = 120/7$?

**(a)** Compare the distribution and sample means.

**(b)** Compare the distribution and sample variances.

**(c)** Compare $P(X < 35)$ with the proportion of times that are less than 35.

**(d)** If possible, make some graphical comparisons.

**(e)** What is your conclusion?

**3.3-8** If $X$ is $\chi^2(5)$, determine the constants $c$ and $d$ so that $P(c < X < d) = 0.95$ and $P(X < c) = 0.025$. That is, find $c = \chi^2_{0.975}(5)$ and $d = \chi^2_{0.025}(5)$.

**3.3-9** Let $X$ have a gamma distribution with parameters $\alpha = 2$ and $\theta$. If $x = 2$ is the mode of the distribution, find $\theta$ and $P(X < 9.488)$.

**3.3-10** Let $X$ have the uniform distribution $U(0, 1)$. Find the distribution function and then the p.d.f. of $Y = -2 \ln X$. HINT: Find $P(Y \leq y) = P(X \geq e^{-y/2})$ when $0 < y < \infty$.

**3.3-11** Find the uniform distribution $U(a, b)$ that has the same mean and variance as a $\chi^2(8)$ distribution.

**3.3-12** Let $f(x)$ be the p.d.f. of a gamma distribution. Show that $\int_0^\infty f(x)\, dx = 1$ by changing variables $y = x/\theta$.

## 3.4 THE NORMAL DISTRIBUTION

The normal distribution is perhaps the most important distribution in statistical applications since many estimators, like $Y/n$ for $p$ and $\overline{X}$ for $\mu$ have approximate normal distributions. One explanation of this fact is the role of the normal distribution in the Central Limit Theorem. One form of this theorem will be considered in Section 3.6. In addition, some measurements, but not all, have approximate normal distributions. As a matter of fact, the authors do not like it when instructors "grade on the curve" because usually test scores are not normally distributed!

The random variable $X$ has a **normal distribution** if its p.d.f. is defined by

$$f(x) = \frac{1}{b\sqrt{2\pi}} \exp\left[ -\frac{(x-a)^2}{2b^2} \right], \qquad -\infty < x < \infty,$$

where $a$ and $b$ are parameters satisfying $-\infty < a < \infty, 0 < b < \infty$, and also where $\exp[v]$ means $e^v$. In the formula for $f(x)$ for this normal p.d.f., the symbols $a$ and $b$ are the mean and the standard deviation of $X$. In Example 5.2-10 we use *Maple* to show this and also that $f(x)$ is indeed a p.d.f.

Thus we usually write the normal p.d.f. as

$$f(x) = \frac{1}{\sigma\sqrt{2\pi}} \exp\left[ -\frac{(x-\mu)^2}{2\sigma^2} \right], \qquad -\infty < x < \infty,$$

and say that the random variable $X$ with this p.d.f. is $N(\mu, \sigma^2)$. If $Z$ is $N(0, 1)$, we shall say that $Z$ has a **standard normal distribution**. Moreover, the distribution function of $Z$ is

$$\Phi(z) = P(Z \leq z) = \int_{-\infty}^{z} \frac{1}{\sqrt{2\pi}} e^{-w^2/2}\, dw.$$

It is not possible to evaluate this integral by finding an antiderivative that can be expressed as an elementary function. However, numerical approximations for integrals of this type have been tabulated and are given in Tables Va and Vb in the Appendix. The bell-shaped curved in Figure 3.4-1 represents the graph of the p.d.f. of $Z$, and the shaded area equals $\Phi(z_0)$.

Values of $\Phi(z)$ for $z \geq 0$ are given in Appendix Table Va. Because of the symmetry of the standard normal p.d.f., it is true that $\Phi(-z) = 1 - \Phi(z)$ for all real $z$. Thus Appendix Table Va is sufficient. However, it is sometimes convenient to be able to read $\Phi(-z)$, for $z > 0$, directly from a table. This can be done using values in Appendix Table Vb which lists right tail probabilities. Again, because of the symmetry of the standard normal p.d.f., when $z > 0$, $\Phi(-z) = P(Z \leq -z) = P(Z > z)$ can be read directly from Appendix Table Vb.

**EXAMPLE 3.4-1**   If $Z$ is $N(0,1)$, then, using Appendix Table Va, we obtain

$$P(Z \leq 1.24) = \Phi(1.24) = 0.8925,$$

$$P(1.24 \leq Z \leq 2.37) = \Phi(2.37) - \Phi(1.24) = 0.9911 - 0.8925 = 0.0986,$$

$$P(-2.37 \leq Z \leq -1.24) = P(1.24 \leq Z \leq 2.37) = 0.0986.$$

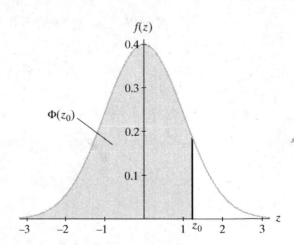

**Figure 3.4-1** Standard normal p.d.f.

Using Appendix Table Vb, we find that

$$P(Z > 1.24) = 0.1075,$$

$$P(Z \leq -2.14) = P(Z \geq 2.14) = 0.0162.$$

We obtain

$$P(-2.14 \leq Z \leq 0.77) = P(Z \leq 0.77) - P(Z \leq -2.14)$$

$$= 0.7794 - 0.0162 = 0.7632$$

using both tables.

There are times when we want to read the normal probability table in the opposite way, essentially finding the inverse of the standard normal distribution function. That is, given a probability $p$, find a constant $a$ so that $P(Z \leq a) = p$. This is illustrated in the next example, and we denote $a$ by $z_{1-p}$; that is, $1 - p$ is the probability to the right of $z_{1-p}$. In general, $\alpha$ is the probability to the right of $z_\alpha$ when using $N(0, 1)$.

**EXAMPLE 3.4-2**    If the distribution of $Z$ is $N(0, 1)$, to find constants $a$ and $b$ such that

$$P(Z \leq a) = 0.9147 \qquad \text{and} \qquad P(Z \geq b) = 0.0526,$$

find the given probabilities in Appendix Tables Va and Vb, respectively, and read off the corresponding values of $z$. From Appendix Table Va we see that $a = z_{0.0853} = 1.37$ and from Appendix Table Vb we see that $b = z_{0.0526} = 1.62$.

If $X$ is $N(\mu, \sigma^2)$, the next theorem shows that the random variable $(X - \mu)/\sigma$ is $N(0, 1)$. Thus Tables Va and Vb in the Appendix can be used to find probabilities concerning $X$.

**Theorem 3.4-1**    If $X$ is $N(\mu, \sigma^2)$, then $Z = (X - \mu)/\sigma$ is $N(0, 1)$.

***Proof***    The distribution function of $Z$ is

$$P(Z \leq z) = P\left(\frac{X - \mu}{\sigma} \leq z\right) = P(X \leq z\sigma + \mu)$$

$$= \int_{-\infty}^{z\sigma+\mu} \frac{1}{\sigma\sqrt{2\pi}} \exp\left[-\frac{(x - \mu)^2}{2\sigma^2}\right] dx.$$

In the integral representing $P(Z \leq z)$, use the change of variable of integration given by $w = (x - \mu)/\sigma$ (i.e., $x = w\sigma + \mu$) to obtain

$$P(Z \leq z) = \int_{-\infty}^{z} \frac{1}{\sqrt{2\pi}} e^{-w^2/2} dw.$$

But this is the expression for $\Phi(z)$, the distribution function of a standardized normal random variable. Hence $Z$ is $N(0, 1)$.

This theorem can be used to find probabilities about $X$, which is $N(\mu, \sigma^2)$, as follows:

$$P(a \leq X \leq b) = P\left(\frac{a - \mu}{\sigma} \leq \frac{X - \mu}{\sigma} \leq \frac{b - \mu}{\sigma}\right) = \Phi\left(\frac{b - \mu}{\sigma}\right) - \Phi\left(\frac{a - \mu}{\sigma}\right),$$

since $(X - \mu)/\sigma$ is $N(0, 1)$.

**EXAMPLE 3.4-3**    If $X$ is $N(3, 16)$, then

$$P(4 \leq X \leq 8) = P\left(\frac{4 - 3}{4} \leq \frac{X - 3}{4} \leq \frac{8 - 3}{4}\right)$$

$$= \Phi(1.25) - \Phi(0.25) = 0.8944 - 0.5987 = 0.2957,$$

$$P(0 \le X \le 5) = P\left(\frac{0-3}{4} \le Z \le \frac{5-3}{4}\right)$$

$$= \Phi(0.5) - \Phi(-0.75) = 0.6915 - 0.2266 = 0.4649.$$

Similarly,

$$P(-2 \le X \le 1) = \Phi(-0.5) - \Phi(-1.25) = 0.3085 - 0.1056 = 0.2029.$$

**EXAMPLE 3.4-4** If $X$ is $N(25, 36)$, we find a constant $c$ such that

$$P(|X - 25| \le c) = 0.9544.$$

We want

$$P\left(\frac{-c}{6} \le \frac{X - 25}{6} \le \frac{c}{6}\right) = 0.9544.$$

Thus

$$\Phi\left(\frac{c}{6}\right) - \left[1 - \Phi\left(\frac{c}{6}\right)\right] = 0.9544$$

and

$$\Phi\left(\frac{c}{6}\right) = 0.9772.$$

Hence $c/6 = 2$ and $c = 12$. That is, the probability that $X$ falls within two standard deviations of its mean is the same as the probability that the standard normal variable $Z$ falls within two units (standard deviations) of zero.

In the next theorem we give a relationship between the chi-square and normal distributions.

**Theorem 3.4-2**    If the random variable $X$ is $N(\mu, \sigma^2)$, $\sigma^2 > 0$, then the random variable $V = (X - \mu)^2 / \sigma^2 = Z^2$ is $\chi^2(1)$.

***Proof***    Because $V = Z^2$, the distribution function $G(v)$ of $V$ is, for $v \ge 0$,

$$G(v) = P(Z^2 \le v) = P(-\sqrt{v} \le Z \le \sqrt{v}).$$

That is, with $v \ge 0$, and since $Z = (X - \mu)/\sigma$ is $N(0, 1)$,

$$G(v) = \int_{-\sqrt{v}}^{\sqrt{v}} \frac{1}{\sqrt{2\pi}} e^{-z^2/2}\, dz = 2 \int_0^{\sqrt{v}} \frac{1}{\sqrt{2\pi}} e^{-z^2/2}\, dz.$$

If we change the variable of integration by writing $z = \sqrt{y}$, then, since $dz/dy = 1/(2\sqrt{y})$, we have

$$G(v) = \int_0^v \frac{1}{\sqrt{2\pi y}} e^{-y/2}\, dy, \qquad 0 < v,$$

Of course, $G(v) = 0$, when $v < 0$. Hence the p.d.f. $g(v) = G'(v)$ of the continuous type random variable $V$ is, by one form of the fundamental theorem of calculus,

$$g(v) = \frac{1}{\sqrt{\pi}\sqrt{2}}\, v^{1/2-1}\, e^{-v/2}, \qquad 0 < v < \infty.$$

From its form it must be a gamma p.d.f. and thus $\sqrt{\pi} = \Gamma(1/2)$.

**EXAMPLE 3.4-5**    If $Z$ is $N(0,1)$, then $P(|Z| < 1.96 = \sqrt{3.842}) = P(Z^2 < 3.842) = 0.95$ from the chi-square table with $r = 1$.

Given a set of observations of a random variable $X$, it is a challenge to determine the distribution of $X$. In particular, how can we decide whether or not $X$ has an approximate normal distribution? If we have a large number of observations of $X$, a histogram or a stem-and-leaf diagram of the observations can often be helpful. (See Exercises 3.4-7 and 3.6-9.) For small samples a **quantile-quantile** ($q$-$q$) **plot** can be used to check on whether it is reasonable to assume that the sample arises from a normal distribution.

The $100p$th percentile of a distribution is also called the quantile of order $p$ and is denoted by $q_p$. Moreover if $y_1 \leq y_2 \leq \cdots \leq y_n$ are the order statistics of a sample of size $n$, then $y_r$ is called the sample quantile of order $r/(n+1)$ as well as the $100r/(n+1)$ sample percentile.

Suppose the quantiles of a certain normal distribution were plotted against the corresponding quantiles of the sample and these pairs of points were on a straight line with slope 1 and intercept 0. Of course, we would then believe that we have an ideal sample from that normal distribution with that certain mean and standard deviation. This plot, however, requires that we know the mean and the standard deviation of this normal distribution, and we usually do not. However, since the quantile, $q_p$, of $N(\mu, \sigma^2)$ is related to the corresponding one, $z_{1-p}$, of $N(0,1)$ by $q_p = \mu + \sigma z_{1-p}$, we can always plot the quantiles of $N(0,1)$ against the corresponding sample quantiles and get the needed information. That is, if the sample quantiles are plotted as the $x$ coordinate of the pair and the $N(0,1)$ quantiles as the $y$ coordinate and if the graph is almost a straight line, then it is reasonable to assume that the sample arises from a normal distribution. Moreover, the reciprocal of the slope of that straight line is a good estimate of the standard deviation $\sigma$ because $z_{1-p} = (q_p - \mu)/\sigma$.

**EXAMPLE 3.4-6**    When researching ground water it is often important to know the characteristics of the soil at a certain site. Many of these characteristics, such as porosity, are at least partially dependent upon the grain size. The diameter of individual grains of soil can be measured. Here are the diameters (in mm) of 30 randomly selected grains. (Also see Exercise 3.1-4.)

| | | | | | | | | | |
|------|------|------|------|------|------|------|------|------|------|
| 1.24 | 1.36 | 1.28 | 1.31 | 1.35 | 1.20 | 1.39 | 1.35 | 1.41 | 1.31 |
| 1.28 | 1.26 | 1.37 | 1.49 | 1.32 | 1.40 | 1.33 | 1.28 | 1.25 | 1.39 |
| 1.38 | 1.34 | 1.40 | 1.27 | 1.33 | 1.36 | 1.43 | 1.33 | 1.29 | 1.34 |

For these data, $\bar{x} = 1.33$ and $s^2 = 0.0040$. May we assume that these are observations of a random variable $X$ that has an approximate normal distribution with $\mu = 1.33$ and $\sigma^2 = 0.0040$? To help answer this question, we shall construct a $q$-$q$ plot of the standard normal quantiles that correspond to $p = 1/31, 2/31, \ldots, 30/31$ versus the ordered observations. To find these quantiles, it is helpful to use the computer. Minitab could be used. We found the quantiles using *Maple*.

| k | Diameters in mm ($x$) | $p = k/31$ | $z_{1-p}$ | k | Diameters in mm ($x$) | $p = k/31$ | $z_{1-p}$ |
|---|---|---|---|---|---|---|---|
| 1 | 1.20 | 0.0323 | −1.85 | 16 | 1.34 | 0.5161 | 0.04 |
| 2 | 1.24 | 0.0645 | −1.52 | 17 | 1.34 | 0.5484 | 0.12 |
| 3 | 1.25 | 0.0968 | −1.30 | 18 | 1.35 | 0.5806 | 0.20 |
| 4 | 1.26 | 0.1290 | −1.13 | 19 | 1.35 | 0.6129 | 0.29 |
| 5 | 1.27 | 0.1613 | −0.99 | 20 | 1.36 | 0.6452 | 0.37 |
| 6 | 1.28 | 0.1935 | −0.86 | 21 | 1.36 | 0.6774 | 0.46 |
| 7 | 1.28 | 0.2258 | −0.75 | 22 | 1.37 | 0.7097 | 0.55 |
| 8 | 1.28 | 0.2581 | −0.65 | 23 | 1.38 | 0.7419 | 0.65 |
| 9 | 1.29 | 0.2903 | −0.55 | 24 | 1.39 | 0.7742 | 0.75 |
| 10 | 1.31 | 0.3226 | −0.46 | 25 | 1.39 | 0.8065 | 0.86 |
| 11 | 1.31 | 0.3548 | −0.37 | 26 | 1.40 | 0.8387 | 0.99 |
| 12 | 1.32 | 0.3871 | −0.29 | 27 | 1.40 | 0.8710 | 1.13 |
| 13 | 1.33 | 0.4194 | −0.20 | 28 | 1.41 | 0.9032 | 1.30 |
| 14 | 1.33 | 0.4516 | −0.12 | 29 | 1.43 | 0.9355 | 1.52 |
| 15 | 1.33 | 0.4839 | −0.04 | 30 | 1.49 | 0.9677 | 1.85 |

A $q$-$q$ plot of these data is shown in Figure 3.4-2. Note that the points do fall close to a straight line so the normal probability model seems to be appropriate based on these few data.

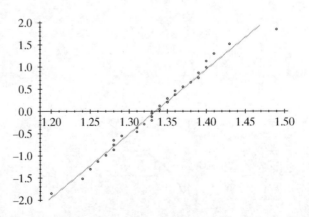

Figure 3.4-2  $q$-$q$ Plot, $N(0,1)$ quantiles versus grain diameters

EXAMPLE 3.4-7    At this point it would be useful to see how $q$-$q$ plots can help us say something about the distribution from which a sample arose. Using six different distributions for which the p.d.f.s have different shapes, we have simulated a random sample of size $n = 500$ from each. In Figure 3.4-3 are displayed six $q$-$q$ plots with the

$N(0,1)$ quantiles on the $y$-axis and the ordered observations (quantiles of the sample) on the $x$-axis. The distributions from which the samples were taken are:

| | | |
|---|---|---|
| **(a)** | $N(2,1)$ | $\mu = 2,\ \sigma^2 = 1$ |
| **(b)** | $U(0,4)$ | $\mu = 2,\ \sigma^2 = 16/12$ |
| **(c)** | $f(x) = e^{-x},\ 0 < x < \infty$ | $\mu = 1,\ \sigma^2 = 1$ |
| **(d)** | $f(x) = (3/64)x^2,\ 0 < x < 4$ | $\mu = 3,\ \sigma^2 = 3/5$ |
| **(e)** | $f(x) = (3/2)x^2,\ -1 < x < 1$ | $\mu = 0,\ \sigma^2 = 3/5$ |
| **(f)** | $f(x) = \dfrac{15}{16\sqrt{6}}\dfrac{1}{(1+x^2/6)^{7/2}},\ -\infty < x < \infty$ | $\mu = 6,\ \sigma^2 = 6/4$ |

($t$-distribution with 6 degrees of freedom defined in Section 4.1)

The reader should sketch the graphs of each of these p.d.f.s and compare them to the $q$-$q$ plots. Note in particular that the sample from the normal distribution yielded a linear plot. The uniform distribution, **(b)** $U(0,4)$, has tails that are "lighter" than the normal tails and note how the $q$-$q$ plot turns downward on the left and upward on the right. The same can be said about the $U$-shaped distribution, **(e)**, but more so. The distributions **(c)** and **(d)** are skewed ones: **(c)** to the right and **(d)** to the left; note how the ends of the corresponding $q$-$q$ plots respond. Finally, we study the $t$-distribution later in Section 4.1 . It has heavier tails than the normal and the $q$-$q$ plot tends to become more horizontal at the extreme left and extreme right.

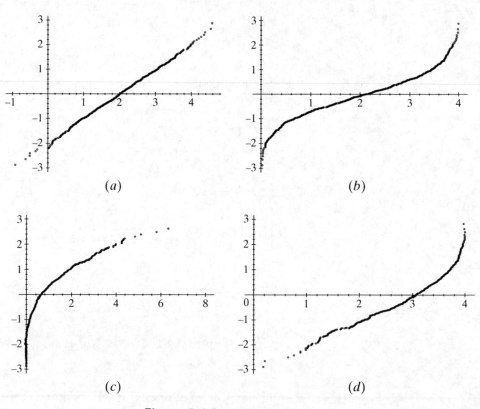

Figure 3.4-3 Examples of $q$-$q$ plots

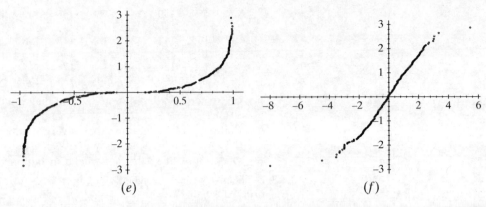

$(e)$                              $(f)$

**Figure 3.4-3** *(Continued)*

## EXERCISES 3.4

**3.4-1** If $Z$ is $N(0,1)$, find

    **(a)** $P(0 \leq Z \leq 0.87)$.         **(b)** $P(-2.64 \leq Z \leq 0)$.

    **(c)** $P(-2.13 \leq Z \leq -0.56)$.     **(d)** $P(|Z| > 1.39)$.

    **(e)** $P(Z < -1.62)$.            **(f)** $P(|Z| > 1)$.

    **(g)** $P(|Z| > 2)$.             **(h)** $P(|Z| > 3)$.

**3.4-2** Find the values of **(a)** $z_{0.10}$, **(b)** $-z_{0.05}$, **(c)** $-z_{0.0485}$, and **(d)** $z_{0.9656}$.

**3.4-3** If $X$ is normally distributed with a mean of 6 and a variance of 25, find

    **(a)** $P(6 \leq X \leq 12)$.     **(b)** $P(0 \leq X \leq 8)$.

    **(c)** $P(-2 < X \leq 0)$.     **(d)** $P(X > 21)$.

    **(e)** $P(|X - 6| < 5)$.     **(f)** $P(|X - 6| < 10)$.

    **(g)** $P(|X - 6| < 15)$.     **(h)** $P(|X - 6| < 12.41)$.

**3.4-4** If $X$ is $N(650, 625)$, find

    **(a)** $P(600 \leq X < 660)$.

    **(b)** A constant $c > 0$ such that $P(|X - 650| \leq c) = 0.9544$.

**3.4-5** If $X$ is $N(\mu, \sigma^2)$, show that $Y = aX + b$ is $N(a\mu + b, a^2\sigma^2)$, $a \neq 0$. HINT: Find the distribution function $P(Y \leq y)$ of $Y$ and, in the resulting integral, let $w = ax + b$ or, equivalently, $x = (w - b)/a$.

**3.4-6** Consider the weights of carrots in a prepackaged "1-pound" bag. Let $X$ equal the weight of such a bag of carrots. Would you expect $X$ to have a normal distribution? These are the weights of $n = 12$ observations of $X$:

$$
\begin{array}{cccccc}
1.12 & 1.13 & 1.19 & 1.25 & 1.06 & 1.31 \\
1.12 & 1.23 & 1.29 & 1.17 & 1.20 & 1.11
\end{array}
$$

Construct a $q$-$q$ plot of the corresponding $N(0,1)$ versus the ordered carrot weights to answer this question.

**3.4-7** A company manufactures windows that are then inserted into an automobile. Each window has 5 studs for attaching it. A pull-out test is used to determine the force required to pull a stud out of a window. (Note that this is an example of destructive testing.) Let $X$ equal the force required for pulling studs out of position 4. Sixty observations of $X$ were as follows:

$$
\begin{array}{rrrrrrrrrr}
159 & 150 & 147 & 160 & 155 & 142 & 143 & 151 & 154 & 133 \\
151 & 146 & 140 & 146 & 137 & 148 & 154 & 157 & 142 & 153 \\
135 & 144 & 135 & 165 & 118 & 158 & 126 & 147 & 123 & 140 \\
125 & 151 & 153 & 158 & 144 & 163 & 150 & 150 & 137 & 164 \\
137 & 156 & 139 & 134 & 171 & 144 & 160 & 147 & 155 & 175 \\
162 & 160 & 149 & 149 & 158 & 152 & 165 & 131 & 150 & 120 \\
\end{array}
$$

**(a)** Construct an ordered stem-and-leaf diagram using as stems 11•, 12*, 12•, and so on.

**(b)** Construct a q-q plot using about every 5th observation in the ordered array and the corresponding quantiles of $N(0, 1)$.

**(c)** Does it look like $X$ has a normal distribution?

**3.4-8** A chemistry major weighed 19 plain m&m's® (in grams) on a ±0.0001 scale. The ordered weights are

$$
\begin{array}{ccccccc}
0.7938 & 0.8032 & 0.8089 & 0.8222 & 0.8268 & 0.8383 & 0.8442 \\
0.8490 & 0.8528 & 0.8572 & 0.8674 & 0.8734 & 0.8786 & \\
0.8850 & 0.8873 & 0.8920 & 0.9069 & 0.9150 & 0.9243 & \\
\end{array}
$$

**(a)** Construct a stem-and-leaf diagram using three-digit leaves, with stems 0.7•, 0.8*, 0.8t, etc.

**(b)** Construct a q-q plot of these data and the corresponding quantiles for the standard normal distribution, $N(0, 1)$.

**(c)** Do these data look like observations from a normal distribution? Why?

**3.4-9** If $X$ is $N(7, 4)$, find $P[15.364 \le (X - 7)^2 \le 20.096]$.

**3.4-10** Let the distribution of $X$ be $N(\mu, \sigma^2)$ Show that the points of inflection of the graph of the p.d.f. of $X$ occur at $x = \mu \pm \sigma$.

**3.4-11** If $X$ is $N(75, 100)$, compute the conditional probability $P(X > 85 \mid X > 80)$.

**3.4-12** If $X$ is $N(\mu, \sigma^2)$, show that $E(|X - \mu|) = \sigma \sqrt{2/\pi}$. HINT: Write the integral representing $E(|X - \mu|)$ as $2 \int_{\mu}^{\infty} (x - \mu) f(x) \, dx$, where $f(x)$ is the p.d.f. of $N(\mu, \sigma^2)$.

**3.4-13** Show that the constant $c$ can be selected so that

$$
f(x) = c3^{-x^2}, \qquad -\infty < x < \infty,
$$

satisfies the conditions of a normal p.d.f. HINT: Write $3 = e^{\ln 3}$ or use a CAS. (See Section 5.2 .)

**3.4-14** Let $X$ have the p.d.f.

$$
f(x) = \frac{2}{\sqrt{2\pi}} e^{-x^2/2}, \qquad 0 < x < \infty.
$$

Find the mean and the variance of $X$. HINT: Compute $E(X)$ directly and evaluate $E(X^2)$ by comparing its integral to the variance of $N(0, 1)$, or use a CAS. (See Section 5.2 .)

## 3.5  ESTIMATION IN THE CONTINUOUS CASE

We introduced estimation in Section 2.4, particularly maximum likelihood estimation. We review that technique by considering a continuous-type distribution, namely the exponential distribution, having p.d.f.

$$
f(x; \theta) = \frac{1}{\theta} e^{-x/\theta}, \qquad 0 < x < \infty,
$$

where the **parameter space** is given by $\Omega = \{\theta : 0 < \theta < \infty\}$. Recall that the mean of $X$ is $\mu = \theta$. Thus if we have a random sample $X_1, X_2, \ldots, X_n$ from this distribution, the sample mean $\overline{X}$ could be considered an **estimator** of $\theta$ using the **method of moments**. That is, here we equate the first moment of the empirical distribution to that of the theoretical distribution. Once the sample is observed to be the values $x_1, x_2, \ldots, x_n$, the computed mean $\bar{x}$ is an **estimate** of $\theta$ and hopefully is close to $\theta$.

The **likelihood function** in the continuous case is defined in a fashion similar to that in the discrete case, namely

$$L(\theta) = f(x_1; \theta)f(x_2; \theta) \cdots f(x_n; \theta), \qquad \theta \, \varepsilon \, \Omega,$$

even though the p.d.f. $f(x; \theta)$ no longer equals probability. In our special situation, it is

$$L(\theta) = \left(\frac{1}{\theta} e^{-x_1/\theta}\right)\left(\frac{1}{\theta} e^{-x_2/\theta}\right) \cdots \left(\frac{1}{\theta} e^{-x_n/\theta}\right)$$

$$= \left(\frac{1}{\theta}\right)^n e^{-\Sigma x_i/\theta}, \qquad \text{where } 0 < \theta < \infty.$$

Again we find it to our advantage to take logarithms to obtain

$$\ln L(\theta) = -n \ln \theta - \sum_{i=1}^{n} \frac{x_i}{\theta}$$

and

$$\frac{d[\ln L(\theta)]}{d\theta} = -\frac{n}{\theta} + \frac{\sum_{i=1}^{n} x_i}{\theta^2} = 0.$$

This latter equation has the solution

$$\theta = \frac{1}{n} \sum_{i=1}^{n} x_i = \bar{x},$$

and the second derivative of $\ln[L(\theta)]$ at $\bar{x}$ is

$$\left[\frac{n}{\theta^2} - \frac{2 \sum_{i=1}^{n} x_i}{\theta^3}\right]_{\theta=\bar{x}} = \frac{n}{\bar{x}^2} - \frac{2n\bar{x}}{\bar{x}^3} = \frac{-n}{\bar{x}^2} < 0.$$

Since it is negative, we have that $\widehat{\theta} = \overline{X}$ is the **maximum likelihood estimator** of $\theta$. This is the same estimator as that found by the method of moments.

EXAMPLE 3.5-1   Let $X_1, X_2, \ldots, X_n$ be a random sample from $N(\mu = \theta, \sigma^2)$, where the parameter space $\Omega = \{\theta: -\infty < \theta < \infty\}$ and $\sigma^2$ is known. Then

$$L(\theta) = \prod_{i=1}^{n} \frac{1}{\sqrt{2\pi}\,\sigma} e^{-(x_i-\theta)^2/2\sigma^2}$$

$$= \left(\frac{1}{\sqrt{2\pi}\,\sigma}\right)^n e^{-\Sigma(x_i-\theta)^2/2\sigma^2}, \qquad -\infty < \theta < \infty.$$

Thus

$$\ln L(\theta) = -n \ln(\sqrt{2\pi}\,\sigma) - \sum_{i=1}^{n} \frac{(x_i - \theta)^2}{2\sigma^2}$$

and

$$\frac{d[\ln L(\theta)]}{d\theta} = + \sum_{i=1}^{n} \frac{2(x_i - \theta)}{2\sigma^2} = 0.$$

The latter equation can be written

$$\sum_{i=1}^{n} x_i - n\theta = 0,$$

which has the solution $\theta = \bar{x}$. Since the second derivative of $\ln L(\theta)$ is $-n/\sigma^2 < 0$, $\hat{\theta} = \overline{X}$ is the maximum likelihood estimator of $\theta$.

**EXAMPLE 3.5-2**    Let $X_1, X_2, \ldots, X_n$ be a random sample from a gamma distribution in which $\alpha$ is known, but $\theta$ is an unknown positive value. The likelihood function is

$$L(\theta) = \prod_{i=1}^{n} \frac{1}{\Gamma(\alpha)\theta^\alpha} x_i^{\alpha-1} e^{-x_i/\theta}$$

$$= \left[\frac{1}{\Gamma(\alpha)}\right]^n \frac{1}{\theta^{n\alpha}} (x_1 x_2 \cdots x_n)^{\alpha-1} e^{-\Sigma x_i/\theta}, \qquad 0 < \theta < \infty.$$

Thus

$$\ln L(\theta) = -n \ln \Gamma(\alpha) - n\alpha \ln(\theta) + (\alpha - 1) \ln(x_1 x_2 \cdots x_n) - \frac{\sum_{i=1}^{n} x_i}{\theta}$$

and

$$\frac{d[\ln L(\theta)]}{d\theta} = \frac{-n\alpha}{\theta} + \frac{\sum_{i=1}^{n} x_i}{\theta^2} = 0,$$

the solution of which is $\theta = \sum_{i=1}^{n} x_i/n\alpha = \bar{x}/\alpha$. It is easy to show that the second derivative is negative at this solution; so $\hat{\theta} = \overline{X}/\alpha$ is the maximum likelihood estimator of $\theta$. This is actually a generalization of the illustration involving the exponential distribution; we see this by letting $\alpha = 1$.

In both of these examples, 3.5-1 and 3.5-2, the maximum likelihood estimators of $\theta$ are unbiased because $\mu_{\overline{X}} = E(\overline{X}) = \mu$, the mean of the underlying distribution; hence, in those examples,

$$E(\overline{X}) = \theta \qquad \text{and} \qquad E\left(\frac{\overline{X}}{\alpha}\right) = \frac{\alpha\theta}{\alpha} = \theta,$$

respectively.

The preceding discussion can be extended to two or more unknown parameters. For two parameters, say $\theta_1$ and $\theta_2$, the p.d.f. of $X$ is $f(x; \theta_1, \theta_2)$ with parameter space $\Omega$. With a random sample from this distribution the **likelihood function** is

$$L(\theta_1, \theta_2) = f(x_1; \theta_1, \theta_2) f(x_2; \theta_1, \theta_2) \cdots f(x_n; \theta_1, \theta_2)$$

where $(\theta_1, \theta_2) \; \varepsilon \; \Omega$. Say the values

$$\theta_1 = u_1(x_1, x_2, \ldots, x_n) \qquad \text{and} \qquad \theta_2 = u_2(x_1, x_2, \ldots, x_n)$$

maximize $L(\theta_1, \theta_2)$. Then

$$\widehat{\theta_1} = u_1(X_1, X_2, \ldots, X_n) \qquad \text{and} \qquad \widehat{\theta_2} = u_2(X_1, X_2, \ldots, X_n)$$

are the **maximum likelihood estimators** of $\theta_1$ and $\theta_2$, respectively. In the method of moments, we equate the first two moments of the empirical distribution to those of the theoretical distribution and solve these equations for $\theta_1$ and $\theta_2$ to get the method of moments estimators.

EXAMPLE 3.5-3   Let $X_1, X_2, \ldots, X_n$ be a random sample from $N(\theta_1, \theta_2)$, where

$$\Omega = \{(\theta_1, \theta_2): \ -\infty < \theta_1 = \mu < \infty, 0 < \theta_2 = \sigma^2 < \infty\}.$$

Then

$$L(\theta_1, \theta_2) = \prod_{i=1}^{n} \frac{1}{\sqrt{2\pi\theta_2}} e^{-(x_i - \theta_1)^2 / 2\theta_2}$$

$$= \left( \frac{1}{\sqrt{2\pi\theta_2}} \right)^n e^{-\sum (x_i - \theta_1)^2 / 2\theta_2}.$$

Thus

$$\ln[L(\theta_1, \theta_2)] = -\frac{n}{2} \ln(2\pi\theta_2) - \frac{\sum_{i=1}^{n} (x_i - \theta_1)^2}{2\theta_2}.$$

The first partial derivatives of $\ln[L(\theta_1, \theta_2)]$ are

$$\frac{\partial \ln[L(\theta_1, \theta_2)]}{\partial \theta_1} = \frac{1}{\theta_2} \sum_{i=1}^{n} (x_i - \theta_1) = 0$$

and

$$\frac{\partial \ln[L(\theta_1, \theta_2)]}{\partial \theta_2} = -\frac{n}{2\theta_2} + \frac{1}{2\theta_2^2} \sum_{i=1}^{n} (x_i - \theta_1)^2 = 0.$$

Solving these equations simultaneously, we see that the maximum likelihood estimators of $\theta_1$ and $\theta_2$ are

$$\widehat{\theta_1} = \overline{X} \qquad \text{and} \qquad \widehat{\theta_2} = \frac{1}{n} \sum_{i=1}^{n} (X_i - \overline{X})^2 = V,$$

respectively. (We need results from multivariate calculus to prove that these solutions actually provide a maximum for $L$.) Here $E(\widehat{\theta_1}) = \theta_1$; however we show that $E(\widehat{\theta_2}) \neq \theta_2$ and hence $\widehat{\theta_2}$ is a biased estimator of $\theta_2$. This inequality is true because

$$\sum_{i=1}^{n} (X_i - \theta_1)^2 = \sum_{i=1}^{n} (X_i - \overline{X})^2 + n(\overline{X} - \theta_1)^2;$$

so

$$E\left[ \sum_{i=1}^{n} (X_i - \theta_1)^2 \right] = E\left[ \sum_{i=1}^{n} (X_i - \overline{X})^2 \right] + E[n(\overline{X} - \theta_1)^2].$$

Since $\sigma^2 = \theta_2$, we have

$$n\theta_2 = E\left[\sum_{i=1}^{n}(X_i - \overline{X})^2\right] + \theta_2,$$

because $E[(\overline{X} - \theta_1)^2/(\theta_2/n)] = 1$. Accordingly

$$E\left[\sum_{i=1}^{n}(X_i - \overline{X})^2\right] = (n-1)\theta_2;$$

thus

$$E\left[\frac{\sum_{i=1}^{n}(X_i - \overline{X})^2}{n-1}\right] = \theta_2.$$

Note that

$$E(\widehat{\theta_2}) = E\left[\frac{\sum_{i=1}^{n}(X_i - \overline{X})^2}{n}\right]$$

$$= \left(\frac{n-1}{n}\right)E\left[\frac{\sum_{i=1}^{n}(X_i - \overline{X})^2}{n-1}\right] = \left(\frac{n-1}{n}\right)\theta_2.$$

That is, $S^2 = \sum_{i=1}^{n}(X_i - \overline{X})^2/(n-1)$ is *not* the maximum likelihood estimator but is the unbiased estimator of $\sigma^2 = \theta_2$. This has been alluded to earlier.

**EXAMPLE 3.5-1**    Let $X_1, X_2, \ldots, X_n$ be a random sample from a gamma distribution with $\alpha = \theta_1$ and $\theta = \theta_2$, where $\theta_1 > 0, \theta_2 > 0$. It is difficult to maximize

$$L(\theta_1, \theta_2) = \frac{1}{[\Gamma(\theta_1)]^n \theta_2^{n\theta_1}}(x_1 x_2 \cdots x_n)^{\theta_1 - 1}e^{-\Sigma x_i/\theta_2}$$

with respect to $\theta_1$ and $\theta_2$ due to the presence of the gamma function $\Gamma(\theta_1)$. We can, however, use the method of moments by equating the first two theoretical moments, namely $\mu = \theta_1\theta_2$ and $\sigma^2 = \theta_1\theta_2^2$, to the first two moments of the empirical distribution. That is,

$$\theta_1\theta_2 = \overline{X} \qquad \text{and} \qquad \theta_1\theta_2^2 = V,$$

the solutions of which are

$$\widetilde{\theta}_1 = \frac{\overline{X}^2}{V} \qquad \text{and} \qquad \widetilde{\theta}_2 = \frac{V}{\overline{X}},$$

the method of moments estimators of $\theta_1$ and $\theta_2$.

There is a huge advantage of using maximum likelihood estimators in what is called the **regular case**. There are some mathematical conditions needed to have regular cases, but the main one is that the parameters do not appear in an endpoint of the support (space) of $X$. For illustration of a p.d.f. that is not a regular case is

$$f(x; \theta) = \frac{1}{\theta}, \qquad 0 \le x \le \theta,$$

where $\Omega = \{\theta: \ 0 < \theta < \infty\}$. Here $\theta$ is the upper endpoint of the support. While we can maximize

$$L(\theta) = \prod_{i=1}^{n} \left(\frac{1}{\theta}\right)^n, \qquad 0 \le x_i \le \theta$$

by making $\theta$ as small as possible, namely $\widehat{\theta} = \max(X_i)$; for if it were any smaller, the likelihood function would equal zero. That is, $L(\theta) = 0$ if $\theta < x_i$ for some $x_i$, and zero would not be a maximum for $L(\theta)$. The p.d.f.s given in Examples 3.5-1 through 3.5-4 are all regular cases.

In regular cases, the maximum likelihood estimators have approximate normal distributions provided $n$ is large enough. We do not prove this now, but this is the real reason that the normal distribution is so important to statisticians. It is *not* the fact that some underlying distributions are close to being normal; remember that no model is exactly right but many are useful. But it is the fact that many estimators, including maximum likelihood estimators, have approximate normal distributions. As you will see shortly, this is a most important fact and is even true in regular cases of discrete distributions in which the parameters do not enter the support of $X$.

What does this mean in the examples that we have studied thus far? In our first illustration of the exponential distribution, $\widehat{\theta} = \overline{X}$ is approximately $N(\theta, \theta^2/n)$ because there $\mu_{\overline{X}} = \theta$ and $\sigma^2_{\overline{X}} = \sigma^2/n = \theta^2/n$. In Example 3.5-1, we will see later that $\overline{X}$ has an exact normal distribution, $N(\theta, \sigma^2/n)$. In Example 3.5-2, $\widehat{\theta} = \overline{X}/\alpha$ is approximately $N(\theta, \theta^2/\alpha n)$ because

$$E(\widehat{\theta}) = E\left(\frac{\overline{X}}{\alpha}\right) = \frac{\alpha\theta}{\alpha} = \theta$$

and

$$\text{Var}(\widehat{\theta}) = \left(\frac{1}{\alpha}\right)^2 \text{Var}(\overline{X}) = \left(\frac{1}{\alpha^2}\right)\left(\frac{\alpha\theta^2}{n}\right) = \frac{\theta^2}{\alpha n}.$$

In Example 3.5-3, it is true that $\widehat{\theta}_1 = \overline{X}$ is $N(\theta_1, \theta_2/n)$ and $\widehat{\theta}_2 = V$ is approximately $N(\theta_2, 2\theta_2^2/n)$ for large $n$ even though $V$ is a biased estimator of $\theta_2$. The statement about $V$ being approximately normal with that mean and variance can not be proved at this time; so we simply accept it. In the discrete case in which $Y = X_1 + X_2 + \cdots + X_n$ is $b(n, p = \theta)$, the m.l.e. of $\theta, \widehat{\theta} = \overline{X} = Y/n$, has an approximate $N[\theta, \theta(1-\theta)/n]$ distribution because

$$E\left(\frac{Y}{n}\right) = \frac{E(Y)}{n} = \frac{n\theta}{n} = \theta \quad \text{and} \quad \text{Var}\left(\frac{Y}{n}\right) = \left(\frac{1}{n}\right)^2 n\theta(1-\theta) = \frac{\theta(1-\theta)}{n}.$$

Not only are the maximum likelihood estimators in regular case approximately normal, but it is also true that the mean $\overline{X}$ of a random sample of size $n$ from any distribution with mean $\mu$ and finite variance $\sigma^2$ is approximately $N(\mu, \sigma^2/n)$, provided the sample size $n$ is large enough. This latter fact is essentially a statement of what is called the **Central Limit Theorem**. Right now we accept these statements as facts, and investigate one important use of them, namely that of finding confidence intervals for unknown parameters. This is why these approximate normal distributions are so important.

Let us say that an estimator $U = u(X_1, X_2, \ldots, X_n)$ of $\theta$ has a normal or approximate normal distribution with unknown mean $\theta$ and variance $\sigma^2_U$, which at the moment we assume is known. Then

$$\frac{U - \theta}{\sigma_U} \quad \text{is (approximately)} \quad N(0,1).$$

From normal tables, we have

$$P\left(-2 \leq \frac{U - \theta}{\sigma_U} \leq 2\right) \approx 0.95.$$

This is actually 0.9544 if $U$ is exactly normal. Solving these inequalities for $\theta$, we have the equivalent probability statement that

$$P(U - 2\sigma_U \leq \theta \leq U + 2\sigma_U) \approx 0.95.$$

Say the random sample is observed to be $x_1, x_2, \ldots, x_n$ and the statistic $U$ computed to be $u = u(x_1, x_2, \ldots, x_n)$. At this point, if $\sigma_U$ is known, we are fairly confident that the computed interval

$$u(x_1, x_2, \ldots, x_n) - 2\sigma_U \quad \text{to} \quad u(x_1, x_2, \ldots, x_n) + 2\sigma_U$$

covers $\theta$ because the probability of that occurring before the sample was taken was very high, namely 0.95. Thus we call the interval, written for simplicity as $u \pm 2\sigma_U$, an approximate **95 percent confidence interval** for $\theta$ and the 95 percent is the **confidence coefficient**.

Now if $\sigma_U$ is unknown and depends upon parameters itself, we estimate those parameters appropriately and obtain an estimate of $\sigma_U$, say $\tilde{\sigma}_U$. For illustration, in Section 2.5, with $Y$ being $b(n, p = \theta)$, we noted that the standard error

$$\tilde{\sigma}_{Y/n} = \sqrt{\frac{(Y/n)(1 - Y/n)}{n}}$$

was an estimate of $\sqrt{\theta(1 - \theta)/n}$, the standard deviation of $Y/n$. Hence we used

$$\frac{y}{n} \pm 2\sqrt{\frac{(y/n)(1 - y/n)}{n}}$$

as an approximate 95 percent confidence interval for $p = \theta$.

The other 95 percent confidence interval given in Section 2.5 was for $\mu$, namely

$$\bar{x} \pm 2\left(\frac{s}{\sqrt{n}}\right),$$

where the standard error $s/\sqrt{n}$ is an estimate of the standard deviation, $\sigma/\sqrt{n}$, of $\bar{X}$. Clearly, by using values other than 2 from the normal table, we may obtain an 80, a 90, a 99, or a 99.73 percent confidence interval for the unknown parameter by using 1.282, 1.645, 2.576, or 3, respectively.

At this point, there is always a question about how large $n$ must be to have these intervals be fairly good in maintaining that advertised percentage. There is no easy answer, because it depends upon the situation. For illustration, let us say with a continuous-type symmetric distribution with single mode and the variance known, $n$ can be quite small, maybe as little as $n = 4$ or 5, for a confidence interval for $\mu$. As the distribution becomes skewed, more observations are needed. As a matter of fact, with the standard deviation being estimated by the standard error, we would like $n$ to be at least 30 and much more if the underlying distribution is badly skewed. Moreover, with discrete distributions, like the binomial, $n = 50$ or more is not unreasonable, particularly if $p$ is not close to 1/2. We can think of highly skewed discrete distributions in which we would prefer the sample size to be several hundred. With these warnings, a rough rule would be to have $n$ at least 30, and we are more comfortable with 50 without additional assumptions. Now, of course, one can always give an interval like $\bar{x} \pm 2s/\sqrt{n}$ as a confidence interval for $\mu$ as long as it is noted that if $n$ is not large enough, this may have a confidence coefficient much different from 95 percent. Most of the time these confidence coefficients would be in the 90s, but we have seen them as low as 65 percent.

## EXERCISES 3.5

**3.5-1** Find the maximum likelihood estimates for $\theta_1 = \mu$ and $\theta_2 = \sigma^2$ if a random sample of size 15 from $N(\mu, \sigma^2)$ yielded the following values:

| | | | | |
|---|---|---|---|---|
| 31.5 | 36.9 | 33.8 | 30.1 | 33.9 |
| 35.2 | 29.6 | 34.4 | 30.5 | 34.2 |
| 31.6 | 36.7 | 35.8 | 34.5 | 32.7 |

**3.5-2** A random sample $X_1, X_2, \ldots, X_n$ of size $n$ is taken from $N(\mu, \sigma^2)$, where the variance $\theta = \sigma^2$ is such that $0 < \theta < \infty$ and $\mu$ is a known real number. Show that the maximum likelihood estimator for $\theta$ is $\hat{\theta} = (1/n) \sum_{i=1}^{n} (X_i - \mu)^2$ and that this estimator is an unbiased estimator of $\theta$.

**3.5-3** Let $f(x; \theta) = \theta x^{\theta-1}$, $0 < x < 1$, $\theta \in \Omega = \{\theta : 0 < \theta < \infty\}$. Let $X_1, X_2, \ldots, X_n$ denote a random sample of size $n$ from this distribution.

  **(a)** Sketch the p.d.f. of $X$ for **(i)** $\theta = 1/2$, **(ii)** $\theta = 1$, and **(iii)** $\theta = 2$.

  **(b)** Show that $\hat{\theta} = -n/\ln\left(\prod_{i=1}^{n} X_i\right)$ is the maximum likelihood estimator of $\theta$.

  **(c)** For each of the following three sets of 10 observations from this distribution, calculate the values of the maximum likelihood estimate and the method of moments estimate for $\theta$.

| | | | | | |
|---|---|---|---|---|---|
| **(i)** | 0.0256 | 0.3051 | 0.0278 | 0.8971 | 0.0739 |
| | 0.3191 | 0.7379 | 0.3671 | 0.9763 | 0.0102 |
| **(ii)** | 0.9960 | 0.3125 | 0.4374 | 0.7464 | 0.8278 |
| | 0.9518 | 0.9924 | 0.7112 | 0.2228 | 0.8609 |
| **(iii)** | 0.4698 | 0.3675 | 0.5991 | 0.9513 | 0.6049 |
| | 0.9917 | 0.1551 | 0.0710 | 0.2110 | 0.2154 |

**3.5-4** Let $f(x; \theta) = (1/\theta)x^{(1-\theta)/\theta}$, $0 < x < 1$, $0 < \theta < \infty$.

  **(a)** Show that the maximum likelihood estimator of $\theta$ is $\hat{\theta} = -(1/n) \sum_{i=1}^{n} \ln X_i$.

  **(b)** Show that $E(\hat{\theta}) = \theta$ and thus $\hat{\theta}$ is an unbiased estimator of $\theta$.

**3.5-5** Let $X_1, X_2, \ldots, X_n$ be a random sample from a distribution with p.d.f.

$$f(x; \theta) = e^{-(x-\theta)}, \qquad \theta < x < \infty,$$

where $-\infty < \theta < \infty$. Find the maximum likelihood estimator of $\theta$. HINT: This is not a regular case of estimation

**3.5-6** Let $X_1, X_2, \ldots, X_n$ be a random sample from a distribution with distribution function

$$F(x) = \begin{cases} 0, & x \leq 1, \\ 1 - \left(\dfrac{1}{x}\right)^{\theta}, & 1 < x < \infty, \end{cases}$$

where $\theta > 0$. Find the maximum likelihood estimator of $\theta$.

**3.5-7** The sample mean $\overline{X}$ is the maximum likelihood estimator of $\theta$ if the underlying p.d.f. is $f(x; \theta) = (1/\theta)e^{-x/\theta}$, $0 < x < \infty$, where $\theta > 0$. For this distribution, $E(\overline{X}) = \theta$ and $\text{Var}(\overline{X}) = \theta^2/n$, where $n$ is the sample size. Use the fact that $\overline{X}$ is approximately $N(\theta, \theta^2/n)$ to construct an approximate 95 percent confidence interval for $\theta$.

**3.5-8** Let $\overline{X}$ be the mean of a random sample of size $n$ from a distribution with unknown mean $\mu$ and known variance $\sigma^2$. Then $\bar{x} \pm 2\sigma/\sqrt{n}$ can serve as an approximate 95 percent confidence interval for $\mu$. If $\sigma = 4$, find $n$ so that this confidence interval is $\bar{x} \pm 0.5$.

## 3.6 THE CENTRAL LIMIT THEOREM

In Section 2.5 we found that the mean $\overline{X}$ of a random sample of size $n$ from a discrete distribution with mean $\mu$ and variance $\sigma^2 > 0$ is a random variable with the properties that

$$E(\overline{X}) = \mu \qquad \text{and} \qquad \text{Var}(\overline{X}) = \frac{\sigma^2}{n}.$$

Thus, as $n$ increases, the variance of $\overline{X}$ decreases. Consequently, the distribution of $\overline{X}$ clearly depends on $n$, and we see that we are dealing with sequences of distributions. Also note that the probability becomes concentrated in a small interval centered at $\mu$. That is, as $n$ increases, $\overline{X}$ tends to converge to $\mu$, or $(\overline{X} - \mu)$ tends to converge to 0 in a probability sense. When sampling from a normal distribution, we will show, in Chapter 6, that the distribution of $\overline{X}$ is $N(\mu, \sigma^2/n)$.

In general, if we let

$$W = \frac{\sqrt{n}}{\sigma}(\overline{X} - \mu) = \frac{\overline{X} - \mu}{\sigma/\sqrt{n}} = \frac{Y - n\mu}{\sqrt{n}\,\sigma},$$

where $Y$ is the sum of a random sample of size $n$ from some distribution with mean $\mu$ and variance $\sigma^2$, then for each positive integer $n$,

$$E(W) = E\left[\frac{\overline{X} - \mu}{\sigma/\sqrt{n}}\right] = \frac{E(\overline{X}) - \mu}{\sigma/\sqrt{n}} = \frac{\mu - \mu}{\sigma/\sqrt{n}} = 0$$

and

$$\text{Var}(W) = E(W^2) = E\left[\frac{(\overline{X} - \mu)^2}{\sigma^2/n}\right] = \frac{E[(\overline{X} - \mu)^2]}{\sigma^2/n} = \frac{\sigma^2/n}{\sigma^2/n} = 1.$$

Thus, while $\overline{X} - \mu$ "degenerates" to zero, the factor $\sqrt{n}/\sigma$ in $\sqrt{n}(\overline{X} - \mu)/\sigma$ "spreads out" the probability enough to prevent this degeneration. What then is the distribution of $W$ as $n$ increases? One observation that might shed some light on the answer to this question can be made immediately. If the sample arises from a normal distribution, we show in Example 6.2-4 that $\overline{X}$ is $N(\mu, \sigma^2/n)$ and hence $W$ is $N(0, 1)$ for each positive $n$. Thus, in the limit, the distribution of $W$ must be $N(0, 1)$. So if the solution of the question does not depend on the underlying distribution (i.e., it is unique), the answer must be $N(0, 1)$. As we will see, this is exactly the case, and this result is so important it is called the Central Limit Theorem, the proof of which is given in Section 6.3 .

| Theorem 3.6-1 | **(Central Limit Theorem)** If $\overline{X}$ is the mean of a random sample $X_1, X_2, \ldots, X_n$ of size $n$ from a distribution with a finite mean $\mu$ and a finite positive variance $\sigma^2$, then the distribution of |

$$W = \frac{\overline{X} - \mu}{\sigma/\sqrt{n}} = \frac{\sum_{i=1}^{n} X_i - n\mu}{\sqrt{n}\,\sigma} = \frac{Y - n\mu}{\sqrt{n}\,\sigma},$$

where $Y = \sum_{i=1}^{n} X_i$, is $N(0,1)$ in the limit as $n \to \infty$.

A practical use of the Central Limit Theorem is approximating, when $n$ is "sufficiently large," the distribution function of $W$, namely

$$P(W \le w) \approx \int_{-\infty}^{w} \frac{1}{\sqrt{2\pi}}\, e^{-z^2/2}\, dz = \Phi(w).$$

We present some illustrations of this application, discuss "sufficiently large," and try to give an intuitive feeling for the Central Limit Theorem.

**EXAMPLE 3.6-1**   Let $\overline{X}$ be the mean of a random sample of $n = 25$ currents (in milliamperes) in a strip of wire in which each measurement has a mean of 15 and a variance of 4. Then $\overline{X}$ has an approximate $N(15, 4/25)$ distribution. For illustration,

$$P(14.4 < \overline{X} < 15.6) = P\left(\frac{14.4 - 15}{0.4} < \frac{\overline{X} - 15}{0.4} < \frac{15.6 - 15}{0.4}\right)$$

$$\approx \Phi(1.5) - \Phi(-1.5) = 0.9332 - 0.0668 = 0.8664.$$

**EXAMPLE 3.6-2**   Let $X_1, X_2, \ldots, X_{20}$ denote a random sample of size 20 from the uniform distribution $U(0,1)$. Here $E(X_i) = 1/2$ and $\text{Var}(X_i) = 1/12$, for $i = 1, 2, \ldots, 20$. If $Y = X_1 + X_2 + \cdots + X_{20}$, then $\mu_Y = 20(1/2)$ and $\sigma_Y^2 = 20(1/12)$ and

$$P(Y \le 9.1) = P\left(\frac{Y - 20(1/2)}{\sqrt{20/12}} \le \frac{9.1 - 10}{\sqrt{20/12}}\right) = P(W \le -0.697)$$

$$\approx \Phi(-0.697) = 0.2423.$$

Also,

$$P(8.5 \le Y \le 11.7) = P\left(\frac{8.5 - 10}{\sqrt{5/3}} \le \frac{Y - 10}{\sqrt{5/3}} \le \frac{11.7 - 10}{\sqrt{5/3}}\right)$$

$$= P(-1.162 \le W \le 1.317)$$

$$\approx \Phi(1.317) - \Phi(-1.162)$$

$$= 0.9061 - 0.1226 = 0.7835.$$

**EXAMPLE 3.6-3**   Let $\overline{X}$ denote the mean of a random sample of size 25 from the distribution whose p.d.f. is $f(x) = x^3/4, 0 < x < 2$. It is easy to show that $\mu = 8/5 = 1.6$ and $\sigma^2 = 8/75$.

Thus

$$P(1.5 \leq \overline{X} \leq 1.65) = P\left(\frac{1.5 - 1.6}{\sqrt{8/75}/\sqrt{25}} \leq \frac{\overline{X} - 1.6}{\sqrt{8/75}/\sqrt{25}} \leq \frac{1.65 - 1.6}{\sqrt{8/75}/\sqrt{25}}\right)$$

$$= P(-1.531 \leq W \leq 0.765)$$

$$\approx \Phi(0.765) - \Phi(-1.531)$$

$$= 0.7779 - 0.0629 = 0.7150.$$

These examples have shown how the Central Limit Theorem can be used for approximating certain probabilities concerning the mean $\overline{X}$ or the sum $Y = \sum_{i=1}^{n} X_i$ of a random sample. That is, $\overline{X}$ is approximately $N(\mu, \sigma^2/n)$, and $Y$ is approximately $N(n\mu, n\sigma^2)$ when $n$ is "sufficiently large," where $\mu$ and $\sigma^2$ are the mean and the variance of the underlying distribution from which the sample arose. Generally, if $n$ is greater than 25 or 30, these approximations will be good. However, if the underlying distribution is symmetric, unimodal, and of the continuous type, a value of $n$ as small as 4 or 5 can yield a very adequate approximation. Moreover, if the original distribution is approximately normal, $\overline{X}$ would have a distribution very close to normal when $n$ equals 2 or 3. In fact, we know that if the sample is taken from $N(\mu, \sigma^2)$, $\overline{X}$ is exactly $N(\mu, \sigma^2/n)$ for every $n = 1, 2, 3, \ldots$.

The following examples will help to illustrate the previous remarks and will give the reader a better intuitive feeling about the Central Limit Theorem. In particular, we shall see how the size of $n$ affects the distribution of $\overline{X}$ and $Y = \sum_{i=1}^{n} X_i$ for samples from several underlying distributions.

EXAMPLE 3.6-4    Let $X_1, X_2, X_3, X_4$ be a random sample of size 4 from the uniform distribution $U(0, 1)$ with p.d.f. $f(x) = 1, 0 < x < 1$. Then $\mu = 1/2$ and $\sigma^2 = 1/12$. We shall compare the graph of the p.d.f. of

$$Y = \sum_{i=1}^{n} X_i$$

with the graph of the $N[n(1/2), n(1/12)]$ p.d.f. for $n = 2$ and 4, respectively.

By methods given in Section 5.2 we can determine that the p.d.f. of $Y = X_1 + X_2$ is

$$g(y) = \begin{cases} y, & 0 < y \leq 1, \\ 2 - y, & 1 < y < 2. \end{cases}$$

This is the triangular p.d.f. that is graphed in Figure 3.6-1(a). In this figure the $N[2(1/2), 2(1/12)]$ p.d.f. is also graphed.

Moreover, the p.d.f. of $Y = X_1 + X_2 + X_3 + X_4$ is

$$g(y) = \begin{cases} \dfrac{y^3}{6}, & 0 \leq y < 1, \\[2mm] \dfrac{-3y^3 + 12y^2 - 12y + 4}{6}, & 1 \leq y < 2, \\[2mm] \dfrac{3y^3 - 24y^2 + 60y - 44}{6}, & 2 \leq y < 3, \\[2mm] \dfrac{-y^3 + 12y^2 - 48y + 64}{6}, & 3 \leq y \leq 4. \end{cases}$$

This p.d.f. is graphed in Figure 3.6-1(b) along with the $N[4(1/2), 4(1/12)]$ p.d.f. If we are interested in finding $P(1.7 \leq Y \leq 3.2)$, this could be done by evaluating

$$\int_{1.7}^{3.2} g(y)\, dy,$$

which is tedious (see Exercise 3.6-8). It is much easier to use a normal approximation, which results in a number very close to the exact value.

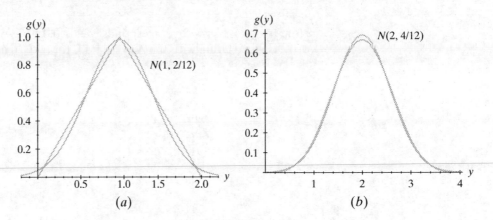

**Figure 3.6-1**  p.d.f.s of sums of uniform random variables

In Example 3.6-4 and Exercise 3.6-8 we show that even for a small value of $n$, like $n = 4$, the sum of the sample items has an approximate normal distribution. The following example illustrates that for some underlying distributions (particularly skewed ones) $n$ must be quite large to obtain a satisfactory approximation. In order to keep the scale on the horizontal axis the same for each value of $n$, we will use the following result.

Let $f(x)$ and $F(x)$ be the p.d.f. and distribution function of a random variable, $X$, of the continuous type having mean $\mu$ and variance $\sigma^2$. Let $W = (X - \mu)/\sigma$. The distribution function of $W$ is given by

$$G(w) = P(W \leq w) = P\left(\frac{X - \mu}{\sigma} \leq w\right)$$

$$= P(X \leq \sigma w + \mu) = F(\sigma w + \mu).$$

Thus the p.d.f. of $W$ is given by the derivative of $F(\sigma w + \mu)$, namely

$$g(w) = \sigma f(\sigma w + \mu).$$

**EXAMPLE 3.6-5**   Let $X_1, X_2, \ldots, X_n$ be a random sample of size $n$ from a chi-square distribution with one degree of freedom. If

$$Y = \sum_{i=1}^{n} X_i,$$

then we show (Chapter 6 ) that $Y$ is $\chi^2(n)$, and $E(Y) = n$, $\text{Var}(Y) = 2n$. Let

$$W = \frac{Y - n}{\sqrt{2n}}.$$

The p.d.f. of $W$ is given by

$$g(w) = \sqrt{2n}\,\frac{(\sqrt{2n}\,w + n)^{n/2-1}}{\Gamma\!\left(\dfrac{n}{2}\right)2^{n/2}}\,e^{-(\sqrt{2n}\,w+n)/2}, \qquad -n/\sqrt{2n} < w < \infty.$$

Note that $w > -n/\sqrt{2n}$ corresponds to $y > 0$. In Figure 3.6-2(a) and (b), the graph of $W$ is given along with the $N(0,1)$ p.d.f. for $n = 20$ and $100$, respectively.

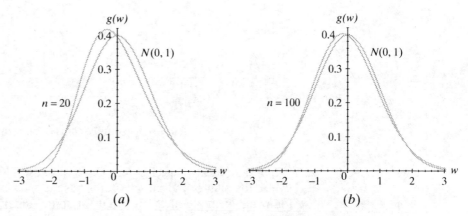

**Figure 3.6-2** p.d.f.s of $W = (Y - n)/\sqrt{2n}$

So far all the illustrations have concerned distributions of the continuous type. However, the hypotheses for the Central Limit Theorem do not require the distribution to be continuous. We shall consider applications of the Central Limit Theorem for discrete-type distributions in the next section.

**EXERCISES 3.6**

**3.6-1** Let $\overline{X}$ be the mean of a random sample of size 12 from the uniform distribution on the interval $(0, 1)$. Approximate $P(1/2 \le \overline{X} \le 2/3)$.

**3.6-2** Let $Y = X_1 + X_2 + \cdots + X_{15}$ be the sum of a random sample of size 15 from the distribution whose p.d.f. is $f(x) = (3/2)x^2$, $-1 < x < 1$. Approximate

$$P(-0.3 \le Y \le 1.5).$$

**3.6-3** Let $\overline{X}$ be the mean of a random sample of size 36 from an exponential distribution with mean 3. Approximate $P(2.5 \le \overline{X} \le 4)$.

**3.6-4** A random sample of size $n = 18$ is taken from the distribution with p.d.f $f(x) = 1 - x/2$, $0 \le x \le 2$.

  **(a)** Find $\mu$ and $\sigma^2$.   **(b)** Find, approximately, $P(2/3 \le \overline{X} \le 5/6)$.

**3.6-5** Let $X$ equal the maximal oxygen intake of a human on a treadmill, where the measurements are in milliliters of oxygen per minute per kilogram of weight.

Assume that for a particular population the mean of $X$ is $\mu = 54.030$ and the standard deviation is $\sigma = 5.8$. Let $\overline{X}$ be the sample mean of a random sample of size $n = 47$. Find $P(52.761 \leq \overline{X} \leq 54.453)$, approximately.

**3.6-6** Let $X$ equal the weight in grams of a miniature candy bar. Assume that $\mu = E(X) = 24.43$ and $\sigma^2 = \text{Var}(X) = 2.20$. Let $\overline{X}$ be the sample mean of a random sample of $n = 30$ candy bars. Find

(a) $E(\overline{X})$.   (b) $\text{Var}(\overline{X})$.   (c) $P(24.17 \leq \overline{X} \leq 24.82)$, approximately.

**3.6-7** Let $X$ equal the birth weight in grams of a baby born in the Sudan. Assume that $E(X) = 3320$ and $\text{Var}(X) = 660^2$. Let $\overline{X}$ be the sample mean of a random sample of size $n = 225$. Find $P(3233.76 \leq \overline{X} \leq 3406.24)$, approximately.

**3.6-8** In Example 3.6-4, with $n = 4$, compute $P(1.7 \leq Y \leq 3.2)$ and compare this answer with the normal approximation of this probability.

**3.6-9** Five measurements in "ohms per square" of the electronic conductive coating (that allows light to pass through it) on a thin, clear piece of glass were made and the average was calculated. This was repeated 150 times, yielding the following averages:

| | | | | | | | | | |
|---|---|---|---|---|---|---|---|---|---|
| 83.86 | 75.86 | 79.65 | 90.57 | 95.37 | 97.97 | 77.00 | 80.80 | 83.53 | 83.17 |
| 80.20 | 81.42 | 89.14 | 88.68 | 85.15 | 90.11 | 89.03 | 85.00 | 82.57 | 79.46 |
| 81.20 | 82.80 | 81.28 | 74.64 | 69.85 | 75.60 | 76.60 | 78.30 | 88.51 | 86.32 |
| 84.03 | 95.27 | 90.05 | 77.50 | 75.14 | 81.33 | 86.12 | 78.30 | 80.30 | 84.96 |
| 79.30 | 88.96 | 82.76 | 83.13 | 79.60 | 86.20 | 85.16 | 87.86 | 91.00 | 91.10 |
| 84.04 | 92.10 | 79.20 | 83.76 | 87.87 | 86.71 | 81.89 | 85.72 | 75.84 | 74.27 |
| 93.08 | 81.75 | 75.66 | 75.35 | 76.55 | 84.86 | 90.68 | 91.02 | 90.97 | 98.30 |
| 91.84 | 97.41 | 73.60 | 90.65 | 80.20 | 74.75 | 90.35 | 79.66 | 86.88 | 83.00 |
| 86.24 | 80.50 | 74.25 | 91.20 | 70.16 | 78.40 | 85.60 | 80.82 | 75.95 | 80.75 |
| 81.86 | 82.18 | 82.98 | 84.00 | 76.85 | 85.00 | 79.50 | 86.56 | 83.30 | 72.40 |
| 79.20 | 86.20 | 82.36 | 84.08 | 86.11 | 88.25 | 88.93 | 93.12 | 78.30 | 77.24 |
| 82.52 | 81.37 | 83.72 | 86.90 | 84.37 | 92.60 | 95.01 | 78.95 | 81.40 | 88.40 |
| 76.10 | 85.33 | 82.95 | 80.20 | 88.21 | 83.49 | 81.00 | 82.72 | 81.12 | 83.62 |
| 91.18 | 85.90 | 79.01 | 77.56 | 81.13 | 80.60 | 81.65 | 70.70 | 69.36 | 79.09 |
| 71.35 | 67.20 | 67.43 | 69.95 | 66.76 | 76.35 | 69.45 | 80.13 | 84.26 | 88.13 |

(a) Construct a frequency table for these 150 observations using 10 intervals of equal length, and using $(66.495, 69.695)$ for the first class interval. Which is the modal class?

(b) Construct a relative frequency histogram for the grouped data.

(c) Superimpose a normal p.d.f. using the mean 82.661 and variance 42.2279 of these 150 values as the true mean and variance. Do these 150 means of five measurements seem to be normally distributed?

**3.6-10** A church has pledges (in dollars) with a mean of 2000 and a standard deviation of 500. A random sample of size $n = 25$ is taken from all of this church's pledges and the sample mean $\overline{X}$ is considered. Approximate $P(\overline{X} > 2050)$.

**3.6-11** The tensile strength $X$ of paper has $\mu = 30$ and $\sigma = 3$ (pounds per square inch). A random sample of size $n = 100$ is taken from the distribution of tensile strengths. Compute the probability that the sample mean $\overline{X}$ is greater than 29.5 pounds per square inch.

**3.6-12** At certain times during the year, a bus company runs a special van holding ten passengers from Iowa City to Chicago. After the opening of sales of the tickets, the time (in minutes) between sales of tickets for the trip has a gamma distribution with $\alpha = 3$ and $\theta = 2$. Approximate the probability of being sold out within one hour.

**3.6-13** Let $X_1, X_2, X_3, X_4$ represent the random times in days needed to complete four steps of a project. These times are independent and have gamma distributions with common $\theta = 2$ and common $\alpha = 4$. One step must be completed before the next can be started. Let $Y$ equal the total time needed to complete the project. Approximate $P(Y \leq 35)$. Does it seem appropriate to use the Central Limit Theorem when $n = 4$ in this exercise?

**3.6-14** Assume that the sick leave taken by the typical worker per year has $\mu = 10$, $\sigma = 2$, measured in days. A firm has $n = 20$ employees. Assuming independence, how many sick days should the firm budget if the financial officer wants the probability of exceeding the budgeted days to be less than 20%?

## 3.7 APPROXIMATIONS FOR DISCRETE DISTRIBUTIONS

In this section we illustrate how the normal distribution can be used to approximate probabilities for certain discrete-type distributions. One of the more important discrete distributions is the binomial distribution. To see how the Central Limit Theorem can be applied, recall that a binomial random variable can be described as the sum of Bernoulli random variables. That is, let $X_1, X_2, \ldots, X_n$ be a random sample from a Bernoulli distribution with a mean $\mu = p$ and a variance $\sigma^2 = p(1-p)$, where $0 < p < 1$. Then $Y = \sum_{i=1}^{n} X_i$ is $b(n, p)$. The Central Limit Theorem states that the distribution of

$$W = \frac{Y - np}{\sqrt{np(1-p)}} = \frac{\overline{X} - p}{\sqrt{p(1-p)/n}}$$

is $N(0, 1)$ in the limit as $n \to \infty$. Thus, if $n$ is sufficiently large, the distribution of $Y$ is approximately $N[np, np(1-p)]$, and probabilities for the binomial distribution $b(n, p)$ can be approximated using this normal distribution. A rule often stated is that $n$ is "sufficiently large" if $np \geq 5$ and $n(1-p) \geq 5$. This can be used as a rough guide although as $p$ deviates more and more from 0.5, we need larger and larger sample sizes, because the underlying Bernoulli distribution becomes more skewed.

Note that we shall be approximating probabilities from a discrete distribution with probabilities for a continuous distribution. Let us discuss a reasonable procedure in this situation. Let $Y$ be $b(n, p)$. Recall that the probability

$$P(Y = k) = f(k) = \frac{n!}{k!(n-k)!} p^k (1-p)^{n-k}, \qquad k = 0, 1, 2, \ldots, n$$

can be represented graphically as the area of a rectangle with a base of length one centered at $k$ (that is, the base goes from $k - 1/2$ to $k + 1/2$) and a height equal to $P(X = k)$. The collection of these rectangles for $k = 0, 1, \ldots, n$ is the probability histogram for $Y$. Figures 3.7-1($a$) and ($b$) show the graphs of the probability histograms for the $b(10, 1/2)$ and $b(18, 1/6)$ distributions, respectively. When using the normal distribution to approximate probabilities for the binomial distribution, areas under the p.d.f. for the normal distribution will be used to approximate areas of rectangles in the probability histogram for the binomial distribution. That is, the area under the $N[np, np(1-p)]$ p.d.f. between $k - 1/2$ and $k + 1/2$ will be used to approximate $P(Y = k)$.

**EXAMPLE 3.7-1**    Let $Y$ be $b(10, 1/2)$. Then, using the Central Limit Theorem, $P(a < Y < b)$ can be approximated using the normal distribution with mean $10(1/2) = 5$ and variance $10(1/2)(1/2) = 5/2$. Figure 3.7-1($a$) shows the graph of the probability histogram for $b(10, 1/2)$ and the graph of the p.d.f. of the normal distribution $N(5, 5/2)$. Note that the area of the rectangle whose base is

$$\left( k - \frac{1}{2}, k + \frac{1}{2} \right)$$

and the area under the normal curve between $k - 1/2$ and $k + 1/2$ are approximately equal for each integer $k$.

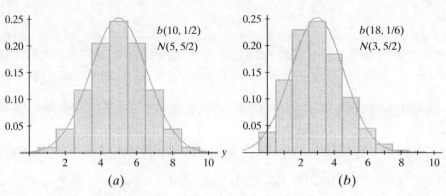

Figure 3.7-1 Normal approximation for the binomial distribution

EXAMPLE 3.7-2 Let $Y$ be $b(18, 1/6)$. Because $np = 18(1/6) = 3 < 5$, the normal approximation is not as good here. Figure 3.7-1($b$) illustrates this by depicting the skewed probability histogram for $b(18, 1/6)$ and the symmetric p.d.f. of the normal distribution $N(3, 5/2)$.

EXAMPLE 3.7-3 Let $Y$ have the binomial distribution of Example 3.7-1 and Figure 3.7-1($a$), namely $b(10, 1/2)$. Then

$$P(3 \le Y < 6) = P(2.5 \le Y \le 5.5)$$

because $P(Y = 6)$ is not in the desired answer. But the latter equals

$$P\left(\frac{2.5 - 5}{\sqrt{10/4}} \le \frac{Y - 5}{\sqrt{10/4}} \le \frac{5.5 - 5}{\sqrt{10/4}}\right) \approx \Phi(0.316) - \Phi(-1.581)$$

$$= 0.6240 - 0.0570 = 0.5670.$$

Using Appendix Table II, we find that $P(3 \le Y < 6) = 0.5683$.

EXAMPLE 3.7-4 Let $Y$ be $b(36, 1/2)$. Then, since

$$\mu = (36)(1/2) = 18 \quad \text{and} \quad \sigma^2 = (36)(1/2)(1/2) = 9,$$

$$P(12 < Y \le 18) = P(12.5 \le Y \le 18.5)$$

$$= P\left(\frac{12.5 - 18}{\sqrt{9}} \le \frac{Y - 18}{\sqrt{9}} \le \frac{18.5 - 18}{\sqrt{9}}\right)$$

$$\approx \Phi(0.167) - \Phi(-1.833)$$

$$= 0.5329.$$

Note that 12 was increased to 12.5 because $P(Y = 12)$ is not included in the desired probability. Using the binomial formula, we find that (you may verify this answer using your calculator)

$$P(12 < Y \le 18) = P(13 \le Y \le 18) = 0.5334.$$

Also,

$$P(Y = 20) = P(19.5 \leq Y \leq 20.5)$$

$$= P\left(\frac{19.5 - 18}{\sqrt{9}} \leq \frac{Y - 18}{\sqrt{9}} \leq \frac{20.5 - 18}{\sqrt{9}}\right)$$

$$\approx \Phi(0.833) - \Phi(0.5)$$

$$= 0.1060.$$

Using the binomial formula, we have $P(Y = 20) = 0.1063$. So, in this situation, the approximation is extremely good.

Note that, in general, if $Y$ is $b(n, p)$, $n$ is large, and $k = 0, 1, \ldots, n$,

$$P(Y \leq k) \approx \Phi\left(\frac{k + 1/2 - np}{\sqrt{npq}}\right)$$

and

$$P(Y < k) \approx \Phi\left(\frac{k - 1/2 - np}{\sqrt{npq}}\right),$$

because in the first case $k$ is included and in the second it is not.

We now show how the Poisson distribution with large enough mean can be approximated using a normal distribution.

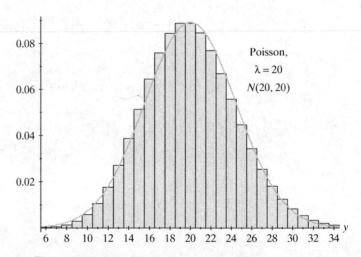

**Figure 3.7-2** Normal approximation of Poisson, $\lambda = 20$

EXAMPLE 3.7-5   A random variable having a Poisson distribution with mean 20 can be thought of as the sum $Y$ of the observations of a random sample of size 20 from a Poisson distribution with mean 1 (see Chapter 6 ). Thus

$$W = \frac{Y - 20}{\sqrt{20}}$$

has a distribution that is approximately $N(0,1)$, and the distribution of $Y$ is approximately $N(20, 20)$ (see Figure 3.7-2). So, for illustration,

$$P(16 < Y \le 21) = P(16.5 \le Y \le 21.5)$$

$$= P\left( \frac{16.5 - 20}{\sqrt{20}} \le \frac{Y - 20}{\sqrt{20}} \le \frac{21.5 - 20}{\sqrt{20}} \right)$$

$$\approx \Phi(0.335) - \Phi(-0.783) = 0.4142.$$

Note that 16 is increased to 16.5 because $Y = 16$ is not included in the event $\{16 < Y \le 21\}$. The answer using the Poisson formula is 0.4226.

In general, if $Y$ has a Poisson distribution with mean $\lambda$, then the distribution of

$$W = \frac{Y - \lambda}{\sqrt{\lambda}}$$

is approximately $N(0,1)$ when $\lambda$ is sufficiently large.

## EXERCISES 3.7

**3.7-1** Let the distribution of $Y$ be $b(25, 1/2)$. Find the following probabilities in two ways: exactly using Appendix Table II and approximately using the Central Limit Theorem. Compare the two results in each of the three cases.

(a) $P(10 < Y \le 12)$.   (b) $P(12 \le Y < 15)$.   (c) $P(Y = 12)$.

**3.7-2** Among the gifted 7th-graders who score very high on a mathematics exam, approximately 20% are left-handed or ambidextrous. Let $X$ equal the number of left-handed or ambidextrous students among a random sample of $n = 25$ gifted 7th-graders. Find $P(2 < X < 9)$

(a) Using Appendix Table II.

(b) Approximately, using the Central Limit Theorem.

**3.7-3** A public opinion poll in Southern California was conducted to determine whether Southern Californians are prepared for the big earthquake that experts predict will devastate the region sometime in the next 50 years. It was learned that "60% have not secured objects in their homes that might fall and cause injury and damage during a temblor." In a random sample of $n = 864$ Southern Californians, let $X$ equal the number who "have not secured objects in their homes." Find $P(496 \le X \le 548)$, approximately.

**3.7-4** Let $X$ equal the number out of $n = 48$ mature aster seeds that will germinate when $p = 0.75$ is the probability that a particular seed germinates. Approximate $P(35 \le X \le 40)$.

**3.7-5** Let $p$ equal the proportion of all college students who would say yes to the question, "Would you drink from the same glass as your friend if you suspected that this friend were an AIDS virus carrier?" Assume that $p = 0.10$. Let $X$ equal the number of students out of a random sample of size $n = 100$ who would say yes to this question. Approximate $P(X \le 11)$.

> **Remark**   The value used for $p$ was based on a poll conducted in a class at San Diego State University. It would be interesting for you to conduct a survey at your college to estimate the value of $p$.

**3.7-6** In adults the pneumococcus bacterium causes 70% of pneumonia cases. In a random sample of $n = 84$ adults who have pneumonia, let $X$ equal the number whose pneumonia was caused by the pneumococcus bacterium. Use the normal distribution to find, approximately, $P(X \le 52)$.

**3.7-7** Let $X_1, X_2, \ldots, X_{48}$ be a random sample of size 48 from the distribution with p.d.f. $f(x) = 1/x^2$, $1 < x < \infty$. Approximate the probability that at most 10 of these random variables have values greater than 4. HINT: Let the $i$th trial be a success if $X_i > 4$, $i = 1, 2, \ldots, 48$ and let $Y$ equal the number of successes.

**3.7-8** A candy maker produces mints that have a label weight of 20.4 grams. Assume that the distribution of the weights of these mints is $N(21.37, 0.16)$.

  **(a)** Let $X$ denote the weight of a single mint selected at random from the production line. Find $P(X < 20.857)$.

  **(b)** During a particular shift 100 mints are selected at random and weighed. Let $Y$ equal the number of these mints that weigh less than 20.857 grams. Approximate $P(Y \leq 5)$.

  **(c)** Let $\overline{X}$ equal the sample mean of the 100 mints selected and weighed on a particular shift. Find $P(21.31 \leq \overline{X} \leq 21.39)$.

**3.7-9** Let $X$ equal the number of alpha particles emitted by barium-133 per second and counted by a Geiger counter. Assume that $X$ has a Poisson distribution with $\lambda = 49$. Approximate $P(45 < X < 60)$.

**3.7-10** Let $X$ equal the number of alpha particles counted by a Geiger counter during 30 seconds. Assume that the distribution of $X$ is Poisson with a mean of 4829. Determine (approximately) $P(4776 \leq X \leq 4857)$.

**3.7-11** Let $X_1, X_2, \ldots, X_{30}$ be a random sample of size 30 from a Poisson distribution with a mean of 2/3. Approximate

  **(a)** $P\left(15 < \sum_{i=1}^{30} X_i \leq 22\right)$.    **(b)** $P\left(21 \leq \sum_{i=1}^{30} X_i < 27\right)$.

**3.7-12** In the casino game roulette, the probability of winning with a bet on red is $p = 18/38$. Let $Y$ equal the number of winning bets out of 1000 independent bets that are placed. Find $P(Y > 500)$ approximately.

**3.7-13** About 60% of all Americans have a sedentary life style. Select $n = 96$ Americans at random (assume independence). What is the probability that between 50 and 60, inclusive, do not exercise regularly?

**3.7-14** If $X$ is $b(100, 0.1)$, find the approximate value of $P(12 \leq X \leq 14)$ using

  **(a)** The normal approximation,

  **(b)** The Poisson approximation.

  **(c)** Find the exact probability using the binomial.

**3.7-15** Let $X_1, X_2, \ldots, X_{36}$ be a random sample of size 36 from the geometric distribution with p.m.f. $f(x) = (1/4)^{x-1}(3/4)$, $x = 1, 2, 3, \ldots$. Approximate

  **(a)** $P\left(46 \leq \sum_{i=1}^{36} X_i \leq 49\right)$.    **(b)** $P(1.25 \leq \overline{X} \leq 1.50)$.

  HINT: Observe that the distribution of the sum is of the discrete type.

**3.7-16** A die is rolled 24 independent times. Let $Y$ be the sum of the 24 resulting values. Recalling that $Y$ is a random variable of the discrete type, approximate

  **(a)** $P(Y \geq 86)$.    **(b)** $P(Y < 86)$.    **(c)** $P(70 < Y \leq 86)$.

**3.7-17** In the United States the probability that a child dies in his or her first year of life is about $p = 0.01$ (it is slightly less than this in truth). Consider a group of 5000 such infants. What is the probability that between 45 and 53 (including 45 and 53) die in the first year of life?

**3.7-18** Let $Y$ equal the sum of $n = 100$ Bernoulli trials. That is, $Y$ is $b(100, p)$. For each of **(i)** $p = 0.1$, **(ii)** $p = 0.5$, and **(iii)** $p = 0.8$,

  **(a)** Draw on the same graph the approximating normal p.d.f.s.

  **(b)** Find, approximately, $P(\,|\,Y/100 - p\,| \leq 0.015)$.

**3.7-19** The number of trees in one acre has a Poisson distribution with mean 60. Assuming independence, compute approximately $P(5950 \leq X \leq 6100)$, where $X$ is the number of trees in 100 acres.

**3.7-20** A communication system for a company has 35 outside lines. If the number $X$ of lines in use at a given time follows a Poisson distribution with mean 25, compute the probability that an incoming call cannot find an open line.

**3.7-21** The number of flaws in a certain unit of material has a Poisson distribution with the mean of two. Fifty of these units are taken at random.

   **(a)** What is the probability that the total number of flaws in the 50 is less that 110?

   **(b)** What is the probability that at least two of the 50 have more than 6 flaws?

**3.7-22** The number $X$ of flaws on a certain tape of length one yard follows a Poisson distribution with mean 0.3. We examine $n = 100$ such tapes and count the total number $Y$ of flaws. Approximate $P(Y \leq 25)$.

# CHAPTER
# THREE COMMENTS

Here we give examples of how p.d.f.s are created starting with a probability background. Let us first consider the gamma distribution.

Say we have a Poisson process (see comments at the end of Chapter 2) involving time intervals with mean $\lambda$, and let $T$ be the waiting time until the $r$th change occurs. With $t > 0$, the distribution function of $T$ is given by

$$F(t) = P(T \le t) = 1 - P(T > t) = 1 - \sum_{k=0}^{r-1} \frac{(\lambda t)^k e^{-\lambda t}}{k!},$$

since the latter summation is the probability of having less than $r$ changes in the interval $[0, t]$. The derivative of $F(t)$ is the p.d.f. of $T$, namely

$$f(t) = \lambda^r t^{r-1} e^{-\lambda t}/(r - 1)!, \qquad 0 < t < \infty.$$

This is a gamma p.d.f. with $\theta = 1/\lambda$ and $\alpha = r$. So we can think of this p.d.f. as an appropriate model for a waiting time under certain assumptions.

Of course, if $\alpha = 1$, then we obtain the exponential p.d.f.

$$f(t) = (1/\theta)e^{-t/\theta}, \qquad 0 < t < \infty.$$

It is interesting that the failure rate (hazard rate, force of mortality),

$$\frac{f(x)}{1 - F(x)} = \frac{(1/\theta)e^{-t/\theta}}{1 - (1 - e^{-t/\theta})} = \frac{1}{\theta},$$

is a constant. This means that if the length of life of people had an exponential distribution, the probability of a young man dying in the next year would be exactly the same as that of an old man. Hopefully $\theta$ would be large so that $1/\theta$ would be small, and we would have found a mathematical "fountain of youth." Unfortunately, the failure rate usually increases for products and people.

Engineers often find that the hazard (failure) rate is equal to $\alpha x^{\alpha-1}/\beta^\alpha$, $\alpha > 1$, $\beta > 0$, which, by solving a simple differential equation, results in the **Weibull** p.d.f.

$$f(x) = \frac{\alpha x^{\alpha-1}}{\beta^\alpha} e^{-(x/\beta)^\alpha}, \qquad 0 < x < \infty.$$

A value of $\alpha$ around three frequently provides a good model for the life of many engineering products. On the other hand, actuaries find that the force of mortality of humans is exponential, like $ae^{bx}$, $a > 0$, $b > 0$, which gives the **Gompertz** p.d.f.

$$f(x) = ab^{bx}e^{-(1/b)(e^{bx}-1)}, \qquad 0 < x < \infty.$$

For illustration, they say that the force of mortality of males often increases about 10 percent per year, meaning the probability of a male dying within one year doubles about every $72/10 = 7.2$ years, by the "Rule of 72." Thank goodness the force of mortality is extremely low for a five-year old boy, possibly about 0.00004; that is, the probability that such a boy will die in the next year is about 4 out of 100,000. This, however, is much higher than the probability winning "power ball."

# CHAPTER

# 4

# APPLICATIONS OF STATISTICAL INFERENCE

## 4.1 SUMMARY OF NECESSARY THEORETICAL RESULTS

In order to carry out many more statistical inferences other than the simple confidence intervals that we have considered, we need some additional theoretical results which we give in this section. Some of these will be proved in Chapter 6; so we state a few of them as theorems and we give an example of each.

**Remark** If the instructor so chooses, Section 6.2 on the moment-generating function (m.g.f.) of linear functions can be studied at this time.

### Theorem 4.1-1

Let $X_1, X_2, \ldots, X_n$ be $n$ independent chi-square random variables with $r_1, r_2, \ldots, r_n$ degrees of freedom, respectively. Then the random variable $Y = X_1 + X_2 + \cdots + X_n$ has a chi-square distribution with $r_1 + r_2 + \cdots + r_n$ degrees of freedom.

### EXAMPLE 4.1-1

Let $X_1, X_2, X_3$ be three independent chi-square random variables with $r_1 = 2$, $r_2 = 5$, $r_3 = 4$ degrees of freedom. Then the random variable $Y = X_1 + X_2 + X_3$ is $\chi^2(11)$ and $P(Y < 19.68) = 0.95$ from Table IV.

### Theorem 4.1-2

If $X_1, X_2, \ldots, X_n$ are independent normal random variables with means $\mu_1, \mu_2, \ldots, \mu_n$ and variances $\sigma_1^2, \sigma_2^2, \ldots, \sigma_n^2$, respectively, then the random variable $Y = \sum_{i=1}^{n} a_i X_i$ has a normal distribution with mean $\mu_Y = \sum_{i=1}^{n} a_i \mu_i$ and variance $\sigma_Y^2 = \sum_{i=1}^{n} a_i^2 \sigma_i^2$.

EXAMPLE 4.1-2

Let $X_1, X_2, X_3$ be three independent normal random variables with means $\mu_1 = 4$, $\mu_2 = -5$, $\mu_3 = 2$ and variances $\sigma_1^2 = 4$, $\sigma_2^2 = 16$, $\sigma_3^2 = 9$, respectively. Then the random variable $Y = X_1 - X_2$ is normal with $\mu_Y = (1)(4) + (-1)(-5) = 9$ and $\sigma_Y^2 = (1)^2(4) + (-1)^2(16) = 20$. Moreover, the random variable $Z = X_1 + X_2 + X_3$ is normal with mean $\mu_Z = 4 + (-5) + 2 = 1$ and variance $\sigma_Z^2 = 4 + 16 + 9 = 29x$.

---

**Theorem 4.1-3**

Let $\overline{X}$ and $S^2$ be the mean and the variance of a random sample of size $n$ from a distribution which is $N(\mu, \sigma^2)$. Then $\overline{X}$ and $S^2$ are independent random variables with distributions such that

$$\overline{X} \quad \text{is} \quad N(\mu, \sigma^2/n)$$

and

$$\frac{(n-1)S^2}{\sigma^2} \quad \text{is} \quad \chi^2(n-1).$$

---

EXAMPLE 4.1-3

Let $X_1, X_2, \ldots, X_n$ be a random sample from a distribution which is $N(\mu, \sigma^2)$. We know from Section 3.4 that $(X_i - \mu)^2/\sigma^2$ is $\chi^2(1)$, $i = 1, 2, \ldots, n$. From Theorem 4.1-1, it must be true that $\sum_{i=1}^{n}(X_i - \mu)^2/\sigma^2$ is $\chi^2(n)$ as we are adding $n$ independent chi-square variables together. From Theorem 4.1-3, note that if we replace $\mu$ by its estimator $\overline{X}$ we obtain $(n-1)S^2/\sigma^2 = \sum_{i=1}^{n}(X_i - \overline{X})/\sigma^2$, which is $\chi^2(n-1)$. That is, we have lost one degree of freedom by replacing a parameter in a chi-square variable by its estimator. Later, we see that this is generalized: If $p$ parameters in a $\chi^2(r)$ random variable, $p < r$, are replaced by "reasonable" estimators, the resulting chi-square variable has $r - p$ degrees of freedom.

Before we close this section on results needed to make a number of statistical inferences, we consider two important distributions. We do not try to develop them here, but only give the definitions of these two random variables and refer the reader to Tables VI and VII in which a few probabilities concerning these $T$ and $F$ random variables are given. Also see Examples 5.2-12 and 5.2-13 for the p.d.f.s and some graphs for the $T$ and $F$ distributions.

Let

$$T = \frac{Z}{\sqrt{U/r}},$$

where $Z$ has a $N(0, 1)$ distribution, $U$ has a $\chi^2(r)$ distribution, and $Z$ and $U$ are independent. We say that $T$ has a Student's $t$ distribution with $r$ degrees of freedom. For illustration, if $r = 10$, $P[T \leq 2.764] = 0.99$ from Table VI in the Appendix. In general, from Table VI, we can find $t_\alpha(r)$ so that $P[T > t_\alpha(r)] = \alpha$.

EXAMPLE 4.1-4

(continuation of Example 4.1-3) With Theorem 4.1-3, $\overline{X}$ has a $N(\mu, \sigma^2/n)$ distribution so that

$$\frac{\overline{X} - \mu}{\sigma/\sqrt{n}} \quad \text{is} \quad N(0, 1).$$

Of course $(n-1)S^2/\sigma^2$ is $\chi^2(n-1)$ and $\overline{X}$ and $S^2$ are independent. Hence

$$T = \frac{\dfrac{\overline{X} - \mu}{\sigma/\sqrt{n}}}{\sqrt{\dfrac{(n-1)S^2}{\sigma^2}\Big/(n-1)}} = \frac{\overline{X} - \mu}{S/\sqrt{n}}$$

has a Student's $t$ distribution with $n-1$ degree of freedom.

Let

$$F = \frac{U_1/r_1}{U_2/r_2},$$

where $U_i$ is $\chi^2(r_i)$, $i = 1, 2$, and $U_1$ and $U_2$ are independent. Then we say that $F$ has a Fisher's $F$ distribution (some call it Snedecor's $F$) with $r_1$ and $r_2$ degrees of freedom, denoted by $F(r_1, r_2)$. For illustration, if $r_1 = 5$ and $r_2 = 10$, then $P(F \leq 4.24) = 0.975$ and $P(F > 5.64) = 0.01$ from Table VII in the Appendix. In general from that table, we can find $F_\alpha$ so that $P(F > F_\alpha) = \alpha$ for selected values of $\alpha$.

**EXAMPLE 4.1-5**    Let $X_1, X_2, \ldots, X_n$ and $Y_1, Y_2, \ldots, Y_m$ be independent random samples from $N(\mu_X, \sigma_X^2)$ and $N(\mu_Y, \sigma_Y^2)$, respectively. We know, from Theorem 4.1-3, that $(n-1)S_X^2/\sigma_X^2$ is $\chi^2(n-1)$ and $(m-1)S_Y^2/\sigma_Y^2$ is $\chi^2(m-1)$ and the two are independent since the $X$s and $Y$s are independent. Thus

$$F = \frac{\dfrac{(n-1)S_X^2}{\sigma_X^2(n-1)}}{\dfrac{(m-1)S_Y^2}{\sigma_Y^2(m-1)}} = \frac{S_X^2 \sigma_Y^2}{S_Y^2 \sigma_X^2}$$

has an $F$ distribution with $n-1$ and $m-1$ degrees of freedom.

## EXERCISES 4.1

**4.1-1** Let $X_1$ and $X_2$ be independent chi-square variables with $r_1 = 4$ and $r_2 = 2$ degrees of freedom, respectively.

(a) Determine $P(2.204 < X_1 + X_2 < 16.81)$.

(b) Find $P(12.59 < X_1 + X_2)$.

**4.1-2** Let $X_1$ and $X_2$ be independent random variables. Let $Y = X_1 + X_2$ be $\chi^2(14)$ and let $X_1$ be $\chi^2(3)$.

(a) Guess the distribution of $X_2$. Note that we prove this in Section 6.2.

(b) Determine $P(3.053 < X_2 < 24.72)$.

**4.1-3** Let the independent random variables $X_1$ and $X_2$ be $N(\mu_1 = 3, \sigma_1^2 = 9)$ and $N(\mu_2 = 6, \sigma_2^2 = 16)$, respectively. Determine $P(-10 < Y < 5)$, where $Y = X_1 - X_2$.

**4.1-4** Three random steps in series are needed to complete a certain procedure. The means and the standard deviations of the respective steps are $\mu_1 = 6$ hours, $\mu_2 = 4$ hours, $\mu_3 = 5$ hours and $\sigma_1 = 2$ hours, $\sigma_2 = 2$ hours, $\sigma_3 = 3$ hours. Assuming independence and normal distributions, compute the probability that the procedure will be completed in less than 20 hours.

**4.1-5** Let $\overline{X}$ and $S^2$ be the mean and the variance of a random sample of size $n = 16$ from the normal distribution $N(\mu, \sigma^2)$.

**(a)** Find $d$ from Table VI such that

$$P\left(-d < \frac{\overline{X} - \mu}{S/\sqrt{16}} < d\right) = 0.95.$$

**(b)** Rewrite the inequalities in part (a) so that

$$P[u(\overline{X}, S) < \mu < v(\overline{X}, S)] = 0.95.$$

Find $u(\overline{X}, S)$ and $v(\overline{X}, S)$ so that once $\overline{x}$ and $s$ are computed the interval from $u(\overline{x}, s)$ to $v(\overline{x}, s)$ provides a 95% confidence interval for $\mu$.

**4.1-6** (continuation of Example 4.1-5.) In Example 4.1-5 let $\sigma_X^2 = \sigma_Y^2 = \sigma^2$, which is unknown.

**(a)** We know that $\overline{X}$ is $N(\mu_X, \sigma^2/n)$ and $\overline{Y}$ is $N(\mu_Y, \sigma^2/m)$; so argue that $\overline{X} - \overline{Y}$ is $N[\mu_X - \mu_Y, \sigma^2(1/n + 1/m)]$.

**(b)** Argue that $[(n - 1)S_X^2 + (m - 1)S_Y^2]/\sigma^2$ is $\chi^2(n+m-2)$.

**(c)** Show that

$$T = \cfrac{\cfrac{\overline{X} - \overline{Y} - (\mu_X - \mu_Y)}{\sqrt{\sigma^2(1/n + 1/m)}}}{\sqrt{\cfrac{[(n-1)S_X^2 + (m-1)S_Y^2]/\sigma^2}{n + m - 2}}}$$

has a Student's $t$ distribution with $n + m - 2$ degrees of freedom and eliminate $\sigma$ in that $T$.

**4.1-7** (continuation of Exercise 4.1-6.) In Exercise 4.1-6, let $n = 8$ and $m = 10$.

**(a)** Find $d$ from Table VI so that

$$P(-d < T < d) = 0.90.$$

**(b)** Rewrite the inequalities in this probability statement in part (a) so that $\mu_X - \mu_Y$ is "trapped" in the middle.

**(c)** If $\overline{x} = 7.3$, $s_x = 12.7$ and $\overline{y} = 6.4$, $s_y = 10.3$, using the result of part (b) compute a 90% confidence interval for $\mu_X - \mu_Y$.

**(d)** Assuming that $\sigma_X^2$ might not equal $\sigma_Y^2$, use the $F$ of Example 4.1-5 to find $c$ and $d$ from Table VII so that

$$P(c < F < d) = 0.95 \qquad \text{and} \qquad P(F < d) = 0.975.$$

HINT: Note that $P(c < F) = P(1/c > 1/F) = 0.975$ and that $1/F$ has an $F(m-1, n-1)$ distribution.

**(e)** Rewrite $P(c < F < d) = 0.95$ so that

$$P[u(S_X^2, S_Y^2) < \sigma_Y^2/\sigma_X^2 < v(S_X^2, S_Y^2)] = 0.95.$$

**(f)** Use the values in part (c) to find a 95% confidence interval for $\sigma_Y^2/\sigma_X^2$.

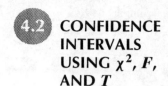

## CONFIDENCE INTERVALS USING $\chi^2$, $F$, AND $T$

In this section, we assume that all samples arise from normal distributions. However, in actual practice, we must always question the assumption of normality and make certain that those underlying distributions are at least approximately normal before using this theory.

We use the fact (Theorem 4.1-3) that $W = (n-1)S^2/\sigma^2$ is $\chi^2(n-1)$ to find a confidence interval for $\sigma^2$. From Table IV in the Appendix with $n-1$ degrees of freedom select $a$ and $b$ such that

$$P\left(a \leq \frac{(n-1)S^2}{\sigma^2} \leq b\right) = 1 - \alpha,$$

where $P(W \leq a) = P(W \geq b) = \alpha/2$ so that $a = \chi^2_{1-\alpha/2}(n-1)$ and $b = \chi^2_{\alpha/2}(n-1)$. (Note that $\chi^2_p(r)$ is a number that cuts off $p$ probability to the *right* of it for a $\chi^2(r)$ random variable.) Then, solving the inequalities, we have

$$1 - \alpha = P\left(\frac{a}{(n-1)S^2} \leq \frac{1}{\sigma^2} \leq \frac{b}{(n-1)S^2}\right)$$

$$= P\left(\frac{(n-1)S^2}{b} \leq \sigma^2 \leq \frac{(n-1)S^2}{a}\right).$$

Thus the probability that the random interval $[(n-1)S^2/b, (n-1)S^2/a]$ contains the unknown $\sigma^2$ is $1 - \alpha$. Once the values of $X_1, X_2, \ldots, X_n$ are observed to be $x_1, x_2, \ldots, x_n$ and $s^2$ computed, then the interval $[(n-1)s^2/b, (n-1)s^2/a]$ is a $100(1-\alpha)\%$ confidence interval for $\sigma^2$. It follows that a $100(1-\alpha)\%$ confidence interval for $\sigma$, the standard deviation, is given by

$$\left[\sqrt{\frac{(n-1)s^2}{b}}, \sqrt{\frac{(n-1)s^2}{a}}\right] = \left[\sqrt{\frac{n-1}{b}}\, s, \sqrt{\frac{n-1}{a}}\, s\right]$$

$$= \left[\sqrt{\frac{n-1}{\chi^2_{\alpha/2}(n-1)}}\, s, \sqrt{\frac{n-1}{\chi^2_{1-\alpha/2}(n-1)}}\, s\right].$$

**EXAMPLE 4.2-1**    Assume that the time in days required for maturation of seeds of a species of Guardiola, a flowering plant found in Mexico, is $N(\mu, \sigma^2)$. A random sample of $n = 13$ seeds, both parents having narrow leaves, yielded $\bar{x} = 18.97$ days and

$$12s^2 = \sum_{i=1}^{13} (x_i - \bar{x})^2 = 128.41.$$

A 90% confidence interval for $\sigma^2$ is

$$\left[\frac{128.41}{21.03}, \frac{128.41}{5.226}\right] = [6.11, 24.57]$$

because $5.226 = \chi^2_{0.95}(12)$ and $21.03 = \chi^2_{0.05}(12)$ from Table IV in the Appendix. The corresponding 90% confidence interval for $\sigma$ is

$$[\sqrt{6.11}, \sqrt{24.57}] = [2.47, 4.96].$$

There are occasions when it is of interest to compare the variances of two normal distributions. We do this by finding a confidence interval for $\sigma^2_X/\sigma^2_Y$ using the ratio of $S^2_X/\sigma^2_X$ and $S^2_Y/\sigma^2_Y$, where $S^2_X$ and $S^2_Y$ are the two sample variances based on two independent samples of sizes $n$ and $m$ from $N(\mu_X, \sigma^2_X)$ and $N(\mu_Y, \sigma^2_Y)$, respectively. However the reciprocal of that ratio can be rewritten as follows:

$$\frac{\dfrac{S^2_Y}{\sigma^2_Y}}{\dfrac{S^2_X}{\sigma^2_X}} = \frac{\left[\dfrac{(m-1)S^2_Y}{\sigma^2_Y}\right]\Big/(m-1)}{\left[\dfrac{(n-1)S^2_X}{\sigma^2_X}\right]\Big/(n-1)}.$$

Since $(m-1)S^2_Y/\sigma^2_Y$ and $(n-1)S^2_X/\sigma^2_X$ are independent chi-square variables with $(m-1)$ and $(n-1)$ degrees of freedom, respectively, we know from Example 4.1-5 that the distribution of this ratio is $F(m-1, n-1)$. That is,

$$F = \frac{\dfrac{(m-1)S^2_Y}{\sigma^2_Y(m-1)}}{\dfrac{(n-1)S^2_X}{\sigma^2_X(n-1)}} = \frac{\dfrac{S^2_Y}{\sigma^2_Y}}{\dfrac{S^2_X}{\sigma^2_X}}$$

has an $F$ distribution with $r_1 = m - 1$ and $r_2 = n - 1$ degrees of freedom. This is the ratio that we want to use to find a confidence interval for $\sigma^2_X/\sigma^2_Y$.

To form the confidence interval, select constants $c$ and $d$ from Table VII in the Appendix so that

$$1 - \alpha = P\left(c \le \frac{S^2_Y/\sigma^2_Y}{S^2_X/\sigma^2_X} \le d\right)$$

$$= P\left(c\,\frac{S^2_X}{S^2_Y} \le \frac{\sigma^2_X}{\sigma^2_Y} \le d\,\frac{S^2_X}{S^2_Y}\right).$$

Because of the limitations of Table VII, we generally let $c = F_{1-\alpha/2}(m-1, n-1) = 1/F_{\alpha/2}(n-1, m-1)$ and $d = F_{\alpha/2}(m-1, n-1)$. (See Exercise 4.1-7(d).) If $s^2_x$ and $s^2_y$ are the observed values of $S^2_X$ and $S^2_Y$, respectively, then

$$\left[\frac{1}{F_{\alpha/2}(n-1, m-1)}\,\frac{s^2_x}{s^2_y}, F_{\alpha/2}(m-1, n-1)\,\frac{s^2_x}{s^2_y}\right]$$

is a $100(1 - \alpha)\%$ confidence interval for $\sigma^2_X/\sigma^2_Y$. By taking square roots of both endpoints, we would obtain a $100(1 - \alpha)\%$ confidence interval for $\sigma_X/\sigma_Y$.

**EXAMPLE 4.2-2**    In Example 4.2-1, denote $\sigma^2$ by $\sigma^2_X$. There $(n-1)s^2_x = 12s^2 = 128.41$. Assume that the time in days required for maturation of seeds of a species of Guardiola, both parents having broad leaves, is $N(\mu_1, \sigma^2_Y)$. A random sample of size $m = 9$ seeds yielded $\bar{y} = 23.20$ and

$$8s_y^2 = \sum_{i=1}^{9} (y_i - \bar{y})^2 = 36.72.$$

A 98% confidence interval for $\sigma_X^2/\sigma_Y^2$ is given by

$$\left[ \left( \frac{1}{5.67} \right) \frac{(128.41)/12}{(36.72)/8}, (4.50) \frac{(128.41)/12}{(36.72)/8} \right] = [0.41, 10.49]$$

because $F_{0.01}(12, 8) = 5.67$ and $F_{0.01}(8, 12) = 4.50$. It follows that a 98% confidence interval for $\sigma_X/\sigma_Y$ is

$$[\sqrt{0.41}, \sqrt{10.49}] = [0.64, 3.24].$$

Although we are able to formally find a confidence interval for the ratio of two distribution variances and/or standard deviations, we should point out that these intervals are generally not too useful because they are often very wide. Moreover, these intervals are not very robust. That is, the confidence coefficients are not very accurate if we deviate much from underlying normal distributions, because, in those instances, the distribution of $(n - 1)S^2/\sigma^2$ could deviate greatly from $\chi^2(n-1)$.

We have found a confidence interval for the mean $\mu$ of a normal distribution, assuming that the value of the standard deviation $\sigma$ is known or when $\sigma$ is unknown but the sample size is large. However, in many applications, the sample sizes are small and we do not know the value of the standard deviation, although in some cases we might have a very good idea about its value. For illustration, a manufacturer of light bulbs probably has a good notion from past experience of the value of the standard deviation of the length of life of different types of light bulbs. But certainly, most of the time, the investigator will not have any more idea about the standard deviation than about the mean—and frequently less. Let us consider how to proceed under these circumstances.

If the random sample arises from a normal distribution, we use the fact (Example 4.1-4) that

$$T = \frac{\dfrac{\overline{X} - \mu}{\sigma/\sqrt{n}}}{\sqrt{\dfrac{(n-1)S^2}{\sigma^2} \Big/ (n-1)}} = \frac{\overline{X} - \mu}{S/\sqrt{n}}$$

has a $t$ distribution with $r = n-1$ degrees of freedom, because $Z = (\overline{X}-\mu)/(\sigma/\sqrt{n})$ is $N(0, 1)$, $W = (n - 1)S^2/\sigma^2$ is $\chi^2(n-1)$, and $Z$ and $W$ are independent. Select $t_{\alpha/2}(n-1)$ so that $P[T \geq t_{\alpha/2}(n-1)] = \alpha/2$. Then

$$1 - \alpha = P\left[ -t_{\alpha/2}(n-1) \leq \frac{\overline{X} - \mu}{S/\sqrt{n}} \leq t_{\alpha/2}(n-1) \right]$$

$$= P\left[ -t_{\alpha/2}(n-1)\left( \frac{S}{\sqrt{n}} \right) \leq \overline{X} - \mu \leq t_{\alpha/2}(n-1)\left( \frac{S}{\sqrt{n}} \right) \right]$$

$$= P\left[ -\overline{X} - t_{\alpha/2}(n-1)\left(\frac{S}{\sqrt{n}}\right) \le -\mu \le -\overline{X} + t_{\alpha/2}(n-1)\left(\frac{S}{\sqrt{n}}\right) \right]$$

$$= P\left[ \overline{X} - t_{\alpha/2}(n-1)\left(\frac{S}{\sqrt{n}}\right) \le \mu \le \overline{X} + t_{\alpha/2}(n-1)\left(\frac{S}{\sqrt{n}}\right) \right].$$

The observations of a random sample provide computed values of $\overline{x}$ and $s^2$ and

$$\left[ \overline{x} - t_{\alpha/2}(n-1)\left(\frac{s}{\sqrt{n}}\right), \overline{x} + t_{\alpha/2}(n-1)\left(\frac{s}{\sqrt{n}}\right) \right]$$

is a $100(1 - \alpha)\%$ confidence interval for $\mu$.

**EXAMPLE 4.2-3**  Let $X$ equal the amount of butterfat in pounds produced by a typical cow during a 305-day milk production period between her first and second calves. Assume that the distribution of $X$ is $N(\mu, \sigma^2)$. To estimate $\mu$ a farmer measured the butterfat production for $n = 20$ cows yielding the following data:

$$\begin{array}{cccccccccc}
481 & 537 & 513 & 583 & 453 & 510 & 570 & 500 & 457 & 555 \\
618 & 327 & 350 & 643 & 499 & 421 & 505 & 637 & 599 & 392
\end{array}$$

For these data, $\overline{x} = 507.50$ and $s = 89.75$. Thus a point estimate of $\mu$ is $\overline{x} = 507.50$. Since $t_{0.05}(19) = 1.729$, a 90% confidence interval for $\mu$ is

$$507.50 \pm 1.729\left(\frac{89.75}{\sqrt{20}}\right)$$

$$507.50 \pm 34.70,$$

or equivalently, $[472.80, 542.20]$.

Let $T$ have a $t$ distribution with $n - 1$ degrees of freedom. Then $t_{\alpha/2}(n-1) > z_{\alpha/2}$. Consequently, we would expect the interval $\overline{x} \pm z_{\alpha/2}(\sigma/\sqrt{n})$ to be shorter than the interval $\overline{x} \pm t_{\alpha/2}(n-1)(s/\sqrt{n})$. After all, we have more information, namely the value of $\sigma$, in constructing the first interval. However, the length of the second interval is very much dependent on the value of $s$. If the observed $s$ is smaller than $\sigma$, a shorter confidence interval could result by the second procedure. But on the average, $\overline{x} \pm z_{\alpha/2}(\sigma/\sqrt{n})$ is the shorter of the two confidence intervals. (See Exercise 4.2-10.)

**EXAMPLE 4.2-4**  To compare confidence intervals when $\sigma$ is known or when $\sigma$ is unknown, 50 samples of $n = 5$ observations were simulated from a $N(50, 16)$ distribution. For each sample of size 5, a 90% confidence interval was calculated using the known $\sigma = 4$, namely, $\overline{x} \pm 1.645(4/\sqrt{5})$. Those 50 confidence intervals are depicted in Figure 4.2-1($a$). For those same data a 90% confidence interval was calculated for $\mu$ assuming that $\sigma$ was unknown and using $\overline{x} \pm 2.132(s/\sqrt{5})$. These are depicted in Figure 4.2-1($b$). Note the different lengths of the latter intervals while the length of the $z$ intervals are all equal. Some of the $t$ intervals are longer and some are shorter than the corresponding $z$ intervals. The average length of these intervals is 7.399, while the length of a $z$ interval is 5.885. It can be shown that the expected length of a $t$ interval with the given characteristics is 7.169. For the $z$ intervals, 43 (86%) contain the mean $\mu = 50$, while 45 (90%) of these $t$ intervals contain the mean. If this simulation were repeated, the results will be different but it should always be true that approximately 90% of each set of intervals contain the mean.

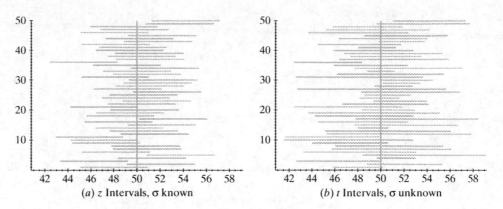

(a) z Intervals, σ known                (b) t Intervals, σ unknown

**Figure 4.2-1** 90% confidence intervals for $\mu$

If we are not able to assume that the underlying distribution is normal but $\mu$ and $\sigma$ are both unknown, approximate confidence intervals for $\mu$ can still be constructed using

$$T = \frac{\overline{X} - \mu}{S/\sqrt{n}},$$

which now only has an approximate $t$ distribution. Generally, this approximation is quite good for many nonnormal distributions (i.e., it is robust), in particular, if the underlying distribution is symmetric, unimodal, and of the continuous type. However, if the distribution is highly skewed, there is great danger using this approximation. (See Exercise 5.3-11.) In such a situation, it would be safer to use certain nonparametric methods for finding a confidence interval for the median of the distribution, one of which is given in Section 4.9.

There is one other aspect of confidence intervals that should be mentioned. So far we have created only what are called **two-sided confidence intervals** for the mean $\mu$. Sometimes it happens that you might want only a lower (or upper) bound on $\mu$. We proceed as follows.

Say $\overline{X}$ is the mean of a random sample of size $n$ from the normal distribution $N(\mu, \sigma^2)$, where say for the moment that $\sigma^2$ is known. Then

$$P\left(\frac{\overline{X} - \mu}{\sigma/\sqrt{n}} \leq z_\alpha\right) = 1 - \alpha,$$

or equivalently,

$$P\left[\overline{X} - z_\alpha\left(\frac{\sigma}{\sqrt{n}}\right) \leq \mu\right] = 1 - \alpha.$$

Once $\overline{X}$ is observed to be equal to $\overline{x}$, then $[\overline{x} - z_\alpha(\sigma/\sqrt{n}), \infty)$ is a $100(1 - \alpha)\%$ **one-sided confidence interval** for $\mu$. That is, with the confidence coefficient of $1 - \alpha$, $\overline{x} - z_\alpha(\sigma/\sqrt{n})$ is a lower bound for $\mu$. Similarly, $(-\infty, \overline{x} + z_\alpha(\sigma/\sqrt{n})]$ is a one-sided confidence interval for $\mu$ and $\bar{x} + z_\alpha(\sigma/\sqrt{n})$ provides an upper bound for $\mu$ with confidence coefficient $1 - \alpha$.

When $\sigma$ is unknown, we would use $T = (\bar{X} - \mu)/(S/\sqrt{n})$ to find the corresponding lower or upper bounds for $\mu$, namely $\bar{x} - t_\alpha(n-1)(s/\sqrt{n})$ and $\bar{x} + t_\alpha(n-1)(s/\sqrt{n})$.

Now consider the problem of constructing confidence intervals for the difference of the means of two normal distributions when the variances are unknown but the sample sizes are small. Let $X_1, X_2, \ldots, X_n$ and $Y_1, Y_2, \ldots, Y_m$ be two independent random samples from the distributions $N(\mu_X, \sigma_X^2)$ and $N(\mu_Y, \sigma_Y^2)$, respectively. If the sample sizes are not large (say considerably smaller than 30), this problem can be a difficult one. However, even in these cases, if we can assume common, but unknown, variances, say $\sigma_X^2 = \sigma_Y^2 = \sigma^2$, there is a way out of our difficulty.

We know that

$$Z = \frac{\bar{X} - \bar{Y} - (\mu_X - \mu_Y)}{\sqrt{\sigma^2/n + \sigma^2/m}}$$

is $N(0, 1)$. Moreover, since the random samples are independent,

$$U = \frac{(n-1)S_X^2}{\sigma^2} + \frac{(m-1)S_Y^2}{\sigma^2}$$

is the sum of two independent chi-square random variables; and thus the distribution of $U$ is $\chi^2(n+m-2)$. In addition, the independence of the sample means and sample variances implies that $Z$ and $U$ are independent. According to the definition of a $T$ random variable,

$$T = \frac{Z}{\sqrt{U/(n+m-2)}}$$

has a $t$ distribution with $n + m - 2$ degrees of freedom. That is,

$$T = \frac{\dfrac{\bar{X} - \bar{Y} - (\mu_X - \mu_Y)}{\sqrt{\sigma^2/n + \sigma^2/m}}}{\sqrt{\left[\dfrac{(n-1)S_X^2}{\sigma^2} + \dfrac{(m-1)S_Y^2}{\sigma^2}\right]\Big/(n+m-2)}}$$

$$= \frac{\bar{X} - \bar{Y} - (\mu_X - \mu_Y)}{\sqrt{\left[\dfrac{(n-1)S_X^2 + (m-1)S_Y^2}{n+m-2}\right]\left[\dfrac{1}{n} + \dfrac{1}{m}\right]}}$$

has a $t$ distribution with $r = n + m - 2$ degrees of freedom. Thus, with $t_0 = t_{\alpha/2}(n+m-2)$,

$$P(-t_0 \leq T \leq t_0) = 1 - \alpha.$$

Solving the inequality for $\mu_X - \mu_Y$ yields

$$P\left(\bar{X} - \bar{Y} - t_0 S_P\sqrt{\frac{1}{n} + \frac{1}{m}} \leq \mu_X - \mu_Y \leq \bar{X} - \bar{Y} + t_0 S_P\sqrt{\frac{1}{n} + \frac{1}{m}}\right)$$

where the pooled estimator of the common standard deviation is

$$S_P = \sqrt{\frac{(n-1)S_X^2 + (m-1)S_Y^2}{n+m-2}}.$$

If $\bar{x}, \bar{y}$, and $s_p$ are the observed values of $\overline{X}, \overline{Y}$, and $S_P$, then

$$\left[ \bar{x} - \bar{y} - t_0 s_p \sqrt{\frac{1}{n} + \frac{1}{m}}, \bar{x} - \bar{y} + t_0 s_p \sqrt{\frac{1}{n} + \frac{1}{m}} \right]$$

is a $100(1-\alpha)\%$ confidence interval for $\mu_X - \mu_Y$.

**EXAMPLE 4.2-5**    Suppose that scores on a standardized test in mathematics taken by students from large and small high schools are $N(\mu_X, \sigma^2)$ and $N(\mu_Y, \sigma^2)$, respectively, where $\sigma^2$ is unknown. If a random sample of $n = 9$ students from large high schools yielded $\bar{x} = 81.31$, $s_x^2 = 60.76$ and a random sample of $m = 15$ students from small high schools yielded $\bar{y} = 78.61$, $s_y^2 = 48.24$, the endpoints for a 95% confidence interval for $\mu_X - \mu_Y$ are given by

$$81.31 - 78.61 \pm 2.074 \sqrt{\frac{8(60.76) + 14(48.24)}{22}} \sqrt{\frac{1}{9} + \frac{1}{15}}$$

because $t_{0.025}(22) = 2.074$. The 95% confidence interval is $[-3.65, 9.05]$.

In the case that the variances $\sigma_X^2$ and $\sigma_Y^2$ are known or the sample sizes large so that $s_x^2 \approx \sigma_X^2$ and $s_y^2 \approx \sigma_Y^2$, we would use the facts that

$$\frac{\overline{X} - \overline{Y} - (\mu_X - \mu_Y)}{\sqrt{\sigma_X^2/n + \sigma_Y^2/m}} \quad \text{and} \quad \frac{\overline{X} - \overline{Y} - (\mu_X - \mu_Y)}{\sqrt{S_X^2/n + S_Y^2/m}}$$

have $N(0,1)$ and approximate $N(0,1)$ distributions, respectively, to find confidence intervals for $\mu_X - \mu_Y$. The respective intervals are

$$\bar{x} - \bar{y} \pm z_{\alpha/2} \sqrt{\frac{\sigma_X^2}{n} + \frac{\sigma_Y^2}{m}} \quad \text{and} \quad \bar{x} - \bar{y} \pm z_{\alpha/2} \sqrt{\frac{s_x^2}{n} + \frac{s_y^2}{m}}.$$

**Remark**    It is interesting to consider the two-sample $T$ in more detail. It is

$$T = \frac{\overline{X} - \overline{Y} - (\mu_X - \mu_Y)}{\sqrt{\frac{(n-1)S_X^2 + (m-1)S_Y^2}{n+m-2}\left(\frac{1}{n} + \frac{1}{m}\right)}}$$

$$= \frac{\overline{X} - \overline{Y} - (\mu_X - \mu_Y)}{\sqrt{\left[\frac{(n-1)S_X^2}{nm} + \frac{(m-1)S_Y^2}{nm}\right]\left[\frac{n+m}{n+m-2}\right]}}.$$

Now since $(n-1)/n \approx 1$, $(m-1)/m \approx 1$, and $(n+m)/(n+m-2) \approx 1$, we have that

$$T \approx \frac{\overline{X} - \overline{Y} - (\mu_X - \mu_Y)}{\sqrt{\dfrac{S_X^2}{m} + \dfrac{S_Y^2}{n}}}.$$

In this form we note that each variance is divided by the wrong sample size! That is, as noted following Example 4.2-5, if the sample sizes are large or the variances known, we would like

$$\sqrt{\frac{S_X^2}{n} + \frac{S_Y^2}{m}} \quad \text{or} \quad \sqrt{\frac{\sigma_X^2}{n} + \frac{\sigma_Y^2}{m}}$$

in the denominator; so $T$ seems to use the wrong sample sizes. Thus, using this $T$ is particularly bad when the sample sizes and the variances are unequal; and thus caution must be taken in using that $T$ in constructing a confidence interval for $\mu_X - \mu_Y$. That is, if $n < m$ and $\sigma_X^2 < \sigma_Y^2$, then $T$ does not have a distribution which is close to that of a Student $t$-distribution with $n + m - 2$ degrees of freedom: its spread is much less than the Student $t$'s as the term $s_y^2/n$ in the denominator is much larger than it should be. On the other hand, if $m < n$ and $\sigma_X^2 < \sigma_Y^2$, then $s_x^2/m + s_y^2/n$ is generally smaller than it should be and the distribution of $T$ is spread out more than that of the Student $t$.

There is a way out of this difficulty, however. When the underlying distributions are close to normal, but the sample sizes and the variances are seemingly much different, we suggest the use of

$$U = \frac{\overline{X} - \overline{Y} - (\mu_X - \mu_Y)}{\sqrt{\dfrac{S_X^2}{n} + \dfrac{S_Y^2}{m}}}$$

where it has been proved that $U$ has an approximate $t$-distribution with $\lfloor v \rfloor$ degrees of freedom, with

$$v = \frac{\left(\dfrac{s_x^2}{n} + \dfrac{s_y^2}{m}\right)^2}{\dfrac{1}{n-1}\left(\dfrac{s_x^2}{n}\right)^2 + \dfrac{1}{m-1}\left(\dfrac{s_y^2}{m}\right)^2}.$$

Here $\lfloor v \rfloor$ is the "floor" or greatest integer in $v$ so the number of degrees of freedom equals $v$ rounded down. This type of approximation was first suggested by B. L. Welch.

**EXAMPLE 4.2-6** To help understand the above remark, a simulation was done using *Maple*. In order to obtain a $q$-$q$ plot of the quantiles of a $t$-distribution, a CAS or some type of computer program is very important because of the challenge in finding these quantiles.

*Maple* was used to simulate $N = 500$ observations of $T$ and $N = 500$ observations of $U$. In Figure 4.2-2, $n = 6$, $m = 18$, the $X$ observations were generated from the $N(0, 1)$ distribution, and the $Y$ observations were generated from the $N(0, 36)$ distribution. For the value of $v$ for Welch's approximate $t$ distribution, we used the distribution variances rather than the sample variances.

For the simulation results shown in Figure 4.2-3, $n = 18$, $m = 6$, the $X$ observations were generated from the $N(0, 1)$ distribution, and the $Y$ observations were generated from the $N(0, 36)$ distribution.

Remember that in the development of the $T$ statistic, it was assumed that the variances of the two distributions are equal.

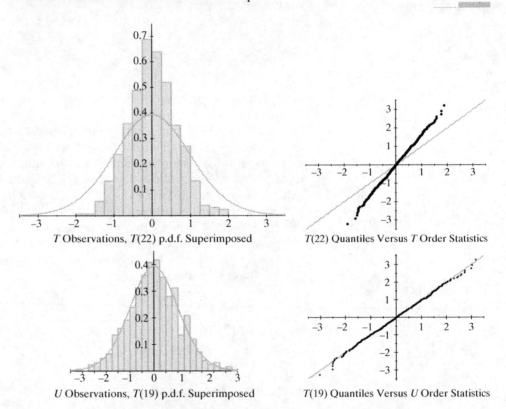

*T* Observations, $T(22)$ p.d.f. Superimposed

*T(22)* Quantiles Versus *T* Order Statistics

*U* Observations, $T(19)$ p.d.f. Superimposed

*T(19)* Quantiles Versus *U* Order Statistics

**Figure 4.2-2** Observations of $T$ and of $U$, $n = 6$, $m = 18$, $\sigma_X^2 = 1$, $\sigma_Y^2 = 36$

## EXERCISES 4.2

**4.2-1** Let $X$ equal the length (in centimeters) of a certain species of fish when caught in the spring. Assume that the distribution of $X$ is $N(\mu, \sigma^2)$. A random sample of $n = 13$ observations of $X$ are

$$
\begin{array}{ccccccc}
13.1 & 5.1 & 18.0 & 8.7 & 16.5 & 9.8 & 6.8 \\
12.0 & 17.8 & 25.4 & 19.2 & 15.8 & 23.0 &
\end{array}
$$

**(a)** Give a point estimate of the standard deviation $\sigma$ of this species of fish.

**(b)** Find a 95% confidence interval for $\sigma$.

**4.2-2** A student who works in a blood lab tested 25 men for cholesterol levels and found the following values:

$$
\begin{array}{ccccccccc}
164 & 272 & 261 & 248 & 235 & 192 & 203 & 278 & 268 \\
230 & 242 & 305 & 286 & 310 & 345 & 289 & 326 & \\
335 & 297 & 328 & 400 & 228 & 194 & 338 & 252 &
\end{array}
$$

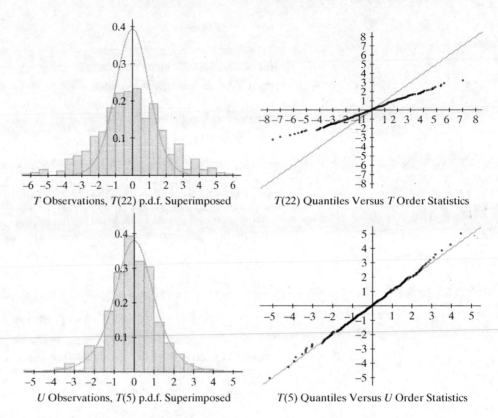

Figure 4.2-3  Observations of $T$ and of $U$, $n = 18$, $m = 6$, $\sigma_X^2 = 1$, $\sigma_Y^2 = 36$

Assume that these values represent observations of a random sample taken from $N(\mu, \sigma^2)$.

**(a)** Calculate the sample mean and sample variance for these data.

**(b)** Find a 90% confidence interval for $\sigma^2$.

**(c)** Find a 90% confidence interval for $\sigma$.

**(d)** Does the assumption of normality seem to be valid?

**4.2-3** Let $X_1, X_2, X_3, \ldots, X_n$ be a random sample from $N(\mu, \sigma^2)$, with known mean $\mu$. Describe how you would construct a confidence interval for the unknown variance $\sigma^2$. HINT: Use the fact that $\sum_{i=1}^{n}(X_i - \mu)^2/\sigma^2$ is $\chi^2(n)$.

**4.2-4** Let $X$ and $Y$ equal the weights of a phosphorus-free laundry detergent in a "6-pound" box and a "12-pound" box, respectively. Assume that the distributions of $X$ and $Y$ are $N(\mu_X, \sigma_X^2)$ and $N(\mu_Y, \sigma_Y^2)$, respectively. A random sample of $n = 10$ observations of $X$ yielded a sample mean of $\bar{x} = 6.10$ pounds with a sample variance of $s_x^2 = 0.0040$, while an independent random sample of $m = 9$ observations of $Y$ yielded a sample mean of $\bar{y} = 12.10$ pounds with a sample variance of $s_y^2 = 0.0076$.

**(a)** Give a point estimate of $\sigma_X^2/\sigma_Y^2$.

**(b)** Find a 95% confidence interval for $\sigma_X^2/\sigma_Y^2$.

**4.2-5** Let $X$ and $Y$ equal the concentration in parts per billion of chromium in the blood for healthy persons and for persons with a suspected disease, respectively. Assume that the distributions of $X$ and $Y$ are $N(\mu_X, \sigma_X^2)$ and $N(\mu_Y, \sigma_Y^2)$, respectively. Using $n = 8$ observations of $X$:

15  23  12  18  9  28  11  10

and $m = 10$ observations of $Y$:

$$25 \quad 20 \quad 35 \quad 15 \quad 40 \quad 16 \quad 10 \quad 22 \quad 18 \quad 32$$

(a) Give a point estimate of $\sigma_X^2/\sigma_Y^2$.

(b) Find a one-sided 95% confidence interval which is an upper bound for $\sigma_X^2/\sigma_Y^2$.

**4.2-6** Let $X_1, X_2, \ldots, X_n$ be a random sample of size $n$ from a normal distribution, $N(\mu, \sigma^2)$. Select $a$ and $b$ so that

$$P\left(a \le \frac{(n-1)S^2}{\sigma^2} \le b\right) = 1 - \alpha.$$

So a $100(1-\alpha)$% confidence interval for $\sigma$ is $[\sqrt{(n-1)/b}\,s, \sqrt{(n-1)/a}\,s]$. Find values of $a$ and $b$ that minimize the length of this confidence interval. That is, minimize

$$k = s\sqrt{n-1}\left(\frac{1}{\sqrt{a}} - \frac{1}{\sqrt{b}}\right)$$

under the restriction

$$G(b) - G(a) = \int_a^b g(u)\,du = 1 - \alpha,$$

where $G(u)$ and $g(u)$ are the distribution function and p.d.f. of a $\chi^2(n-1)$ distribution, respectively. HINT: Due to the restriction, $b$ is a function of $a$. In particular, by taking derivatives of the restricting equation with respect to $a$, show that $\dfrac{db}{da} = \dfrac{g(a)}{g(b)}$. Determine $\dfrac{dk}{da}$. By setting $\dfrac{dk}{da} = 0$, show that $a$ and $b$ must satisfy

$$a^{n/2}e^{-a/2} - b^{n/2}e^{-b/2} = 0.$$

NOTE: It is possible to solve for the values of $a$ and $b$. See Exercise 5.2-1.

**4.2-7** Thirteen tons of cheese are stored in some old gypsum mines, including "22-pound" wheels (label weight). A random sample of $n = 9$ of these wheels yielded the following weights in pounds:

$$\begin{array}{ccccc}
21.50 & 18.95 & 18.55 & 19.40 & 19.15 \\
22.35 & 22.90 & 22.20 & 23.10 &
\end{array}$$

Assuming that the distribution of the weights of the wheels of cheese is $N(\mu, \sigma^2)$, find a 95% confidence interval for $\mu$.

**4.2-8** In a study of maximal aerobic capacity (*Journal of Applied Physiology 65*, 6 [December 1988], pp. 2696-2708), 12 women were used as subjects, and one measurement that was made was blood plasma volume. The following data give their blood plasma volumes in liters:

$$\begin{array}{cccccc}
3.15 & 2.99 & 2.77 & 3.12 & 2.45 & 3.85 \\
2.99 & 3.87 & 4.06 & 2.94 & 3.53 & 3.20
\end{array}$$

Assume that these are observations of a normally distributed random variable $X$ that has mean $\mu$ and standard deviation $\sigma$.

(a) Give the value of a point estimate of $\mu$.

(b) Determine point estimates of $\sigma^2$ and $\sigma$.

(c) Find a 90% confidence interval for $\mu$.

**4.2-9** A leakage test was conducted to determine the effectiveness of a seal designed to keep the inside of a plug air tight. An air needle was inserted in the plug and this was placed under water. The pressure was then increased until leakage was observed. Let $X$ equal the pressure in pounds per square inch. Assume that the distribution of

$X$ is $N(\mu, \sigma^2)$. Using the following $n = 10$ observations of $X$:

$$3.1 \quad 3.3 \quad 4.5 \quad 2.8 \quad 3.5 \quad 3.5 \quad 3.7 \quad 4.2 \quad 3.9 \quad 3.3$$

(a) Find a point estimate of $\mu$.

(b) Find a point estimate of $\sigma$.

(c) Find a 95% one-sided confidence interval for $\mu$ that provides an upper bound for $\mu$.

**4.2-10** Let $X_1, X_2, \ldots, X_n$ be a random sample of size $n$ from the normal distribution $N(\mu, \sigma^2)$. Calculate the expected length of a 95% confidence interval for $\mu$ assuming that $n = 5$ and the variance is

(a) known,

(b) unknown. HINT: To find $E(S)$, first determine $E[\sqrt{(n-1)S^2/\sigma^2}\,]$, recalling that $(n-1)S^2/\sigma^2$ is $\chi^2(n-1)$.

**4.2-11** An interior automotive supplier places several electrical wires in a harness. A pull test measures the force required to pull spliced wires apart. A customer requires that each wire that is spliced into the harness must withstand a pull force of 20 pounds. Let $X$ equal the pull force required to pull 20 gauge wires apart. Assume that the distribution of $X$ is $N(\mu, \sigma^2)$. The following data give 20 observations of $X$.

| | | | | | | | | | |
|---|---|---|---|---|---|---|---|---|---|
| 28.8 | 24.4 | 30.1 | 25.6 | 26.4 | 23.9 | 22.1 | 22.5 | 27.6 | 28.1 |
| 20.8 | 27.7 | 24.4 | 25.1 | 24.6 | 26.3 | 28.2 | 22.2 | 26.3 | 24.4 |

(a) Find point estimates for $\mu$ and $\sigma$.

(b) Find a 99% one-sided confidence interval for $\mu$ that provides a lower bound for $\mu$.

**4.2-12** Independent random samples of the heights of adult males living in two countries yielded the following results: $n = 12$, $\bar{x} = 65.7$ inches, $s_x = 4$ inches and $m = 15$, $\bar{y} = 68.2$ inches, $s_y = 3$ inches. Find an approximate 98% confidence interval for the difference $\mu_X - \mu_Y$ of the means of the populations of heights. Assume that $\sigma_X^2 = \sigma_Y^2$.

**4.2-13** Consider the butterfat production (in pounds) for a cow during a 305-day milk production period following the birth of a calf. Let $X$ and $Y$ equal the butterfat production for such cows on a farm in Wisconsin and a farm in Michigan. Twelve observations of $X$ are:

| | | | | | |
|---|---|---|---|---|---|
| 649 | 657 | 714 | 877 | 975 | 468 |
| 567 | 849 | 721 | 791 | 874 | 405 |

Sixteen observations of $Y$ are:

| | | | | | | | |
|---|---|---|---|---|---|---|---|
| 699 | 891 | 632 | 815 | 589 | 764 | 524 | 727 |
| 597 | 868 | 652 | 978 | 479 | 733 | 549 | 790 |

(a) Assuming that $X$ is $N(\mu_X, \sigma^2)$ and $Y$ is $N(\mu_Y, \sigma^2)$, find a 95% confidence interval for $\mu_X - \mu_Y$.

(b) Construct box-and-whisker diagrams for these two sets of data on the same graph.

(c) Does there seem to be a significant difference in butterfat production for cows on these two farms?

**4.2-14** A test was conducted to determine if a wedge on the end of a plug fitting designed to hold a seal onto that plug was doing its job. The data taken were in the form of measurements of the force required to remove a seal from the plug first with the wedge in place, say $X$, and the force required without the plug, say $Y$. Assume that the distributions of $X$ and $Y$ are $N(\mu_X, \sigma^2)$ and $N(\mu_Y, \sigma^2)$. Ten independent observations of $X$ are:

$$3.26 \quad 2.26 \quad 2.62 \quad 2.62 \quad 2.36 \quad 3.00 \quad 2.62 \quad 2.40 \quad 2.30 \quad 2.40$$

Ten independent observations of $Y$ are:

$$1.80 \quad 1.46 \quad 1.54 \quad 1.42 \quad 1.32 \quad 1.56 \quad 1.36 \quad 1.64 \quad 2.00 \quad 1.54$$

**(a)** Find a 95% confidence interval for $\mu_X - \mu_Y$.

**(b)** Construct box-and-whisker diagrams of these data on the same figure.

**(c)** Is the wedge necessary?

**4.2-15** Let $\overline{X}$, $\overline{Y}$, $S_X^2$, and $S_Y^2$ be the respective sample means and unbiased estimates of the variances using independent samples of sizes $n$ and $m$ from the normal distributions $N(\mu_X, \sigma_X^2)$ and $N(\mu_Y, \sigma_Y^2)$, where $\mu_X$, $\mu_Y$, $\sigma_X^2$, and $\sigma_Y^2$ are unknown. If, however, $\sigma_X^2/\sigma_Y^2 = d$, a known constant, argue that

**(a)** $\dfrac{(\overline{X} - \overline{Y}) - (\mu_X - \mu_Y)}{\sqrt{d\sigma_Y^2/n + \sigma_Y^2/m}}$ is $N(0, 1)$.

**(b)** $\dfrac{(n-1)S_X^2}{d\sigma_Y^2} + \dfrac{(m-1)S_Y^2}{\sigma_Y^2}$ is $\chi^2(n+m-2)$.

**(c)** The two random variables in (a) and (b) are independent.

**(d)** With these results, construct a random variable (not depending upon $\sigma_Y^2$) that has a $t$ distribution and can be used to construct a confidence interval for $\mu_X - \mu_Y$.

**4.2-16** Let $\overline{X}$ denote the mean of a random sample of size $n$ from a distribution that has mean $\mu$ and variance $\sigma^2 = 10$. Find $n$ so that the probability is approximately 0.95 that the random interval $\overline{X} - 1/2$ to $\overline{X} + 1/2$ includes $\mu$.

**4.2-17** Let $\overline{X}$ and $\overline{Y}$ be the means of two independent random samples, each of size $n$, from the respective distributions $N(\mu_X, \sigma^2)$ and $N(\mu_Y, \sigma^2)$, where the common variance $\sigma^2$ is known. Find $n$ such that

$$P(\overline{X} - \overline{Y} - \sigma/5 < \mu_X - \mu_Y < \overline{X} - \overline{Y} + \sigma/5) = 0.90.$$

## 4.3 CONFIDENCE INTERVALS AND TESTS OF HYPOTHESES

The first major area of statistical inference involves the estimation of parameters. We have introduced both point estimation through maximum likelihood estimation and interval estimation with confidence intervals. We now consider a second major area of statistical inference, namely **tests of statistical hypotheses**, in which many such tests are closely related to confidence intervals for parameters. To see this, let us begin with an illustration.

Many statisticians are involved in a reform movement in teaching introductory statistics. This concerns mainly those statistics courses that do not have calculus as a prerequisite. This reform requires students to be more actively involved through projects, discussions, computer work, and analyses of statistical reports found in the media. There is much less lecturing by the instructor and hopefully more thinking about the use of good statistical methods by the student. Initially, we find that in such a course both the instructor and the student really work harder, but there seems to be more satisfaction for both in doing so. That is, the claim is the students learn statistics better. To be honest, there has not been a good assessment of that claim. How might we investigate such a situation?

Suppose all persons involved, some of whom might be for the reform and some against it, agree on some type of final test to be given to all students taking a course using the reform method. Moreover, let us assume that many students, who have taken the traditional statistics course that uses the lecture system, have taken this type of test with scores that average about 75 with a standard deviation of 10. Later on we consider the possibility of comparing two groups of students, one using the reform and one using the lecture methods, by giving a newly developed test to each group. However, at this moment, let us assume that we have enough information about that new test and expect students who are taught through the lecture system to obtain scores with mean about 75 and standard deviation about 10.

Statisticians supporting the reform believe that students using this new method will have a mean greater than 75. That is, if $\mu$ is that mean, they believe that $\mu > 75$. This conjecture is called a **statistical hypothesis** because it is a statement about the parameter of a distribution. Here it is about the mean $\mu$ of the distribution of scores of the students taking a reform statistics course, namely $\mu > 75$.

There are some statisticians who believe that the reform movement is a waste of time and effort, and they think that this mean $\mu$ will not change much and even possibly decrease. That is, they think that at best $\mu = 75$. Of course, this is also a statistical hypothesis; it is called the **null hypothesis** or no change hypothesis, and it is denoted by $H_0: \mu = 75$. The other hypothesis which we denote by $H_1: \mu > 75$ is called the **alternative hypothesis**, and it is often the hypothesis of a research worker who has proposed a new way of doing something. Let us say that both groups agree that the standard deviation will remain about $\sigma = 10$ for the scores of those taking reform statistics. Later we discuss the situation in which we are unwilling to make that assumption about $\sigma$.

How do we test $H_0: \mu = 75$ against $H_1: \mu > 75$? We need to select some students at random out of the population of students who might take introductory statistics. We would not just take volunteers, because these might be the better students that would bias our findings. Say we need to fill a mid-size class so we select $n = 64$ students at random from a population of students who want (or must) take elementary statistics. Everyone involved has agreed on a certain instructor to teach this reform statistics class, and this final assessment test will be given at the end of the semester.

Let us also agree on the method to evaluate the success or failure of the reform method. We will obtain 64 scores, $X_1, X_2, \ldots, X_{64}$, which we assume to be a random sample from all possible scores of the many students who might take this reform course. With the sample mean, $\overline{X}$, we could construct a one-sided confidence interval for $\mu$. To do this, recall that $\overline{X}$ has an approximate $N(\mu, \sigma^2/n)$ distribution and consider

$$P\left(\frac{\overline{X} - \mu}{\sigma/\sqrt{n}} \leq z_\alpha\right) \approx 1 - \alpha$$

or, equivalently,

$$P\left(\overline{X} - z_\alpha \frac{\sigma}{\sqrt{n}} \leq \mu\right) \approx 1 - \alpha.$$

That is, we are $100(1 - \alpha)$ percent confident that this new $\mu$ is above $\bar{x} - z_\alpha(\sigma/\sqrt{n})$, where $\bar{x}$ is the observed sample mean. In our case, say $\alpha = 0.05$ and, of course, $n = 64$ and $\sigma = 10$ so that the lower bound on the one-sided confidence interval is $\bar{x} - 1.645(10/\sqrt{64}) = \bar{x} - 2.06$.

Therefore, if the sample mean $\bar{x} = 77.47$, for example, this lower bound would be $77.47 - 2.06 = 75.41$. That is, we would be 95% confident that $\mu$ of these test scores is above 75.41 and thus we would believe that the reform method has increased the mean and we would accept the alternative hypotheses $H_1: \mu > 75$. If, on the other hand, the sample mean equaled something like 75.92, then that lower bound for $\mu$ equals $75.92 - 2.06 = 73.86$ and we certainly have no reason to reject $H_0: \mu = 75$ based upon these data.

Several observations must be made at this point. Note in these two cases, if we compute

$$Z = \frac{\overline{X} - 75}{\sigma/\sqrt{n}}$$

we get, respectively,

$$z = \frac{77.47 - 75}{10/\sqrt{64}} = 1.976 \quad \text{and} \quad z = \frac{75.92 - 75}{10/\sqrt{64}} = 0.736.$$

The first value $z = 1.976 > 1.645$ and the second value $z = 0.736 < 1.645$. That is, in the first case, we see that 75 is not in the one-sided confidence interval if $(\overline{x} - 75)/(\sigma/\sqrt{n}) \geq 1.645$ and in the second case, 75 is in that confidence interval if $(\overline{x} - 75)/(\sigma/\sqrt{n}) < 1.645$. The reason for this is very clear when we note the equivalence of the inequalities

$$\frac{\overline{x} - \mu}{\sigma/\sqrt{n}} < z_\alpha \quad \text{and} \quad \overline{x} - z_\alpha \frac{\sigma}{\sqrt{n}} < \mu,$$

with $\mu$ being replaced by 75 and the two observed $\overline{x}$'s by 77.47 and 75.92, respectively. Accordingly, we can simply compute the statistic $Z$ and compare it to an appropriate $z_\alpha$, where $\alpha$ is selected according to the desired confidence coefficient of the one-sided confidence interval.

Thus we could reject $H_0$: $\mu = 75$ if

$$Z = \frac{\overline{X} - 75}{\sigma/\sqrt{n}} \geq z_\alpha$$

for then 75 would not be in that one-sided confidence interval. At this point we recognize that using this **test of a statistical hypothesis** could lead to two types of errors. It would be possible for this $\mu$ to equal 75, and still the $\overline{X}$ be such that the computed $Z \geq z_\alpha$ so that we reject $H_0$: $\mu = 75$; this is a **Type I error**, rejecting the null hypothesis when it is true. The other type of error could occur if the mean $\mu$ is actually greater than 75 and the $\overline{X}$ is such that the computed $Z < z_\alpha$ and we would not reject $H_0$: $\mu = 75$ in favor of $H_1$: $\mu > 75$. That is, suppose $\mu$ is actually equal to 76.5 and $\overline{x}$ is such that we do not reject $H_0$: $\mu = 75$; this is a **Type II error**, accepting the null hypothesis when in fact the alternative hypothesis is true.

Let us compute the probabilities of these two types of errors in our particular situation when we have assumed $\sigma = 10$ and taken $n = 64$ and $\alpha = 0.05$. The probability of the Type I error is

$$P\left(\frac{\overline{X} - 75}{10/\sqrt{64}} \geq 1.645; \mu = 75\right) \approx 0.05 = \alpha$$

because $(\overline{X} - 75)/(10/\sqrt{64})$ has an approximate $N(0, 1)$ distribution when $\mu = 75$. That is, that $\alpha = 0.05$ associated with the $100(1 - \alpha)\% = 95\%$ confidence interval is the probability of the Type I error and this probability $\alpha = 0.05$ is called the **significance level of the test**.

Say we compute the probability of the Type II error when $\mu = 76.5$. That is,

$$P\left(\frac{\overline{X} - 75}{10/\sqrt{64}} < 1.645; \mu = 76.5\right)$$

$$= P\left(\frac{\overline{X} - 76.5}{10/\sqrt{64}} < 1.645 - \frac{76.5 - 75}{10/\sqrt{64}} = 0.445; \mu = 76.5\right) \approx 0.67$$

since $(\overline{X} - 76.5)/(10/\sqrt{64})$ has an approximate $N(0, 1)$ distribution when $\mu = 76.5$. This is a fairly large probability of accepting $H_0: \mu = 75$ when in fact $\mu$ is actually equal to 76.5. Often this probability of the Type II error is denoted by $\beta = 0.67$.

While $\alpha = 0.05$ is reasonably low, $\beta = 0.67$ is too high; and we might want to rethink this test so that $\beta$ is much smaller, say $\beta = 0.10$. We will see that this requires a much larger sample size than $n = 64$. Note that the test is really this: if $\overline{X}$ gets large, say $\overline{X} \geq c$, we reject $H_0: \mu = 75$; but if $\overline{X} < c$, then we do not reject $H_0: \mu = 75$. So we want to find $n$ and $c$ such that

$$\alpha = 0.05 = P(\overline{X} \geq c; \mu = 75) = P\left(\frac{\overline{X} - 75}{10/\sqrt{n}} \geq \frac{c - 75}{10/\sqrt{n}}; \mu = 75\right)$$

and

$$\beta = 0.10 = P(\overline{X} < c; \mu = 76.5) = P\left(\frac{\overline{X} - 76.5}{10/\sqrt{n}} < \frac{c - 76.5}{10/\sqrt{n}}; \mu = 76.5\right).$$

Since in each case $\overline{X}$ has been standardized and has an approximate $N(0, 1)$ distribution, it must be true that

$$\frac{c - 75}{10/\sqrt{n}} = 1.645 \qquad \text{and} \qquad \frac{c - 76.5}{10/\sqrt{n}} = -1.282. \tag{4.3-1}$$

If the second equation is subtracted from the first, we obtain

$$\frac{1.5\sqrt{n}}{10} = 2.927$$

and thus

$$\sqrt{n} = \frac{(2.927)(10)}{1.5} = 19.51.$$

That is, $n = 380.8$ or, in fact, $n = 381$ since $n$ must be an integer. This means that the constant $c$ is about

$$c = 75 + 1.645 \frac{10}{\sqrt{n}} = 75.84.$$

That is, the test statistic is now

$$\frac{\overline{X} - 75}{10/\sqrt{381}} \geq 1.645$$

to achieve the desired $\alpha = 0.05$ and the desired $\beta = 0.10$, when $\mu = 76.5$.

The practical meaning of this is that we must have a large lecture of about 400 students, probably with discussion sections of reasonable size handled by teaching assistants. Is this the best way to try this reform method? Probably not. However, can one university afford to have about six to eight mid-size classes with different instructors using the reform method? Probably not. Even ideally it would be quite impossible to have one university create about 16 sections of 25 students, each with a qualified instructor. We believe that the best solution would be to have several universities or colleges working together on this project. The various instructors, all of whom believe in the reform movement, would need to agree on some common ground rules of selecting students at random and then on reform teaching techniques in covering certain topics. Hopefully this would be a reasonable way to get a sample size of around 400 students, who would take this common final assessment test on the same day, hopefully about the same hour, or it would be possible for students at one college to pass some information on to others at another college.

Another illustration of these new concepts concerning statistical tests is the following. Let $X$ equal the breaking strength of a steel bar. If the steel bar is manufactured by process I, $X$ is $N(50, 36)$. It is hoped that if process II (a new process) is used, $X$ will be $N(55, 36)$. Given a large number of steel bars manufactured by process II, how could we test whether the increase in the mean breaking strength was realized?

That is, we are assuming $X$ is $N(\mu, 36)$ and $\mu$ is equal to 50 or 55. We want to test the null hypothesis $H_0$: $\mu = 50$ against alternative hypothesis $H_1$: $\mu = 55$. Note that each of these hypotheses completely specifies the distribution of $X$. That is, $H_0$ states that $X$ is $N(50, 36)$, and $H_1$ states that $X$ is $N(55, 36)$. A hypothesis that completely specifies the distribution of $X$ is called a **simple hypothesis**; otherwise it is called a **composite hypothesis** (composed of at least two simple hypotheses). For example, $H_1$: $\mu > 50$ would be a composite hypothesis because it is composed of all normal distributions with $\sigma^2 = 36$ and means greater than 50. In order to test which of the two hypotheses, $H_0$ or $H_1$, is true, we shall set up a rule based on the breaking strengths $x_1, x_2, \ldots, x_n$ of $n$ bars (the observed values of a random sample of size $n$ from this new normal distribution). The rule leads to a decision to accept or reject $H_0$; so it is necessary to partition the sample space into two parts, say $C$ and $C'$, so that if $(x_1, x_2, \ldots, x_n) \in C$, $H_0$ is rejected, and if $(x_1, x_2, \ldots, x_n) \in C'$, $H_0$ is accepted (not rejected). The rejection region $C$ for $H_0$ is called the **critical region** for the test. Often the partitioning of the sample space is specified in terms of the values of a statistic called the **test statistic**. In this illustration we could let $\overline{X}$ be the test statistic and, for example, take $C = \{(x_1, x_2, \ldots, x_n) : \overline{x} \geq 53\}$. We could then define the critical region as those values of the test statistic for which $H_0$ is rejected. That is, the given critical region is equivalent to defining $C = \{\overline{x} : \overline{x} \geq 53\}$ in the $\overline{x}$ space. If $(x_1, x_2, \ldots, x_n) \in C$ when $H_0$ is true, $H_0$ would be rejected when it is true, a Type I error. If $(x_1, x_2, \ldots, x_n) \in C'$ when $H_1$ is true, $H_0$ would be accepted, that is, not rejected, when in fact $H_1$ is true, a Type II error. Recall that the probability of a Type I error is called the significance level of the test and is denoted by $\alpha$. That is,

$$\alpha = P[(X_1, X_2, \ldots, X_n) \in C; H_0]$$

is the probability that $(X_1, X_2, \ldots, X_n)$ falls in $C$ when $H_0$ is true. The probability of a Type II error is denoted by $\beta$; that is,

$$\beta = P[(X_1, X_2, \ldots, X_n) \in C'; \, H_1]$$

is the probability of accepting (failing to reject) $H_0$ when it is false. For illustration, suppose $n = 16$ bars were tested and $C = \{\bar{x} : \bar{x} \geq 53\}$. Then $\overline{X}$ is $N(50, 36/16)$ when $H_0$ is true and is $N(55, 36/16)$ when $H_1$ is true. Thus

$$\alpha = P(\overline{X} \geq 53; \, H_0) = P\left(\frac{\overline{X} - 50}{6/4} \geq \frac{53 - 50}{6/4}; \, H_0\right)$$

$$= 1 - \Phi(?) = 0.0228$$

and

$$\beta = P(\overline{X} < 53; \, H_1) = P\left(\frac{\overline{X} - 55}{6/4} < \frac{53 - 55}{6/4}; \, H_1\right)$$

$$= \Phi\left(-\frac{4}{3}\right) = 1 - 0.9087 = 0.0913.$$

See Figure 4.3-1 for the graphs of the probability density functions of $\overline{X}$ when $H_0$ and $H_1$ are true, respectively. Note that a decrease in the size of $\alpha$ leads to an increase in the size of $\beta$, and vice versa. Both $\alpha$ and $\beta$ can be decreased if the sample size $n$ is increased.

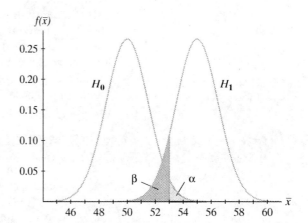

**Figure 4.3-1** p.d.f. of $\overline{X}$ under $H_0$ and $H_1$

The two examples that we have considered illustrate what we call **one-sided tests**: In each case we tested $H_0$: $\mu = \mu_0$ against $H_1$: $\mu > \mu_0$ by rejecting $H_0$ if $\overline{X} \geq c$ or $(\overline{X} - \mu_0)/(\sigma/\sqrt{n}) \geq z_\alpha$. Of course, we sometimes want to decrease the mean (say, as in golf) and we would consider $H_0$: $\mu = \mu_0$ against $H_1$: $\mu < \mu_0$ by the one-sided test of rejecting $H_0$ if $\overline{X} \leq c$ or $(\overline{X} - \mu_0)/(\sigma/\sqrt{n}) \leq -z_\alpha$. There are times in which two-sided alternatives $H_1$: $\mu \neq \mu_0$ occur for we honestly do not know in which direction $\mu$ will go after our changes (often from research). Our **two-sided tests** will be of the form of rejecting $H_0$: $\mu = \mu_0$ if $|\overline{X} - \mu_0| \geq c$ or $|(\overline{X} - \mu_0)/(\sigma/\sqrt{n})| \geq z_{\alpha/2}$. Note $z_{\alpha/2}$ is used so that $P(|Z| \geq z_{\alpha/2}) = \alpha$, where $Z$ is the standardized value of $\overline{X}$ when $H_0$: $\mu = \mu_0$ is true. Again note that this rule is equivalent to rejecting $H_0$: $\mu = \mu_0$ and accepting $H_1$: $\mu \neq \mu_0$ if $\mu_0$ is not in the open $100(1 - \alpha)$% confidence interval

$$(\bar{x} - z_{\alpha/2}\,\sigma/\sqrt{n}, \bar{x} + z_{\alpha/2}\,\sigma/\sqrt{n}).$$

These tests are summarized in Table 4.3-1.

### Table 4.3-1  Tests of Hypotheses About One Mean, Variance Known

| $H_0$ | $H_1$ | Critical Region |
|---|---|---|
| $\mu = \mu_0$ | $\mu > \mu_0$ | $z \geq z_\alpha$ or $\bar{x} \geq \mu_0 + z_\alpha\sigma/\sqrt{n}$ |
| $\mu = \mu_0$ | $\mu < \mu_0$ | $z \leq -z_\alpha$ or $\bar{x} \leq \mu_0 - z_\alpha\sigma/\sqrt{n}$ |
| $\mu = \mu_0$ | $\mu \neq \mu_0$ | $|z| \geq z_{\alpha/2}$ or $|\bar{x} - \mu_0| \geq z_{\alpha/2}\sigma/\sqrt{n}$ |

The point to stress in this section is that all the confidence intervals that we have considered thus far in this text can be converted to tests of hypotheses about those parameters. Say the parameter under consideration is $\theta$: examples of $\theta$ are $\theta = \mu$, $\theta = p$, $\theta = \sigma^2$, $\theta = \sigma_1^2/\sigma_2^2$, and $\theta = \mu_1 - \mu_2$; and we could be using approximate normal, $\chi^2$, $t$, or $F$ random variables to construct those confidence intervals for $\theta$. If the alternative hypothesis $H_1$ is two-sided, we construct a two-sided $100(1 - \alpha)$ percent confidence interval for $\theta$. If $\theta_0$ of $H_0$: $\theta = \theta_0$ is not in that interval, we reject $H_0$ and accept $H_1$: $\theta \neq \theta_0$. For example, in the two-sample case we often want to test $H_0$: $\mu_1 = \mu_2$ or $\theta = \mu_1 - \mu_2 = 0$; if zero is not in the $100(1-\alpha)$ percent confidence interval for $\mu_1 - \mu_2$, we reject $H_0$ and the significance level of the test is that $\alpha$. In particular, if it is a 95 percent confidence interval, then $\alpha = 0.05$ is the significance level of the test.

If the alternative is one-sided like $H_1$: $\theta > \theta_0$, we would construct a one-sided $100(1-\alpha)$ percent confidence interval having a lower bound for $\theta$. If $\theta_0$ is not in that confidence interval, we reject $H_0$: $\theta = \theta_0$ and accept $H_1$: $\theta > \theta_0$ and the significance level is $\alpha$. On the other hand, $H_0$: $\theta = \theta_0$ is rejected in favor of $H_0$: $\theta < \theta_0$ if a one-sided $100(1 - \alpha)$ percent confidence interval with an upper bound does not include $\theta_0$. These ideas are illustrated with examples and exercises throughout this chapter.

## EXERCISES 4.3

**4.3-1** Assume that IQ scores for a certain population are approximately $N(\mu, 100)$. To test $H_0$: $\mu = 110$ against the one-sided alternative hypothesis $H_1$: $\mu > 110$, we take a random sample of size $n = 16$ from this population and observe $\bar{x} = 113.5$. Do we accept or reject $H_0$ at the

**(a)** 5% significance level?

**(b)** 10% significance level?

**4.3-2** Let $X$ equal the Brinell hardness measurement of ductile iron subcritically annealed. Assume that the distribution of $X$ is $N(\mu, 100)$. We shall test the null hypothesis $H_0$: $\mu = 170$ against the alternative hypothesis $H_1$: $\mu > 170$ using $n = 25$ observations of $X$.

**(a)** Define the test statistic and a critical region that has a significance level of $\alpha = 0.05$. Sketch a figure showing this critical region.

**(b)** A random sample of $n = 25$ observations of $X$ yielded the following measurements:

| | | | | | | | | |
|---|---|---|---|---|---|---|---|---|
| 170 | 167 | 174 | 179 | 179 | 156 | 163 | 156 | 187 |
| 156 | 183 | 179 | 174 | 179 | 170 | 156 | 187 | |
| 179 | 183 | 174 | 187 | 167 | 159 | 170 | 179 | |

Calculate the value of the test statistic and clearly give your conclusion.

**4.3-3** A certain size bag is designed to hold 25 pounds of potatoes. A farmer fills such bags in the field. Assume that the weight $X$ of potatoes in a bag is $N(\mu, 9)$. We shall test the null hypothesis $H_0: \mu = 25$ against the alternative hypothesis $H_1: \mu < 25$. Let $X_1, X_2, X_3, X_4$ be a random sample of size 4 from this distribution, and let the critical region $C$ for this test be defined by $\bar{x} \le 22.5$, where $\bar{x}$ is the observed value of $\overline{X}$.

**(a)** What is the significance level, $\alpha$, for your test?

**(b)** If the random sample of four bags of potatoes yielded the values $x_1 = 21.24$, $x_2 = 24.81$, $x_3 = 23.62$, $x_4 = 26.82$, would you accept or reject $H_0$ using your test?

**4.3-4** Assume that SAT mathematics scores of students who attend small liberal arts colleges are $N(\mu, 8100)$. We shall test $H_0: \mu = 530$ against the alternative hypothesis $H_1: \mu < 530$. Given a random sample of size $n = 36$ SAT mathematics scores, let the critical region be defined by $C = \{\bar{x}: \bar{x} < 510.77\}$, where $\bar{x}$ is the observed mean of the sample. What is the value of the significance level of this test?

**4.3-5** Let $X$ equal the yield of alfalfa in tons per acre per year. Assume that $X$ is $N(1.5, 0.09)$. It is hoped that a new fertilizer will increase the average yield. We shall test the null hypothesis $H_0: \mu = 1.5$ against the alternative hypothesis $H_1: \mu > 1.5$. Assume that the variance continues to equal $\sigma^2 = 0.09$ with the new fertilizer. Using $\overline{X}$, the mean of a random sample of size $n$, as the test statistic, reject $H_0$ if $\bar{x} \ge c$. Find $n$ and $c$ so that $P(\overline{X} \ge c; \mu = 1.5) = 0.05$ and $P(\overline{X} \ge c; \mu = 1.7) = 0.95$; that is, $\alpha = 0.05$ and $\beta = 1 - 0.95 = 0.05$ when $\mu = 1.7$.

**4.3-6** Let $X$ equal the number of pounds of butterfat produced by a Holstein cow during the 305-day milking period following the birth of a calf. Assume that the distribution of $X$ is $N(\mu, 140^2)$. To test the null hypothesis $H_0: \mu = 715$ against the alternative hypothesis $H_1: \mu < 715$, let the critical region be defined by $C = \{\bar{x}: \bar{x} \le 668.94\}$, where $\bar{x}$ is the sample mean of $n = 25$ butterfat weights from 25 cows selected at random.

**(a)** What is your conclusion using the following 25 observations of $X$?

| | | | | | | | | |
|---|---|---|---|---|---|---|---|---|
| 425 | 710 | 661 | 664 | 732 | 714 | 934 | 761 | 744 |
| 653 | 725 | 657 | 421 | 573 | 535 | 602 | 537 | 405 |
| 874 | 791 | 721 | 849 | 567 | 468 | 975 | | |

**(b)** What is the significance level of this test?

**4.3-7** Let $\overline{X}$ be the mean of a random sample of size $n$ from $N(\mu, \sigma^2 = 25)$. To test $H_0: \mu = 20$ against $H_1: \mu < 20$, we reject $H_0$ if $\overline{X} \le c$. Find $n$ and $c$ so that $P(\overline{X} \le c; \mu = 20) = 0.10$ and $P(\overline{X} \le c; \mu = 19) = 0.90$; that is, $\alpha = 0.10$ and $\beta = 1 - 0.90 = 0.10$ when $\mu = 19$.

## 4.4  BASIC TESTS CONCERNING ONE PARAMETER

In Section 4.3 we stressed the relationship between confidence intervals for parameters and tests of statistical hypotheses about those parameters. In this section we summarize many of the resulting tests, but first we introduce another term that is simply a different way of considering the value of a test statistic. Most computer programs automatically print this out; it is called the **probability value** or, for brevity, the **p-value**. The p-value associated with a test is the probability that we obtain a value of the test statistic that is at least as extreme (in the direction of the alternative) as the observed value of our test statistic; this probability is calculated when $H_0$ is true. Rather than select the critical region ahead of time, the p-value of a test can be reported and the reader then makes a decision.

To understand the p-value clearly, consider the test statistic

$$Z - \frac{\overline{X} - \mu_0}{\sigma/\sqrt{n}}$$

used in the illustration in Section 4.3. There to test $H_0$: $\mu = 75 = \mu_0$ against $H_1$: $\mu > 75 = \mu_0$, we assumed that we knew $\sigma \approx 10$ and we found that we needed about $n = 400$ observations to have a reasonable test. That is, we needed about $n = 400$ observations to have the probabilities of the two types of errors to be about $\alpha = 0.05$ and $\beta \approx 0.10$. So the test was to reject $H_0$: $\mu = 75$ and accept $H_1$: $\mu > 75$ if the test statistic

$$\frac{\overline{X} - 75}{10/\sqrt{400}} = \frac{\overline{X} - 75}{0.5} \geq 1.645$$

or, equivalently, if $\overline{X} \geq 75.8225$. Say we observe $\overline{X}$ to be $\bar{x} = 76$ so that by this test we clearly reject $H_0$: $\mu = 75$ and accept $H_1$: $\mu > 75$. However, note that we could have computed the test statistic to be $(76 - 75)/0.5 = 2$. Since $(\overline{X} - 75)/0.5$ is $N(0, 1)$ when $H_0$: $\mu = 75$, the probability of a standard normal random variable being greater than or equal to two is

$$p\text{-value} = P(Z \geq 2) = 0.0228 < 0.05.$$

If the $p$-value is less than $\alpha$, we reject $H_0$ and accept $H_1$. Please observe that, in this case, the $p$-value $= 0.0228 \leq \alpha = 0.05$ is exactly the same as the test statistic $z = 2 \geq 1.645$. And if the $p$-value would be greater than 0.05 then the test statistic would be less than 1.645. The $p$-value is simply another statistic which is a one-to-one transformation of the test statistic: given one of these statistics, we can compute the other one. However, many statisticians today like to use the $p$-values to make decisions associated with various statistical hypotheses because they feel as if the value (tail end probability) of the $p$-value is more meaningful that that of $z$.

If in the previous discussion the alternative hypothesis had been the two-sided one $H_1$: $\mu \neq 75$, then we would need to double the previous $p$-value for that would be the probability that the absolute value of a standard normal random variable exceeded two, namely

$$p\text{-value} = P(|Z| \geq 2) = 2(0.0228) = 0.0456,$$

which would still lead to rejection if $\alpha = 0.05$.

**EXAMPLE 4.4-1**    Assume that the underlying distribution is normal with unknown mean $\mu$ but known variance $\sigma^2 = 100$. Say we are testing $H_0$: $\mu = 60$ against $H_1$: $\mu > 60$ with a sample mean $\overline{X}$ based on $n = 52$ observations. Suppose that we obtain the observed sample mean of $\bar{x} = 62.75$. If we compute the probability of obtaining an $\bar{x}$ of that value of 62.75 or greater when $\mu = 60$, then we obtain the $p$-value associated with $\bar{x} = 62.75$. That is,

$$p\text{-value} = P(\overline{X} \geq 62.75; \mu = 60)$$

$$= P\left(\frac{\overline{X} - 60}{10/\sqrt{52}} \geq \frac{62.75 - 60}{10/\sqrt{52}}; \mu = 60\right)$$

$$= 1 - \Phi\left(\frac{62.75 - 60}{10/\sqrt{52}}\right) = 1 - \Phi(1.983) = 0.0237.$$

If this $p$-value is small, we tend to reject the hypothesis $H_0$: $\mu = 60$. For example, rejection of $H_0$: $\mu = 60$ if the $p$-value is less than or equal to $\alpha = 0.05$ is exactly the same as rejection if

$$\bar{x} \geq 60 + (1.645)\left(\frac{10}{\sqrt{52}}\right) = 62.718.$$

Here

$$p\text{-value} = 0.0237 \leq \alpha = 0.05 \qquad \text{and} \qquad z = \frac{\bar{x} - 60}{10/\sqrt{52}} = 1.983 \geq 1.645.$$

To help the reader keep the definition of $p$-value in mind, we note that it can be thought of as that **tail-end probability**, under $H_0$, of the distribution of the statistic, here $Z$, beyond the observed value of the statistic. See Figure 4.4-1 for the $p$-value associated with $z = 1.983$.

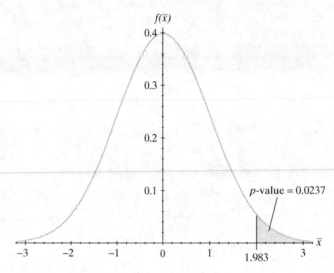

**Figure 4.4-1** Illustration of $p$-value

Let us begin the summary of the various tests that are associated with the confidence intervals that we have studied thus far. The first has the assumption that the underlying distribution is $N(\mu, \sigma^2)$, where $\sigma^2$ is known, and the test statistic is

$$Z = \frac{\overline{X} - \mu_0}{\sigma/\sqrt{n}}.$$

The three different tests of $H_0$: $\mu = \mu_0$ for the three different alternatives were given in Table 4.3-1.

We now consider an underlying distribution which is $N(\mu, \sigma^2)$, but $\sigma^2$ is now unknown, and the test statistic is

$$T = \frac{\overline{X} - \mu_0}{S/\sqrt{n}}.$$

$T$ has a $T(n-1)$ distribution. The tests are given in Table 4.4-1. The third test in the table is associated with the two-sided confidence interval given in Section 4.2.

**EXAMPLE 4.4-2**  Let $X$ (in millimeters) equal the growth in 15 days of a tumor induced in a mouse. Assume that the distribution of $X$ is $N(\mu, \sigma^2)$. We shall test the null hypothesis $H_0$: $\mu = \mu_0 = 4.0$ millimeters against the two-sided alternative hypothesis $H_1$: $\mu \neq 4.0$. If we use $n = 9$ observations and a significance level of $\alpha = 0.10$, the critical region,

## Table 4.4-1 Tests of Hypotheses for One Mean, Variance Unknown

| $H_0$ | $H_1$ | Critical Region |
|-------|-------|-----------------|
| $\mu = \mu_0$ | $\mu > \mu_0$ | $t \geq t_\alpha(n-1)$ or $\bar{x} \geq \mu_0 + t_\alpha(n-1)s/\sqrt{n}$ |
| $\mu = \mu_0$ | $\mu < \mu_0$ | $t \leq -t_\alpha(n-1)$ or $\bar{x} \leq \mu_0 - t_\alpha(n-1)s/\sqrt{n}$ |
| $\mu = \mu_0$ | $\mu \neq \mu_0$ | $|t| \geq t_{\alpha/2}(n-1)$ or $|\bar{x} - \mu_0| \geq t_{\alpha/2}(n-1)s/\sqrt{n}$ |

shown in Figure 4.4-2($a$), is

$$|t| = \frac{|\bar{x} - 4.0|}{s/\sqrt{9}} \geq t_{\alpha/2}(8) = 1.860.$$

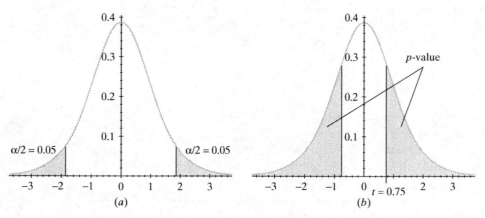

Figure 4.4-2  Test about the mean of tumor growths

If we are given that $n = 9$, $\bar{x} = 4.3$, and $s = 1.2$, we see that

$$t = \frac{4.3 - 4.0}{1.2/\sqrt{9}} = \frac{0.3}{0.4} = 0.75.$$

Thus

$$|t| = |0.75| < 1.860$$

and we accept (do not reject) $H_0$: $\mu = 4.0$ at the $\alpha = 10\%$ significance level. See Figure 4.4-2($b$). The $p$-value is

$$p\text{-value} = P(|T| \geq 0.75) = 2P(T \geq 0.75).$$

With our $t$-tables with 8 degrees of freedom, we can not find this $p$-value exactly. It is about 0.50 because

$$P(|T| \geq 0.706) = 2P(T \geq 0.706) = 0.50,$$

and the computer gives the $p$-value to be 0.4748. See Figure 4.4-2.

**Remark**  In discussing the test of a statistical hypothesis, the word *accept $H_0$* might better be replaced by *do not reject $H_0$* or by *fail to reject $H_0$*. That is, if, in Example 4.4-2, $\bar{x}$ is close enough to 4.0 so that we accept $\mu = 4.0$, we do not want that acceptance to imply that $\mu$ is actually equal to 4.0. We want to say that the data do not deviate enough from $\mu = 4.0$ for us to reject that hypothesis; that is, we do not reject $\mu = 4.0$ with these observed data. With this understanding, we sometimes use accept, and sometimes fail to reject or do not reject, the null hypothesis.

There is, oftentimes, interest in comparing the means of two different distributions or populations. We must consider two situations: that in which $X$ and $Y$ are dependent, and that in which $X$ and $Y$ are independent. If $X$ and $Y$ are dependent, let $W = X - Y$, and the hypothesis that $\mu_X = \mu_Y$ would be replaced with the hypothesis $H_0: \mu_W = 0$. For example, suppose that $X$ and $Y$ equal the resting pulse rate for a person before and after taking an 8-week program in aerobic dance. We would be interested in testing $H_0: \mu_W = 0$ (no change) against $H_1: \mu_W > 0$ (the aerobic dance program decreased resting pulse rate). Because $X$ and $Y$ are measurements on the same person, $X$ and $Y$ are clearly dependent. If we can assume that the distribution of $W$ is (approximately) $N(\mu_W, \sigma^2)$, then the appropriate $t$-test for a single mean could be used, selecting from Table 4.4-1. This is often called a **paired $t$ test**.

EXAMPLE 4.4-3   Twenty-four girls in the 9th and 10th grades were put on an ultra-heavy rope-jumping program. Someone thought that such a program would increase their speed when running the 40-yard dash. Let $W$ equal the difference in time to run the 40-yard dash-the "before program time" minus the "after program time." Assume that the distribution of $W$ is (approximately) $N(\mu_W, \sigma_W^2)$. We shall test the null hypothesis $H_0: \mu_W = 0$ against the alternative hypothesis $H_1: \mu_W > 0$. The test statistic and critical region that has an $\alpha = 0.05$ significance level are given by

$$t = \frac{\overline{w} - 0}{s_w / \sqrt{24}} \geq t_{0.05}(23) = 1.714.$$

The following data give the difference in time that it took each girl to run the 40-yard dash, with positive numbers indicating a faster time after the program:

| | | | | | | | |
|---|---|---|---|---|---|---|---|
| 0.28 | 0.01 | 0.13 | 0.33 | −0.03 | 0.07 | −0.18 | −0.14 |
| −0.33 | 0.01 | 0.22 | 0.29 | −0.08 | 0.23 | 0.08 | 0.04 |
| −0.30 | −0.08 | 0.09 | 0.70 | 0.33 | −0.34 | 0.50 | 0.06 |

For these data, $\overline{w} = 0.079$ and $s_w = 0.255$. Thus the observed value of the test statistic is

$$t = \frac{0.079 - 0}{0.255 / \sqrt{24}} = 1.518.$$

Since $1.518 < 1.714$, the null hypothesis is not rejected. Note, however, that $t_{0.10}(23) = 1.319$ and $t = 1.518 > 1.319$. Thus, the null hypothesis would be rejected at an $\alpha = 0.10$ significance level. Another way of saying this is that

$$0.05 < p\text{-value} < 0.10 \qquad \text{or} \qquad p\text{-value} = 0.0713 \ (\text{using the computer}).$$

It would be instructive for you to draw a figure illustrating this.

Let us continue with the $N(\mu, \sigma^2)$ assumption, but concern ourselves with tests about the variance $\sigma^2$, in which the null hypothesis is $H_0: \sigma^2 = \sigma_0^2$. The test statistic is

$$\chi^2 = \frac{(n-1)S^2}{\sigma_0^2},$$

which, under $H_0$, has a chi-square distribution with $r = n - 1$. The tests are summarized in Table 4.4-2.

| $H_0$ | $H_1$ | Critical Region |
|---|---|---|
| **Table 4.4-2 Tests About the Variance** | | |
| $\sigma^2 = \sigma_0^2$ | $\sigma^2 > \sigma_0^2$ | $s^2 \geq \dfrac{\sigma_0^2 \chi_\alpha^2(n-1)}{n-1}$ or $\chi^2 \geq \chi_\alpha^2(n-1)$ |
| $\sigma^2 = \sigma_0^2$ | $\sigma^2 < \sigma_0^2$ | $s^2 \leq \dfrac{\sigma_0^2 \chi_{1-\alpha}^2(n-1)}{n-1}$ or $\chi^2 \leq \chi_{1-\alpha}^2(n-1)$ |
| $\sigma^2 = \sigma_0^2$ | $\sigma^2 \neq \sigma_0^2$ | $s^2 \leq \dfrac{\sigma_0^2 \chi_{1-\alpha/2}^2(n-1)}{n-1}$ or $s^2 \geq \dfrac{\sigma_0^2 \chi_{\alpha/2}^2(n-1)}{n-1}$ |
| | | or $\chi^2 \leq \chi_{1-\alpha/2}^2(n-1)$ or $\chi^2 \geq \chi_{\alpha/2}^2(n-1)$ |

EXAMPLE 4.4-4    Suppose that $n = 23$ and $\alpha = 0.05$. To test $H_0$: $\sigma^2 = 100$ against $H_1$: $\sigma^2 \neq 100$, we let $\chi^2 = (n-1)S^2/\sigma_0^2 = 22S^2/100$. Because

$$\chi_{0.975}^2(22) = 10.98 \quad \text{and} \quad \chi_{0.025}^2(22) = 36.78,$$

$H_0$ will be rejected if

$$\chi^2 = \frac{22s^2}{100} \leq 10.98 \quad \text{or} \quad \chi^2 = \frac{22s^2}{100} \geq 36.78$$

or, equivalently, if

$$s^2 \leq c_1 = \frac{100(10.98)}{22} = 49.91 \quad \text{or} \quad s^2 \geq c_2 = \frac{100(36.78)}{22} = 167.18.$$

Given that the observed value of the sample variance was $s^2 = 147.82$, the hypothesis $H_0$: $\sigma^2 = 100$ was not rejected. Note that the 95% confidence interval for $\sigma^2$,

$$\left[ \frac{(22)(147.82)}{36.78}, \frac{(22)(147.82)}{10.98} \right] = [88.42, 296.18],$$

contains $\sigma^2 = 100$. Also, the observed value of the chi-square test statistic is

$$\chi^2 = \frac{22(147.82)}{100} = 32.52.$$

Because

$$10.98 < 32.52 < 36.78,$$

we would again accept $H_0$ as expected. See Figure 4.4-3.

If $H_1$: $\sigma^2 > 100$ had been the alternative hypothesis, $H_0$: $\sigma^2 = 100$ would have been rejected if

$$\chi^2 = \frac{22s^2}{100} \geq \chi_{0.05}^2(22) = 33.92$$

or, equivalently,

$$s^2 \geq \frac{100\chi_{0.05}^2(22)}{22} = \frac{(100)(33.92)}{22} = 154.18.$$

Because

$$\chi^2 = 32.52 < 33.92 \qquad \text{and} \qquad s^2 = 147.82 < 154.18,$$

$H_0$ would not be rejected in favor of this one-sided alternative hypothesis for $\alpha = 0.05$, although we observe that a slightly larger data set might lead to rejection.

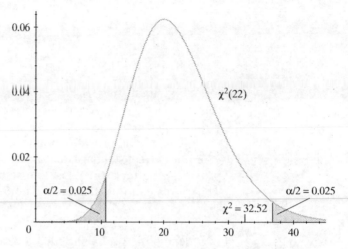

**Figure 4.4-3** Test about the variance of IQ scores

Finally, we turn our attention to the model in which $Y$ is binomial $b(n, p)$, where $p$ is unknown. To test $H_0: p = p_0$ we begin with a procedure that is consistent with the confidence intervals for $p$. Let $\widehat{p} = Y/n$. Then, if the null hypothesis is $H_0: p = p_0$, then the alternative hypothesis $H_1: p < p_0$ is rejected if

$$z = \frac{\widehat{p} - p_0}{\sqrt{\dfrac{\widehat{p}(1 - \widehat{p})}{n}}} \leq -z_\alpha.$$

This is equivalent to the statement that

$$p_0 \notin \left[ 0, \widehat{p} + z_\alpha \sqrt{\dfrac{\widehat{p}(1 - \widehat{p})}{n}} \right),$$

where the latter is a one-sided confidence interval providing an upper bound for $p$. Or if the alternative hypothesis is $H_1: p \neq p_0$, then $H_0$ is rejected if

$$\frac{|\widehat{p} - p_0|}{\sqrt{\dfrac{\widehat{p}(1 - \widehat{p})}{n}}} \geq z_{\alpha/2}.$$

This is equivalent to

$$p_0 \notin \left( \widehat{p} - z_{\alpha/2} \sqrt{\dfrac{\widehat{p}(1 - \widehat{p})}{n}} , \widehat{p} + z_{\alpha/2} \sqrt{\dfrac{\widehat{p}(1 - \widehat{p})}{n}} \right),$$

where the latter is a two-sided confidence interval for $p$.

We summarize tests for $H_0: p = p_0$ in Table 4.4-3.

### Table 4.4-3 Tests of Hypotheses for One Proportion

| $H_0$ | $H_1$ | Critical Region |
|---|---|---|
| $p = p_0$ | $p > p_0$ | $z = \dfrac{y/n - p_0}{\sqrt{\hat{p}(1-\hat{p})/n}} \geq z_\alpha$ |
| $p = p_0$ | $p < p_0$ | $z = \dfrac{y/n - p_0}{\sqrt{\hat{p}(1-\hat{p})/n}} \leq -z_\alpha$ |
| $p = p_0$ | $p \neq p_0$ | $|z| = \dfrac{|y/n - p_0|}{\sqrt{\hat{p}(1-\hat{p})/n}} \geq z_{\alpha/2}$ |

**Remark**  In testing $H_0$: $p = p_0$, some (maybe most) statisticians use $p_0$ rather than $\hat{p}$ in the denominator of $z$. That is, $\sqrt{\hat{p}(1-\hat{p})/n}$ is replaced by $\sqrt{p_0(1-p_0)/n}$. We do not have a strong preference one way or the other since the two methods provide about the same numerical result. The substitution gives the correct standard deviation of $\hat{p}$ when $H_0$: $p = p_0$ is true, but the former provides a better estimate of that standard deviation when $H_0$ is clearly false. The estimate of this standard deviation is often called the standard error of $\hat{p}$.

**EXAMPLE 4.4-5**    Let $p$ denote the probability that a certain tennis player is successful on her first serve. Since $p$ has been about 0.4 in the past, she decides to take lessons to improve that probability. After the lessons, she wants to test $H_0$: $p = 0.4$ (no improvement) against $H_1$: $p > 0.4$ (improvement) by keeping track of her next 200 first serves in future games. It turns out that 92 of them are successful. Thus, with the $\hat{p} = 92/200 = 0.46$ in the estimate of the standard deviation, we obtain

$$z = \frac{0.46 - 0.40}{\sqrt{(0.46)(0.54)/200}} = 1.70 > 1.645$$

and, using $p_0 = 0.4$ in that standard deviation, we obtain

$$z = \frac{0.46 - 0.40}{\sqrt{(0.4)(0.6)/200}} = 1.73 > 1.645.$$

In either case, we would reject $H_0$ in favor of $H_1$ at the $\alpha = 0.05$ significance level. Thus these data seem to indicate that the lessons have helped this tennis player achieve greater success on her first serves.

One of the more interesting statistical problems is the "fill" problem or variations of it. For example, some statistics books consider the situation of filling a 12 ounce bottle with some product and testing $H_0$: $\mu = 12$ against $H_1$: $\mu > 12$. What we really want to know is something about the percent of bottles that are filled with more than 12 ounces of the product. This topic is covered in Section 4.9.

### EXERCISES 4.4

**4.4-1** Assume that the weight of cereal in a "10-ounce box" is $N(\mu, \sigma^2)$. To test $H_0$: $\mu = 10.1$ against $H_1$: $\mu > 10.1$, we take a random sample of size $n = 16$ and observe that $\bar{x} = 10.4$ and $s = 0.4$.

(a) Do we accept or reject $H_0$ at the 5% significance level?

(b) What is the approximate $p$-value of this test?

**4.4-2** Let $X$ equal the forced vital capacity (FVC) in liters for a female college student. (This is the amount of air that a student can force out of her lungs.) Assume that the distribution of $X$ is (approximately) $N(\mu, \sigma^2)$. Suppose it is known that $\mu = 3.4$ liters. A volleyball coach claims that the FVC of volleyball players is greater than 3.4. She plans to test her claim using a random sample of size $n = 9$.

   **(a)** Define the null hypothesis.

   **(b)** Define the alternative (coach's) hypothesis.

   **(c)** Define the test statistic.

   **(d)** Define a critical region for which $\alpha = 0.05$. Draw a figure illustrating your critical region.

   **(e)** Calculate the value of the test statistic given that the random sample yielded the following forced vital capacities:

$$3.4 \quad 3.6 \quad 3.8 \quad 3.3 \quad 3.4 \quad 3.5 \quad 3.7 \quad 3.6 \quad 3.7$$

   **(f)** What is your conclusion?

   **(g)** What is the approximate $p$-value of this test?

**4.4-3** Vitamin $B_6$ is one of the vitamins in a multiple vitamin pill manufactured by a pharmaceutical company. The pills are produced with a mean of 50 milligrams of vitamin $B_6$ per pill. The company believes that there is a deterioration of 1 milligram per month, so that after 3 months they expect that $\mu = 47$. A consumer group suspects that $\mu < 47$ after 3 months.

   **(a)** Define a critical region to test $H_0$: $\mu = 47$ against $H_1$: $\mu < 47$ at an $\alpha = 0.05$ significance level based on a random sample of size $n = 20$.

   **(b)** If the 20 pills yielded a mean of $\bar{x} = 46.94$ with a standard deviation of $s = 0.15$, what is your conclusion?

   **(c)** What is the approximate $p$-value of this test?

**4.4-4** Each of 51 golfers hit three golf balls of brand $X$ and three golf balls of brand $Y$ in a random order. Let $X_i$ and $Y_i$ equal the averages of the distances traveled by the brand $X$ and brand $Y$ golf balls hit by the $i$th golfer, $i = 1, 2, \ldots, 51$. Let $W_i = X_i - Y_i$, $i = 1, 2, \ldots, 51$. Test $H_0$: $\mu_W = 0$ against $H_1$: $\mu_W > 0$, where $\mu_W$ is the mean of the differences. If $\bar{w} = 2.07$ and $s_w^2 = 84.63$, would $H_0$ be accepted or rejected at an $\alpha = 0.05$ significance level?

**4.4-5** A vendor of milk products produces and sells low-fat dry milk to a company that uses it to produce baby formula. In order to determine the fat content of the milk, both the company and the vendor take a sample from each lot and test it for fat content in percent. Ten sets of paired test results are

| Lot Number | Company Test Results ($X$) | Vendor Test Results ($Y$) |
|---|---|---|
| 1 | 0.50 | 0.79 |
| 2 | 0.58 | 0.71 |
| 3 | 0.90 | 0.82 |
| 4 | 1.17 | 0.82 |
| 5 | 1.14 | 0.73 |
| 6 | 1.25 | 0.77 |
| 7 | 0.75 | 0.72 |
| 8 | 1.22 | 0.79 |
| 9 | 0.74 | 0.72 |
| 10 | 0.80 | 0.91 |

Let $\mu_W$ denote the mean of the difference $X - Y$. Test $H_0: \mu_W = 0$ against $H_1: \mu_W > 0$ using a paired $t$-test with the differences. Let $\alpha = 0.05$.

**4.4-6** Let $X$ equal the number of pounds of butterfat produced by a Holstein cow during the 305-day milking period following the birth of a calf. We shall test the null hypothesis $H_0: \sigma^2 = 140^2$ against the alternative hypothesis $H_1: \sigma^2 > 140^2$.

(a) Give the test statistic and a critical region that has a significance level of $\alpha = 0.05$, assuming that there are $n = 25$ observations.  •

(b) Calculate the value of the test statistic and give your conclusion using the following 25 observations of $X$.

$$
\begin{array}{ccccccccc}
425 & 710 & 661 & 664 & 732 & 714 & 934 & 761 & 744 \\
653 & 725 & 657 & 421 & 573 & 535 & 602 & 537 & 405 \\
874 & 791 & 721 & 849 & 567 & 468 & 975 & & \\
\end{array}
$$

(c) Find a 98% one-sided confidence interval that gives an upper bound for $\mu$.

**4.4-7** In May the fill weights of 6-pound boxes of laundry soap had a mean of 6.13 pounds with a standard deviation of 0.095. The goal was to decrease the standard deviation. The company decided to adjust the filling machines and then test $H_0: \sigma = 0.095$ against $H_1: \sigma < 0.095$. In June a random sample of size $n = 20$ yielded $\bar{x} = 6.10$ and $s = 0.065$.

(a) At an $\alpha = 0.05$ significance level, was the company successful?

(b) What is the approximate $p$-value of your test?

**4.4-8** Bowl A contains 100 red balls and 200 white balls; bowl B contains 200 red balls and 100 white balls. Let $p$ denote the probability of drawing a red ball from a bowl, but say $p$ is unknown, since it is unknown whether bowl A or bowl B is being used. We shall test the simple null hypothesis $H_0: p = 1/3$ against the simple alternative hypothesis $H_1: p = 2/3$. Draw three balls at random, one at a time and with replacement from the selected bowl. Let $X$ equal the number of red balls drawn. Then let the critical region be $C = \{x: x = 2, 3\}$. What are the values of $\alpha$ and $\beta$, the probabilities of Type I and Type II errors, respectively?

**4.4-9** Let $Y$ be $b(100, p)$. To test $H_0: p = 0.08$ against $H_1: p < 0.08$, we reject $H_0$ and accept $H_1$ if and only if $Y \leq 6$.

(a) Determine the significance level $\alpha$ of the test.

(b) Find the probability of the Type II error if in fact $p = 0.04$.

**4.4-10** Let $p$ denote the probability that, for a particular tennis player, the first serve is good. Since $p = 0.40$, this player decided to take lessons in order to increase $p$. When the lessons are completed, the hypothesis $H_0: p = 0.40$ will be tested against $H_1: p > 0.40$ based on $n = 25$ trials. Let $y$ equal the number of first serves that are good, and let the critical region be defined by $C = \{y: y \geq 13\}$.

(a) Determine $\alpha = P(Y \geq 13; p = 0.40)$. Use Table II in the Appendix.

(b) Find $\beta = P(Y < 13)$ when $p = 0.60$; that is, $\beta = P(Y \leq 12; p = 0.60)$. Use Table II.

(c) What is the $p$-value associated with $y = 15$?

**4.4-11** Let $Y$ be $b(192, p)$. We reject $H_0: p = 0.75$ and accept $H_1: p > 0.75$ if and only if $Y \geq 152$. Use the normal approximation to determine

(a) $\alpha = P(Y \geq 152; p = 0.75)$.

(b) $\beta = P(Y < 152)$ when $p = 0.80$.

**4.4-12** To determine whether the "one" side on a commercially manufactured die is heavy and the "six" side is light, some students kept track of the number of observed 1s when observing $n = 8000$ rolls of the dice. Let $p$ equal the probability of rolling a one with such a die. We shall test the null hypothesis $H_0: p = 1/6$ against the alternative hypothesis $H_1: p < 1/6$.

(a) Define the test statistic and an $\alpha = 0.05$ critical region.

(b) If $y = 1265$ ones were observed in 8000 rolls, calculate the value of the test statistic and state your conclusion.

(c) Is 1/6 in the 95% one-sided confidence interval providing an upper bound for $p$?

**4.4-13** Let $p$ equal the fraction defective of a certain manufactured item. To test $H_0$: $p = 1/26$ against $H_1$: $p > 1/26$, we inspect $n$ items selected at random and let $Y$ be the number of defective items in this sample. We reject $H_0$ if the observed $y \geq c$. Find $n$ and $c$ so that $\alpha = P(Y \geq c; p = 1/26) = 0.05$ and $P(Y \geq c; p = 1/10) = 0.90$ approximately; that is, $\beta = 0.10$ when $p = 1/10$. Hint: Use either the normal or Poisson approximation to help solve this exercise.

## 4.5   TESTS OF THE EQUALITY OF TWO PARAMETERS

Let independent random variables $X$ and $Y$ have normal distributions $N(\mu_X, \sigma_X^2)$ and $N(\mu_Y, \sigma_Y^2)$, respectively. There are times when we are interested in testing whether the distributions of $X$ and $Y$ are the same. So if the assumption of normality is valid, we would be interested in testing whether the two means are equal and whether the two variances are equal.

We first consider a test of the equality of the two means. When $X$ and $Y$ are independent and normally distributed, we can test hypotheses about their means using the same $t$-statistic that was used for constructing a confidence interval for $\mu_X - \mu_Y$ in Section 4.2. Recall that the $t$-statistic used for constructing the confidence interval assumed that the variances of $X$ and $Y$ are equal. That is why we shall later consider a test for the equality of two variances.

We begin with an example and then give a table that lists some hypotheses and critical regions. A botanist is interested in comparing the growth response of dwarf pea stems to two different levels of the hormone indoleacetic acid (IAA). Using 16-day-old pea plants, the botanist obtains 5-millimeter sections and floats these sections on solutions with different hormone concentrations to observe the effect of the hormone on the growth of the pea stem. Let $X$ and $Y$ denote, respectively, the independent growths that can be attributed to the hormone during the first 26 hours after sectioning for $(0.5)(10)^{-4}$ and $10^{-4}$ levels of concentration of IAA. The botanist would like to test the null hypothesis $H_0$: $\mu_X - \mu_Y = 0$ against the alternative hypothesis $H_1$: $\mu_X - \mu_Y < 0$. If we can assume $X$ and $Y$ are independent and normally distributed with common variance, respective random samples of sizes $n$ and $m$ give a test based on the statistic

$$T = \frac{\overline{X} - \overline{Y}}{\sqrt{\dfrac{(n-1)S_X^2 + (m-1)S_Y^2}{n+m-2}\left(\dfrac{1}{n} + \dfrac{1}{m}\right)}} \tag{4.5-1}$$

$$= \frac{\overline{X} - \overline{Y}}{S_P\sqrt{\dfrac{1}{n} + \dfrac{1}{m}}}$$

where

$$S_P = \sqrt{\frac{(n-1)S_X^2 + (m-1)S_Y^2}{n+m-2}}. \tag{4.5-2}$$

$T$ has a $t$ distribution with $r = n + m - 2$ degrees of freedom when $H_0$ ($\mu_X = \mu_Y$) is true and the variances are (approximately) equal. The hypothesis $H_0$ will be rejected in favor of $H_1$ if the observed value of $T$ is less than $-t_\alpha(n+m-2)$.

**EXAMPLE 4.5-1**   In the preceding discussion, the botanist measured the growths of pea stem segments, in millimeters, for $n = 11$ observations of $X$:

0.8  1.8  1.0  0.1  0.9  1.7  1.0  1.4  0.9  1.2  0.5

and $m = 13$ observations of $Y$:

$$1.0 \quad 0.8 \quad 1.6 \quad 2.6 \quad 1.3 \quad 1.1 \quad 2.4$$
$$1.8 \quad 2.5 \quad 1.4 \quad 1.9 \quad 2.0 \quad 1.2$$

For these data, $\bar{x} = 1.03$, $s_x^2 = 0.24$, $\bar{y} = 1.66$, and $s_y^2 = 0.35$. The critical region for testing $H_0: \mu_X - \mu_Y = 0$ against $H_1: \mu_X - \mu_Y < 0$ is $t \leq -t_{0.05}(22) = -1.717$, where $t$ is the two-sample $t$ found in Equation 4.5-1. Since

$$t = \frac{1.03 - 1.66}{\sqrt{\dfrac{10(0.24) + 12(0.35)}{11 + 13 - 2} \left( \dfrac{1}{11} + \dfrac{1}{13} \right)}}$$

$$= -2.81 \; < \; -1.717,$$

$H_0$ is clearly rejected at an $\alpha = 0.05$ significance level. Notice that the approximate $p$-value of this test is 0.005 because $-t_{0.005}(22) = -2.819$. See Figure 4.5-1. Also, the sample variances do not differ too much; thus most statisticians would use this two-sample $t$-test.

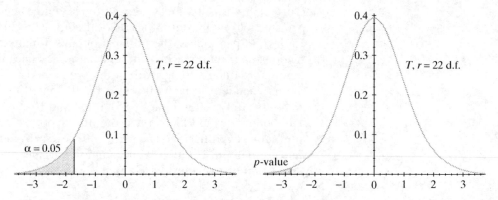

**Figure 4.5-1** Critical region and $p$-value for pea stem growths

It is also instructive to construct box-and-whisker diagrams to gain a visual comparison of the two samples. For these two sets of data, the five-number summaries (minimum, three quartiles, maximum) are

$$0.1 \quad 0.8 \quad 1.0 \quad 1.4 \quad 1.8$$

for the $X$ sample and

$$0.8 \quad 1.15 \quad 1.6 \quad 2.2 \quad 2.6$$

for the $Y$ sample. The two box plots are shown in Figure 4.5-2.

Table 4.5-1 summarizes the different tests using the $T$-statistic. Of course, if we know the variances, we would use the statistic

$$Z = \frac{\overline{X} - \overline{Y}}{\sqrt{\dfrac{\sigma_X^2}{n} + \dfrac{\sigma_Y^2}{m}}}$$

**Figure 4.5-2**  Box plots for pea stem growths

| Table 4.5-1 | Tests of Hypotheses for Equality of Two Means | |
|---|---|---|
| $H_0$ | $H_1$ | **Critical Region** |
| $\mu_X = \mu_Y$ | $\mu_X > \mu_Y$ | $t \geq t_\alpha(n+m-2)$ or $\bar{x} - \bar{y} \geq t_\alpha(n+m-2)s_p\sqrt{1/n + 1/m}$ |
| $\mu_X = \mu_Y$ | $\mu_X < \mu_Y$ | $t \leq -t_\alpha(n+m-2)$ or $\bar{x} - \bar{y} \leq -t_\alpha(n+m-2)s_p\sqrt{1/n + 1/m}$ |
| $\mu_X = \mu_Y$ | $\mu_X \neq \mu_Y$ | $|t| \geq t_{\alpha/2}(n+m-2)$ or $|\bar{x} - \bar{y}| \geq t_{\alpha/2}(n+m-2)s_p\sqrt{1/n + 1/m}$ |

to test $H_0$: $\mu_X = \mu_Y$; and, under $H_0$, $Z$ would be $N(0,1)$ or at least approximately so if the underlying distributions are not normal. If the variances are not known but the sample sizes large enough, we would use $Z$ with $\sigma_X^2$ and $\sigma_Y^2$ replaced by $S_X^2$ and $S_Y^2$, respectively, and the resulting $Z$ would be approximately $N(0,1)$.

With the previous normal and independent assumptions, we test $H_0$: $\sigma_X^2 = \sigma_Y^2$ using the statistic

$$F = \frac{\dfrac{(n-1)S_X^2}{\sigma_X^2(n-1)}}{\dfrac{(m-1)S_Y^2}{\sigma_Y^2(m-1)}} = \frac{S_X^2}{S_Y^2}$$

which has an $F$ distribution with $r_1 = n-1$ and $r_2 = m-1$ degrees of freedom provided $H_0$ is true. The tests are summarized in Table 4.5-2. Recall that $1/F$, the reciprocal of $F$, has an $F$ distribution with $m-1$ and $n-1$ degrees of freedom so all critical regions may be written in terms of right-tail rejection regions so that the critical values can be selected easily from Appendix Table VII. The first and second tests in the table are associated with one-sided confidence intervals for $\sigma_X^2/\sigma_Y^2$.

**EXAMPLE 4.5-2**   A biologist who studies spiders believes that not only do female green lynx spiders tend to be longer than their male counterparts but also that the lengths of the female spiders seem to vary more than those of the male spiders. We shall test

## Table 4.5-2 Tests of Hypotheses of the Equality of Variances

| $H_0$ | $H_1$ | Critical Region |
|---|---|---|
| $\sigma_X^2 = \sigma_Y^2$ | $\sigma_X^2 > \sigma_Y^2$ | $\dfrac{s_x^2}{s_y^2} \geq F_\alpha(n-1, m-1)$ |
| $\sigma_X^2 = \sigma_Y^2$ | $\sigma_X^2 < \sigma_Y^2$ | $\dfrac{s_y^2}{s_x^2} \geq F_\alpha(m-1, n-1)$ |
| $\sigma_X^2 = \sigma_Y^2$ | $\sigma_X^2 \neq \sigma_Y^2$ | $\dfrac{s_x^2}{s_y^2} \geq F_{\alpha/2}(n-1, m-1)$ or |
| | | $\dfrac{s_y^2}{s_x^2} \geq F_{\alpha/2}(m-1, n-1)$ |

whether this latter belief is true. Suppose that the distribution of the length $X$ of male spiders is $N(\mu_X, \sigma_X^2)$ and the length $Y$ of female spiders is $N(\mu_Y, \sigma_Y^2)$, and $X$ and $Y$ are independent. We shall test $H_0: \sigma_X^2/\sigma_Y^2 = 1$ (i.e., $\sigma_X^2 = \sigma_Y^2$) against the alternative hypothesis $H_1: \sigma_X^2/\sigma_Y^2 < 1$ (i.e., $\sigma_X^2 < \sigma_Y^2$.) If we use $n = 30$ and $m = 30$ observations of $X$ and $Y$, respectively, a critical region that has a significance level of $\alpha = 0.01$ is

$$\frac{s_y^2}{s_x^2} \geq F_{0.01}(29, 29) = 2.42,$$

approximately, using interpolation in Appendix Table VII. For the following $n = 30$ observations of $X$

| | | | | | |
|---|---|---|---|---|---|
| 5.20 | 4.70 | 5.75 | 7.50 | 6.45 | 6.55 |
| 4.70 | 4.80 | 5.95 | 5.20 | 6.35 | 6.95 |
| 5.70 | 6.20 | 5.40 | 6.20 | 5.85 | 6.80 |
| 5.65 | 5.50 | 5.65 | 5.85 | 5.75 | 6.35 |
| 5.75 | 5.95 | 5.90 | 7.00 | 6.10 | 5.80 |

and $m = 30$ observations of $Y$

| | | | | | |
|---|---|---|---|---|---|
| 8.25 | 9.95 | 5.90 | 7.05 | 8.45 | 7.55 |
| 9.80 | 10.80 | 6.60 | 7.55 | 8.10 | 9.10 |
| 6.10 | 9.30 | 8.75 | 7.00 | 7.80 | 8.00 |
| 9.00 | 6.30 | 8.35 | 8.70 | 8.00 | 7.50 |
| 9.50 | 8.30 | 7.05 | 8.30 | 7.95 | 9.60 |

where the measurements are in millimeters, $\bar{x} = 5.917$, $s_x^2 = 0.4399$, $\bar{y} = 8.153$ and $s_y^2 = 1.4100$. Since

$$\frac{s_y^2}{s_x^2} = \frac{1.4100}{0.4399} = 3.2053 > 2.42,$$

the null hypothesis is rejected in favor of the biologist's belief. This is illustrated in Figure 4.5-3.

There are two graphs of the data that can give some insight into the comparison of these two sets of data. In Figure 4.5-4(a) a quantile-quantile (q-q) plot is shown

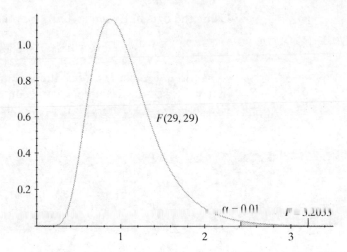

**Figure 4.5-3** Test of variances for green lynx spiders

that graphs the order statistics of the lengths of the female spiders against the order statistics of the male spiders. From this $q$-$q$ plot, it can be seen that the standard deviation of the male spiders is less than that of the female spiders and also that the male spiders are not as long as the female spiders. This can also be seen in Figure 4.5-4($b$), where box plots of the lengths of the male ($X$) and female ($Y$) spiders are given.

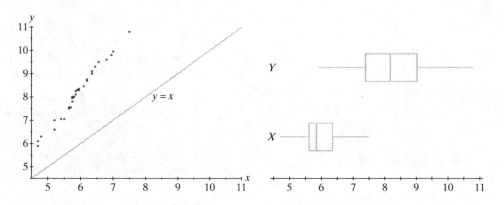

**Figure 4.5-4** Quantile-quantile and box plots for lengths of male ($X$) and female ($Y$) spiders

In Example 4.5-1 we used a $t$-statistic for testing the equality of means that assumed the variances were equal. In the next example we shall test whether that assumption is valid.

**EXAMPLE 4.5-3**   For Example 4.5-1, given $n = 11$ observations of $X$ and $m = 13$ observations of $Y$, where $X$ is $N(\mu_X, \sigma_X^2)$ and $Y$ is $N(\mu_Y, \sigma_Y^2)$, we shall test the null hypothesis $H_0$: $\sigma_X^2/\sigma_Y^2 = 1$ against a two-sided alternative hypothesis. At an $\alpha = 0.05$ significance level, $H_0$ is rejected if

$$s_x^2/s_y^2 \geq F_{0.025}(10, 12) = 3.37 \qquad \text{or} \qquad s_y^2/s_x^2 \geq F_{0.025}(12, 10) = 3.62.$$

Using the data in Example 4.5-1, we obtain

$$s_x^2/s_y^2 = 0.24/0.35 = 0.686 \qquad \text{and} \qquad s_y^2/s_x^2 = 1.458,$$

so we do not reject $H_0$. Thus the assumption of equal variances for the $t$-statistic that was used in Example 4.5-1 seems to be valid. See Figure 4.5-5, noting that $F_{0.975}(10, 12) = 1/F_{0.025}(12, 10) = 1/3.62 = 0.276.$

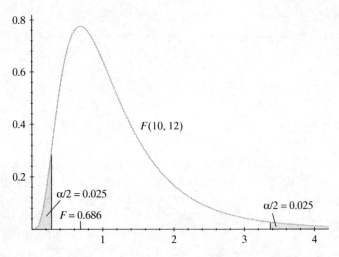

**Figure 4.5-5** Test of variances for pea stem growths

Often there is interest in testing the equality of the probability of success for two different methods of doing some task. Let $Y_1$ and $Y_2$ be the numbers of successes in $n_1$ and $n_2$ independent trials with probabilities $p_1$ and $p_2$ of success, respectively. We know that $\widehat{p_1} = Y_1/n_1$ and $\widehat{p_2} = Y_2/n_2$ have approximate normal distributions so that

$$Z = \frac{Y_1/n_1 - Y_2/n_2 - (p_1 - p_2)}{\sqrt{p_1(1 - p_1)/n_1 + p_2(1 - p_2)/n_2}}$$

$$= \frac{\widehat{p_1} - \widehat{p_2} - (p_1 - p_2)}{\sqrt{p_1(1 - p_1)/n_1 + p_2(1 - p_2)/n_2}}$$

is approximately $N(0, 1)$. We used this fact to establish a confidence interval for $p_1 - p_2$, although we did need to substitute $\widehat{p_1}$ for $p_1$ and $\widehat{p_2}$ for $p_2$ in that denominator. So if we are testing $H_0$: $p_1 - p_2 = 0$, we could use the test statistic

$$Z = \frac{\widehat{p_1} - \widehat{p_2} - 0}{\sqrt{\dfrac{\widehat{p_1}(1 - \widehat{p_1})}{n_1} + \dfrac{\widehat{p_2}(1 - \widehat{p_2})}{n_2}}}$$

to give tests that are equivalent to one-sided or two-sided confidence intervals for $p_1 - p_2$. These are summarized in Table 4.5-3.

**Remark**  In testing $H_0$: $p_1 = p_2$, some (again maybe most) statisticians replace both $\widehat{p_1}$ and $\widehat{p_2}$ in the denominator of $Z$ by an estimate of the common $p_1 = p_2$,

| **Table 4.5-3 Tests of Hypotheses for Two Proportions** | | |
|---|---|---|
| $H_0$ | $H_1$ | **Critical Region** |
| $p_1 = p_2$ | $p_1 > p_2$ | $z \geq z_\alpha$ |
| $p_1 = p_2$ | $p_1 < p_2$ | $z \leq -z_\alpha$ |
| $p_1 = p_2$ | $p_1 \neq p_2$ | $|z| \geq z_{\alpha/2}$ |

namely $\widehat{p} = (Y_1 + Y_2)/(n_1 + n_2)$. The resulting test statistic is

$$Z = \frac{\widehat{p_1} - \widehat{p_2} - 0}{\sqrt{\widehat{p}(1 - \widehat{p})\left(\dfrac{1}{n_1} + \dfrac{1}{n_2}\right)}}.$$

Again we do not have a strong preference one way or the other; the numerical results are about the same. With this common estimate $\widehat{p}$, the estimate of the standard deviation of $\widehat{p_1} - \widehat{p_2}$ is better if $H_0$ is true; but the estimate of this standard deviation is better using the individual $\widehat{p_1}$ and $\widehat{p_2}$ in the denominator if $H_0$ is clearly false. In either case, the estimate of the standard deviation is called the standard error of $\widehat{p_1} - \widehat{p_2}$.

**EXAMPLE 4.5-4** A machine shop that manufactures toggle levers has both a day and a night shift. A toggle lever is defective if a standard nut cannot be screwed onto the threads. Let $p_1$ and $p_2$ be the respective proportion of defective levers among those manufactured by the day and night shifts. We test the null hypothesis $H_0$: $p_1 = p_2$ against the one-sided alternative $H_1$: $p_1 < p_2$ based on two random samples, each of size $n = 1000$, taken from the production of the respective shifts. Let $\alpha = 0.05$. Say $y_1 = 37$ and $y_2 = 53$ defectives were observed for the day and night shifts. We calculate the test statistic in those two ways. If we use the individual $\widehat{p_1}$ and $\widehat{p_2}$ in the denominator, we obtain

$$z = \frac{0.037 - 0.053}{\sqrt{\dfrac{(0.037)(0.963)}{1000} + \dfrac{(0.053)(0.947)}{1000}}} = -1.727 < -1.645.$$

With the estimate of the common $p_1 = p_2$, namely $\widehat{p} = (37 + 53)/2000 = 0.045$, we have

$$z = \frac{0.037 - 0.053}{\sqrt{(0.045)(0.955)\left(\dfrac{1}{1000} + \dfrac{1}{1000}\right)}} = -1.726 < -1.645.$$

We reject $H_0$: $p_1 = p_2$ and accept $H_1$: $p_1 < p_2$ and look for ways to improve the production of the night shift.

**EXERCISES 4.5**

**4.5-1** Let $X$ and $Y$ denote the weights in grams of male and female gallinules, respectively. Assume that $X$ is $N(\mu_X, \sigma_X^2)$ and $Y$ is $N(\mu_Y, \sigma_Y^2)$.

(a) Given $n = 16$ observations of $X$ and $m = 13$ observations of $Y$, define a test statistic and critical region for testing the null hypothesis $H_0$: $\mu_X = \mu_Y$ against the one-sided alternative hypothesis $H_1$: $\mu_X > \mu_Y$. Let $\alpha = 0.01$. (Assume the variances are equal.)

**(b)** Given that $\bar{x} = 415.16$, $s_x^2 = 1356.75$, $\bar{y} = 347.40$, and $s_y^2 = 692.21$, calculate the value of the test statistic and state your conclusion.

**(c)** Test whether the assumption of equal variances is valid. Let $\alpha = 0.05$.

**4.5-2** Among the data collected for the World Health Organization air quality monitoring project is a measure of suspended particles in $\mu g/m^3$. Let $X$ and $Y$ equal the concentration of suspended particles in $\mu g/m^3$ in the city center (commercial district), for Melbourne and Houston, respectively. Using $n = 13$ observations of $X$ and $m = 16$ observations of $Y$, we shall test $H_0: \mu_X = \mu_Y$ against $H_1: \mu_X < \mu_Y$.

**(a)** Define the test statistic and critical region, assuming that the variances are equal. Let $\alpha = 0.05$.

**(b)** If $\bar{x} = 72.9$, $s_x = 25.6$, $\bar{y} = 81.7$, and $s_y = 28.3$, calculate the value of the test statistic and state your conclusion.

**(c)** Give limits for the $p$-value of this test.

**(d)** Test whether the assumption of equal variances is valid. Let $\alpha = 0.05$.

**4.5-3** Some nurses in County Public Health conducted a survey of women who had received inadequate prenatal care. They used information from birth certificates to select mothers for the survey. The mothers that were selected were divided into two groups: 14 mothers who said they had 5 or fewer prenatal visits and 14 mothers who said they had 6 or more prenatal visits. Let $X$ and $Y$ equal the respective birthweights of the babies from these two sets of mothers and assume that the distribution of $X$ is $N(\mu_X, \sigma^2)$ and the distribution of $Y$ is $N(\mu_Y, \sigma^2)$.

**(a)** Define the test statistic and critical region for testing $H_0: \mu_X - \mu_Y = 0$ against $H_1: \mu_X - \mu_Y < 0$. Let $\alpha = 0.05$.

**(b)** Given that the observations of $X$ were

$$
\begin{array}{ccccccc}
49 & 108 & 110 & 82 & 93 & 114 & 134 \\
114 & 96 & 52 & 101 & 114 & 120 & 116
\end{array}
$$

and the observations of $Y$ were

$$
\begin{array}{ccccccc}
133 & 108 & 93 & 119 & 119 & 98 & 106 \\
131 & 87 & 153 & 116 & 129 & 97 & 110
\end{array}
$$

calculate the value of the test statistic and state your conclusion.

**(c)** Approximate the $p$-value.

**(d)** Construct box plots on the same figure for these two sets of data. Do the box plots support your conclusion?

**(e)** Test whether the assumption of equal variances is valid. Let $\alpha = 0.05$.

**4.5-4** Let $X$ and $Y$ equal the forces required to pull stud No. 3 and stud No. 4 out of a window that has been manufactured for an automobile. Assume that the distributions of $X$ and $Y$ are $N(\mu_X, \sigma_X^2)$ and $N(\mu_Y, \sigma_Y^2)$, respectively.

**(a)** If $m = n = 10$ observations are selected randomly, define a test statistic and critical region for testing $H_0: \mu_X - \mu_Y = 0$ against a two-sided alternative hypothesis. Let $\alpha = 0.05$. Assume that the variances are equal.

**(b)** Given n = 10 observations of X:

$$111 \quad 120 \quad 139 \quad 136 \quad 138 \quad 149 \quad 143 \quad 145 \quad 111 \quad 123$$

and $m = 10$ observations of Y:

$$152 \quad 155 \quad 133 \quad 134 \quad 119 \quad 155 \quad 142 \quad 146 \quad 157 \quad 149$$

calculate the value of the test statistic and clearly state your conclusion.

**(c)** What is the approximate $p$-value of this test?

**(d)** Construct box plots on the same figure for these two sets of data. Do the box plots confirm your decision in part (b)?

**(e)** Test whether the assumption of equal variances is valid.

**4.5-5** Let $X$ and $Y$ denote the tarsus lengths of male and female grackles, respectively. Assume that $X$ is $N(\mu_X, \sigma_X^2)$ and $Y$ is $N(\mu_Y, \sigma_Y^2)$. Given that $n = 25$, $\bar{x} = 33.80$, $s_x^2 = 4.88$, $m = 29$, $\bar{y} = 31.66$, and $s_y^2 = 5.81$, test

(a) $H_0$: $\sigma_X^2 / \sigma_Y^2 = 1$ against a two-sided alternative with $\alpha = 0.02$.

(b) $H_0$: $\mu_X = \mu_Y$ against $H_1$: $\mu_X > \mu_Y$ with $\alpha = 0.01$.

**4.5-6** Weight checks are done on the scales for an automatic bagger for water softener pellets. Each bagger has two scales, say south and north, that operate alternately to fill 80# bags of pellets. Let $X$ and $Y$ equal the weights of the bags of pellets from these two scales. Assume that the distributions of $X$ and $Y$ are $N(\mu_X, \sigma_X^2)$ and $N(\mu_Y, \sigma_Y^2)$, respectively. Ten bags were selected randomly from each scale and weighed by hand on a scale with a "high" degree of accuracy yielding the following weights:

| $X$: | 80.51 | 80.46 | 80.75 | 80.50 | 80.36 |
|------|-------|-------|-------|-------|-------|
|      | 80.32 | 80.36 | 80.78 | 80.26 | 80.34 |
| $Y$: | 80.31 | 80.28 | 80.40 | 80.35 | 80.38 |
|      | 80.28 | 80.27 | 80.16 | 80.59 | 80.56 |

(a) Test $H_0$: $\mu_X = \mu_Y$ against $H_1$: $\mu_X \neq \mu_Y$. Assume that $\sigma_X^2 = \sigma_Y^2$. Give limits for the $p$-value of the test and state your conclusion.

(b) Test the assumption of equal variances against a two-sided alternative hypothesis.

(c) Draw box-and-whisker diagrams on the same graph. Does this figure confirm your answers?

**4.5-7** Let $p_m$ and $p_f$ be the respective proportions of male and female white-crowned sparrows that return to their hatching site. Give the endpoints for a 95% confidence interval for $p_m - p_f$, given that 124 out of 894 males and 70 out of 700 females returned. (*The Condor*, 1992, pp. 117–133.) Does this agree with the conclusion of a test of $H_0$: $p_m = p_f$ against $H_1$: $p_m \neq p_f$ with $\alpha = 0.05$?

**4.5-8** *TIME*, April 18, 1994, reported the results of a telephone poll of 800 adult Americans, 605 of them nonsmokers, who were asked the following question: "Should the federal tax on cigarettes be raised by \$1.25 to pay for health care reform?" Let $p_1$ and $p_2$ equal the proportions of nonsmokers and smokers, respectively, who would say yes to this question. Given that $y_1 = 351$ nonsmokers and $y_2 = 41$ smokers said yes,

(a) With $\alpha = 0.05$, test $H_0$: $p_1 = p_2$ against $H_1$: $p_1 \neq p_2$.

(b) Find a 95% confidence interval for $p_1 - p_2$. Is this in agreement with the conclusion of part (a)?

**4.5-9** For developing countries in Africa and the Americas, let $p_1$ and $p_2$ be the respective proportions of babies with a low birth weight (below 2500 grams). We shall test $H_0$: $p_1 = p_2$ against the alternative hypothesis $H_1$: $p_1 > p_2$.

(a) Define a critical region that has an $\alpha = 0.05$ significance level.

(b) If respective random samples of sizes $n_1 = 900$ and $n_2 = 700$ yielded $y_1 = 135$ and $y_2 = 77$ babies with a low birth weight, what is your conclusion?

(c) What would your decision be with a significance level of $\alpha = 0.01$?

(d) What is the $p$-value of your test?

## 4.6   SIMPLE LINEAR REGRESSION

There is often interest in the relation between two variables, for example, a student's scholastic aptitude test score in mathematics and this same student's grade in calculus. Frequently, one of these variables, say $x$, is known in advance of the other, and hence there is interest in predicting a future random variable $Y$. Since $Y$ is a random variable, we cannot predict its future observed value $Y = y$ with certainty. Thus let us first concentrate on the problem of estimating the mean of $Y$, that is, $E(Y)$. Now $E(Y)$ is usually a function of $x$; for example, in our illustration with the calculus grade, say $Y$, we would expect $E(Y)$ to increase with increasing mathematics aptitude score $x$. Sometimes $E(Y) = \mu(x)$

is assumed to be of a given form, such as linear or quadratic or exponential; that is, $\mu(x)$ could be assumed to be equal to $\alpha + \beta x$ or $\alpha + \beta x + \gamma x^2$ or $\alpha e^{\beta x}$. To estimate $E(Y) = \mu(x)$, or equivalently the parameters $\alpha$, $\beta$, and $\gamma$, we observe the random variable $Y$ for each of $n$ different values of $x$, say $x_1, x_2, \ldots, x_n$. Once the $n$ independent experiments have been performed, we have $n$ pairs of known numbers $(x_1, y_1), (x_2, y_2), \ldots, (x_n, y_n)$. These pairs are then used to estimate the mean $E(Y)$. Problems like this are often classified under **regression** because $E(Y) = \mu(x)$ is frequently called a regression curve.

> **Remark**   A model for the mean like $\alpha + \beta x + \gamma x^2$, is called a **linear model** because it is linear in the parameters, $\alpha$, $\beta$, and $\gamma$ although it is not linear in $x$. Thus $\alpha e^{\beta x}$ is not a linear model because it is not linear in $\alpha$ and $\beta$.

Let us begin with the case in which $E(Y) = \mu(x)$ is a linear function in $x$. The data points are $(x_1, y_1), (x_2, y_2), \ldots, (x_n, y_n)$; so the first problem is that of fitting a straight line to the set of data (see Figure 4.6-1). In addition to assuming that the mean of $Y$ is a linear function, we assume that for a particular value of $x$, the value of $Y$ will differ from its mean by a random amount $\varepsilon$. We further assume that the distribution of $\varepsilon$ is $N(0, \sigma^2)$. So we have for our linear model

$$Y_i = \alpha_1 + \beta x_i + \varepsilon_i,$$

where $\varepsilon_i$, for $i = 1, 2, \ldots, n$, are independent and $N(0, \sigma^2)$.

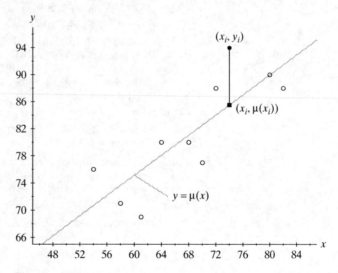

**Figure 4.6-1**  Scatter plot and least squares regression line

We shall now find point estimates for $\alpha_1$, $\beta$, and $\sigma^2$. For convenience we let $\alpha_1 = \alpha - \beta \bar{x}$ so that

$$Y_i = \alpha + \beta(x_i - \bar{x}) + \varepsilon_i, \text{ where } \bar{x} = \frac{1}{n} \sum_{i=1}^{n} x_i.$$

Then $Y_i$ is equal to a constant, $\alpha + \beta(x_i - \bar{x})$, plus a normal random variable $\varepsilon_i$. Hence $Y_1, Y_2, \ldots, Y_n$ are mutually independent normal variables with respective means $\alpha + \beta(x_i - \bar{x})$, $i = 1, 2, \ldots, n$, and unknown variance $\sigma^2$. Their joint p.d.f. is therefore the product of the individual probability density functions; that is, the likelihood function equals

$$L(\alpha, \beta, \sigma^2) = \prod_{i=1}^{n} \frac{1}{\sqrt{2\pi\sigma^2}} \exp\left\{-\frac{[y_i - \alpha - \beta(x_i - \bar{x})]^2}{2\sigma^2}\right\}$$

$$= \left(\frac{1}{2\pi\sigma^2}\right)^{n/2} \exp\left\{-\frac{\sum_{i=1}^{n}[y_i - \alpha - \beta(x_i - \bar{x})]^2}{2\sigma^2}\right\}.$$

To maximize $L(\alpha, \beta, \sigma^2)$, or, equivalently, to minimize

$$-\ln L(\alpha, \beta, \sigma^2) = \frac{n}{2}\ln(2\pi\sigma^2) + \frac{\sum_{i=1}^{n}[y_i - \alpha - \beta(x_i - \bar{x})]^2}{2\sigma^2},$$

we must select $\alpha$ and $\beta$ to minimize

$$H(\alpha, \beta) = \sum_{i=1}^{n}[y_i - \alpha - \beta(x_i - \bar{x})]^2.$$

Since $|y_i - \alpha - \beta(x_i - \bar{x})| = |y_i - \mu(x_i)|$ is the vertical distance from the point $(x_i, y_i)$ to the line $y = \mu(x)$, we note that $H(\alpha, \beta)$ represents the sum of the squares of those distances. Thus selecting $\alpha$ and $\beta$ so that the sum of the squares is minimized means that we are fitting the straight line to the data by the **method of least squares**.

To minimize $H(\alpha, \beta)$, we find the two first partial derivatives

$$\frac{\partial H(\alpha, \beta)}{\partial \alpha} = 2\sum_{i=1}^{n}[y_i - \alpha - \beta(x_i - \bar{x})](-1)$$

and

$$\frac{\partial H(\alpha, \beta)}{\partial \beta} = 2\sum_{i=1}^{n}[y_i - \alpha - \beta(x_i - \bar{x})][-(x_i - \bar{x})].$$

Setting $\partial H(\alpha, \beta)/\partial \alpha = 0$, we obtain

$$\sum_{i=1}^{n} y_i - n\alpha - \beta\sum_{i=1}^{n}(x_i - \bar{x}) = 0.$$

Since

$$\sum_{i=1}^{n}(x_i - \bar{x}) = 0,$$

we have that

$$\sum_{i=1}^{n} y_i - n\alpha = 0$$

and thus the estimate of $\alpha$ is

$$\hat{\alpha} = \bar{y}.$$

The equation $\partial H(\alpha, \beta)/\partial \beta = 0$ yields, with $\alpha$ replaced by $\bar{y}$,

$$\sum_{i=1}^{n} (y_i - \bar{y})(x_i - \bar{x}) - \beta \sum_{i=1}^{n} (x_i - \bar{x})^2 = 0$$

or, equivalently, the estimate of $\beta$ is

$$\widehat{\beta} = \frac{\displaystyle\sum_{i=1}^{n} (y_i - \bar{y})(x_i - \bar{x})}{\displaystyle\sum_{i=1}^{n} (x_i - \bar{x})^2} = \frac{\displaystyle\sum_{i=1}^{n} y_i(x_i - \bar{x})}{\displaystyle\sum_{i=1}^{n} (x_i - \bar{x})^2}.$$

Thus, to find the mean line of best fit, $\mu(x) = \alpha + \beta(x_i - \bar{x})$, we use

$$\widehat{\alpha} = \bar{y} \tag{4.6-1}$$

and

$$\widehat{\beta} = \frac{\displaystyle\sum_{i=1}^{n} y_i(x_i - \bar{x})}{\displaystyle\sum_{i=1}^{n} (x_i - \bar{x})^2} = \frac{\sum_{i=1}^{n} x_i y_i - \left(\dfrac{1}{n}\right)\left(\displaystyle\sum_{i=1}^{n} x_i\right)\left(\displaystyle\sum_{i=1}^{n} y_i\right)}{\displaystyle\sum_{i=1}^{n} x_i^2 - \left(\dfrac{1}{n}\right)\left(\displaystyle\sum_{i=1}^{n} x_i\right)^2}. \tag{4.6-2}$$

To find the maximum likelihood estimate of $\sigma^2$, consider the partial derivative

$$\frac{\partial[-\ln L(\alpha, \beta, \sigma^2)]}{\partial(\sigma^2)} = \frac{n}{2\sigma^2} - \frac{\displaystyle\sum_{i=1}^{n} [y_i - \alpha - \beta(x_i - \bar{x})]^2}{2(\sigma^2)^2}.$$

Setting this equal to zero and replacing $\alpha$ and $\beta$ by their solutions $\widehat{\alpha}$ and $\widehat{\beta}$, we obtain the estimate

$$\widehat{\sigma^2} = \frac{1}{n} \sum_{i=1}^{n} [y_i - \widehat{\alpha} - \widehat{\beta}(x_i - \bar{x})]^2. \tag{4.6-3}$$

A formula for $n\widehat{\sigma^2}$ useful in its calculation is

$$n\widehat{\sigma^2} = \sum_{i=1}^{n} y_i^2 - \frac{1}{n}\left(\sum_{i=1}^{n} y_i\right)^2 - \widehat{\beta}\sum_{i=1}^{n} x_i y_i + \widehat{\beta}\left(\frac{1}{n}\right)\left(\sum_{i=1}^{n} x_i\right)\left(\sum_{i=1}^{n} y_i\right). \tag{4.6-4}$$

Note that the summand in Equation 4.6-3 for $\widehat{\sigma^2}$ is the square of the difference between the value of $y_i$ and the predicted mean of $y_i$. Let $\widehat{y}_i = \widehat{\alpha} + \widehat{\beta}(x_i - \bar{x})$, the predicted mean value of $y_i$. The difference

$$y_i - \widehat{y}_i = y_i - \widehat{\alpha} - \widehat{\beta}(x_i - \bar{x})$$

is called the $i$th **residual**, $i = 1, 2, \ldots, n$. The maximum likelihood estimate of $\sigma^2$ is then the sum of the squares of the residuals divided by $n$. It should always be true that the sum of the residuals is equal to zero. However, in practice, due to round off, the sum of the observed residuals, $y_i - \widehat{y}_i$, sometimes differs slightly from zero.

If we have $n$ bivariate observations $(x_1, y_1), (x_2, y_2), \ldots, (x_n, y_n)$, it is sometimes convenient to have available what is called the **sample correlation coefficient**

$$r = \frac{s_{xy}}{s_x s_y},$$

where the **sample covariance** $s_{xy}$ is defined as

$$s_{xy} = \frac{1}{n-1} \sum_{i=1}^{n} (x_i - \bar{x})(y_i - \bar{y}) = \frac{1}{n-1} \sum_{i=1}^{n} (x_i - \bar{x}) y_i.$$

That is, the sample correlation coefficient equals

$$r = \frac{s_{xy}}{s_x s_y} = \left(\frac{s_{xy}}{s_x^2}\right)\left(\frac{s_x}{s_y}\right) = \widehat{\beta}\frac{s_x}{s_y},$$

because $\widehat{\beta} = \dfrac{s_{xy}}{s_x^2}$; or, equivalently,

$$\widehat{\beta} = r\frac{s_y}{s_x}.$$

Thus the straight line fitted by the method of least squares is

$$\widehat{y} = \widehat{\alpha} + \widehat{\beta}(x - \bar{x}) = \bar{y} + r\frac{s_y}{s_x}(x - \bar{x}).$$

Note that the sample mean of $\widehat{y}_1, \widehat{y}_2, \ldots, \widehat{y}_n$, is

$$\bar{\widehat{y}} = \frac{1}{n} \sum_{i=1}^{n} \widehat{y}_i$$

$$= \frac{1}{n} \sum_{i=1}^{n} [\bar{y} + r\frac{s_y}{s_x}(x_i - \bar{x})]$$

$$= \bar{y}.$$

The sample variance of $\widehat{y}_1, \widehat{y}_2, \ldots, \widehat{y}_n$, is

$$s_{\widehat{y}}^2 = \frac{1}{n-1} \sum_{i=1}^{n} [\widehat{y}_i - \bar{y}]^2$$

$$= \frac{1}{n-1} \sum_{i=1}^{n} [\bar{y} + r\frac{s_y}{s_x}(x_i - \bar{x}) - \bar{y}]^2$$

$$= r^2 \frac{s_y^2}{s_x^2} \frac{1}{n-1} \sum_{i=1}^{n} (x_i - \bar{x})^2$$

$$= r^2 s_y^2.$$

Thus

$$r^2 = \frac{s_{\widehat{y}}^2}{s_y^2}.$$

Note that $r^2$ is the fraction of the variation in the values of $y_1, y_2, \ldots, y_n$ that is explained by the regression of $y$ on $x$.

### Table 4.6-1 Calculations for Test Score Data

| $x$ | $y$ | $x^2$ | $xy$ | $y^2$ | $\widehat{y}$ | $y - \widehat{y}$ | $(y - \widehat{y})^2$ |
|---|---|---|---|---|---|---|---|
| 70 | 77 | 4,900 | 5,390 | 5,929 | 82.561566 | −5.561566 | 30.931016 |
| 74 | 94 | 5,476 | 6,956 | 8,836 | 85.529956 | 8.470044 | 71.741645 |
| 72 | 88 | 5,184 | 6,336 | 7,744 | 84.045761 | 3.954239 | 15.636006 |
| 68 | 80 | 4,624 | 5,440 | 6,400 | 81.077371 | −1.077371 | 1.160728 |
| 58 | 71 | 3,364 | 4,118 | 5,041 | 73.656395 | −2.656395 | 7.056434 |
| 54 | 76 | 2,916 | 4,104 | 5,776 | 70.688004 | 5.311996 | 28.217302 |
| 82 | 88 | 6,724 | 7,216 | 7,744 | 91.466737 | −3.466737 | 12.018265 |
| 64 | 80 | 4,096 | 5,120 | 6,400 | 78.108980 | 1.891020 | 3.575957 |
| 80 | 90 | 6,400 | 7,200 | 8,100 | 89.982542 | 0.017458 | 0.000305 |
| 61 | 69 | 3,721 | 4,209 | 4,761 | 75.882687 | −6.882687 | 47.371380 |
| 683 | 813 | 47,405 | 56,089 | 66,731 | | 0.000001 | 217.709038 |

**EXAMPLE 4.6-1**    The data plotted in Figure 4.6-1 are 10 pairs of test scores of 10 students in a psychology class, $x$ being the score on a preliminary test and $y$ the score on the final examination. The values of $x$ and $y$ are shown in Table 4.6-1. The sums that are needed to calculate estimates of the parameters are also given. Of course, the estimates of $\alpha$ and $\beta$ have to be found before the residuals can be calculated.

Thus $\widehat{\alpha} = 813/10 = 81.3$,

$$\widehat{\beta} = \frac{56,089 - (683)(813)/10}{47,405 - (683)(683)/10} = \frac{561.1}{756.1} = 0.742.$$

Since $\bar{x} = 683/10 = 68.3$, the least squares regression line is

$$\widehat{y} = 81.3 + (0.742)(x - 68.3).$$

The maximum likelihood estimate of $\sigma^2$ is

$$\widehat{\sigma^2} = \frac{217.709038}{10} = 21.7709.$$

A residual plot for these data is shown in Figure 4.6-2. For the data in Table 4.6-1, $s_y^2 = 70.4556$, $s_{\widehat{y}}^2 = 46.2657$, and $r = 0.81035$. We note that

$$\frac{s_{\widehat{y}}^2}{s_y^2} = \frac{46.2657}{70.4556} = 0.6567 = r^2$$

and thus 65.67% of the variation in the final exam scores is due to the regression of $y$ on $x$. If there is no trend or pattern among the residuals (that is, the points seem like a random sample from a distribution with mean zero), the estimate of the mean of the $y$ values seems to be a fairly good fit.

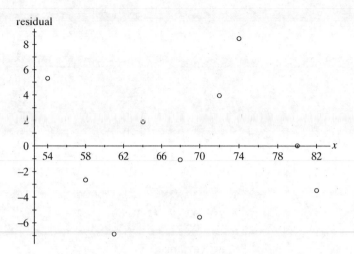

**Figure 4.6-2** Residuals plot for data in Table 4.6-1

We shall now consider the problem of finding the distributions of $\widehat{\alpha}$, $\widehat{\beta}$, and $\widehat{\sigma^2}$ (or distributions of functions of these estimators). We would like to be able to say something about the error of the estimates to find confidence intervals for the parameters.

During the preceding discussion we have treated $x_1, x_2, \ldots, x_n$ as constants. Of course, many times they can be set by the experimenter; for example, a chemist in experimentation might produce a compound at many different temperatures. But it is also true that these numbers might be observations on an earlier random variable, such as an SAT score or preliminary test grade (as in Example 4.6-1), but we consider the problem on the condition that these $x$ values are given in either case. Thus, in finding the distributions of the estimators $\widehat{\alpha}$, $\widehat{\beta}$, and $\widehat{\sigma^2}$, the only random variables are $Y_1, Y_2, \ldots, Y_n$.

Since the estimator $\widehat{\alpha}$ is a linear function of independent and normally distributed random variables, $\widehat{\alpha}$ has a normal distribution (see Theorem 4.1-2) with mean

$$E(\widehat{\alpha}) = E\left(\frac{1}{n}\sum_{i=1}^{n} Y_i\right) = \frac{1}{n}\sum_{i=1}^{n} E(Y_i)$$

$$= \frac{1}{n}\sum_{i=1}^{n} [\alpha + \beta(x_i - \overline{x})] = \alpha,$$

and variance

$$\operatorname{Var}(\widehat{\alpha}) = \sum_{i=1}^{n} \left(\frac{1}{n}\right)^2 \operatorname{Var}(Y_i) = \frac{\sigma^2}{n}.$$

The estimator $\widehat{\beta}$ is also a linear function of $Y_1, Y_2, \ldots, Y_n$ and hence has a normal distribution with mean

$$
E(\widehat{\beta}) = \frac{\sum_{i=1}^{n}(x_i - \bar{x})E(Y_i)}{\sum_{i=1}^{n}(x_i - \bar{x})^2}
$$

$$
= \frac{\sum_{i=1}^{n}(x_i - \bar{x})[\alpha + \beta(x_i - \bar{x})]}{\sum_{i=1}^{n}(x_i - \bar{x})^2}
$$

$$
= \frac{\alpha \sum_{i=1}^{n}(x_i - \bar{x}) + \beta \sum_{i=1}^{n}(x_i - \bar{x})^2}{\sum_{i=1}^{n}(x_i - \bar{x})^2} = \beta
$$

and variance

$$
\mathrm{Var}(\widehat{\beta}) = \sum_{i=1}^{n}\left[\frac{x_i - \bar{x}}{\sum_{j=1}^{n}(x_j - \bar{x})^2}\right]^2 \mathrm{Var}(Y_i)
$$

$$
= \frac{\sum_{i=1}^{n}(x_i - \bar{x})^2}{\left[\sum_{i=1}^{n}(x_i - \bar{x})^2\right]^2}\sigma^2 = \frac{\sigma^2}{\sum_{i=1}^{n}(x_i - \bar{x})^2}.
$$

It can be shown that

$$
\sum_{i=1}^{n}[Y_i - \alpha - \beta(x_i - \bar{x})]^2 = \sum_{i=1}^{n}\{(\widehat{\alpha} - \alpha) + (\widehat{\beta} - \beta)(x_i - \bar{x})
$$

$$
+ [Y_i - \widehat{\alpha} - \widehat{\beta}(x_i - \bar{x})]\}^2
$$

$$
= n(\widehat{\alpha} - \alpha)^2 + (\widehat{\beta} - \beta)^2 \sum_{i=1}^{n}(x_i - \bar{x})^2
$$

$$
+ \sum_{i=1}^{n}[Y_i - \widehat{\alpha} - \widehat{\beta}(x_i - \bar{x})]^2. \tag{4.6-5}
$$

From the fact that $Y_i$, $\widehat{\alpha}$, and $\widehat{\beta}$ have normal distributions, we know that each of

$$
\frac{[Y_i - \alpha - \beta(x_i - \bar{x})]^2}{\sigma^2}, \quad \frac{(\widehat{\alpha} - \alpha)^2}{\left[\dfrac{\sigma^2}{n}\right]}, \quad \frac{(\widehat{\beta} - \beta)^2}{\left[\dfrac{\sigma^2}{\sum_{i=1}^{n}(x_i - \bar{x})^2}\right]}
$$

has a chi-square distribution with one degree of freedom. Since $Y_1, Y_2, \ldots, Y_n$ are mutually independent, then (Theorem 4.1-1)

$$\frac{\sum_{i=1}^{n} [Y_i - \alpha - \beta(x_i - \overline{x})]^2}{\sigma^2}$$

is $\chi^2(n)$. That is, the left-hand member of Equation 4.6-5 divided by $\sigma^2$ is $\chi^2(n)$ and is equal to the sum of two $\chi^2(1)$ variables and

$$\frac{\sum_{i=1}^{n} [Y_i - \widehat{\alpha} - \widehat{\beta}(x_i - \overline{x})]^2}{\sigma^2} = \frac{n\widehat{\sigma^2}}{\sigma^2} \geq 0.$$

Thus we might then guess that $n\widehat{\sigma^2}/\sigma^2$ is $\chi^2(n-2)$. This is true, and, moreover, $\widehat{\alpha}$, $\widehat{\beta}$, and $\widehat{\sigma^2}$ are mutually independent. [For a proof, see Hogg, McKean, and Craig, *Introduction to Mathematical Statistics*, 6th ed. (Upper Saddle River, NJ: Prentice Hall, 2005).]

Suppose now that we are interested in forming a confidence interval for $\beta$, the slope of the line. We can use the fact that

$$T_1 = \frac{\sqrt{\sum_{i=1}^{n} (x_i - \overline{x})^2} \left( \frac{\widehat{\beta} - \beta}{\sigma} \right)}{\sqrt{\frac{n\widehat{\sigma^2}}{\sigma^2(n-2)}}} = \frac{\widehat{\beta} - \beta}{\sqrt{\frac{n\widehat{\sigma^2}}{(n-2) \sum_{i=1}^{n} (x_i - \overline{x})^2}}}$$

has a $t$ distribution with $n-2$ degrees of freedom. Thus

$$P \left[ -t_{\gamma/2}(n-2) \leq \frac{\widehat{\beta} - \beta}{\sqrt{\frac{n\widehat{\sigma^2}}{(n-2) \sum_{i=1}^{n} (x_i - \overline{x})^2}}} \leq t_{\gamma/2}(n-2) \right] = 1 - \gamma.$$

It follows that

$$\left[ \widehat{\beta} - t_{\gamma/2}(n-2) \sqrt{\frac{n\widehat{\sigma^2}}{(n-2) \sum_{i=1}^{n} (x_i - \overline{x})^2}}, \right.$$

$$\left. \widehat{\beta} + t_{\gamma/2}(n-2) \sqrt{\frac{n\widehat{\sigma^2}}{(n-2) \sum_{i=1}^{n} (x_i - \overline{x})^2}} \right]$$

is a $100(1 - \gamma)\%$ confidence interval for $\beta$.

Similarly,

$$T_2 = \frac{\frac{\sqrt{n}(\widehat{\alpha} - \alpha)}{\sigma}}{\sqrt{\frac{n\widehat{\sigma^2}}{\sigma^2(n-2)}}} = \frac{\widehat{\alpha} - \alpha}{\sqrt{\frac{\widehat{\sigma^2}}{n-2}}}$$

has a $t$ distribution with $n - 2$ degrees of freedom. Thus $T_2$ can be used to make inferences about $\alpha$. The fact that $n\widehat{\sigma^2}/\sigma^2$ has a chi-square distribution with $n - 2$ degrees of freedom can be used to make inferences about the variance $\sigma^2$.

If we were to test $H_0: \beta = \beta_0$ against the two-sided alternative $H_1: \beta \neq \beta_0$ at significance level $\gamma$, we could see if $\beta_0$ is in the two-sided $100(1 - \gamma)$ percent confidence interval for $\beta$. If we were testing $H_0: \beta = \beta_0$ against a one-sided alternative $H_0: \beta > \beta_0$ we could see if $\beta_0$ is in or out of the one-sided $100(1 - \gamma)$ percent confidence interval having lower bound

$$\widehat{\beta} - t_\gamma(n{-}2)\sqrt{\frac{n\widehat{\sigma^2}}{(n - 2)\sum_{i=1}^{n}(x_i - \overline{x})^2}}.$$

Equivalently, we could reject $H_0$ in the one-sided case if the test statistic

$$T_1 = \frac{\widehat{\beta} - \beta_0}{\sqrt{\dfrac{n\widehat{\sigma^2}}{(n - 2)\sum_{i=1}^{n}(x_i - \overline{x})^2}}} \geq t_\gamma(n{-}2).$$

Alternatively, we can compute $T_1 = t_0$, say, and determine the

$$p\text{-value} = P[T(n{-}2) \geq t_0]$$

and reject $H_0: \beta = \beta_0$ and accept $H_1: \beta > \beta_0$ if $p$-value $\leq \gamma$.

| Table 4.6-2 Tests About the Slope of the Regression Line | | |
|---|---|---|
| $H_0$ | $H_1$ | Critical Region |
| $\beta = \beta_0$ | $\beta > \beta_0$ | $t_1 \geq t_\gamma(n{-}2)$ |
| $\beta = \beta_0$ | $\beta < \beta_0$ | $t_1 \leq -t_\gamma(n{-}2)$ |
| $\beta = \beta_0$ | $\beta \neq \beta_0$ | $|t_1| \geq t_{\gamma/2}(n{-}2)$ |

The null hypothesis along with three possible alternative hypotheses are given in Table 4.6-2; these tests are equivalent to stating that we reject $H_0$ if $\beta_0$ is not in certain confidence intervals. For example, the first test is equivalent to rejecting $H_0$ if $\beta_0$ is not in the one-sided confidence interval with lower bound

$$\widehat{\beta} - t_\gamma(n{-}2)\sqrt{\frac{n\widehat{\sigma^2}}{(n - 2)\displaystyle\sum_{i=1}^{n}(x_i - \overline{x})^2}}.$$

Often $\beta_0 = 0$ and we test the hypothesis $H_0: \beta = 0$. That is, we test the null hypothesis that the slope is equal to zero.

**EXAMPLE 4.6-2**   Let $x$ equal a student's preliminary test score in a psychology course and $y$ is the same student's score on the final examination. With $n = 10$ students, we shall test $H_0: \beta = 0$ against $H_1: \beta \neq 0$. At the 0.01 significance level, the critical region is $|t_1| \geq t_{0.005}(8) = 3.355$. Using the data in Example 4.6-1, the observed value of $T_1$ is

$$t_1 = \frac{0.742 - 0}{\sqrt{10(21.7709)/8(756.1)}} = \frac{0.742}{0.1897} = 3.911.$$

Thus we reject $H_0$.

## EXERCISES 4.6

**4.6-1** The midterm and final exam scores of 10 students in a statistics course are tabulated as shown.

(a) Calculate the least squares regression line for these data.

(b) Plot the points and the least squares regression line on the same graph.

(c) Find the value of $\widehat{\sigma^2}$.

| Midterm | Final | Midterm | Final |
|---------|-------|---------|-------|
| 70 | 87 | 67 | 73 |
| 74 | 79 | 70 | 83 |
| 80 | 88 | 64 | 79 |
| 84 | 98 | 74 | 91 |
| 80 | 96 | 82 | 94 |

**4.6-2** Students' scores on the mathematics portion of the ACT examination, $x$, and on the final examination in first semester calculus (200 points possible), $y$, are given.

(a) Calculate the least squares regression line for these data.

(b) Plot the points and the least squares regression line on the same graph.

| x | y | x | y |
|---|---|---|---|
| 25 | 138 | 20 | 100 |
| 20 | 84 | 25 | 143 |
| 26 | 104 | 26 | 141 |
| 26 | 112 | 28 | 161 |
| 28 | 88 | 25 | 124 |
| 28 | 132 | 31 | 118 |
| 29 | 90 | 30 | 168 |
| 32 | 183 | | |

**4.6-3** Chemists often use ion sensitive electrodes to measure the ion concentration of aqueous solutions. These devices measure the migration of the charge of these ions and give a reading in millivolts (mV). A standard curve is produced by measuring known concentrations (in ppm) and fitting a line to the millivolt data. The following table gives the concentrations in ppm and the voltage in mV for calcium ISE.

| ppm | mV | ppm | mV | ppm | mV |
|-----|------|-----|------|-----|------|
| 0 | 1.72 | 75 | 2.40 | 150 | 4.47 |
| 0 | 1.68 | 75 | 2.32 | 150 | 4.51 |
| 0 | 1.74 | 75 | 2.33 | 150 | 4.43 |
| 50 | 2.04 | 100 | 2.91 | 200 | 6.67 |
| 50 | 2.11 | 100 | 3.00 | 200 | 6.66 |
| 50 | 2.17 | 100 | 2.89 | 200 | 6.57 |

**(a)** Calculate the correlation coefficient and the least squares regression line for mV vs. ppm.

**(b)** Plot the points and the least squares regression line on the same graph.

**(c)** Calculate and plot the residuals. Does linear regression seem to be appropriate?

**4.6-4** Show that the endpoints for a $100(1 - \gamma)\%$ confidence interval for $\alpha$ are

$$\widehat{\alpha} \pm t_{\gamma/2}(n-2)\sqrt{\frac{\widehat{\sigma^2}}{n-2}} \,.$$

**4.6-5** Show that a $100(1 - \gamma)\%$ confidence interval for $\sigma^2$ is

$$\left[ \frac{n\widehat{\sigma^2}}{\chi^2_{\gamma/2}(n-2)}, \frac{n\widehat{\sigma^2}}{\chi^2_{1-\gamma/2}(n-2)} \right].$$

**4.6-6** Find 95% confidence intervals for $\alpha$, $\beta$, and $\sigma^2$ for the data in Exercise 4.6-1.

**4.6-7** Let $x$ and $y$ equal the ACT scores in social science and natural science for a student who is applying for admission to a small liberal arts college. A sample of $n = 15$ such students yielded the following data.

| x | y | x | y | x | y |
|----|----|----|----|----|----|
| 32 | 28 | 30 | 27 | 26 | 32 |
| 23 | 25 | 17 | 23 | 16 | 22 |
| 23 | 24 | 20 | 30 | 21 | 28 |
| 23 | 32 | 17 | 18 | 24 | 31 |
| 26 | 31 | 18 | 18 | 30 | 26 |

**(a)** Calculate the least squares regression line for these data.

**(b)** Plot the points and the least squares regression line on the same graph.

**(c)** Find point estimates for $\alpha$, $\beta$, and $\sigma^2$.

**(d)** Find 95% confidence intervals for $\alpha$, $\beta$, and $\sigma^2$ under the usual assumptions.

**4.6-8** For the data given in Exercise 4.6-1, test $H_0$: $\beta = 0$ against $H_1$: $\beta > 0$ at the $\gamma = 0.025$ significance level using a $t$-test.

**4.6-9** For the data given in Exercise 4.6-2, test $H_0$: $\beta = 0$ against $H_1$: $\beta > 0$ at the $\gamma = 0.025$ significance level using a $t$-test.

**4.6-10** Show that

$$\sum_{i=1}^{n} [Y_i - \alpha - \beta(x_i - \bar{x})]^2 = n(\widehat{\alpha} - \alpha)^2 + (\widehat{\beta} - \beta)^2 \sum_{i=1}^{n} (x_i - \bar{x})^2$$

$$+ \sum_{i=1}^{n} [Y_i - \widehat{\alpha} - \widehat{\beta}(x_i - \bar{x})]^2.$$

**4.6-11** Let the independent random variables $Y_1, Y_2, \ldots, Y_n$ have the respective p.d.f.s $N(\beta x_i, \gamma^2 x_i^2)$, $i = 1, 2, \ldots, n$, where the given numbers $x_1, x_2, \ldots, x_n$ are not all equal and no one is zero. Find the maximum likelihood estimators of $\beta$ and $\gamma^2$.

**4.6-12** The "Golden Ratio" is $\phi = (1 + \sqrt{5})/2$. A mathematician who was interested in music analyzed Mozart's sonata movements that are divided into two distinct sections, both of which are repeated in performance. The length of the Exposition in measures is represented by $a$ and the length of the Development and Recapitulation is represented by $b$. His conjecture was that Mozart divided his movements near to the golden ratio. That is, he was interested in studying whether a scatter plot of $a + b$ against $b$ would not only be linear but would actually fall along the line $y = \phi x$. Here are the data in which the first column identifies the piece and movement by the Köchel cataloging system.

| Köchel | a | b | a + b | Köchel | a | b | a + b |
|---|---|---|---|---|---|---|---|
| 279, I | 38 | 62 | 100 | 279, II | 28 | 46 | 74 |
| 279, III | 56 | 102 | 158 | 280, I | 56 | 88 | 144 |
| 280, II | 24 | 36 | 60 | 280, III | 77 | 113 | 190 |
| 281, I | 40 | 69 | 109 | 281, II | 46 | 60 | 106 |
| 282, I | 15 | 18 | 33 | 282, III | 39 | 63 | 102 |
| 283, I | 53 | 67 | 120 | 283, II | 14 | 23 | 37 |
| 283, III | 102 | 171 | 273 | 284, I | 51 | 76 | 127 |
| 309, I | 58 | 97 | 155 | 311, I | 39 | 73 | 112 |
| 310, I | 49 | 84 | 133 | 330, I | 58 | 92 | 150 |
| 330, III | 68 | 103 | 171 | 332, I | 93 | 136 | 229 |
| 332, III | 90 | 155 | 245 | 333, I | 63 | 102 | 165 |
| 333, II | 31 | 50 | 81 | 457, I | 74 | 93 | 167 |
| 533, I | 102 | 137 | 239 | 533, II | 46 | 76 | 122 |
| 545, I | 28 | 45 | 73 | 547a, I | 78 | 118 | 196 |
| 570, I | 79 | 130 | 209 | | | | |

**(a)** Make a scatter plot of the points $a + b$ against the points $b$. Is this plot linear?

**(b)** Find the equation of the least squares regression line. Superimpose it on the scatter plot.

**(c)** On the scatter plot, superimpose the line $y = \phi x$. Compare this line with the least squares regression line. This could be done graphically.

**(d)** Find the sample mean of the points $(a + b)/b$. Is this close to $\phi$?

Remark  This exercise is based on the article "The Golden Section and the Piano Sonatas of Mozart" by John Putz and appears in Volume 68, No. 4, October, 1995, of *Mathematics Magazine*, pp. 275–282.

## 4.7  MORE ON LINEAR REGRESSION

In this section we use the notation and assumptions of Section 4.6. We have noted that $\widehat{Y} = \widehat{\alpha} + \widehat{\beta}(x - \overline{x})$ is a point estimate for the mean of $Y$ for some given $x$, or we could think of this as a prediction of the value of $Y$ for this given $x$. But how close is $\widehat{Y}$ to the mean of $Y$ or to $Y$ itself? We shall now find a confidence interval for $\alpha + \beta(x - \overline{x})$ and a prediction interval for $Y$, given a particular value of $x$.

To find a confidence interval for

$$E(Y) = \mu(x) = \alpha + \beta(x - \overline{x}),$$

let

$$\widehat{Y} = \widehat{\alpha} + \widehat{\beta}(x - \overline{x}).$$

Recall that $\widehat{Y}$ is a linear combination of normally and independently distributed random variables, $\widehat{\alpha}$ and $\widehat{\beta}$, so that $\widehat{Y}$ has a normal distribution. Furthermore

$$E(\widehat{Y}) = E[\widehat{\alpha} + \widehat{\beta}(x - \overline{x})]$$

$$= \alpha + \beta(x - \overline{x})$$

and

$$\mathrm{Var}(\widehat{Y}) = \mathrm{Var}[\widehat{\alpha} + \widehat{\beta}(x - \overline{x})]$$

$$= \frac{\sigma^2}{n} + \frac{\sigma^2}{\displaystyle\sum_{i=1}^{n} (x_i - \overline{x})^2}(x - \overline{x})^2$$

$$= \sigma^2 \left[ \frac{1}{n} + \frac{(x - \overline{x})^2}{\displaystyle\sum_{i=1}^{n} (x_i - \overline{x})^2} \right].$$

Recall that the distribution of $n\widehat{\sigma^2}/\sigma^2$ is $\chi^2(n{-}2)$. Since $\widehat{\alpha}$ and $\widehat{\beta}$ are independent of $\widehat{\sigma^2}$, we can form the $t$ statistic

$$T = \frac{\widehat{\alpha} + \widehat{\beta}(x - \overline{x}) - [\alpha + \beta(x - \overline{x})]}{\sigma\sqrt{\dfrac{1}{n} + \dfrac{(x - \overline{x})^2}{\sum_{i=1}^{n}(x_i - \overline{x})^2}} \Bigg/ \sqrt{\dfrac{n\widehat{\sigma^2}}{(n-2)\sigma^2}}}$$

which has a $t$ distribution with $r = n - 2$ degrees of freedom. Select $t_{\gamma/2}(n{-}2)$ from Table VI in the Appendix so that

$$P[-t_{\gamma/2}(n{-}2) \leq T \leq t_{\gamma/2}(n{-}2)] = 1 - \gamma.$$

This becomes

$$P[\widehat{\alpha} + \widehat{\beta}(x - \overline{x}) - c\, t_{\gamma/2}(n-2) \leq \alpha + \beta(x - \overline{x}) \leq$$

$$\widehat{\alpha} + \widehat{\beta}(x - \overline{x}) + c\, t_{\gamma/2}(n-2)] = 1 - \gamma,$$

where

$$c = \sqrt{\frac{n\widehat{\sigma^2}}{n-2}} \sqrt{\frac{1}{n} + \frac{(x - \overline{x})^2}{\sum_{i=1}^{n}(x_i - \overline{x})^2}}.$$

Thus the endpoints for a $100(1-\gamma)\%$ confidence interval for $\mu(x) = \alpha + \beta(x - \overline{x})$ are

$$\widehat{\alpha} + \widehat{\beta}(x - \overline{x}) \pm c\, t_{\gamma/2}(n-2).$$

 Note that the width of this interval depends on the particular value of $x$ because $c$ depends on $x$ (see Example 4.7-1).

   We have used $(x_1, y_1), (x_2, y_2), \ldots, (x_n, y_n)$ to estimate $\alpha$ and $\beta$. Suppose that we are given a value of $x$, say $x_{n+1}$. A point estimate of the corresponding value of $Y$ is

$$\widehat{y}_{n+1} = \widehat{\alpha} + \widehat{\beta}(x_{n+1} - \overline{x}).$$

However, $\widehat{y}_{n+1}$ is just one possible value of the random variable

$$Y_{n+1} = \alpha + \beta(x_{n+1} - \overline{x}) + \varepsilon_{n+1},$$

What can we say about possible values for $Y_{n+1}$? We shall now obtain a **prediction interval** for $Y_{n+1}$ when $x = x_{n+1}$ that is similar to the confidence interval for the mean of $Y$ when $x = x_{n+1}$.

   We have that

$$Y_{n+1} = \alpha + \beta(x_{n+1} - \overline{x}) + \varepsilon_{n+1},$$

where $\varepsilon_{n+1}$ is $N(0, \sigma^2)$ and $\overline{x} = (1/n)\sum_{i=1}^{n} x_i$. Now

$$W = Y_{n+1} - \widehat{\alpha} - \widehat{\beta}(x_{n+1} - \overline{x})$$

is a linear combination of normally and independently distributed random variables, so $W$ has a normal distribution. The mean of $W$ is

$$E(W) = E[Y_{n+1} - \widehat{\alpha} - \widehat{\beta}(x_{n+1} - \overline{x})]$$

$$= \alpha + \beta(x_{n+1} - \overline{x}) - \alpha - \beta(x_{n+1} - \overline{x}) = 0.$$

Since $Y_{n+1}, \widehat{\alpha}$ and $\widehat{\beta}$ are independent, the variance of $W$ is

$$\text{Var}(W) = \sigma^2 + \frac{\sigma^2}{n} + \frac{\sigma^2}{\displaystyle\sum_{i=1}^{n}(x_i - \overline{x})^2}(x_{n+1} - \overline{x})^2$$

$$= \sigma^2 \left[ 1 + \frac{1}{n} + \frac{(x_{n+1} - \overline{x})^2}{\displaystyle\sum_{i=1}^{n}(x_i - \overline{x})^2} \right].$$

Recall that $\widehat{n\sigma^2}/[(n-2)\sigma^2]$ is $\chi^2(n-2)$. Since $Y_{n+1}, \widehat{\alpha}$, and $\widehat{\beta}$ are independent of $\widehat{\sigma^2}$, we can form the $t$-statistic

$$T = \frac{\dfrac{Y_{n+1} - \widehat{\alpha} - \widehat{\beta}\,(x_{n+1} - \overline{x})}{\sigma\sqrt{1 + \dfrac{1}{n} + \dfrac{(x_{n+1} - \overline{x})^2}{\sum_{i=1}^{n}(x_i - \overline{x})^2}}}}{\sqrt{\dfrac{\widehat{n\sigma^2}}{(n-2)\sigma^2}}}$$

which has a $t$-distribution with $r = n-2$ degrees of freedom. Select a constant $t_{\gamma/2}(n-2)$ from Table VI in the Appendix so that

$$P[-t_{\gamma/2}(n-2) \leq T \leq t_{\gamma/2}(n-2)] = 1 - \gamma.$$

Solving this inequality for $Y_{n+1}$, we have

$$P[\widehat{\alpha} + \widehat{\beta}\,(x_{n+1} - \overline{x}) - d\,t_{\gamma/2}(n-2) \leq Y_{n+1} \leq$$
$$\widehat{\alpha} + \widehat{\beta}(x_{n+1} - \overline{x}) + d\,t_{\gamma/2}(n-2)] = 1 - \gamma,$$

where

$$d = \sqrt{\frac{\widehat{n\sigma^2}}{n-2}}\sqrt{1 + \frac{1}{n} + \frac{(x_{n+1} - \overline{x})^2}{\sum_{i=1}^{n}(x_i - \overline{x})^2}}.$$

Thus the endpoints for a $100(1 - \gamma)\%$ prediction interval for $Y_{n+1}$ are

$$\widehat{\alpha} + \widehat{\beta}(x_{n+1} - \overline{x}) \pm d\,t_{\gamma/2}(n-2).$$

We shall now illustrate a 95% confidence interval for $\mu(x)$ and a 95% prediction interval for $Y$ for a given value of $x$ using the data in Example 4.6-1. To find such intervals, the calculating formulas that are given in Equations 4.6-1, 4.6-2, and 4.6-4 can be used.

**EXAMPLE 4.7-1** To find a 95% confidence interval for $\mu(x)$ using the data in Example 4.6-1, note that we have already found that $\overline{x} = 68.3, \widehat{\alpha} = 81.3, \widehat{\beta} = 561.1/756.1 = 0.7421$, and $\widehat{\sigma^2} = 21.7709$. We also need

$$\sum_{i=1}^{n}(x_i - \overline{x})^2 = \sum_{i=1}^{n}x_i^2 - \left(\frac{1}{n}\right)\left(\sum_{i=1}^{n}x_i\right)^2$$

$$= 47,405 - \frac{683^2}{10} = 756.1.$$

For 95% confidence, $t_{0.025}(8) = 2.306$. When $x = 60$, the endpoints for a 95% confidence interval for $\mu(60)$ are

$$81.3 + 0.7421(60 - 68.3) \pm \left[\sqrt{\frac{10(21.7709)}{8}}\sqrt{\frac{1}{10} + \frac{(60 - 68.3)^2}{756.1}}\right](2.306)$$

$$75.1406 \pm 5.2589.$$

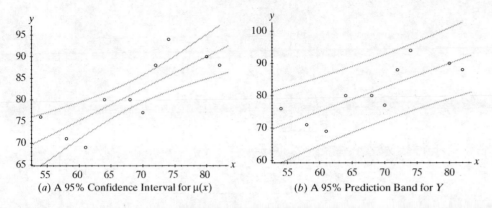

(a) A 95% Confidence Interval for $\mu(x)$      (b) A 95% Prediction Band for $Y$

**Figure 4.7-1** Data from Example 4.6-1

Similarly, when $x = 70$ the endpoints for a 95% confidence interval for $\mu(70)$ are

$$82.5616 \pm 3.8761.$$

Note that the lengths of these intervals depend on the particular value of $x$. A 95% confidence band for $\mu(x)$ is graphed in Figure 4.7-1($a$) along with the scatter diagram and $\widehat{y} = \widehat{\alpha} + \widehat{\beta}(x - \overline{x})$.

The endpoints for a 95% prediction interval for $Y$ when $x = 60$ are

$$81.3 + 0.7421(60 - 68.3) \pm \left[ \sqrt{\frac{10(21.7709)}{8}} \sqrt{1.1 + \frac{(60 - 68.3)^2}{756.1}} \right](2.306)$$

$$75.1406 \pm 13.1289.$$

Note that this interval is much wider than the confidence interval for $\mu(60)$. In Figure 4.7-1($b$) the 95% prediction band for $Y$ is graphed along with the scatter diagram and the least squares regression line.

In using any statistical method, it is always important to check on the adequacy of the fitted model. This is certainly true in linear regression; and, in Section 4.6, we have already suggested considering a residual plot, namely $y_i - \widehat{y_i}$ against $x_i$. For example, say we plot $y_i$ against $x_i$ and $y_i - \widehat{y_i}$ against $x_i$, $i = 1, 2, \ldots, 100$, and obtain the two graphs in Figure 4.7-2. It really seems as if the straight line is not the best fitting regression curve to these data. Clearly a quadratic regression regression curve, $y = \beta_0 + \beta_1 x + \beta_2 x^2$, looks to be a better fit, or possibly some transformation of either $y$ or $x$ would lead to a better fit.

**Remark**   The data used in Figure 4.7-2 were simulated as follows. For each pair, $(x_i, y_i)$, $x_i$ was selected randomly from the interval $(1, 9)$. Given the value $x_i$, $y_i$ was selected randomly from the interval $[g(x_i) - 1, g(x_i) + 1]$, where $g(x) = -x^2/4 + 3x$. For the given 100 observations, the best fitting quadratic regression curve is given by $y = 0.124 + 2.989x - 0.249x^2$.

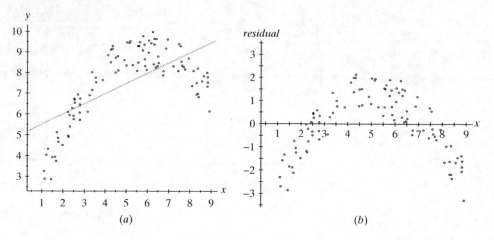

**Figure 4.7-2** Scatter plot with linear regression line and a residual plot

Example 4.7-2 illustrates a transformation with the data set in Table 4.7-1 that comes from Exercise 9.3-6, *Applied Statistics for Engineers and Physical Scientists*, 2nd ed., New York: Macmillan, 1992, by R. V. Hogg and J. Ledolter.

| Table 4.7-1 Weights and Miles Per Gallon (mpg) for 38 Automobiles | | | | | |
|---|---|---|---|---|---|
| **Weight** | **mpg** | **Weight** | **mpg** | **Weight** | **mpg** |
| 4.360 | 16.9 | 3.830 | 18.2 | 4.054 | 15.5 |
| 2.585 | 26.5 | 3.605 | 19.2 | 2.910 | 21.9 |
| 3.940 | 18.5 | 1.975 | 34.1 | 2.155 | 30.0 |
| 1.915 | 35.1 | 2.560 | 27.5 | 2.670 | 27.4 |
| 2.300 | 27.2 | 1.990 | 31.5 | 2.230 | 30.9 |
| 2.135 | 29.5 | 2.830 | 20.3 | 2.670 | 28.4 |
| 3.140 | 17.0 | 2.595 | 28.8 | 2.795 | 21.6 |
| 2.700 | 26.8 | 3.410 | 16.2 | 2.556 | 33.5 |
| 3.380 | 20.6 | 2.200 | 34.2 | 3.070 | 20.8 |
| 2.020 | 31.8 | 3.620 | 18.6 | 2.130 | 37.3 |
| 3.410 | 18.1 | 2.190 | 30.5 | 3.840 | 17.0 |
| 2.815 | 22.0 | 3.725 | 17.6 | 2.600 | 21.5 |
| 3.955 | 16.5 | 1.925 | 31.9 | | |

**EXAMPLE 4.7-2**

Consider the regression model that relates gas mileage and weights of 38 automobiles from the model year 1978–1979. The weights are given in units of 1000 pounds, and the fuel efficiencies are measured by miles per gallon (mpg) of gasoline.

Scatter plots of $y = mpg$ against $x = weight$ and $residual = y - \hat{y}$ against $x$ are given in Figure 4.7-3. The equation of the least squares regression line in Figure 4.7-3($a$) is $y = 48.707 - 8.365x$.

These graphs suggest that a quadratic $y = \beta_0 + \beta_1 x + \beta_2 x^2$ might be a more appropriate model. However, if we think seriously about the situation, it might be better to plot the number of gallons needed to drive each car 100 miles against the weight of that car. The reasoning behind this transformation is that as the

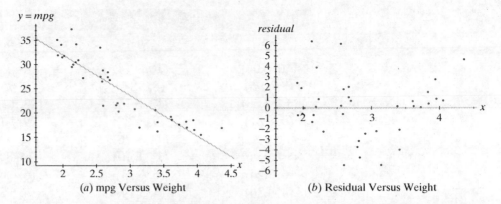

(a) mpg Versus Weight     (b) Residual Versus Weight

Figure 4.7-3 Scatter plot with linear regression line and the residual plot

weight increases, the number of gallons needed to drive the car 100 miles, namely, $100/mpg$, might increase proportionally. That is, if so much gasoline is needed to drive a 2000 pound car 100 miles, then a 4000 pound car might need twice as much. Fitting a straight line to $z = 100/mpg$ against $x = weight$ of the car yields Figure 4.7-4(a) while Figure 4.7-4(b) shows the residual plot, $residual = z - \widehat{z}$ against $x$.

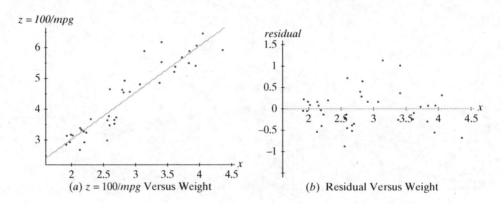

(a) $z = 100/mpg$ Versus Weight     (b) Residual Versus Weight

Figure 4.7-4 Scatter plot of transformed data and the residual plot

In Example 4.7-2, we suggested the transformation $z = 100/y$ because it seemed to be a sensible one. In practice, we often try many transformations. For illustration, if the standard deviation of $y$ seems to increase with the value of $y$, the logarithmic transformation ($\log y$), square root transformation ($\sqrt{y}$), or the more general transformation $y^p$ (for some value of $p$) might be an appropriate one. Plots of these transformed observations against their $x$ values indicate which particular transformation stabilizes the variance the best; and this selection often provides a better linear fit.

Similarly, an investigator often transforms the $x$-values. If, for instance, the $x$-values seemed skewed to the right, we might try the $\log x$, $\sqrt{x}$, or $x^p (p < 1)$ to help achieve a linear fit.

| Table 4.7-2 Anscombe Data | | | |
| a | b | c | d |
| (10, 8.04) | (10, 9.14) | (10, 7.46) | (8, 6.58) |
| (8, 6.95) | (8, 8.14) | (8, 6.77) | (8, 5.76) |
| (13, 7.58) | (13, 8.74) | (13, 12.74) | (8, 7.71) |
| (9, 8.81) | (9, 8.77) | (9, 7.11) | (8, 8.84) |
| (11, 8.33) | (11, 9.26) | (11, 7.81) | (8, 8.47) |
| (14, 9.96) | (14, 8.10) | (14, 8.84) | (8, 7.04) |
| (6, 7.24) | (6, 6.13) | (6, 6.08) | (8, 5.25) |
| (4, 4.26) | (4, 3.10) | (4, 5.39) | (8, 5.56) |
| (12, 10.84) | (12, 9.13) | (12, 8.15) | (8, 7.91) |
| (7, 4.82) | (7, 7.26) | (7, 6.42) | (8, 6.89) |
| (5, 5.68) | (5, 4.74) | (5, 5.73) | (19, 12.50) |

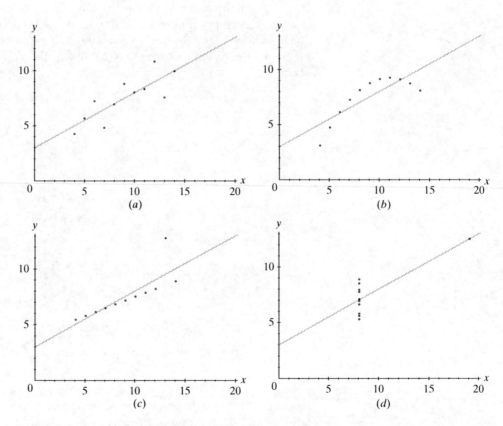

**Figure 4.7-5** For each figure, $\widehat{y} = 3.0 + 0.5x$, $r^2 = 0.67$

EXAMPLE 4.7-3    We close this section with the classical example of Anscombe's (F. J. Anscombe, "Graphs in Statistical Analysis," *The American Statistician, 27*, 17–21 [1973]) that demonstrates dramatically why we should always plot the regression variables. He constructed four data sets, each with $n = 11$ points, that have exactly the same linear fit and the same $r^2$. These data are listed in Table 4.7-2. Clearly only one of them, namely that in part ($a$) of Figure 4.7-5 is really appropriate to fit a linear function. The pattern in part ($b$) is quadratic, at least in the range from $x = 5$ to $x = 15$. In ($c$), ten observations lie on a straight line, but one response at $x = 13$ does not follow the pattern. The plot in ($d$) illustrates the influence of a single observation whose $x$ value is far from all of the others. Jointly, parts ($c$) and ($d$) show the effect of a single observation on the regression estimates. So clearly it is most important to plot the data, if at all possible.

## EXERCISES 4.7

**4.7-1** For the data given in Exercise 4.6-1, with the usual assumptions,
  **(a)** Find a 95% confidence interval for $\mu(x)$ when $x = 68, 75$, and 82.
  **(b)** Find a 95% prediction interval for $Y$ when $x = 68, 75$, and 82.

**4.7-2** For the data given in Exercise 4.6-12, with the usual assumptions,
  **(a)** Find a 95% confidence interval for $\mu(b)$ when $b = 50, 90$, and 130.
  **(b)** Find a 95% prediction interval for $a + b$ when $b = 50, 90$, and 130.

**4.7-3** For the data given in Exercise 4.6-2, with the usual assumptions,
  **(a)** Find a 95% confidence interval for $\mu(x)$ when $x = 20, 25$, and 30.
  **(b)** Find a 95% prediction interval for $Y$ when $x = 20, 25$, and 30.

**4.7-4** For the ACT scores in Exercise 4.6-7, with the usual assumptions,
  **(a)** Find a 95% confidence interval for $\mu(x)$ when $x = 17, 20, 23, 26$, and 29.
  **(b)** Determine a 90% prediction interval for $Y$ when $x = 17, 20, 23, 26$, and 29.

**4.7-5** A computer center recorded the number of programs it maintained during each of 10 consecutive years.
  **(a)** Calculate the least squares regression line for these data.
  **(b)** Plot the points and the line on the same graph.
  **(c)** Find a 95% prediction interval for the number of programs in year 11 under the usual assumptions.

| Year | Number of Programs |
|------|--------------------|
| 1    | 430                |
| 2    | 480                |
| 3    | 565                |
| 4    | 790                |
| 5    | 885                |
| 6    | 960                |
| 7    | 1200               |
| 8    | 1380               |
| 9    | 1530               |
| 10   | 1591               |

**4.7-6** The attendance ($x$) in hundreds at a race track and the amount ($y$) in millions of dollars that was bet on $n = 10$ selected days is given in the following table.

| $x$ | 117 | 128 | 122 | 119 | 131 | 135 | 125 | 120 | 130 | 127 |
|-----|------|------|------|------|------|------|------|------|------|------|
| $y$ | 2.07 | 2.80 | 3.14 | 2.26 | 3.40 | 3.89 | 2.93 | 2.66 | 3.33 | 3.54 |

  **(a)** Plot these 10 points ($x, y$).
  **(b)** Fit a simple linear regression line to the data.
  **(c)** Use these results to construct a 95 percent confidence interval for the mean $\mu(x)$.
  **(d)** Find a 90 percent prediction interval for the amount bet when the attendance is 12,000.

**4.7-7** In the book by G. E. P. Box, W. G. Hunter, and J. S. Hunter: *Statistics for Experimenters*, New York, Wiley, 1978, the authors analyzed the dispersion of an aerosol spray as a function of age. The dispersion $Y$ is measured as the reciprocal of the number of particles in a unit volume. Age $x$ is measured in minutes. The experiments were performed in random order, but they are listed here in increasing order of $x$.

| $x$ | 8 | 22 | 35 | 40 | 57 | 73 | 78 | 87 | 98 |
|---|---|---|---|---|---|---|---|---|---|
| $y$ | 6.16 | 9.88 | 14.35 | 24.06 | 30.34 | 32.17 | 42.18 | 43.23 | 48.76 |

**(a)** Plot these nine points $(x, y)$.

**(b)** Calculate the least-squares regression line and depict it on the graph.

**(c)** Test $H_0: \beta = 0$ against $H_1: \beta > 0$.

**(d)** Estimate the response $y$ at $x = 110$ by constructing a 95 percent prediction interval. Discuss the possible danger of extrapolation when $x = 110$ is outside the range of the observed $x$ values.

## 4.8 ONE-FACTOR ANALYSIS OF VARIANCE

We note here that the usual two-sample problem concerning two means $\mu_1$ and $\mu_2$ is really a regression model

$$Y_j = \beta_1 x_{1j} + \beta_2 x_{2j} + \epsilon_j, \qquad j = 1, 2, \ldots, n + n = 2n,$$

in which $\beta_1 = \mu_1$ and $\beta_2 = \mu_2$ and $(x_{1j}, x_{2j}) = (1, 0)$, for $j = 1, 2, \ldots, n$, and $(x_{1j}, x_{2j}) = (0, 1)$, $j = n + 1, n + 2, \ldots, n + n$. Frequently, experimenters want to compare more than two treatments, say yields of $m$ different corn hybrids with mean yields of $\mu_1, \mu_2, \ldots, \mu_m$ by taking samples of sizes $n_1, n_2, \ldots, n_m$, respectively, where $m > 2$. This situation can be modeled by

$$Y_{ij} = \beta_1 x_{1j} + \beta_2 x_{2j} + \cdots + \beta_m x_{mj} + \epsilon_{ij},$$

where $\epsilon_{ij}$ are independent $N(0, \sigma^2)$ random variables, $\beta_i = \mu_i$, $x_{ij} = 1$ and the other $x$ values equal zero, $i = 1, 2, \ldots, m$ and $j = 1, 2, \ldots, n_i$. Or, equivalently,

$$Y_{ij} = \mu_i + \epsilon_{ij}, \qquad i = 1, 2, \ldots, m, \quad j = 1, 2, \ldots, n_i.$$

That is, $Y_{i1}, Y_{i2}, \ldots, Y_{in_i}$ represents a random sample of size $n_i$ from $N(\mu_i, \sigma^2)$, $i = 1, 2, \ldots, m$. One inference that we wish to consider is the equality of the $m$ means, namely $H_0: \mu_1 = \mu_2 = \cdots = \mu_m = \mu$, $\mu$ unspecified, against all possible alternative hypotheses, $H_1$. In order to test this hypothesis we shall take independent random samples from these distributions. Let $Y_{i1}, Y_{i2}, \ldots, Y_{in_i}$ represent a random sample of size $n_i$ from the normal distribution $N(\mu_i, \sigma^2)$, $i = 1, 2, \ldots, m$. In Table 4.8-1 we have indicated these random samples along with the row means (sample means) where, with $n = n_1 + n_2 + \cdots + n_m$,

$$\overline{Y}_{..} = \frac{1}{n} \sum_{i=1}^{m} \sum_{j=1}^{n_i} Y_{ij} \quad \text{and} \quad \overline{Y}_{i.} = \frac{1}{n_i} \sum_{j=1}^{n_i} Y_{ij}, \qquad i = 1, 2, \ldots, m.$$

The dot in the notation for the means, $\overline{Y}_{..}$ and $\overline{Y}_{i.}$, indicates the index over which the average is taken. Here $\overline{Y}_{..}$ is an average taken over both indices while $\overline{Y}_{i.}$ is just taken over the index $j$.

### Table 4.8-1 One-Factor Random Samples

|  |  |  |  |  | Means |
|---|---|---|---|---|---|
| $Y_1:$ | $Y_{11}$ | $Y_{12}$ | $\cdots$ | $Y_{1n_1}$ | $\overline{Y}_{1\cdot}$ |
| $Y_2:$ | $Y_{21}$ | $Y_{22}$ | $\cdots$ | $Y_{2n_2}$ | $\overline{Y}_{2\cdot}$ |
| $\vdots$ | $\vdots$ | $\vdots$ | $\vdots$ | $\vdots$ | $\vdots$ |
| $Y_m:$ | $Y_{m1}$ | $Y_{m2}$ | $\cdots$ | $Y_{mn_m}$ | $\overline{Y}_{m\cdot}$ |
| Grand Mean: |  |  |  |  | $\overline{Y}_{\cdot\cdot}$ |

To determine a critical region for a test of $H_0$, we shall first partition the sum of squares associated with the variance of the combined samples into two parts. This sum of squares is given by

$$SS(TO) = \sum_{i=1}^{m}\sum_{j=1}^{n_i}(Y_{ij} - \overline{Y}_{\cdot\cdot})^2$$

$$= \sum_{i=1}^{m}\sum_{j=1}^{n_i}(Y_{ij} - \overline{Y}_{i\cdot} + \overline{Y}_{i\cdot} - \overline{Y}_{\cdot\cdot})^2$$

$$= \sum_{i=1}^{m}\sum_{j=1}^{n_i}(Y_{ij} - \overline{Y}_{i\cdot})^2 + \sum_{i=1}^{m}\sum_{j=1}^{n_i}(\overline{Y}_{i\cdot} - \overline{Y}_{\cdot\cdot})^2$$

$$+ 2\sum_{i=1}^{m}\sum_{j=1}^{n_i}(Y_{ij} - \overline{Y}_{i\cdot})(\overline{Y}_{i\cdot} - \overline{Y}_{\cdot\cdot}).$$

The last term of the right-hand member of this identity may be written as

$$2\sum_{i=1}^{m}\left[(\overline{Y}_{i\cdot} - \overline{Y}_{\cdot\cdot})\sum_{j=1}^{n_i}(Y_{ij} - \overline{Y}_{i\cdot})\right] = 2\sum_{i=1}^{m}(\overline{Y}_{i\cdot} - \overline{Y}_{\cdot\cdot})(n_i\overline{Y}_{i\cdot} - n_i\overline{Y}_{i\cdot}) = 0,$$

and the preceding term may be written as

$$\sum_{i=1}^{m}\sum_{j=1}^{n_i}(\overline{Y}_{i\cdot} - \overline{Y}_{\cdot\cdot})^2 = \sum_{i=1}^{m}n_i(\overline{Y}_{i\cdot} - \overline{Y}_{\cdot\cdot})^2.$$

Thus

$$SS(TO) = \sum_{i=1}^{m}\sum_{j=1}^{n_i}(Y_{ij} - \overline{Y}_{i\cdot})^2 + \sum_{i=1}^{m}n_i(\overline{Y}_{i\cdot} - \overline{Y}_{\cdot\cdot})^2.$$

For notation let

$$SS(TO) = \sum_{i=1}^{m}\sum_{j=1}^{n_i}(Y_{ij} - \overline{Y}_{\cdot\cdot})^2, \text{ the total sum of squares.}$$

$$SS(E) = \sum_{i=1}^{m}\sum_{j=1}^{n_i}(Y_{ij} - \overline{Y}_{i\cdot})^2, \quad \text{the sum of squares within treatments, groups, or classes, often called the error sum of squares.}$$

$$SS(T) = \sum_{i=1}^{m} n_i (\overline{Y}_{i.} - \overline{Y}_{..})^2,$$    the sum of squares among the different treatments, groups, or classes, often called the between treatment sum of squares.

Thus

$$SS(TO) = SS(E) + SS(T).$$

When $H_0$ is true, we may regard $Y_{ij}, i = 1, 2, \ldots, m, j = 1, 2, \ldots, n_i$, as a random sample of size $n = n_1 + n_2 + \cdots + n_m$ from the normal distribution $N(\mu, \sigma^2)$. Then $SS(TO)/(n-1)$ is an unbiased estimator of $\sigma^2$ because $SS(TO)/\sigma^2$ is $\chi^2(n-1)$ so that $E[SS(TO)/\sigma^2] = n-1$ and $E[SS(TO)/(n-1)] = \sigma^2$. An unbiased estimator of $\sigma^2$ based only on the sample from the $i$th distribution is

$$W_i = \frac{\sum_{j=1}^{n_i} (Y_{ij} - \overline{Y}_{i.})^2}{n_i - 1} \qquad \text{for } i = 1, 2, \ldots, m,$$

because $(n_i - 1)W_i/\sigma^2$ is $\chi^2(n_i-1)$. Thus

$$E\left[\frac{(n_i - 1)W_i}{\sigma^2}\right] = n_i - 1,$$

and so

$$E(W_i) = \sigma^2, \qquad i = 1, 2, \ldots, m.$$

It follows that the sum of $m$ of these independent chi-square random variables, namely

$$\sum_{i=1}^{m} \frac{(n_i - 1)W_i}{\sigma^2} = \frac{SS(E)}{\sigma^2},$$

is also chi-square with $(n_1 - 1) + (n_2 - 1) + \cdots + (n_m - 1) = n - m$ degrees of freedom. Hence $SS(E)/(n - m)$ is an unbiased estimator of $\sigma^2$. We now have that

$$\frac{SS(TO)}{\sigma^2} = \frac{SS(E)}{\sigma^2} + \frac{SS(T)}{\sigma^2},$$

where

$$\frac{SS(TO)}{\sigma^2} \text{ is } \chi^2(n-1) \text{ and } \frac{SS(E)}{\sigma^2} \text{ is } \chi^2(n-m).$$

Because $SS(T) \geq 0$, there is a theorem (see following remark) that states that $SS(E)$ and $SS(T)$ are independent and the distribution of $SS(T)/\sigma^2$ is $\chi^2(m-1)$.

Remark    The sums of squares, $SS(T)$, $SS(E)$, and $SS(TO)$, are examples of **quadratic forms** in the variables $Y_{ij}, i = 1, 2, \ldots, m, j = 1, 2, \ldots, n_i$. That is, each term in these sums of squares is of second degree in $Y_{ij}$. Furthermore the coefficients of the variables are real numbers, so these sums of squares are called **real quadratic forms**. The following theorem, stated without proof, is used in this chapter. (For a proof, see Hogg, McKean, and Craig, *Introduction to Mathematical Statistics*, 6th ed. [Prentice Hall, 2005].)

| | |
|---|---|
| Theorem 4.8-1 | Let $Q = Q_1 + Q_2 + \cdots + Q_k$, where $Q, Q_1, \ldots, Q_k$ are $k+1$ real quadratic forms in $n$ mutually independent random variables normally distributed with the same variance $\sigma^2$. Let $Q/\sigma^2, Q_1/\sigma^2, \ldots, Q_{k-1}/\sigma^2$ have chi-square distributions with $r, r_1, \ldots, r_{k-1}$ degrees of freedom, respectively. If $Q_k$ is non-negative, then |

(a) $Q_1, \ldots, Q_k$ are mutually independent, and hence,

(b) $Q_k/\sigma^2$ has a chi-square distribution with $r - (r_1 + \cdots + r_{k-1}) = r_k$ degrees of freedom.

Since under $H_0$, $SS(T)/\sigma^2$ is $\chi^2(m-1)$, we have $E[SS(T)/\sigma^2] = m - 1$ and hence $E[SS(T)/(m-1)] = \sigma^2$. Now the estimator of $\sigma^2$, which is based on $SS(E)$, namely $SS(E)/(n-m)$, is always unbiased whether $H_0$ is true or false. However, if the means $\mu_1, \mu_2, \ldots, \mu_m$ are not equal, the expected value of the estimator based on $SS(T)$ will be greater than $\sigma^2$. To make this last statement clear, we have

$$E[SS(T)] = E\left[\sum_{i=1}^{m} n_i(\overline{Y}_{i.} - \overline{Y}_{..})^2\right] = E\left[\sum_{i=1}^{m} n_i\overline{Y}_{i.}^2 - n\overline{Y}_{..}^2\right]$$

$$= \sum_{i=1}^{m} n_i\{\mathrm{Var}(\overline{Y}_{i.}) + [E(\overline{Y}_{i.})]^2\} - n\{\mathrm{Var}(\overline{Y}_{..}) + [E(\overline{Y}_{..})]^2\}$$

$$= \sum_{i=1}^{m} n_i\left\{\frac{\sigma^2}{n_i} + \mu_i^2\right\} - n\left\{\frac{\sigma^2}{n} + \overline{\mu}^2\right\}$$

$$= (m-1)\sigma^2 + \sum_{i=1}^{m} n_i(\mu_i - \overline{\mu})^2,$$

where $\overline{\mu} = (1/n)\sum_{i=1}^{m} n_i\mu_i$. If $\mu_1 = \mu_2 = \cdots = \mu_m = \mu$,

$$E\left(\frac{SS(T)}{m-1}\right) = \sigma^2.$$

If the means are not all equal,

$$E\left[\frac{SS(T)}{m-1}\right] = \sigma^2 + \sum_{i=1}^{m} n_i\frac{(\mu_i - \overline{\mu})^2}{m-1} > \sigma^2.$$

Exercise 4.8-2 also illustrates the fact that the estimator using $SS(T)$ is usually greater than that using $SS(E)$ when $H_0$ is false.

We can base our test of $H_0$ on the ratio of $SS(T)/(m-1)$ and $SS(E)/(n-m)$, both of which are unbiased estimators of $\sigma^2$, provided that $H_0: \mu_1 = \mu_2 = \cdots = \mu_m$ is true so that, under $H_0$, the ratio would assume values near one. However, as the means $\mu_1, \mu_2, \ldots, \mu_m$ begin to differ, this ratio tends to become large, since

$E[SS(T)/(m-1)]$ gets larger. Under $H_0$, the ratio

$$\frac{SS(T)/(m-1)}{SS(E)/(n-m)} = \frac{[SS(T)/\sigma^2]/(m-1)}{[SS(E)/\sigma^2]/(n-m)} = F$$

has an $F$ distribution with $m-1$ and $n-m$ degrees of freedom because $SS(T)/\sigma^2$ and $SS(E)/\sigma^2$ are independent chi-square variables. We would reject $H_0$ if the observed value of $F$ is too large because this would indicate that we have a relatively large $SS(T)$, which suggests that the means are unequal. Thus the critical region is of the form $F \geq F_\alpha(m-1, n-m)$.

The information for tests of the equality of several means is often summarized in an **analysis-of-variance table** or **ANOVA** table like that given in Table 4.8-2, where the mean square (MS) is the sum of squares (SS) divided by its degrees of freedom.

### Table 4.8-2 Analysis-of-Variance Table

| Source | Sum of Squares (SS) | Degrees of Freedom | Mean Square (MS) | F-Ratio |
|---|---|---|---|---|
| Treatment | SS(T) | $m-1$ | $MS(T) = \dfrac{SS(T)}{m-1}$ | $\dfrac{MS(T)}{MS(E)}$ |
| Error | SS(E) | $n-m$ | $MS(E) = \dfrac{SS(E)}{n-m}$ | |
| Total | SS(TO) | $n-1$ | | |

**EXAMPLE 4.8-1** Let $Y_1, Y_2, Y_3, Y_4$ be independent random variables that have normal distributions $N(\mu_i, \sigma^2)$, $i = 1, 2, 3, 4$. We shall test

$$H_0: \mu_1 = \mu_2 = \mu_3 = \mu_4 = \mu$$

against all alternatives based on a random sample of size $n_i = 3$ from each of the four distributions. A critical region of size $\alpha = 0.05$ is given by

$$F = \frac{SS(T)/(4-1)}{SS(E)/(12-4)} \geq 4.07 = F_{0.05}(3, 8).$$

The observed data are given in Table 4.8-3. (Clearly, these data are not observations from normal distributions. They were selected to illustrate the calculations.)

For these data, the calculated SS(TO), SS(E), and SS(T) are

$$SS(TO) = (13-11)^2 + (8-11)^2 + \cdots + (15-11)^2 + (10-11)^2 = 80;$$

$$SS(E) = (13-10)^2 + (8-10)^2 + \cdots + (15-12)^2 + (10-12)^2 = 50;$$

$$SS(T) = 3[(10-11)^2 + (13-11)^2 + (9-11)^2 + (12-11)^2] = 30.$$

| Table 4.8-3 Illustrative Data | | | | $\overline{Y}_i.$ |
|---|---|---|---|---|
| | **Observations** | | | |
| $Y_1$: | 13 | 8 | 9 | 10 |
| $Y_2$: | 15 | 11 | 13 | 13 |
| $Y_3$: | 8 | 12 | 7 | 9 |
| $Y_4$: | 11 | 15 | 10 | 12 |
| $\overline{Y}_{..}$ | | | | 11 |

Note that since SS(TO) = SS(E) + SS(T), only two of the three values need to be calculated from the data directly. Here the computed value of $F$ is

$$\frac{30/3}{50/8} = 1.6 < 4.07,$$

and $H_0$ is not rejected. The $p$-value is the probability, under $H_0$, of observing an $F$ that is at least as large as this observed $F$. It is often given by computer programs.

The information for this example is summarized in the ANOVA table, Table 4.8-4. Again we note that (here and elsewhere) the $F$ statistic is the ratio of two appropriate mean squares.

| Table 4.8-4 ANOVA Table for Illustrative Data | | | | | |
|---|---|---|---|---|---|
| **Source** | **Sum of Squares (SS)** | **Degrees of Freedom** | **Mean Square (MS)** | **$F$-Ratio** | **$p$-value** |
| Treatment | 30 | 3 | 30/3 | 1.6 | 0.264 |
| Error | 50 | 8 | 50/8 | | |
| Total | 80 | 11 | | | |

Formulas that sometimes simplify the calculations of SS(TO), SS(T), and SS(E) are

$$SS(TO) = \sum_{i=1}^{m}\sum_{j=1}^{n_i} Y_{ij}^2 - \frac{1}{n}\left[\sum_{i=1}^{m}\sum_{j=1}^{n_i} Y_{ij}\right]^2,$$

$$SS(T) = \sum_{i=1}^{m}\frac{1}{n_i}\left[\sum_{j=1}^{n_i} Y_{ij}\right]^2 - \frac{1}{n}\left[\sum_{i=1}^{m}\sum_{j=1}^{n_i} Y_{ij}\right]^2,$$

and

$$SS(E) = SS(TO) - SS(T).$$

It is interesting to note that in these formulas each square is divided by the number of observations in the sum being squared: $Y_{ij}^2$ by one, $(\sum_{j=1}^{n_i} Y_{ij})^2$ by $n_i$, and $(\sum_{i=1}^{m}\sum_{j=1}^{n_i} Y_{ij})^2$ by $n$. These formulas are used in Example 4.8-2. Although these formulas are useful, you are encouraged to use appropriate statistical packages on a computer to aid you with these calculations. Insight can also be gained by plotting on the same figure box-and-whisker diagrams for each of the samples.

EXAMPLE 4.8-2    A window that is manufactured for an automobile has five studs that are used for attaching it. A company that manufactures these windows performs "pull out tests" to determine the force needed to pull a stud out of the window. Let $Y_i$, $i = 1, 2, 3, 4, 5$, equal the force required at position $i$ and assume that the distribution of $Y_i$ is $N(\mu_i, \sigma^2)$. We shall test the null hypothesis $H_0: \mu_1 = \mu_2 = \mu_3 = \mu_4 = \mu_5$ using 7 independent observations at each position. At an $\alpha = 0.01$ significance level, $H_0$ is rejected if the computed

$$F = \frac{SS(T)/(5-1)}{SS(E)/(35-5)} \geq 4.02 = F_{0.01}(4, 30).$$

The observed data along with certain sums are given in Table 4.8-5. For these data,

| Table 4.8-5  Pull-Out Test Data | | | | | | | | |
|---|---|---|---|---|---|---|---|---|
| | Observations | | | | | | $\sum_{j=1}^{7} Y_{ij}$ | $\sum_{j=1}^{7} Y_{ij}^2$ |
| $Y_1$: | 92 | 90 | 87 | 105 | 86 | 83 | 102 | 645 | 59,847 |
| $Y_2$: | 100 | 108 | 98 | 110 | 114 | 97 | 94 | 721 | 74,609 |
| $Y_3$: | 143 | 149 | 138 | 136 | 139 | 120 | 145 | 970 | 134,936 |
| $Y_4$: | 147 | 144 | 160 | 149 | 152 | 131 | 134 | 1017 | 148,367 |
| $Y_5$: | 142 | 155 | 119 | 134 | 133 | 146 | 152 | 981 | 138,415 |
| Totals | | | | | | | | 4334 | 556,174 |

$$SS(TO) = 556,174 - \frac{1}{35}(4334)^2 = 19,500.97$$

$$SS(T) = \frac{1}{7}[645^2 + 721^2 + 970^2 + 1017^2 + 981^2]$$

$$- \frac{1}{35}(4334)^2 = 16,672.11$$

$$SS(E) = 19,500.97 - 16,672.11 = 2828.86.$$

Since the computed $F$ is

$$F = \frac{16,672.11/4}{2828.86/30} = 44.20,$$

the null hypothesis is clearly rejected. This information is summarized in Table 4.8-6.

But why is $H_0$ rejected? The box-and-whisker diagrams shown in Figure 4.8-1 help to answer this question. It looks like the forces required to pull out studs in positions 1 and 2 are similar and those in positions 3, 4, and 5 are quite similar

| Source | Sum of Squares (SS) | Degrees of Freedom | Mean Square (MS) | F | p-value |
|---|---|---|---|---|---|
| Treatment | 16,672.11 | 4 | 4,168.03 | 44.20 | 0.000 |
| Error | 2,828.86 | 30 | 94.30 | | |
| Total | 19,500.97 | 34 | | | |

**Table 4.8-6 ANOVA Table for Pull-Out Tests**

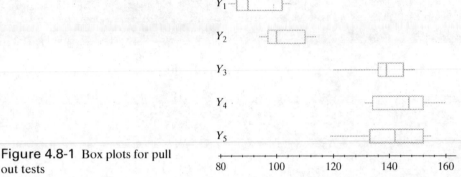

**Figure 4.8-1** Box plots for pull out tests

but different from positions 1 and 2. (See Exercise 4.8-8.) An examination of the window would confirm that this is the case.

As with the two sample $t$-test, the $F$-test works quite well even if the underlying distributions are non-normal, unless they are highly skewed or the variances are quite different. In these latter cases, we might need to transform the observations to make the data more symmetric with about the same variances or to use certain nonparametric methods that are beyond the scope of this course.

### EXERCISES 4.8

**4.8-1** Let $\mu_1, \mu_2, \mu_3$ be, respectively, the means of three normal distributions with a common but unknown variance $\sigma^2$. In order to test, at the $\alpha = 0.05$ significance level, the hypothesis $H_0: \mu_1 = \mu_2 = \mu_3$ against all possible alternative hypotheses, we take a random sample of size 4 from each of these distributions. Determine whether we accept or reject $H_0$ if the observed values from these three distributions are, respectively,

$$
\begin{array}{lcccc}
Y_1: & 5 & 9 & 6 & 8 \\
Y_2: & 11 & 13 & 10 & 12 \\
Y_3: & 10 & 6 & 9 & 9
\end{array}
$$

**4.8-2** For the following set of data show that the computed $SS(E)/(n - m) = 1$ and $SS(T)/(m - 1) = 75$. This suggests that the unbiased estimate of $\sigma^2$ based on $SS(T)$ is usually greater than $\sigma^2$ when the true means are unequal.

$$
\begin{array}{lccc}
Y_1: & 4 & 5 & 6 \\
Y_2: & 9 & 10 & 11 \\
Y_3: & 14 & 15 & 16
\end{array}
$$

**4.8-3** Hogg and Ledolter (see R. V. Hogg and J. Ledolter, *Applied Statistics for Engineers and Physical Scientists*, 2nd ed. [New York: Macmillan, 1992]) report that a civil engineer wishes to compare the strengths of three different types of beams, one (A) made of steel and two (B and C) made of different and more expensive alloys. A certain deflection (in units of 0.001 inch) was measured for each beam when submitted to a given force; thus a small deflection would indicate a beam of great strength. The order statistics for the three samples of sizes $n_1 = 8, n_2 = 6$, and $n_3 = 6$ are the following:

| | | | | | | | | |
|---|---|---|---|---|---|---|---|---|
| A: | 79 | 82 | 83 | 84 | 85 | 86 | 86 | 87 |
| B: | 74 | 75 | 76 | 77 | 78 | 82 | | |
| C: | 77 | 78 | 79 | 79 | 79 | 82 | | |

Use these data to test the equality of the three means. Use $\alpha = 0.05$.

**4.8-4** Montgomery considers the strengths of a synthetic fiber that is possibly affected by the percentage of cotton in the fiber. Five levels of this percentage are considered with five observations taken at each level.

| Percentage of Cotton | Tensile Strength in Pounds Per Square Inch | | | | |
|---|---|---|---|---|---|
| 15 | 7 | 7 | 15 | 11 | 9 |
| 20 | 12 | 17 | 12 | 18 | 18 |
| 25 | 14 | 18 | 18 | 19 | 19 |
| 30 | 19 | 25 | 22 | 19 | 23 |
| 35 | 7 | 10 | 11 | 15 | 11 |

Use the $F$-test, with $\alpha = 0.05$, to see if there are differences in the breaking strengths due to the percentages of cotton used. (D. C. Montgomery, *Design and Analysis of Experiments*, 2nd ed. [New York: Wiley, 1984], p. 51.)

**4.8-5** Let $Y_1, Y_2, Y_3, Y_4$ equal the cholesterol level of a woman under the age of 50, a man under 50, a woman 50 or older, and a man 50 or older, respectively. Assume that the distribution of $Y_i$ is $N(\mu_i, \sigma^2)$, $i = 1, 2, 3, 4$. We shall test the null hypothesis $H_0$: $\mu_1 = \mu_2 = \mu_3 = \mu_4$ using 7 observations of each $Y_i$.

**(a)** Give a critical region for an $\alpha = 0.05$ significance level.

**(b)** Construct an ANOVA table and state your conclusion using the following data:

| | | | | | | | |
|---|---|---|---|---|---|---|---|
| $Y_1$: | 221 | 213 | 202 | 183 | 185 | 197 | 162 |
| $Y_2$: | 271 | 192 | 189 | 209 | 227 | 236 | 142 |
| $Y_3$: | 262 | 193 | 224 | 201 | 161 | 178 | 265 |
| $Y_4$: | 192 | 253 | 248 | 278 | 232 | 267 | 289 |

**(c)** Give bounds on the $p$-value for this test.

**(d)** Construct box-and-whisker diagrams for each set of data on the same figure and give an interpretation.

**4.8-6** Let $Y_i$, $i = 1, 2, 3, 4$ equal the distance that a golf ball travels when hit from a tee, where $i$ denotes the index of the $i$th manufacturer. Assume that the distribution of $Y_i$ is $N(\mu_i, \sigma^2)$, $i = 1, 2, 3, 4$ when hit by a certain golfer. We shall test the null hypothesis $H_0$: $\mu_1 = \mu_2 = \mu_3 = \mu_4$ using three observations of each random variable.

**(a)** Give a critical region for an $\alpha = 0.05$ significance level.

**(b)** Construct an ANOVA table and state your conclusion using the following data:

| | | | |
|---|---|---|---|
| $Y_1$: | 240 | 221 | 265 |
| $Y_2$: | 286 | 256 | 272 |
| $Y_3$: | 259 | 245 | 232 |
| $Y_4$: | 239 | 215 | 223 |

**(c)** What would your conclusion be if $\alpha = 0.025$?

**(d)** What is the approximate $p$-value of this test?

**4.8-7** The driver of a diesel-powered automobile decided to test the quality of three types of diesel fuel sold in the area based on mpg. Test the null hypothesis that the three means are equal using the following data. Make the usual assumptions and take $\alpha = 0.05$.

| Brand A: | 38.7 | 39.2 | 40.1 | 38.9 | |
| Brand B: | 41.9 | 42.3 | 41.3 | | |
| Brand C: | 40.8 | 41.2 | 39.5 | 38.9 | 40.3 |

**4.8-8** Based on the box-and-whisker diagrams in Figure 4.8-1, it looks like the means of $Y_1$ and $Y_2$ could be equal and also that the means of $Y_3$, $Y_4$, and $Y_5$ could be equal but different from the first two.

   (a) Using the data in Example 4.8-2, test $H_0: \mu_1 = \mu_2$ against a two-sided alternative hypothesis using a $t$-test and an $F$-test. Let $\alpha = 0.05$. Do $F$ and $t$ tests give the same result?

   (b) Using the data in Example 4.8-2, test $H_0: \mu_3 = \mu_4 = \mu_5$. Let $\alpha = 0.05$.

**4.8-9** For an aerosol product, there are three weights: the tare weight (container weight), the concentrate weight, and the propellant weight. Let $Y_1, Y_2, Y_3$ denote the propellant weights on three different days. Assume that each of these independent random variables has a normal distribution with common variance and respective means, $\mu_1, \mu_2, \mu_3$. We shall test the null hypothesis $H_0: \mu_1 = \mu_2 = \mu_3$ using nine observations of each of the random variables.

   (a) Give a critical region for an $\alpha = 0.01$ significance level.

   (b) Construct an ANOVA table and state your conclusion using the following data:

| $Y_1$: | 43.06 | 43.32 | 42.63 | 42.86 | 43.05 |
| | 42.87 | 42.94 | 42.80 | 42.36 | |
| $Y_2$: | 42.33 | 42.81 | 42.13 | 42.41 | 42.39 |
| | 42.10 | 42.42 | 41.42 | 42.52 | |
| $Y_3$: | 42.83 | 42.57 | 42.96 | 43.16 | 42.25 |
| | 42.24 | 42.20 | 41.97 | 42.61 | |

   (c) Construct box-and-whisker diagrams for each set of data on the same figure and give an interpretation.

**4.8-10** A particular process puts a coating on a piece of glass so that it is sensitive to touch. Randomly throughout the day, pieces of glass are selected from the production line and the resistance is measured at 12 different locations on the glass. On each of 3 different days, December 6, December 7, and December 22, the following data give the means of the 12 measurements on each of 11 pieces of glass.

| December 6: | 175.05 | 177.44 | 181.94 | 176.51 | 182.12 | 164.34 |
| | 163.20 | 168.12 | 171.26 | 171.92 | 167.87 | |
| December 7: | 175.93 | 176.62 | 171.39 | 173.90 | 178.34 | 172.90 |
| | 174.67 | 174.27 | 177.16 | 184.13 | 167.21 | |
| December 22: | 167.27 | 161.48 | 161.86 | 173.83 | 170.75 | 172.90 |
| | 173.27 | 170.82 | 170.93 | 173.89 | 177.68 | |

   (a) Use these data to test whether the means on all three days are equal.

   (b) Use box-and-whisker diagrams to confirm your answer.

## 4.9   DISTRIBUTION-FREE CONFIDENCE AND TOLERANCE INTERVALS

We defined the order statistics, percentiles of a sample, and the percentiles $\pi_p$ of a distribution of the continuous type in Section 3.1. In this section we show, using order statistics, how to find confidence intervals for various percentiles $\pi_p$ of continuous distributions. Since little is assumed in finding a confidence interval for $\pi_p$, we often call them **distribution-free confidence intervals**, which tend to be longer than those with stronger model assumptions.

If $Y_1 < Y_2 < Y_3 < Y_4 < Y_5$ are the order statistics of a random sample of size $n = 5$ from a continuous-type distribution, then the sample median $Y_3$ could be thought of as an estimator of the distribution median $\pi_{0.5}$. We shall let $m = \pi_{0.5}$. We could simply use the sample median $Y_3$ as an estimator of the distribution median $m$. However, we are certain that all of us recognize that, with only a sample of size 5, we would be quite lucky if the observed $Y_3 = y_3$ were very close to $m$. Thus we now describe how a confidence interval can be constructed for $m$.

Instead of simply using $Y_3$ as an estimator of $m$, let us also compute the probability that the random interval $(Y_1, Y_5)$ includes $m$. That is, let us determine $P(Y_1 < m < Y_5)$. This is easy if we say that we have success if an individual observation, say $X$, is less than $m$; thus the probability of success on one of the independent trials is $P(X < m) = 0.5$. In order for the first order statistic $Y_1$ to be less than $m$ and the last order statistic $Y_5$ to be greater than $m$, we must have at least one success but not five successes. That is,

$$P(Y_1 < m < Y_5) = \sum_{k=1}^{4} \binom{5}{k} \left(\frac{1}{2}\right)^k \left(\frac{1}{2}\right)^{5-k}$$

$$= 1 - \left(\frac{1}{2}\right)^5 - \left(\frac{1}{2}\right)^5 = \frac{15}{16}.$$

So the probability that the random interval $(Y_1, Y_5)$ includes $m$ is $15/16 \approx 0.94$. Suppose that this random sample is actually taken and the order statistics are observed to equal $y_1 < y_2 < y_3 < y_4 < y_5$, respectively. Then $(y_1, y_5)$ is a 94% confidence interval for $m$.

It is interesting to note what happens as the sample size increases. Let $Y_1 < Y_2 < \cdots < Y_n$ be the order statistics of a random sample of size $n$ from a distribution of the continuous type. Thus $P(Y_1 < m < Y_n)$ is the probability that there is at least one "success" but not $n$ successes, where the probability of success on each trial is $P(X < m) = 0.5$. Consequently,

$$P(Y_1 < m < Y_n) = \sum_{k=1}^{n-1} \binom{n}{k} \left(\frac{1}{2}\right)^k \left(\frac{1}{2}\right)^{n-k}$$

$$= 1 - \left(\frac{1}{2}\right)^n - \left(\frac{1}{2}\right)^n = 1 - \left(\frac{1}{2}\right)^{n-1}.$$

This probability increases as $n$ increases so that the corresponding confidence interval $(y_1, y_n)$ would have a very large confidence coefficient, $1 - (1/2)^{n-1}$. Unfortunately, the interval $(y_1, y_n)$ tends to get wider as $n$ increases, and thus we are not "pinning down" $m$ very well. If we used the interval $(y_2, y_{n-1})$ or $(y_3, y_{n-2})$ we would obtain a shorter interval but also a smaller confidence coefficient. Let us investigate this possibility further.

With the order statistics $Y_1 < Y_2 < \cdots < Y_n$ associated with a random sample of size $n$ from a continuous-type distribution, consider $P(Y_i < m < Y_j)$, where $i < j$. For example, we might want

$$P(Y_2 < m < Y_{n-1}) \quad \text{or} \quad P(Y_3 < m < Y_{n-2}).$$

On each of the $n$ independent trials we say that we have success if that $X$ is less than $m$; thus the probability of success on each trial is $P(X < m) = 0.5$. Consequently, to have the $i$th order statistic $Y_i$ less than $m$ and the $j$th order statistic greater than $m$, we must have at least $i$ successes but fewer than $j$ successes (or else $Y_j < m$). That is,

$$P(Y_i < m < Y_j) = \sum_{k=i}^{j-1} \binom{n}{k} \left(\frac{1}{2}\right)^k \left(\frac{1}{2}\right)^{n-k} = 1 - \alpha.$$

For particular values of $n$, $i$, and $j$, this probability, say $1 - \alpha$, which is the sum of probabilities from a binomial distribution, can be calculated directly or approximated by an area under the normal p.d.f. provided $n$ is large enough. The observed interval $(y_i, y_j)$ could then serve as a $100(1 - a)\%$ confidence interval for the unknown distribution median.

**EXAMPLE 4.9-1**    The lengths in centimeters of $n = 9$ fish of a particular species (*nezumia*) captured off the New England coast were 32.5, 27.6, 29.3, 30.1, 15.5, 21.7, 22.8, 21.2, 19.0. Thus the observed order statistics are

$$15.5 < 19.0 < 21.2 < 21.7 < 22.8 < 27.6 < 29.3 < 30.1 < 32.5.$$

Before the sample is drawn, we know that

$$P(Y_2 < m < Y_8) = \sum_{k-2}^{7} \binom{9}{k}\left(\frac{1}{2}\right)^k\left(\frac{1}{2}\right)^{9-k} = 0.9805 - 0.0195 = 0.9610,$$

from Table II in the Appendix. Thus the confidence interval $(y_2 = 19.0, y_8 = 30.1)$ for $m$, the median of the lengths of all fish of this species, has a 96.1% confidence coefficient.

So that the student need not compute many of these probabilities, we give in Table 4.9-1 the necessary information for constructing confidence intervals of the form $(y_i, y_{n+1-i})$ for the unknown $m$ for sample sizes $n = 5, 6, \cdots, 20$. The subscript $i$ is selected so that the confidence coefficient $P(Y_i < m < Y_{n+1-i})$ is greater than 90% and as close to 95% as possible.

| $n$ | $(i, n + 1 - i)$ | $P(Y_i < m < Y_{n+1-i})$ |
|---|---|---|
| | **Table 4.9-1  Information for Confidence Intervals for $m$** | |
| 5 | $(1, 5)$ | 0.9376 |
| 6 | $(1, 6)$ | 0.9688 |
| 7 | $(1, 7)$ | 0.9844 |
| 8 | $(2, 7)$ | 0.9296 |
| 9 | $(2, 8)$ | 0.9610 |
| 10 | $(2, 9)$ | 0.9786 |
| 11 | $(3, 9)$ | 0.9346 |
| 12 | $(3, 10)$ | 0.9614 |
| 13 | $(3, 11)$ | 0.9776 |
| 14 | $(4, 11)$ | 0.9426 |
| 15 | $(4, 12)$ | 0.9648 |
| 16 | $(5, 12)$ | 0.9232 |
| 17 | $(5, 13)$ | 0.9510 |
| 18 | $(5, 14)$ | 0.9692 |
| 19 | $(6, 14)$ | 0.9364 |
| 20 | $(6, 15)$ | 0.9586 |

We know that when the sample size $n$ gets large, we usually use the fact that the binomial probabilities can be approximated by the normal distribution with mean $np$ and variance $np(1 - p)$. To illustrate how good these approximations are, we compute the probability corresponding to $n = 16$ in Table 4.9-1. Here, using Table II, we have

$$1 - \alpha = P(Y_5 < m < Y_{12}) = \sum_{k=5}^{11} \binom{16}{k} \left(\frac{1}{2}\right)^k \left(\frac{1}{2}\right)^{16-k}$$

$$= P(W = 5, 6, \ldots, 11) = 0.9616 - 0.0384 = 0.9232,$$

where $W$ is $b(16, 1/2)$. The normal approximation gives

$$1 - \alpha = P(4.5 < W < 11.5) = P\left(\frac{4.5 - 8}{2} < \frac{W - 8}{2} < \frac{11.5 - 8}{2}\right)$$

because $W$ has mean $np = 8$ and variance $np(1 - p) = 4$. The standardized variable $Z = (W - 8)/2$ has an approximate normal distribution. Thus

$$1 - \alpha \approx \Phi\left(\frac{3.5}{2}\right) - \Phi\left(\frac{-3.5}{2}\right) = \Phi(1.75) - \Phi(-1.75)$$

$$= 0.9599 - 0.0401 = 0.9198.$$

This compares very favorably with the probability 0.9232 recorded in Table 4.9-1.

The argument used to find a confidence interval for the median $m$ of a distribution of the continuous type can be applied to any percentile $\pi_p$. In this case we say that we have success on a single trial if that $X$ is less than $\pi_p$. Thus the probability of success on each of the independent trials is $P(X < \pi_p) = p$. Accordingly, with $i < j$, $1 - \alpha = P(Y_i < \pi_p < Y_j)$ is the probability that we have at least $i$ successes but fewer than $j$ successes. Thus

$$1 - \alpha = P(Y_i < \pi_p < Y_j) = \sum_{k=i}^{j-1} \binom{n}{k} p^k (1 - p)^{n-k}.$$

Once the sample is observed and the order statistics determined, the known interval $(y_i, y_j)$ could serve as a $100(1 - \alpha)\%$ confidence interval for the unknown distribution percentile $\pi_p$.

**EXAMPLE 4.9-2**    Let the following numbers represent the order statistics of the $n = 27$ observations obtained in a random sample from a certain population of incomes (measured in hundreds of dollars):

| | | | | | |
|---|---|---|---|---|---|
| 161 | 180 | 192 | 205 | 229 | 264 |
| 169 | 183 | 193 | 213 | 241 | 291 |
| 171 | 184 | 196 | 221 | 243 | 317 |
| 174 | 186 | 200 | 222 | 256 | 376 |
| 179 | 187 | 204 | | | |

Say we are interested in estimating the 25th percentile $\pi_{0.25}$ of the population. Since $(n + 1)p = 28(1/4) = 7$, the 7th order statistic, namely $y_7 = 183$, would be a point estimate of $\pi_{0.25}$. To find a confidence interval for $\pi_{0.25}$, let us move down and up a few order statistics from $y_7$, say to $y_4$ and $y_{10}$. What is the confidence coefficient associated with the interval $(y_4, y_{10})$? Before the sample was drawn, we had

$$1 - \alpha = P(Y_4 < \pi_{0.25} < Y_{10}) = \sum_{k=4}^{9} \binom{27}{k}(0.25)^k(0.75)^{27-k}$$

$$= 0.8201.$$

This answer can be found using a calculator or computer. A normal approximation could also be used by letting $W$ be $b(27, 1/4)$ with mean $27/4 = 6.75$ and variance $81/16$. Then

$$1 - \alpha = P(3.5 < W < 9.5)$$

$$\approx \Phi\left(\frac{9.5 - 6.75}{9/4}\right) - \Phi\left(\frac{3.5 - 6.75}{9/4}\right)$$

$$= \Phi\left(\frac{11}{9}\right) - \Phi\left(-\frac{13}{9}\right) = 0.8149.$$

Thus ($y_4 = 174, y_{10} = 187$) serves as an 82.01% or approximate 81.49% confidence interval for $\pi_{0.25}$. It should be noted that we could choose other intervals, such as ($y_3 = 171, y_{11} = 192$), and these would have different confidence coefficients. The persons involved in the study must select the desired confidence coefficient, and then the appropriate order statistics are taken, usually quite symmetrically about the $(n+1)p$th order statistic.

When the number of observations is large, it is important to be able to determine rather easily the order statistics. A stem-and-leaf diagram, as introduced in Section 3.1, can be helpful in determining the needed order statistics. This is illustrated in the next example.

EXAMPLE 4.9-3   The measurements of butterfat produced by $n = 90$ cows during a 305-day milk production period following their first calf are summarized in Table 4.9-2, an ordered stem-and-leaf diagram in which each leaf consists of two digits.

From this display it is quite easy to see that $y_8 = 392$. It takes only a little more work to show that $y_{38} = 494$ and $y_{53} = 526$. The interval (494, 526) serves as a confidence interval for the unknown median $m$ of all butterfat production for the given breed of cows. Its confidence coefficient is

$$P(Y_{38} < m < Y_{53}) = \sum_{k=38}^{52} \binom{90}{k}\left(\frac{1}{2}\right)^k\left(\frac{1}{2}\right)^{90-k}$$

$$\approx \Phi\left(\frac{52.5 - 45}{\sqrt{22.5}}\right) - \Phi\left(\frac{37.5 - 45}{\sqrt{22.5}}\right)$$

$$= \Phi(1.581) - \Phi(-1.581) = 0.8860,$$

because $(90)(1/2) = 45$ and $(90)(1/2)(1/2) = 22.5$. The exact value of 0.8867 can be found with your calculator or computer.

Similarly, ($y_{17} = 437, y_{29} = 470$) is a confidence interval for the first quartile $\pi_{0.25}$ with confidence coefficient

$$P(Y_{17} < \pi_{0.25} < Y_{29}) \approx \Phi\left(\frac{28.5 - 22.5}{\sqrt{16.875}}\right) - \Phi\left(\frac{16.5 - 22.5}{\sqrt{16.875}}\right)$$

$$= \Phi(1.46) - \Phi(-1.46) = 0.8558,$$

### Table 4.9-2 Stem-and-Leaf Display of 90 Butterfat Measurements

| Stems | Leaves | | | | | | | | | Depths |
|-------|----|----|----|----|----|----|----|----|----|--------|
| 2s | 74 | | | | | | | | | 1 |
| 2● | | | | | | | | | | 1 |
| 3* | | | | | | | | | | 1 |
| 3t | 27 | 39 | | | | | | | | 3 |
| 3f | 45 | 50 | | | | | | | | 5 |
| 3s | | | | | | | | | | 5 |
| 3● | 80 | 88 | 92 | 94 | 95 | | | | | 10 |
| 4* | 17 | 18 | | | | | | | | 12 |
| 4t | 21 | 22 | 27 | 34 | 37 | 39 | | | | 18 |
| 4f | 44 | 52 | 53 | 53 | 57 | 58 | | | | 24 |
| 4s | 60 | 64 | 66 | 70 | 70 | 72 | 75 | 78 | | 32 |
| 4● | 81 | 86 | 89 | 91 | 92 | 94 | 96 | 97 | 99 | 41 |
| 5* | 00 | 00 | 01 | 02 | 05 | 09 | 10 | 13 | 13 | 16 | (10) |
| 5t | 24 | 26 | 31 | 32 | 32 | 37 | 37 | 39 | | 39 |
| 5f | 40 | 41 | 44 | 55 | | | | | | 31 |
| 5s | 61 | 70 | 73 | 74 | | | | | | 27 |
| 5● | 83 | 83 | 86 | 93 | 99 | | | | | 23 |
| 6* | 07 | 08 | 11 | 12 | 13 | 17 | 18 | 19 | | 18 |
| 6t | 27 | 28 | 35 | 37 | | | | | | 10 |
| 6f | 43 | 43 | 45 | | | | | | | 6 |
| 6s | 72 | | | | | | | | | 3 |
| 6● | 91 | 96 | | | | | | | | 2 |

because $(90)(1/4) = 22.5$  and  $(90)(1/4)(3/4) = 16.875$. The exact value is 0.8569 using the computer.

It is interesting to compare the lengths of a confidence interval for the mean $\mu$ using $\bar{x} \pm 2(s/\sqrt{n})$ and a 95% one for the median $m$ using the distribution-free techniques of this section. (See Exercise 4.9-6.) Usually, if the sample arises from a distribution that does not deviate too much from the normal, the confidence interval based upon $\bar{x}$ is much shorter. After all, we assume much more creating that confidence interval. With the distribution-free method, all we assume is that the distribution be of the continuous type. So if the distribution is highly skewed or heavy tailed so that outliers could exist, the distribution-free techniques are safer and much more robust. Moreover, the distribution-free technique provides a way to get confidence intervals for various percentiles, and investigators are often interested in these.

This next concept is similar to distribution-free confidence intervals for unknown percentiles, and it really answers "fill" problems better than confidence intervals or tests of hypotheses. It doesn't matter whether the "fill" is cereal, beer, toothpaste, cough syrup, cleaning fluid, nuts, potato chips, etc. It goes something like this. The company wants to place at least 12 ounces of chips, say, in a bag as they have advertised that amount. Clearly the mean weight cannot be $\mu = 12$ or about 50% of the bags would not have enough chips and the FDA would not permit this. Thus the mean must be somewhat higher than 12 and hopefully the standard deviation $\sigma$ small enough so that the contents of almost all bags exceeds 12 ounces in weight. For illustration, if $\mu - 3\sigma = 12$, then, with normal assumptions, only 0.13 percent of the contents of the bags would be slightly underweight and that is certainly acceptable to the FDA (that agency recognizes that variation exists and accepts a very small fraction below the limit).

Thus we see that the "fill" problem is really an overfill one and frequently very expensive. All companies must overfill some so that only a small portion of their products weight less than the advertised amount. Often workers of companies want to be on the safe side and tend to overfill too much. For illustration, if they give an extra ounce of toothpaste on the average, what does this mean to the company? There clearly is the cost of the extra toothpaste; but, more important, the consumer will not buy the toothpaste as often. These two factors account for huge losses to any company in the "fill" business. Hence, companies really want to reduce the variation as much as possible so that the mean can be set only slightly above the advertised weight.

We let $Y_1 < Y_2 < \cdots < Y_n$ be the order statistics of a random sample of size $n$ from a continuous-type distribution with distribution function $F(x)$. Then the random variable $F(Y_i) = Z_i$, $i = 1, 2, \ldots, n$, is the fractional part of the probability for the distribution of the original random variable $X$ which is less than or equal to $Y_i$. In a more advanced course, it can be shown that $Z_i = F(Y_i)$ has a beta distribution with $\alpha = i$ and $\beta = n - i + 1$. (See Section 5.2.)

Consider the $n+1$ random variables $W_1 = Z_1 = F(Y_1)$, $W_2 = Z_2 - Z_1 = F(Y_2) - F(Y_1)$, $W_3 = Z_3 - Z_2 = F(Y_3) - F(Y_2)$, $\cdots$, $W_n = Z_n - Z_{n-1} = F(Y_n) - F(Y_{n-1})$, and $W_{n+1} = 1 - Z_n = 1 - F(Y_n)$. These $W_1, W_2, \cdots, W_n, W_{n+1}$ are called the respective *coverages* of the intervals $(-\infty, Y_1], (Y_1, Y_2], \ldots, (Y_{n-1}, Y_n], (Y_n, \infty)$. It is an interesting result that, if $k \leq n$, the sum of $k$ of these coverages has a beta distribution with $\alpha = k$ and $\beta = n - k + 1$. For illustration, we have already stated that $Z_i = F(Y_i)$, which is the sum of $k = i$ coverages, namely $W_1 + W_2 + \cdots + W_i$, has a beta distribution with $\alpha = i$ and $\beta = n - i + 1$. Moreover, we note that $F(Y_j) - F(Y_i)$, $i < j$, is the sum of $k = j - i$ coverages so that the probability

$$\gamma = P[F(Y_j) - F(Y_i) \geq p] = \int_p^1 \frac{\Gamma(n+1)}{\Gamma(j-i)\Gamma(n-j+i+1)} v^{j-i-1}(1-v)^{n-j+i}\, dv.$$

Once $Y_i$ and $Y_j$ are observed to be $y_i$ and $y_j$, we can refer to the interval $(y_i, y_j)$ as a $100\gamma$ percent **tolerance interval** for $100p$ percent of the probability of the distribution of $X$. Note that there are two proper fractions here: $\gamma$ is the probability that the random interval $(Y_i, Y_j)$ covers the fraction $p$ of the original distribution from which the random sample arose.

Let us consider a simple example.

**EXAMPLE 4.9-4**   Let $Y_1 < Y_2 < \cdots < Y_6$ be the order statistics of a random sample of size $n = 6$ from a continuous-type distribution. Then, if $p = 0.8$, we have

$$\gamma = P[F(Y_6) - F(Y_1) \geq 0.8] = \int_{0.8}^1 \frac{\Gamma(7)}{\Gamma(5)\Gamma(2)} v^4(1-v)\, dv = 0.34.$$

That is, the observed values of $Y_1$ and $Y_6$ provide a 34 percent tolerance interval for 80 percent of the distribution.

Clearly in practice we want $\gamma$ and $p$ to be much larger than 0.34 and 0.8. For illustration, consider a "fill" problem in which the manufacturer states that each container has at least 12 ounces of the product. If $X$ is the amount in a container, the company would be pleased to note that the interval 12.02 to 12.18 ounces is a 98% tolerance interval for 95% of the distribution of $X$. Let us consider an example having larger values of $p$ and $\gamma$.

EXAMPLE 4.9-5    Let $Y_1 < Y_2 < \cdots < Y_{100}$ be the order statistics of a random sample of size $n = 100$ from a continuous-type distribution. Let $p = 0.9$; and thus, since $F(Y_{100}) - F(Y_1)$ is the sum of 99 coverages, we have

$$\gamma = \int_{0.9}^{1} \frac{\Gamma(101)}{\Gamma(99)\Gamma(2)} v^{99-1}(1-v)^{2-1}\, dv$$

$$= \left[100v^{99}(1-v) + v^{100}\right]_{0.9}^{1}$$

$$= 1 - \sum_{k=99}^{100} \binom{100}{k}(0.9)^k(0.1)^{100-k}$$

$$= \sum_{k=0}^{98} \binom{100}{k}(0.9)^k(0.1)^{100-k}.$$

That is, by integrating by parts, we have noted the relationship between the beta and binomial distributions, and we see that we must compute the probability $\gamma$ of 98 or fewer successes for the binomial distribution, $b(100, 0.9)$. Using the computer it is 0.9997. By using the normal approximation it is, with $np = 90$ and $np(1-p) = 9$,

$$P(U \le 98) = P\left(\frac{U - 90}{3} \le \frac{98.5 - 90}{3}\right) = \Phi(2.83) = 0.9977.$$

Thus the observed values of $(y_1, y_{100})$ provide a 99.97 percent tolerance interval for 90 percent of the distribution. By using a shorter interval, say $(y_2, y_{99})$, we would decrease the $\gamma$ somewhat to 0.9922.

Clearly we could find a one-sided tolerance interval based upon the interval $(Y_i, \infty)$ which is associated with the sum of $k = n+1-i$ coverages. In the preceding example, the interval $(y_4, \infty)$ would be a 99.22 percent tolerance interval for 90 percent of the distribution.

# APPLICATIONS

**1.** In *A Data-Based Approach to Statistics* (1994) by Ronald L. Iman, Wadsworth Publishing Co., Belmont, CA 94002, it is reported that a reliability engineer was using an accelerated life test on certain aircraft wheels. In such tests, the conditions are accelerated so as to speed up the time to failure of the item under investigation. Accelerated life tests are used on many products such as automobiles, motors, and tires. In the case of these wheels, a machine is used to roll the wheel under a load for what is equivalent to many miles until failure. The following are the results of 16 wheels in thousands of miles; the values have been ordered.

$$46.7 \quad 47.2 \quad 49.1 \quad 56.5 \quad 56.8 \quad 59.2 \quad 59.9 \quad 63.2$$
$$63.3 \quad 63.4 \quad 63.7 \quad 64.1 \quad 67.1 \quad 67.7 \quad 93.3 \quad 118.5$$

By considering $(16 + 1)(0.5) = 8.5$, $17(0.25) = 4.25$, and $17(0.75) = 12.75$, the estimates of the median and first and third quartiles are, respectively,

$$\widetilde{\pi}_{0.5} = \left(\frac{1}{2}\right)(63.2) + \left(\frac{1}{2}\right)(63.3) = 63.25,$$

$$\widetilde{\pi}_{0.25} = \left(\frac{3}{4}\right)(56.5) + \left(\frac{1}{4}\right)(56.8) = 56.575,$$

$$\widetilde{\pi}_{0.75} = \left(\frac{1}{4}\right)(64.1) + \left(\frac{3}{4}\right)(67.1) = 66.35.$$

Thus the five-number summary is

$$y_1 = 46.7, \quad \widetilde{q}_1 = 56.575, \quad \widetilde{m} = 63.25, \quad \widetilde{q}_3 = 66.35, \quad y_{16} = 118.5,$$

and the corresponding box-plot is show in Figure 4.9-1.

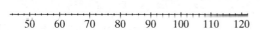

**Figure 4.9-1** Box plot for failures of wheels in thousands of miles

Due to the two large values, these data are very skewed to the right and the engineer did not want to use $\bar{x} \pm 2s/\sqrt{16}$ as a confidence interval for the middle. Accordingly he found a 92.32% confidence interval for the median $m = \pi_{0.5}$ using a table similar to Table 4.9-1 and recorded $y_5 = 56.8$ to $y_{12} = 64.1$ as that interval, along with the point estimate $\tilde{\pi}_{0.5} = 63.25$ of the middle.

**2.** If we refer to the first application at the end of Section 3.1 involving the pull strengths of 98 soldered resistor leads, we have the point estimates of the three quartiles, namely

$$\tilde{\pi}_{0.25} = 63.2, \qquad \tilde{\pi}_{0.5} = 64.4, \qquad \tilde{\pi}_{0.75} = 65.5.$$

Suppose we want to find confidence intervals for $\pi_{0.25}$, $\pi_{0.5}$, and $\pi_{0.75}$ from these data. Let us begin with the median $m = \pi_{0.5}$.

Clearly we would like to find $i$ so that

$$P(Y_i < \pi_{0.5} < Y_{99-i}) = \sum_{k=i}^{98-i} \binom{98}{k} \left(\frac{1}{2}\right)^{98}$$

$$\approx \Phi\left(\frac{98.5 - i - 49}{\sqrt{98(0.5)(0.5)}}\right) - \Phi\left(\frac{i - 0.5 - 49}{\sqrt{98(0.5)(0.5)}}\right) \approx 0.95.$$

Thus

$$\frac{49.5 - i}{4.95} \approx 2, \qquad i \approx 49.5 - 9.9 = 39.6.$$

So if we take $i = 39$, our confidence coefficient will probably be somewhat greater than 0.95. Hence the interval

$$y_{39} = 64.3 \qquad \text{to} \qquad y_{60} = 64.9$$

serves as that confidence interval for $\pi_{0.5}$.

There is not a unique way to finding confidence intervals for $\pi_{0.25}$ and $\pi_{0.75}$. However, we might proceed as follows. Since $y_{25}$ is about the estimate of $\pi_{0.25}$, we could consider

$$P(Y_{25-i} < \pi_{0.25} < Y_{25+i}) = \sum_{k=25-i}^{24+i} \binom{98}{k} \left(\frac{1}{4}\right)^k \left(\frac{3}{4}\right)^{98-k}$$

$$\approx \Phi\left(\frac{24.5 + i - 24.5}{4.29}\right) - \Phi\left(\frac{24.5 - i - 24.5}{4.29}\right) \approx 0.95,$$

since $98(0.25) = 24.5$ and $\sqrt{98(0.25)(0.75)} = 4.29$. Thus

$$\frac{i}{4.29} \approx 2 \qquad \text{or} \qquad i \approx 2(4.29) = 8.58.$$

So if we take $i = 9$, we will increase that confidence coefficient and

$$y_{16} = 62.4 \qquad \text{to} \qquad y_{34} = 64.1$$

serves as a confidence interval for $\pi_{0.25}$. In a similar fashion,

$$y_{65} = 65.1 \qquad \text{to} \qquad y_{83} = 66.2$$

serves as a confidence interval for $\pi_{0.75}$.

**4.9-1** A sample of $n = 9$ electrochromic mirrors was used to measure the following low end reflectivity percentages.

$$7.12 \quad 7.22 \quad 6.78 \quad 6.31 \quad 5.99 \quad 6.58 \quad 7.80 \quad 7.40 \quad 7.05$$

(a) Find the endpoints for an approximate 95% confidence interval for the median, $m$.

(b) The interval $(y_3, y_7)$ could serve as a confidence interval for $m$. Find it and give its confidence coefficient.

**4.9-2** Let $Y_1 < Y_2 < Y_3 < Y_4 < Y_5 < Y_6$ be the order statistics of a random sample of size $n = 6$ from a distribution of the continuous type having $(100p)$th percentile $\pi_p$. Compute

(a) $P(Y_2 < \pi_{0.5} < Y_5)$.

(b) $P(Y_1 < \pi_{0.25} < Y_4)$.

(c) $P(Y_4 < \pi_{0.9} < Y_6)$.

**4.9-3** For $n = 12$ model 1995 automobiles whose horse power is between 290 and 390, the following measurements give the time in seconds for the car to go from 0 to 60 m.p.h.

$$6.2 \quad 7.9 \quad 6.6 \quad 6.4 \quad 5.2 \quad 4.9 \quad 5.5 \quad 7.0 \quad 6.6 \quad 5.4 \quad 5.3 \quad 5.1$$

(a) Find a 96.14% confidence interval for the median, $m$.

(b) The interval $(y_1, y_7)$ could serve as a confidence interval for $\pi_{0.3}$. Find it and give its confidence coefficient.

**4.9-4** Let $m$ denote the median weight of "80-pound" bags of water softener pellets. Use the following random sample of $n = 14$ weights to find an approximate 95% confidence interval for $m$.

$$
\begin{array}{ccccccc}
80.51 & 80.28 & 80.40 & 80.35 & 80.38 & 80.28 & 80.27 \\
80.16 & 80.59 & 80.56 & 80.32 & 80.27 & 80.53 & 80.32
\end{array}
$$

(a) Find a 94.26% confidence interval for $m$.

(b) The interval $(y_6, y_{12})$ could serve as a confidence interval for $\pi_{0.6}$. What is its confidence coefficient?

**4.9-5** A biologist who studies spiders selected a random sample of 20 male green lynx spiders (a spider that does not weave a web but chases and leaps on its prey) and measured in millimeters the lengths of one of the front legs of the 20 spiders. Use the following measurements to construct a confidence interval for $m$ that has a confidence coefficient about equal to 0.95:

$$
\begin{array}{ccccc}
15.10 & 13.55 & 15.75 & 20.00 & 15.45 \\
13.60 & 16.45 & 14.05 & 16.95 & 19.05 \\
16.40 & 17.05 & 15.25 & 16.65 & 16.25 \\
17.75 & 15.40 & 16.80 & 17.55 & 19.05
\end{array}
$$

**4.9-6** The following 25 observations give the time in seconds between submissions of computer programs to a printer queue:

$$
\begin{array}{ccccccccc}
79 & 315 & 445 & 350 & 136 & 723 & 198 & 75 & 161 \\
13 & 215 & 24 & 57 & 152 & 238 & 288 & 272 & \\
9 & 315 & 11 & 51 & 98 & 620 & 244 & 34 &
\end{array}
$$

(a) List the observations in order of magnitude.

(b) Give point estimates of $\pi_{0.25}$, $m$, and $\pi_{0.75}$.

(c) Find the following confidence intervals and give the confidence level (using Table II) for

(i) $(y_3, y_{10})$, a confidence interval for $\pi_{0.25}$.

(ii) $(y_9, y_{17})$, a confidence interval for the median $m$.

(iii) $(y_{16}, y_{23})$, a confidence interval for $\pi_{0.75}$.

(d) Find a confidence interval for $\mu$, whose confidence coefficient corresponds to that of (c), part (ii), using $\bar{x} \pm 2(s/\sqrt{25})$. Compare the lengths of these two confidence intervals of the middles. Is this surprising? Why?

**4.9-7** Let $X$ equal the amount of fluoride in a certain brand of toothpaste. The specifications are $0.85-1.10$ mg/g. The following are 25 observations of $X$.

| | | | | | | | | |
|---|---|---|---|---|---|---|---|---|
| 0.98 | 0.94 | 1.06 | 0.95 | 1.02 | 0.88 | 0.98 | 1.00 | 1.01 |
| 0.95 | 0.99 | 0.92 | 0.93 | 0.92 | 0.89 | 0.92 | 0.98 | |
| 0.90 | 0.98 | 1.00 | 0.97 | 0.87 | 0.95 | 0.91 | 0.90 | |

(a) Construct an ordered stem-and-leaf diagram.

(b) Give a point estimate of the median, $m = \pi_{0.50}$.

(c) Find an approximate 95% confidence interval for the median $m$. Give the exact confidence level.

(d) Give a point estimate for the first quartile.

(e) Find an approximate 90% confidence interval for the first quartile and give the exact confidence level.

(f) Give a point estimate for the third quartile.

(g) Find an approximate 95% confidence interval for the third quartile and give the exact confidence level.

**4.9-8** When placed in solutions of varying ionic strength, paramecia grow blisters in order to counteract the flow of water. The following 60 measurements in microns are blister lengths.

| | | | | | | | |
|---|---|---|---|---|---|---|---|
| 7.42 | 5.73 | 3.80 | 5.20 | 11.66 | 8.51 | 6.31 | 8.49 |
| 10.31 | 6.92 | 7.36 | 5.92 | 6.74 | 8.93 | 9.61 | 11.38 |
| 12.78 | 11.43 | 6.57 | 13.50 | 10.58 | 8.03 | 10.07 | 8.71 |
| 10.09 | 11.16 | 7.22 | 10.10 | 6.32 | 10.30 | 10.75 | 11.51 |
| 11.55 | 11.41 | 9.40 | 4.74 | 6.52 | 12.10 | 6.01 | 5.73 |
| 7.57 | 7.80 | 6.84 | 6.95 | 8.93 | 8.92 | 5.51 | 6.71 |
| 10.40 | 13.44 | 9.33 | 8.57 | 7.08 | 8.11 | 13.34 | 6.58 |
| 8.82 | 7.70 | 12.22 | 7.46 | | | | |

(a) Construct an ordered stem-and-leaf diagram.

(b) Give a point estimate of the median, $m = \pi_{0.50}$.

(c) Find an approximate 95% confidence interval for $m$.

(d) Give a point estimate for the 40th percentile, $\pi_{0.40}$.

(e) Find an approximate 90% confidence interval for $\pi_{0.40}$.

**4.9-9** Let $Y_1 < Y_2 < \cdots < Y_{48}$ be the order statistics of a random sample of size $n = 48$ from a distribution of the continuous type. We want to use $(y_2, y_{47})$ as a $100\gamma$ percent tolerance interval for 75 percent of the distribution.

(a) Determine $\gamma$ using the computer.

(b) Approximate part (a) using the normal approximation to certain binomial probabilities.

**4.9-10** Let $Y_1$ and $Y_n$ be the first and the $n$th order statistics of a random sample of size $n$ from a continuous-type distribution with distribution function $F(x)$. Find the smallest value of $n$ such that $\gamma = P[F(Y_n) - F(Y_1) \geq 0.90]$ is at least 0.95.

**4.9-11** Let $Y_1 < Y_2 < \cdots < Y_{25}$ be the order statistics of a random sample of size $n = 25$ from a continuous-type distribution with distribution function $F(x)$. Using the binomial tables compute $P[F(Y_{21}) - F(Y_5) \geq 0.5]$.

## 4.10 CHI-SQUARE GOODNESS OF FIT TESTS

We now consider applications of the very important chi-square statistic, first proposed by Karl Pearson in 1900. As the reader will see, it is a very adaptable test statistic and can be used for many different types of tests. In particular, one application of it allows us to test the appropriateness of different probabilistic models.

So that the reader can get some idea as to why Pearson first proposed his chi-square statistic, we begin with the binomial case. That is, let $Y_1$ be $b(n, p_1)$, where $0 < p_1 < 1$. According to the central limit theorem,

$$Z = \frac{Y_1 - np_1}{\sqrt{np_1(1 - p_1)}}$$

has a distribution that is approximately $N(0, 1)$ for large $n$, particularly when $np_1 \geq 5$ and $n(1 - p_1) \geq 5$. Thus it is not surprising that $Q_1 = Z^2$ is approximately $\chi^2(1)$. If we let $Y_2 = n - Y_1$ and $p_2 = 1 - p_1$, we see that $Q_1$ may be written as

$$Q_1 = \frac{(Y_1 - np_1)^2}{np_1(1 - p_1)} = \frac{(Y_1 - np_1)^2}{np_1} + \frac{(Y_1 - np_1)^2}{n(1 - p_1)}.$$

Since

$$(Y_1 - np_1)^2 = (n - Y_1 - n[1 - p_1])^2 = (Y_2 - np_2)^2,$$

we have

$$Q_1 = \frac{(Y_1 - np_1)^2}{np_1} + \frac{(Y_2 - np_2)^2}{np_2}.$$

Let us now carefully consider each term in this last expression for $Q_1$. Of course, $Y_1$ is the number of "successes," and $np_1$ is the expected number of "successes"; that is, $E(Y_1) = np_1$. Likewise, $Y_2$ and $np_2$ are, respectively, the number and the expected number of "failures." So each numerator consists of the square of a difference of the observed number and expected number. Note that $Q_1$ can be written as

$$Q_1 = \sum_{i=1}^{2} \frac{(Y_i - np_i)^2}{np_i}, \qquad (4.10\text{-}1)$$

and we have seen intuitively that it has an approximate chi-square distribution with one degree of freedom. In a sense, $Q_1$ measures the "closeness" of the observed numbers to the corresponding expected numbers. For example, if the observed values of $Y_1$ and $Y_2$ equal their expected values, then the computed $Q_1$ is equal to $q_1 = 0$; but if they differ much from them, then the computed $Q_1 = q_1$ is relatively large.

To generalize, we let an experiment have $k$ (instead of only two) mutually exclusive and exhaustive outcomes, say $A_1, A_2, \ldots, A_k$. Let $p_i = P(A_i)$ and thus $\sum_{i=1}^{k} p_i = 1$. The experiment is repeated $n$ independent times, and we let $Y_i$ represent the number of times the experiment results in $A_i, i = 1, 2, \ldots, k$. This joint distribution of $Y_1, Y_2, \ldots, Y_{k-1}$ is a straightforward generalization of the binomial distribution that was given in Section 2.2. The joint p.m.f. of $Y_1, Y_2, \ldots, Y_{k-1}$ is

$$f(y_1, y_2, \ldots, y_{k-1}) = \frac{n!}{y_1! y_2! \cdots y_k!} p_1^{y_1} p_2^{y_2} \cdots p_k^{y_k},$$

recalling that $y_k = n - y_1 - y_2 - \cdots - y_{k-1}$.

Pearson then constructed an expression similar to $Q_1$ (Equation 4.10-1), which involves $Y_1$ and $Y_2 = n - Y_1$, that we denote by $Q_{k-1}$, which involves $Y_1, Y_2, \ldots,$

$Y_{k-1}$, and $Y_k = n - Y_1 - Y_2 - \cdots - Y_{k-1}$, namely,

$$Q_{k-1} = \sum_{i=1}^{k} \frac{(Y_i - np_i)^2}{np_i}.$$

He argued that $Q_{k-1}$ has an approximate chi-square distribution with $k-1$ degrees of freedom in much the same way we argued that $Q_1$ is approximately $\chi^2(1)$. We accept this fact, as the proof is beyond the level of this text.

Some writers suggest that $n$ should be large enough so that $np_i \geq 5, i = 1, 2, \ldots, k$, to be certain that the approximating distribution is adequate. This is probably good advice for the beginner to follow, although we have seen the approximation work very well when $np_i \geq 1$, $i = 1, 2, \ldots, k$. The important thing to guard against is allowing some particular $np_i$ to become so small that the corresponding term in $Q_{k-1}$, namely $(Y_i - np_i)^2/np_i$, tends to dominate the others because of its small denominator. In any case it is important to realize that $Q_{k-1}$ has only an approximate chi-square distribution.

We shall now show how we can use the fact that $Q_{k-1}$ is approximately $\chi^2(k-1)$ to test hypotheses about probabilities of various outcomes. Let an experiment have $k$ mutually exclusive and exhaustive outcomes, $A_1, A_2, \ldots, A_k$. We would like to test whether $p_i = P(A_i)$ is equal to a known number $p_{i0}, i = 1, 2, \ldots, k$. That is, we shall test the hypothesis

$$H_0: p_i = p_{i0}, \qquad i = 1, 2, \ldots, k.$$

In order to test such a hypothesis, we shall take a sample of size $n$, that is repeat the experiment $n$ independent times. We tend to favor $H_0$ if the observed number of times that $A_i$ occurred, say $y_i$, and the number of times $A_i$ was expected to occur if $H_0$ were true, namely $np_{i0}$, are approximately equal. That is, if

$$q_{k-1} = \sum_{i=1}^{k} \frac{(y_i - np_{i0})^2}{np_{i0}}$$

is "small," we tend to favor $H_0$. Since the distribution of $Q_{k-1}$ is approximately $\chi^2(k-1)$, we shall reject $H_0$ if $q_{k-1} \geq \chi_\alpha^2(k-1)$, where $\alpha$ is the desired significance level of the test.

EXAMPLE 4.10-1    If persons are asked to record a string of random digits, such as

$$3 \quad 7 \quad 2 \quad 4 \quad 1 \quad 9 \quad 7 \quad 2 \quad 1 \quad 5 \quad 0 \quad 8 \ldots,$$

we usually find that they are reluctant to record the same or even the two closest numbers in adjacent positions. And yet, in true random digit generation, the probability of the next digit being the same as the preceding one is $p_{10} = 1/10$, the probability of the next being only one away from the preceding (assuming that 0 is one away from 9) is $p_{20} = 2/10$, and the probability of all other possibilities is $p_{30} = 7/10$. We shall test one person's concept of a random sequence by asking him to record a string of 51 digits that seems to represent a random generation. Thus we shall test

$$H_0: p_1 = p_{10} = \frac{1}{10}, \quad p_2 = p_{20} = \frac{2}{10}, \quad p_3 = p_{30} = \frac{7}{10}.$$

The critical region for an $\alpha = 0.05$ significance level is $q_2 \geq \chi^2_{0.05}(2) = 5.991$. The sequence of digits was as follows:

$$
\begin{array}{ccccccccccccc}
5 & 8 & 3 & 1 & 9 & 4 & 6 & 7 & 9 & 2 & 6 & 3 & 0 \\
8 & 7 & 5 & 1 & 3 & 6 & 2 & 1 & 9 & 5 & 4 & 8 & 0 \\
3 & 7 & 1 & 4 & 6 & 0 & 4 & 3 & 8 & 2 & 7 & 3 & 9 \\
8 & 5 & 6 & 1 & 8 & 7 & 0 & 3 & 5 & 2 & 5 & 2 &
\end{array}
$$

We went through this listing and observed how many times the next digit was the same as or was one away from the preceding one.

|  | Frequency | Expected Number |
|---|---|---|
| Same | 0 | $50(1/10) = 5$ |
| One away | 8 | $50(2/10) = 10$ |
| Other | 42 | $50(7/10) = 35$ |
| Total | 50 | 50 |

The computed chi-square statistic is

$$
\frac{(0-5)^2}{5} + \frac{(8-10)^2}{10} + \frac{(42-35)^2}{35} = 6.8 > 5.991 = \chi^2_{0.05}(2).
$$

Thus we would say that this string of 51 digits does not seem to be random.

One major disadvantage in the use of the chi-square test is that it is a many-sided test. That is, the alternative hypothesis is very general, and it would be difficult to restrict alternatives to situations such as, with $k = 3$, $H_1$: $p_1 > p_{10}$, $p_2 > p_{20}$, $p_3 < p_{30}$. As a matter of fact, some statisticians would probably test $H_0$ against this particular alternative $H_1$ by using a linear function of $Y_1, Y_2$, and $Y_3$. However, this sort of discussion is beyond the scope of this book because it involves knowing more about the distributions of linear functions of the dependent random variables $Y_1, Y_2$, and $Y_3$. In any case, the student who truly recognizes that this chi-square statistic tests $H_0$: $p_i = p_{i0}$, $i = 1, 2, \ldots, k$, against all alternatives can usually appreciate the fact that it is more difficult to reject $H_0$ at a given significance level $\alpha$ using the chi-square statistic than it would be if some appropriate "one-sided" test statistic were available.

Many experiments yield a set of data, say $x_1, x_2, \ldots, x_n$, and the experimenter is often interested in determining whether these data can be treated as the observed values of a random sample $X_1, X_2, \ldots, X_n$ from a given distribution. That is, would this proposed distribution be a reasonable probabilistic model for these sample items? To see how the chi-square test can help us answer questions of this sort, consider a very simple example.

**EXAMPLE 4.10-2** Let $X$ denote the number of heads that occur when four coins are tossed at random. Under the assumptions that the four coins are independent and the probability of heads on each coin is 1/2, $X$ is $b(4, 1/2)$. One hundred repetitions of this experiment resulted in 0, 1, 2, 3, and 4 heads being observed on 7, 18, 40, 31, and 4 trials, respectively. Do these results support the assumptions? That is, is $b(4, 1/2)$

a reasonable model for the distribution of $X$? To answer this, we begin by letting $A_1 = \{0\}$, $A_2 = \{1\}$, $A_3 = \{2\}$, $A_4 = \{3\}$, $A_5 = \{4\}$. If $p_{i0} = P(X \in A_i)$ when $X$ is $b(4, 1/2)$, then

$$p_{10} = p_{50} = \binom{4}{0}\left(\frac{1}{2}\right)^4 = \frac{1}{16} = 0.0625,$$

$$p_{20} = p_{40} = \binom{4}{1}\left(\frac{1}{2}\right)^4 = \frac{4}{16} = 0.25,$$

$$p_{30} = \binom{4}{2}\left(\frac{1}{2}\right)^4 = \frac{6}{16} = 0.375.$$

At an approximate $\alpha = 0.05$ significance level, the null hypothesis

$$H_0: p_i = p_{i0}, \qquad i = 1, 2, \ldots, 5,$$

is rejected if the observed value of $Q_4$ is greater than $\chi^2_{0.05}(4) = 9.488$. If we use the 100 repetitions of this experiment that resulted in the observed values of $Y_1, Y_2, \ldots, Y_5$ of $y_1 = 7$, $y_2 = 18$, $y_3 = 40$, $y_4 = 31$, and $y_5 = 4$, the computed value of $Q_4$ is

$$q_4 = \frac{(7 - 6.25)^2}{6.25} + \frac{(18 - 25)^2}{25} + \frac{(40 - 37.5)^2}{37.5} + \frac{(31 - 25)^2}{25} + \frac{(4 - 6.25)^2}{6.25}$$

$$= 4.47.$$

Since $4.47 < 9.488$, the hypothesis is not rejected. That is, the data support the hypothesis that $b(4, 1/2)$ is a reasonable probabilistic model for $X$. Recall that the mean of a chi-square random variable is its number of degrees of freedom. In this example the mean is 4 and the observed value of $Q_4$ is 4.47, just a little greater than the mean.

Thus far all the hypotheses $H_0$ tested with the chi-square statistic $Q_{k-1}$ have been simple ones (i.e., completely specified, namely, in $H_0: p_i = p_{i0}$, $i = 1, 2, \ldots, k$, each $p_{i0}$ has been known). This is not always the case, and it frequently happens that $p_{10}, p_{20}, \ldots, p_{k0}$ are functions of one or more unknown parameters. For example, suppose that the hypothesized model for $X$ in Example 4.10-2 was $H_0: X$ is $b(4, p)$, $0 < p < 1$. Then

$$p_{i0} = P(X \in A_i) = \frac{4!}{(i-1)!(5-i)!}p^{i-1}(1-p)^{5-i}, \qquad i = 1, 2, \ldots, 5,$$

which is a function of the unknown parameter $p$. Of course, if $H_0: p_i = p_{i0}$, $i = 1, 2, \ldots, 5$, is true, for large $n$,

$$Q_4 = \sum_{i=1}^{5} \frac{(Y_i - np_{i0})^2}{np_{i0}}$$

still has an approximate chi-square distribution with four degrees of freedom. The difficulty is that when $Y_1, Y_2, \ldots, Y_5$ are observed to be equal to $y_1, y_2, \ldots, y_5$,

$Q_4$ cannot be computed, since $p_{10}, p_{20}, \ldots, p_{50}$ (and hence $Q_4$) are functions of the unknown parameter $p$.

One way out of the difficulty would be to estimate $p$ from the data and then carry out the computations using this estimate. It is interesting to note the following. Say the estimation of $p$ is carried out by minimizing $Q_4$ with respect to $p$ yielding $\tilde{p}$. This $\tilde{p}$ is sometimes called a **minimum chi-square estimator** of $p$. If then this $\tilde{p}$ is used in $Q_4$, the statistic $Q_4$ still has an approximate chi-square distribution but with only $4 - 1 = 3$ degrees of freedom. That is, the number of degrees of freedom of the approximating chi-square distribution is reduced by one for each parameter estimated by the minimum chi-square technique. We accept this result without proof (as it is a rather difficult one). Although we have considered this when $p_{i0}$, $i = 1, 2, \ldots, k$, is a function of only one parameter, it holds when there is more than one unknown parameter, say $d$. Hence, in a more general situation, the test would be completed by computing $Q_{k-1}$ using $Y_i$ and the estimated $p_{i0}$, $i = 1, 2, \ldots, k$, to obtain $q_{k-1}$ (i.e., $q_{k-1}$ is the minimized chi-square). This value $q_{k-1}$ would then be compared to a critical value $\chi^2_\alpha(k-1-d)$. In our special case, the computed (minimized) chi-square $q_4$ would be compared to $\chi^2_\alpha(3)$.

There is still one trouble with all of this: It is usually very difficult to find minimum chi-square estimators. Hence most statisticians usually use some reasonable method (maximum likelihood is satisfactory) of estimating the parameters. They then compute $q_{k-1}$, recognizing that it is somewhat larger than the minimized chi-square, and compare it with $\chi^2_\alpha(k-1-d)$. Note that this provides a slightly larger probability of rejecting $H_0$ than would the scheme in which the minimized chi-square were used because this computed $q_{k-1}$ is larger than the minimum $q_{k-1}$.

**EXAMPLE 4.10-3**

Let $X$ denote the number of alpha particles emitted by barium-133 in 1/10 of a second. The following 50 observations of $X$ were taken with a Geiger counter in a fixed position:

| | | | | | | | | | |
|---|---|---|---|---|---|---|---|---|---|
| 7 | 4 | 3 | 6 | 4 | 4 | 5 | 3 | 5 | 3 |
| 5 | 5 | 3 | 2 | 5 | 4 | 3 | 3 | 7 | 6 |
| 6 | 4 | 3 | 11 | 9 | 6 | 7 | 4 | 5 | 4 |
| 7 | 3 | 2 | 8 | 6 | 7 | 4 | 1 | 9 | 8 |
| 4 | 8 | 9 | 3 | 9 | 7 | 7 | 9 | 3 | 10 |

The experimenter is interested in determining whether $X$ has a Poisson distribution. To test $H_0$: $X$ is Poisson, we first estimate the mean of $X$, say $\lambda$, with the sample mean, $\bar{x} = 5.4$, of these 50 observations. We then partition the set of outcomes for this experiment into the sets $A_1 = \{0, 1, 2, 3\}$, $A_2 = \{4\}$, $A_3 = \{5\}$, $A_4 = \{6\}$, $A_5 = \{7\}$, and $A_6 = \{8, 9, 10, \ldots\}$. We combined $\{0, 1, 2, 3\}$ into one set $A_1$ and $\{8, 9, 10, \ldots\}$ into another $A_6$ so that the expected number of outcomes for each set is at least 5 when $H_0$ is true. In Table 4.10-1 the data are grouped, and the estimated probabilities specified by the hypothesis that $X$ has a Poisson distribution with an estimated $\widehat{\lambda} = \bar{x} = 5.4$ are given. Since one parameter was estimated, $Q_{6-1}$ has an approximate chi-square distribution with $r = 5 - 1 = 4$ degrees of freedom. Since

$$q_5 = \frac{[13 - 50(0.213)]^2}{50(0.213)} + \cdots + \frac{[10 - 50(0.178)]^2}{50(0.178)}$$

$$= 2.763 < 9.488 = \chi^2_{0.05}(4),$$

$H_0$ is not rejected at the 5% significance level. That is, with only these data, we will not reject the hypothesis that $X$ has a Poisson distribution.

## Table 4.10-1 Grouped Geiger Counter Data

| | Outcome | | | | | |
| | $A_1$ | $A_2$ | $A_3$ | $A_4$ | $A_5$ | $A_6$ |
|---|---|---|---|---|---|---|
| Frequency | 13 | 9 | 6 | 5 | 7 | 10 |
| Probability | 0.213 | 0.160 | 0.173 | 0.156 | 0.120 | 0.178 |
| Expected ($50p_i$) | 10.65 | 8.00 | 8.65 | 7.80 | 6.00 | 8.90 |

Let us now consider the problem of testing a model for the distribution of a random variable $W$ of the continuous type. That is, if $F(w)$ is the distribution function of $W$, we wish to test

$$H_0: F(w) = F_0(w),$$

where $F_0(w)$ is some known distribution function of the continuous type. Recall that we have considered problems of this type using $q$-$q$ plots. In order to use the chi-square statistic, we must partition the set of possible values of $W$ into $k$ sets. One way this can be done is as follows. Partition the interval $[0, 1]$ into $k$ sets with the points $b_0, b_1, b_2, \ldots, b_k$, where

$$0 = b_0 < b_1 < b_2 < \cdots < b_k = 1.$$

Let $a_i = F_0^{-1}(b_i)$, $i = 1, 2, \ldots, k-1$; $A_1 = (-\infty, a_1]$, $A_i = (a_{i-1}, a_i]$ for $i = 2, 3, \ldots, k-1$, and $A_k = (a_{k-1}, \infty)$; and $p_i = P(W \in A_i)$, $i = 1, 2, \ldots, k$. Let $Y_i$ denote the number of times the observed value of $W$ belongs to $A_i$, $i = 1, 2, \ldots, k$, in $n$ independent repetitions of the experiment. Then $Y_1, Y_2, \ldots, Y_k$ have a multinomial distribution with parameters $n, p_1, p_2, \ldots, p_{k-1}$. Also let $p_{i0} = P(W \in A_i)$ when the distribution function of $W$ is $F_0(w)$. The hypothesis that we actually test is a modification of $H_0$, namely

$$H_0': p_i = p_{i0}, \qquad i = 1, 2, \ldots, k.$$

This hypothesis is rejected if the observed value of the chi-square statistic

$$Q_{k-1} = \sum_{i=1}^{k} \frac{(Y_i - np_{i0})^2}{np_{i0}}$$

is at least as great as $\chi_\alpha^2(k-1)$. If the hypothesis $H_0': p_i = p_{i0}$, $i = 1, 2, \ldots, k$, is not rejected, we do not reject the hypothesis $H_0: F(w) = F_0(w)$.

**EXAMPLE 4.10-4**    The following table lists 105 observations of $X$, the times between calls to 911 in a small city. A histogram of these data with an exponential p.d.f. with $\theta = 20$ superimposed is shown in Figure 4.10-1.

| | | | | | | | | | | | | | | |
|---|---|---|---|---|---|---|---|---|---|---|---|---|---|---|
| 30 | 17 | 65 | 8 | 38 | 35 | 4 | 19 | 7 | 14 | 12 | 4 | 5 | 4 | 2 |
| 7 | 5 | 12 | 50 | 33 | 10 | 15 | 2 | 10 | 1 | 5 | 30 | 41 | 21 | 31 |
| 1 | 18 | 12 | 5 | 24 | 7 | 6 | 31 | 1 | 3 | 2 | 22 | 1 | 30 | 2 |
| 1 | 3 | 12 | 12 | 9 | 28 | 6 | 50 | 63 | 5 | 17 | 11 | 23 | 2 | 46 |
| 90 | 13 | 21 | 55 | 43 | 5 | 19 | 47 | 24 | 4 | 6 | 27 | 4 | 6 | 37 |
| 16 | 41 | 68 | 9 | 5 | 28 | 42 | 3 | 42 | 8 | 52 | 2 | 11 | 41 | 4 |
| 35 | 21 | 3 | 17 | 10 | 16 | 1 | 68 | 105 | 45 | 23 | 5 | 10 | 12 | 17 |

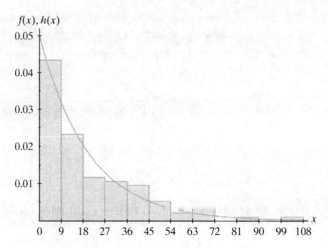

**Figure 4.10-1**  Times between calls to 911

| Table 4.10-2 Summary of Times Between Calls to 911 | | | |
| --- | --- | --- | --- |
| Class | Frequency | Probability | Expected |
| $A_1 = [\ 0,\ 9]$ | 41 | 0.3624 | 38.0490 |
| $A_2 = (\ 9, 18]$ | 22 | 0.2311 | 24.2611 |
| $A_3 = (19, 27]$ | 11 | 0.1473 | 15.4696 |
| $A_4 = (27, 36]$ | 10 | 0.0939 | 9.8638 |
| $A_5 = (36, 45]$ | 9 | 0.0599 | 6.2895 |
| $A_6 = (45, 54]$ | 5 | 0.0382 | 4.0103 |
| $A_7 = (54, 72]$ | 5 | 0.0399 | 4.1876 |
| $A_8 = (72, \infty)$ | 2 | 0.0273 | 2.8690 |

We shall now show how a chi-square goodness of fit test can be used to see whether or not this is an appropriate model for these data. That is, if $X$ is equal to the time between calls to 911, we shall test the null hypothesis that the distribution of $X$ is exponential with a mean of $\theta = 20$. Table 4.10-2 groups the data into eight classes, gives the probabilities and expected values of these classes, and calculates the value of the chi-square statistic with $8 - 1 = 7$ degrees of freedom. The value of the chi-square goodness of fit statistic is

$$q_8 = \frac{(41 - 38.0490)^2}{38.0490} + \frac{(22 - 24.2611)^2}{24.2611} + \cdots + \frac{(2 - 2.8690)^2}{2.8690} = 3.1584.$$

The $p$-value associated with this test is 0.8700, which means that it is an extremely good fit.

Note that we assumed that we knew $\theta = 20$. We could also have run this test letting $\theta = \bar{x} = 20.49$, remembering that we then lose one degree of freedom. For this example, the outcome would be about the same.

It is also true, in dealing with models of random variables of the continuous type, that we must frequently estimate unknown parameters. For example, let $H_0$ be that $W$ is $N(\mu, \sigma^2)$, where $\mu$ and $\sigma^2$ are unknown. With a random sample $W_1, W_2, \ldots, W_n$, we first can estimate $\mu$ and $\sigma^2$, possibly with $\bar{w}$ and $s_w^2$. We partition the space $\{w: -\infty < w < \infty\}$ into $k$ mutually disjoint sets $A_1, A_2, \ldots, A_k$. We then use the estimates of $\mu$ and $\sigma^2$, say $\overline{w}$ and $s^2 = s_w^2$, to provide the estimates

$$p_{i0} = \int_{A_i} \frac{1}{s\sqrt{2\pi}} \exp\left[-\frac{(w - \overline{w})^2}{2s^2}\right] dw,$$

$i = 1, 2, \ldots, k$. Using the observed frequencies $y_1, y_2, \ldots, y_k$ of $A_1, A_2, \ldots, A_k$, respectively, from the observed random sample $w_1, w_2, \ldots, w_n$, and $\widehat{p}_{10}, \widehat{p}_{20}, \ldots, \widehat{p}_{k0}$ estimated with $\overline{w}$ and $s^2 = s_w^2$, we compare the computed

$$q_{k-1} = \sum_{i=1}^{k} \frac{(y_i - n\widehat{p}_{i0})^2}{n\widehat{p}_{i0}}$$

to $\chi_\alpha^2(k-1-2)$. This value $q_{k-1}$ will again be somewhat larger than that which would be found using minimum chi-square estimation, and certain caution should be observed. Several exercises illustrate the procedure in which one or more parameters must be estimated. Finally, it should be noted that the methods given in this section frequently are classified under the more general title of goodness of fit tests. In particular, then, the tests in this section would be **chi-square goodness of fit tests**.

## EXERCISES 4.10

**4.10-1** A 1-pound bag of candy-coated chocolate-covered peanuts contained 224 pieces of candy colored brown, orange, green, and yellow. Test the null hypothesis that the machine filling these bags treats the four colors of candy equally likely; that is, test

$$H_0: p_B = p_O = p_G = p_Y = \frac{1}{4}.$$

The observed values were 42 brown, 64 orange, 53 green, and 65 yellow. You may select the significance level or give an approximate $p$-value.

**4.10-2** A particular brand of candy-coated chocolate comes in five different colors that we shall denote as $A_1 = \{$brown$\}$, $A_2 = \{$yellow$\}$, $A_3 = \{$orange$\}$, $A_4 = \{$green$\}$, and $A_5 = \{$coffee$\}$. Let $p_i$ equal the probability that the color of a piece of candy selected at random belongs to $A_i$, $i = 1, 2, \ldots, 5$. Test the null hypothesis

$$H_0: p_1 = 0.4, \ p_2 = 0.2, \ p_3 = 0.2, \ p_4 = 0.1, \ p_5 = 0.1$$

using a random sample of $n = 580$ pieces of candy whose colors yielded the respective frequencies 224, 119, 130, 48, and 59. You may select the significance level or give an approximate $p$-value.

**4.10-3** In the Arizona Daily Lottery, each weekday a three-digit integer is generated one digit at a time. Let $p_i$ denote the probability of generating digit $i$, $i = 0, 1, \ldots, 9$. Use each of the two sets of digits given in Exercise 2.1-5 to test $H_0: p_0 = p_1 = \cdots = p_9 = 1/10$. Let $\alpha = 0.05$.

**4.10-4** In a biology laboratory students test the Mendelian theory of inheritance using corn. The Mendelian theory of inheritance claims that frequencies of the four categories smooth and yellow, wrinkled and yellow, smooth and purple, and wrinkled and purple will occur in the ratio 9:3:3:1. If a student counted 124, 30, 43, and 11, respectively, for these four categories, would these data support the Mendelian theory? Let $\alpha = 0.05$.

**4.10-5** Let X equal the number of female children in a three-child family. We shall use a chi-square goodness of fit statistic to test the null hypothesis that the distribution of X is $b(3, 0.5)$.

**(a)** Define the test statistic and critical region using an $\alpha = 0.05$ significance level.

**(b)** Among students who were taking statistics, 52 came from families with 3 children. For these families, $x = 0, 1, 2,$ and 3 for 5, 17, 24, and 6 families, respectively. Calculate the value of the test statistic and state your conclusion, considering how the sample was selected.

**4.10-6** It has been claimed that, if a penny minted in 1999 or earlier is balanced on a table and the table is tapped so that the penny falls, the probability of heads is greater than 1/2. Let $X$ equal the number of heads that occur when 5 pennies are balanced and the table is tapped. Students in a statistics class repeated this experiment $n = 85$ times. They observed 0, 1, 2, 3, 4, and 5 heads 0, 6, 13, 30, 28, and 8 times, respectively. Use these data to test the hypothesis that $X$ is $b(5, p)$. Use the data to estimate $p$. Give limits for the $p$-value of this test. In addition, use the 425 individual observations to find a 95% confidence interval for $p$.

**4.10-7** While testing a used tape for bad records, a computer operator counted the number of flaws per 100 feet of tape. Let $X$ equal this random variable. Test the null hypothesis that $X$ has a Poisson distribution with a mean of $\lambda = 2.4$ given that 40 observations of $X$ yielded 5 zeros, 7 ones, 12 two, 9 threes, 5 fours, 1 five, and 1 six. Let $\alpha = 0.05$. HINT: Combine five and six into one set; that is, the last set would be all $x$ values $\geq 5$.

**4.10-8** Let $X$ equal the distance between bad records on a used computer tape. Test the hypothesis that the distribution of $X$ is exponential using the following 90 observations of $X$ and 10 classes of equal probability. Use $\bar{x} = 42.2$ as an estimate of $\theta$. Let $\alpha = 0.05$.

| | | | | | | | | | |
|---|---|---|---|---|---|---|---|---|---|
| 30 | 79 | 38 | 47 | 22 | 52 | 36 | 36 | 7 | 57 |
| 3 | 22 | 30 | 14 | 8 | 32 | 15 | 21 | 12 | 12 |
| 6 | 67 | 6 | 7 | 35 | 78 | 28 | 74 | 5 | 9 |
| 37 | 1 | 3 | 3 | 44 | 160 | 50 | 27 | 61 | 15 |
| 39 | 44 | 130 | 18 | 6 | 1 | 32 | 116 | 23 | 12 |
| 58 | 101 | 68 | 53 | 58 | 21 | 21 | 7 | 79 | 41 |
| 80 | 33 | 71 | 81 | 17 | 10 | 13 | 49 | 21 | 56 |
| 107 | 21 | 17 | 64 | 14 | 36 | 26 | 1 | 54 | 207 |
| 64 | 238 | 25 | 51 | 82 | 8 | 2 | 3 | 43 | 87 |

**4.10-9** A sample of 100 2.2K ohms resistors was tested, yielding the following measurements:

| | | | | | | | | | |
|---|---|---|---|---|---|---|---|---|---|
| 2.17 | 2.18 | 2.17 | 2.19 | 2.23 | 2.18 | 2.16 | 2.23 | 2.29 | 2.20 |
| 2.24 | 2.13 | 2.18 | 2.22 | 2.21 | 2.22 | 2.23 | 2.18 | 2.23 | 2.25 |
| 2.24 | 2.20 | 2.21 | 2.20 | 2.25 | 2.15 | 2.20 | 2.25 | 2.16 | 2.19 |
| 2.20 | 2.18 | 2.18 | 2.19 | 2.22 | 2.19 | 2.20 | 2.19 | 2.22 | 2.21 |
| 2.22 | 2.18 | 2.18 | 2.28 | 2.19 | 2.23 | 2.21 | 2.19 | 2.21 | 2.21 |
| 2.19 | 2.18 | 2.21 | 2.17 | 2.20 | 2.18 | 2.18 | 2.21 | 2.22 | 2.18 |
| 2.22 | 2.23 | 2.19 | 2.23 | 2.18 | 2.19 | 2.18 | 2.21 | 2.18 | 2.15 |
| 2.20 | 2.23 | 2.20 | 2.20 | 2.23 | 2.20 | 2.19 | 2.22 | 2.17 | 2.20 |
| 2.20 | 2.17 | 2.19 | 2.19 | 2.25 | 2.19 | 2.19 | 2.20 | 2.20 | 2.19 |
| 2.21 | 2.14 | 2.24 | 2.21 | 2.19 | 2.23 | 2.18 | 2.22 | 2.23 | 2.19 |

**(a)** Group these data into nine classes of equal length with the class boundaries for the first class being 2.125–2.145.

**(b)** Use a chi-square goodness of fit statistic to test whether these are observations of a normally distributed random variable, after grouping the first two classes and the last two classes to avoid small expected numbers. Estimate $\mu$ and $\sigma$ with the mean and standard deviation of the grouped data. Let $\alpha = 0.05$.

## 4.11  CONTINGENCY TABLES

In this section we demonstrate the flexibility of the chi-square test. We first look at a method for testing whether two or more multinomial distributions are equal, sometimes called a test for homogeneity. Then we consider a test for independence of attributes of classification. Both of these lead to a similar test statistic.

Suppose that each of two independent experiments can terminate in one of the $k$ mutually exclusive and exhaustive events $A_1, A_2, \ldots, A_k$. Let

$$p_{ij} = P(A_i), \qquad i = 1, 2, \ldots, k, \qquad j = 1, 2.$$

That is, $p_{11}, p_{21}, \ldots, p_{k1}$ are the probabilities of the events in the first experiment and $p_{12}, p_{22}, \ldots, p_{k2}$ are those associated with the second experiment. Let the experiments be repeated $n_1$ and $n_2$ independent times, respectively. Also let $Y_{11}, Y_{21}, \ldots, Y_{k1}$ be the frequencies of $A_1, A_2, \ldots, A_k$ associated with the $n_1$ independent trials of the first experiment. Similarly, let $Y_{12}, Y_{22}, \ldots, Y_{k2}$ be the respective frequencies associated with the $n_2$ trials of the second experiment. Of course, $\sum_{i=1}^{k} Y_{ij} = n_j, j = 1, 2$. From the sampling distribution theory corresponding to the basic chi-square test, we know that each of

$$\sum_{i=1}^{k} \frac{(Y_{ij} - n_j p_{ij})^2}{n_j p_{ij}}, \qquad j = 1, 2,$$

has an approximate chi-square distribution with $k - 1$ degrees of freedom. Since the two experiments are independent (and thus the two chi-square variables are independent), the sum

$$\sum_{j=1}^{2} \sum_{i=1}^{k} \frac{(Y_{ij} - n_j p_{ij})^2}{n_j p_{ij}}$$

is approximately chi-square with $k - 1 + k - 1 = 2k - 2$ degrees of freedom.

Usually, $p_{ij}, i = 1, 2, \ldots, k, j = 1, 2$, are unknown, but frequently we wish to test the hypothesis

$$H_0: \ p_{11} = p_{12}, \ p_{21} = p_{22}, \ \ldots, \ p_{k1} = p_{k2};$$

that is, the hypothesis that the corresponding probabilities associated with the two independent experiments are equal. Under $H_0$, we can estimate the unknown

$$p_{i1} = p_{i2}, \qquad i = 1, 2, \ldots, k,$$

by using the relative frequency $(Y_{i1} + Y_{i2})/(n_1 + n_2), i = 1, 2, \ldots, k$. That is, if $H_0$ is true, we can say that the two experiments are actually parts of a large one in which $Y_{i1} + Y_{i2}$ is the frequency of the event $A_i, i = 1, 2, \ldots, k$. Note that we only have to estimate the $k - 1$ probabilities $p_{i1} = p_{i2}$ using

$$\frac{Y_{i1} + Y_{i2}}{n_1 + n_2}, \qquad i = 1, 2, \ldots, k - 1,$$

since the sum of the $k$ probabilities must equal one. That is, the estimator of $p_{k1} = p_{k2}$ is

$$1 - \frac{Y_{11} + Y_{12}}{n_1 + n_2} - \cdots - \frac{Y_{k-1,1} + Y_{k-1,2}}{n_1 + n_2} = \frac{Y_{k1} + Y_{k2}}{n_1 + n_2}.$$

Substituting these estimators, we have that

$$Q = \sum_{j=1}^{2} \sum_{i=1}^{k} \frac{[Y_{ij} - n_j(Y_{i1} + Y_{i2})/(n_1 + n_2)]^2}{n_j(Y_{i1} + Y_{i2})/(n_1 + n_2)}$$

has an approximate chi-square distribution with $2k - 2 - (k - 1) = k - 1$ degrees of freedom. Here $k - 1$ is subtracted from $2k - 2$ because that is the number of estimated parameters. The critical region for testing $H_0$ is of the form

$$q \geq \chi_\alpha^2(k-1).$$

**EXAMPLE 4.11-1**   To test two methods of instruction, 50 students are selected at random from each of two groups. At the end of the instruction period each student is assigned a grade (A, B, C, D, or F) by an evaluating team. The data are recorded as follows:

| | | | Grade | | | |
|---|---|---|---|---|---|---|
| | A | B | C | D | F | Totals |
| Group I | 8 | 13 | 16 | 10 | 3 | 50 |
| Group II | 4 | 9 | 14 | 16 | 7 | 50 |

Accordingly, if the hypothesis $H_0$ that the corresponding probabilities are equal is true, the respective estimates of the probabilities are

$$\frac{8+4}{100} = 0.12, \ 0.22, \ 0.30, \ 0.26, \ \frac{3+7}{100} = 0.10.$$

Thus the estimates of $n_1 p_{i1} = n_2 p_{i2}$ are 6, 11, 15, 13, 5, respectively. Hence the computed value of $Q$ is

$$q = \frac{(8-6)^2}{6} + \frac{(13-11)^2}{11} + \frac{(16-15)^2}{15} + \frac{(10-13)^2}{13} + \frac{(3-5)^2}{5}$$
$$+ \frac{(4-6)^2}{6} + \frac{(9-11)^2}{11} + \frac{(14-15)^2}{15} + \frac{(16-13)^2}{13} + \frac{(7-5)^2}{5}$$
$$= \frac{4}{6} + \frac{4}{11} + \frac{1}{15} + \frac{9}{13} + \frac{4}{5} + \frac{4}{6} + \frac{4}{11} + \frac{1}{15} + \frac{9}{13} + \frac{4}{5} = 5.18.$$

Now, under $H_0$, $Q$ has an approximate chi-square distribution with $k - 1 = 4$ degrees of freedom so the $\alpha = 0.05$ critical region is $q \geq 9.488 = \chi_{0.05}^2(4)$. Here $q = 5.18 < 9.488$, and hence $H_0$ is not rejected at the 5% significance level. Furthermore, the $p$-value for $q = 5.18$ is 0.268, which is greater than most significance levels. Thus with these data, we cannot say there is a difference between the two methods of instruction.

It is fairly obvious how this procedure can be extended to testing the equality of $h$ independent multinomial distributions. That is, let

$$p_{ij} = P(A_i), \qquad i = 1, 2, \ldots, k, \qquad j = 1, 2, \ldots, h,$$

and test

$$H_0: p_{i1} = p_{i2} = \cdots = p_{ih} = p_i, \qquad i = 1, 2, \ldots, k.$$

Repeat the $j$th experiment $n_j$ independent times and let $Y_{1j}, Y_{2j}, \ldots, Y_{kj}$ denote the frequencies of the respective events $A_1, A_2, \ldots, A_k$. Now

$$Q = \sum_{j=1}^{h} \sum_{i=1}^{k} \frac{(Y_{ij} - n_j p_{ij})^2}{n_j p_{ij}}$$

has an approximate chi-square distribution with $h(k-1)$ degrees of freedom. Under $H_0$, we must estimate $k-1$ probabilities using

$$\widehat{p}_i = \frac{\displaystyle\sum_{j=1}^{h} Y_{ij}}{\displaystyle\sum_{j=1}^{h} n_j}, \qquad i = 1, 2, \ldots, k-1,$$

because the estimate of $p_k$ follows from $\widehat{p}_k = 1 - \widehat{p}_1 - \widehat{p}_2 - \cdots - \widehat{p}_{k-1}$. We use these estimates to obtain

$$Q = \sum_{j=1}^{h} \sum_{i=1}^{k} \frac{(Y_{ij} - n_j \widehat{p}_i)^2}{n_j \widehat{p}_i},$$

which has an approximate chi-square distribution with its degrees of freedom given by $h(k-1) - (k-1) = (h-1)(k-1)$.

Let us see how we can use the above procedures to test the equality of two or more independent distributions that are not necessarily multinomial. Suppose first that we are given random variables $U$ and $V$ with distribution functions $F(u)$ and $G(v)$, respectively. It is sometimes of interest to test the hypothesis $H_0: F(x) = G(x)$ for all $x$. We have previously considered tests of $\mu_U = \mu_V, \sigma_U^2 = \sigma_V^2$, using $t$ and $F$, and have used $q$-$q$ plots for graphical comparisons of two random samples. Now we shall only assume that the random variables from the two distributions of the continuous type are independent.

We are interested in testing the hypothesis $H_0: F(x) = G(x)$ for all $x$. This hypothesis will be replaced by another one. Partition the real line into $k$ mutually disjoint sets $A_1, A_2, \ldots, A_k$. Let

$$p_{i1} = P(U \in A_i), \qquad i = 1, 2, \ldots, k,$$

and

$$p_{i2} = P(V \in A_i), \qquad i = 1, 2, \ldots, k.$$

We observe that if $F(x) = G(x)$ for all $x$, then $p_{i1} = p_{i2}, i = 1, 2, \ldots, k$. We replace the hypothesis $H_0: F(x) = G(x)$ with the less restrictive hypothesis $H_0': p_{i1} = p_{i2}$, $i = 1, 2, \ldots, k$. That is, we are now essentially interested in testing the equality of two multinomial distributions.

Let $n_1$ and $n_2$ denote the number of independent observations of $U$ and $V$, respectively. For $i = 1, 2, \ldots, k$, let $Y_{ij}$ denote the number of these observations of $U$ and $V, j = 1, 2$, respectively, that fall into a set $A_i$. At this point, we proceed to make the test of $H_0'$ as described earlier. Of course, if $H_0'$ is rejected at the (approximate) significance level $\alpha$, then $H_0$ is rejected with the same probability. However, if $H_0'$ is true, $H_0$ is not necessarily true. Thus if $H_0'$ is accepted, it is probably better to say that we do not reject $H_0$ than to say $H_0$ is accepted.

In applications, the question of how to select $A_1, A_2, \ldots, A_k$ is frequently raised. Obviously, there is not a single choice for $k$ nor the dividing marks of the partition. But it is interesting to observe that the combined sample can be used in this selection without upsetting the approximate distribution of $Q$. For example, suppose that $n_1 = n_2 = 20$. We could easily select the dividing marks of the partition so that $k = 4$ and one fourth of the combined sample falls in each of the four sets.

EXAMPLE 4.11-2 Select, at random, 20 cars of each of two comparable major-brand models. All 40 cars are submitted to accelerated life testing; that is, they are driven many miles over very poor roads in a short time and their failure times are recorded (in weeks):

| Brand U: | 25 | 31 | 20 | 42 | 39 | 19 | 35 | 36 | 44 | 26 |
|----------|----|----|----|----|----|----|----|----|----|----|
|          | 38 | 31 | 29 | 41 | 43 | 36 | 28 | 31 | 25 | 38 |
| Brand V: | 28 | 17 | 33 | 25 | 31 | 21 | 16 | 19 | 31 | 27 |
|          | 23 | 19 | 25 | 22 | 29 | 32 | 24 | 20 | 34 | 26 |

If we use 23.5, 28.5, and 34.5 as dividing marks, we note that exactly one fourth of the 40 cars fall in each of the resulting four sets. Thus the data can be summarized as follows:

|          | $A_1$ | $A_2$ | $A_3$ | $A_4$ | Totals |
|----------|-------|-------|-------|-------|--------|
| Brand U  | 2     | 4     | 4     | 10    | 20     |
| Brand V  | 8     | 6     | 6     | 0     | 20     |

The estimate of each $p_i$ is $10/40 = 1/4$, which multiplied by $n_j = 20$ gives 5. Hence the computed $Q$ is

$$q = \frac{(2-5)^2}{5} + \frac{(4-5)^2}{5} + \frac{(4-5)^2}{5} + \frac{(10-5)^2}{5} + \frac{(8-5)^2}{5}$$
$$+ \frac{(6-5)^2}{5} + \frac{(6-5)^2}{5} + \frac{(0-5)^2}{5}$$
$$= \frac{72}{5} = 14.4 > 7.815 = \chi_{0.05}^2(3).$$

Also the $p$-value is 0.0028. Hence it seems that the two brands of cars have different distributions for the length of life under accelerated life testing.

Again it should be clear how this can be extended to more than two distributions, and this extension will be illustrated in the exercises.

Now let us suppose that a random experiment results in an outcome that can be classified by two different attributes, such as height and weight. Assume that the first attribute is assigned to one and only one of $k$ mutually exclusive and exhaustive events, say $A_1, A_2, \ldots, A_k$, and the second attribute falls in one and only one of $h$ mutually exclusive and exhaustive events, say $B_1, B_2, \ldots, B_h$. Let the probability of $A_i \cap B_j$ be defined by

$$p_{ij} = P(A_i \cap B_j), \qquad i = 1, 2, \ldots, k, \qquad j = 1, 2, \ldots, h.$$

The random experiment is to be repeated $n$ independent times, and $Y_{ij}$ will denote the frequency of the event $A_i \cap B_j$. Since there are $kh$ such events as $A_i \cap B_j$, the random variable

$$Q_{kh-1} = \sum_{j=1}^{h} \sum_{i=1}^{k} \frac{(Y_{ij} - np_{ij})^2}{np_{ij}}$$

has an approximate chi-square distribution with $kh - 1$ degrees of freedom, provided $n$ is large.

Suppose that we wish to test the hypothesis of the independence of the $A$ and $B$ attributes, namely,

$$H_0: \ P(A_i \cap B_j) = P(A_i)P(B_j), \qquad i = 1, 2, \ldots, k, \qquad j = 1, 2, \ldots, h.$$

Let us denote $P(A_i)$ by $p_{i\cdot}$, and $P(B_j)$ by $p_{\cdot j}$; that is,

$$p_{i\cdot} = \sum_{j=1}^{h} p_{ij} = P(A_i), \qquad \text{and} \qquad p_{\cdot j} = \sum_{i=1}^{k} p_{ij} = P(B_j),$$

and, of course,

$$1 = \sum_{j=1}^{h}\sum_{i=1}^{k} p_{ij} = \sum_{j=1}^{h} p_{\cdot j} = \sum_{i=1}^{k} p_{i\cdot}.$$

Then the hypothesis can be formulated as

$$H_0: \ p_{ij} = p_{i\cdot}p_{\cdot j}, \qquad i = 1, 2, \ldots, k, \qquad j = 1, 2, \ldots, h.$$

To test $H_0$, we can use $Q_{kh-1}$ with $p_{ij}$ replaced by $p_{i\cdot}p_{\cdot j}$. But if $p_{i\cdot}, i = 1, 2, \ldots, k$, and $p_{\cdot j}, j = 1, 2, \ldots, h$, are unknown, as they usually are in the applications, we cannot compute $Q_{kh-1}$ once the frequencies are observed. In such a case we estimate these unknown parameters by

$$\widehat{p}_{i\cdot} = \frac{y_{i\cdot}}{n}, \qquad \text{where} \quad y_{i\cdot} = \sum_{j=1}^{h} y_{ij}$$

is the observed frequency of $A_i, i = 1, 2, \ldots, k$; and

$$\widehat{p}_{\cdot j} = \frac{y_{\cdot j}}{n}, \qquad \text{where} \quad y_{\cdot j} = \sum_{i=1}^{k} y_{ij}$$

is the observed frequency of $B_j, j = 1, 2, \ldots, h$. Since $\sum_{i=1}^{k} p_{i\cdot} = \sum_{j=1}^{h} p_{\cdot j} = 1$, we actually estimate only $k - 1 + h - 1 = k + h - 2$ parameters. So if these estimates are used in $Q_{kh-1}$, with $p_{ij} = p_{i\cdot}p_{\cdot j}$, then, according to the rule stated earlier, the random variable

$$Q = \sum_{j=1}^{h}\sum_{i=1}^{k} \frac{[Y_{ij} - n(Y_{i\cdot}/n)(Y_{\cdot j}/n)]^2}{n(Y_{i\cdot}/n)(Y_{\cdot j}/n)}$$

has an approximate chi-square distribution with $kh - 1 - (k + h - 2) = (k-1)(h-1)$ degrees of freedom, provided that $H_0$ is true. The hypothesis $H_0$ is rejected if the computed value of this statistic exceeds $\chi_\alpha^2[(k-1)(h-1)]$.

**EXAMPLE 4.11-3**    A random sample of 400 undergraduate students at the University of Iowa was taken. The students in the sample were classified according to the college in which they were enrolled and according to their gender. These results are recorded in Table 4.11-1; this table is called a $k \times h$ **contingency table**, where here $k = 2$ and $h = 5$. (Do not be concerned about the numbers in parentheses at this

| | College | | | | | |
|---|---|---|---|---|---|---|
| Gender | Business | Engineering | Liberal Arts | Nursing | Pharmacy | Totals |
| Male | 21 (16.625) | 16 (9.5) | 145 (152) | 2 (7.125) | 6 (4.75) | 190 |
| Female | 14 (18.375) | 4 (10.5) | 175 (168) | 13 (7.875) | 4 (5.25) | 210 |
| Totals | 35 | 20 | 320 | 15 | 10 | 400 |

**Table 4.11-1 Undergraduates at the University of Iowa**

point.) Incidentally, these numbers do actually reflect the composition of the undergraduate colleges at Iowa, but they were modified a little to make the computations easier in this first example.

We desire to test the null hypothesis $H_0: p_{ij} = p_{i.}p_{.j}$, $i = 1, 2$ and $j = 1, 2, 3, 4, 5$, that the college in which a student enrolls is independent of the gender of that student. Under $H_0$, estimates of the probabilities are

$$\widehat{p}_{1.} = \frac{190}{400} = 0.475 \quad \text{and} \quad \widehat{p}_{2.} = \frac{210}{400} = 0.525$$

and

$$\widehat{p}_{.1} = \frac{35}{400} = 0.0875, \widehat{p}_{.2} = 0.05, \widehat{p}_{.3} = 0.8, \widehat{p}_{.4} = 0.0375, \widehat{p}_{.5} = 0.025.$$

The expected numbers $n(y_{i.}/n)(y_{.j}/n)$ are computed as follows:

$$400(0.475)(0.0875) = 16.625,$$

$$400(0.525)(0.0875) = 18.375,$$

$$400(0.475)(0.05) = 9.5,$$

and so on. These are the values recorded in the parentheses in Table 4.11-1. The computed chi-square statistic is

$$q = \frac{(21 - 16.625)^2}{16.625} + \frac{(14 - 18.375)^2}{18.375} + \cdots + \frac{(4 - 5.25)^2}{5.25}$$
$$= 1.15 + 1.04 + 4.45 + 4.02 + 0.32 + 0.29 + 3.69$$
$$+ 3.34 + 0.33 + 0.30 = 18.93.$$

Since the number of degrees of freedom equals $(k-1)(h-1) = 4$, this $q = 18.93 > 13.28 = \chi^2_{0.01}(4)$, and we reject $H_0$ at the $\alpha = 0.01$ significance level. Moreover, since the first two terms of $q$ come from the business college, the next two from engineering, and so on, it is clear that the enrollments in engineering and nursing are more highly dependent on gender than in the other colleges because they have contributed the most to the value of the chi-square statistic. It is also interesting to note that one expected number is less than five, namely 4.75. However, as the associated term in $q$ does not contribute an unusual amount to the chi-square value, it does not concern us.

It is fairly obvious how to extend the testing procedure above to more than two attributes. For example, if the third attribute falls in one and only one of

$m$ mutually exclusive and exhaustive events, say $C_1, C_2, \ldots, C_m$, then we test the independence of the three attributes by using

$$Q = \sum_{r=1}^{m} \sum_{j=1}^{h} \sum_{i=1}^{k} \frac{[Y_{ijr} - n(Y_{i..}/n)(Y_{.j.}/n)(Y_{..r}/n)]^2}{n(Y_{i..}/n)(Y_{.j.}/n)(Y_{..r}/n)},$$

where $Y_{ijr}$, $Y_{i..}$, $Y_{.j.}$, and $Y_{..r}$ are the respective observed frequencies of the events $A_i \cap B_j \cap C_r$, $A_i$, $B_j$, and $C_r$ in $n$ independent trials of the experiment. If $n$ is large and if the three attributes are independent, then $Q$ has an approximate chi-square distribution with $khm - 1 - (k-1) - (h-1) - (m-1) = khm - k - h - m + 2$ degrees of freedom.

Rather than explore this extension further, it is more instructive to note some interesting uses of contingency tables.

**EXAMPLE 4.11-4**   Say we observed 30 values $x_1, x_2, \ldots, x_{30}$ that are claimed to be the values of a random sample. That is, the corresponding random variables $X_1, X_2, \ldots, X_{30}$ were supposed to be mutually independent and to have the same distribution. Say, however, by looking at the 30 values we detect an upward trend that indicates there might have been some dependence and/or the random variables did not actually have the same distribution. One simple way to test if they could be thought of as being observed values of a random sample is the following. Mark each $x$ high (H) or low (L) depending on whether it is above or below the sample median. Then divide the $x$ values into three groups: $x_1, \ldots, x_{10}; x_{11}, \ldots, x_{20}; x_{21}, \ldots, x_{30}$. Certainly if the observations are those of a random sample we would expect five H's and five L's in each group. That is, the attribute classified as H or L should be independent of the group number. The summary of these data provides a $3 \times 2$ contingency table. For example, say the 30 values are

| 5.6 | 8.2 | 7.8 | 4.8 | 5.5 | 8.1 | 6.7 | 7.7 | 9.3 | 6.9 |
|-----|-----|-----|-----|-----|-----|-----|-----|-----|-----|
| 8.2 | 10.1 | 7.5 | 6.9 | 11.1 | 9.2 | 8.7 | 10.3 | 10.7 | 10.0 |
| 9.2 | 11.6 | 10.3 | 11.7 | 9.9 | 10.6 | 10.0 | 11.4 | 10.9 | 11.1 |

The median can be taken to be the average of the two middle observations in magnitude, namely, 9.2 and 9.3. Marking each item H or L after comparing it with this median, we obtain the following $3 \times 2$ contingency table.

| Group | L | H | Totals |
|-------|---|---|--------|
| 1 | 9 | 1 | 10 |
| 2 | 5 | 5 | 10 |
| 3 | 1 | 9 | 10 |
| Totals | 15 | 15 | 30 |

Here each $n(y_{i.}/n)(y_{.j}/n) = 30(10/30)(15/30) = 5$ so that the computed value of $Q$ is

$$q = \frac{(9-5)^2}{5} + \frac{(1-5)^2}{5} + \frac{(5-5)^2}{5} + \frac{(5-5)^2}{5} + \frac{(1-5)^2}{5} + \frac{(9-5)^2}{5}$$

$$= 12.8 > 5.991 = \chi_{0.05}^2(2),$$

since here $(k-1)(h-1) = 2$ degrees of freedom. (The $p$-value is 0.0017.) Hence we reject the conjecture that these 30 values could be the items of a random sample. Obviously, modifications could be made to this scheme: dividing the sample into more (or less) than three groups and rating items differently, such as low (L), middle (M), and high (H).

It cannot be emphasized enough that the chi-square statistic can be used fairly effectively in almost any situation in which there should be independence. For illustration, suppose that we have a group of workers who have essentially the same qualifications (training, experience, etc.). Many believe that salary and gender of the workers should be independent attributes; yet there have been several claims in special cases that there is a dependence—or discrimination—in attributes associated with such a problem.

**EXAMPLE 4.11-5**    Two groups of workers have the same qualifications for a particular type of work. Their experience in salaries is summarized by the following $2 \times 5$ contingency table, in which the upper bound of each salary range is not included in that listing.

| | Salary (thousands of dollars) | | | | | |
|---|---|---|---|---|---|---|
| Group | 20–22 | 22–24 | 24–26 | 26–28 | 28– | Totals |
| 1 | 6 | 11 | 16 | 14 | 13 | 60 |
| 2 | 5 | 9 | 8 | 6 | 2 | 30 |
| Totals | 11 | 20 | 24 | 20 | 15 | 90 |

To test if the group assignment and the salaries seem to be independent with these data at the $\alpha = 0.05$ significance level, we compute

$$q = \frac{[6 - 90(60/90)(11/90)]^2}{90(60/90)(11/90)} + \cdots + \frac{[2 - 90(30/90)(15/90)]^2}{90(30/90)(15/90)}$$

$$= 4.752 < 9.488 = \chi^2_{0.05}(4).$$

With the computer we can calculate that the $p$-value is 0.313. Hence with these limited data, group assignment and salaries seem to be independent. Note that if we had three times as much data such that the frequencies are multiplied by three, the computed chi-square would be three times as great as 4.752 and the hypothesis of independence would be rejected.

Before turning to the exercises, note that we could have thought of the last two examples in this section as testing the equality of two or more multinomial distributions. In Example 4.11-4 the three groups define three binomial distributions; and in Example 4.11-5 the two groups define two multinomial distributions. What would have happened if we had used the computations outlined earlier in this section? It is interesting to note that we obtain exactly the same value of chi-square and in each case the number of degrees of freedom is equal to $(k - 1)(h - 1)$. Hence it makes no difference whether we think of it as a test of independence or a test of the equality of several multinomial distributions. Our advice is to use the terminology that seems most natural for the particular situations.

**4.11-1** We wish to test to see if two groups of nurses distribute their time in six different categories about the same way. That is, the hypothesis under consideration is $H_0$: $p_{i1} = p_{i2}$, $i = 1, 2, \ldots, 6$. To test this, nurses are observed at random throughout several days, each observation resulting in a mark in one of the six categories. The summary is given by the following frequency table:

| | Category | | | | | | |
|---|---|---|---|---|---|---|---|
| | 1 | 2 | 3 | 4 | 5 | 6 | Totals |
| Group I | 95 | 36 | 71 | 21 | 45 | 32 | 300 |
| Group II | 53 | 26 | 43 | 18 | 32 | 28 | 200 |

Use a chi-square test with $\alpha = 0.05$.

**4.11-2** Each of two comparable classes of 15 students responded to two different methods of instructions with the following scores on a standardized test:

| Class U: | 91 | 42 | 39 | 62 | 55 | 82 | 67 | 44 |
|---|---|---|---|---|---|---|---|---|
| | 51 | 77 | 61 | 52 | 76 | 41 | 59 | |
| Class V: | 80 | 71 | 55 | 67 | 61 | 93 | 49 | 78 |
| | 57 | 88 | 79 | 81 | 63 | 51 | 75 | |

Use a chi-square test with $\alpha = 0.05$ to test the equality of the distributions of test scores by dividing the combined sample into three equal parts (low, middle, high).

**4.11-3** Suppose that a third class (W) of 15 students was observed along with classes U and V of Exercise 4.11-2, resulting in scores of

| 91 | 73 | 67 | 83 | 59 | 98 | 87 | 69 |
|---|---|---|---|---|---|---|---|
| 78 | 80 | 65 | 94 | 82 | 74 | 85 | |

Again use a chi-square test with $\alpha = 0.05$ to test the equality of the three distributions by dividing the combined sample into three equal parts.

**4.11-4** In a contingency table, 1015 individuals are classified by gender and by whether they favor, oppose, or have no opinion on a complete ban on smoking in public places. Test the null hypothesis that gender and opinion on smoking in public places are independent. Give the approximate $p$-value of this test.

| | Smoking in Public Places | | | |
|---|---|---|---|---|
| Gender | Favor | Oppose | No Opinion | Totals |
| Male | 262 | 231 | 10 | 503 |
| Female | 302 | 205 | 5 | 512 |
| Totals | 564 | 436 | 15 | 1015 |

**4.11-5** A random sample of 100 students were classified by gender and by the "instrument" that they "played." Test whether the selection of instrument is independent of the gender of the respondent. Approximate the $p$-value of this test.

| Gender | Instrument | | | | | |
|---|---|---|---|---|---|---|
| | Piano | Woodwind | Brass | String | Vocal | Totals |
| Male | 4 | 11 | 15 | 6 | 9 | 45 |
| Female | 7 | 18 | 6 | 6 | 18 | 55 |
| Totals | 11 | 29 | 21 | 12 | 27 | 100 |

**4.11-6** A random sample of 50 women who were tested for cholesterol were classified according to age and cholesterol level and grouped in the following contingency table.

| Age | Cholesterol Level | | | |
|---|---|---|---|---|
| | < 180 | 180–210 | > 210 | Totals |
| < 50 | 5 | 11 | 9 | 25 |
| ≥ 50 | 4 | 3 | 18 | 25 |
| Totals | 9 | 14 | 27 | 50 |

Test the null hypothesis $H_0$: Age and cholesterol level are independent attributes of classification. What is your conclusion if $\alpha = 0.01$?

**4.11-7** Although high school grades and testing scores, such as SAT or ACT, can be used to predict first-year college grade-point average (GPA), many educators claim that a more important factor influencing that GPA is the living conditions of students. In particular, it is claimed that the roommate of the student will have a great influence on his or her grades. To test this, suppose we selected at random 200 students and classified each according to the following two attributes:

**(a)** Ranking of the student's roommate from 1 to 5, from a person who was difficult to live with and discouraged scholarship to one who was congenial but encouraged scholarship.

**(b)** The student's first-year GPA.

Say this gives the following 5 × 4 contingency table.

| Rank of Roommate | Grade-Point Average | | | | |
|---|---|---|---|---|---|
| | Under 2.00 | 2.00–2.69 | 2.70–3.19 | 3.20–4.00 | Totals |
| 1 | 8 | 9 | 10 | 4 | 31 |
| 2 | 5 | 11 | 15 | 11 | 42 |
| 3 | 6 | 7 | 20 | 14 | 47 |
| 4 | 3 | 5 | ?? | 23 | 53 |
| 5 | 1 | 3 | 11 | 12 | 27 |
| Totals | 23 | 35 | 78 | 64 | 200 |

Compute the chi-square statistic used to test the independence of the two attributes and compare it to the critical value associated with $\alpha = 0.05$.

**4.11-8** A study was conducted to determine the media credibility for reporting news. Those surveyed were asked to give their age, gender, education, and the most credible medium. Test whether

(a) Media credibility and age are independent.

(b) Media credibility and gender are independent.

(c) Media credibility and education are independent.

(d) Give the approximate $p$-value for each test.

| | Most Credible Medium | | | |
|---|---|---|---|---|
| Age | Newspaper | Television | Radio | Totals |
| Under 35 | 30 | 68 | 10 | 108 |
| 35–54 | 61 | 79 | 20 | 160 |
| Over 54 | 98 | 43 | 21 | 162 |
| Totals | 189 | 190 | 51 | 430 |

| | Most Credible Medium | | | |
|---|---|---|---|---|
| Gender | Newspaper | Television | Radio | Totals |
| Male | 92 | 108 | 19 | 219 |
| Female | 97 | 81 | 32 | 210 |
| Totals | 189 | 189 | 51 | 429 |

| | Most Credible Medium | | | |
|---|---|---|---|---|
| Education | Newspaper | Television | Radio | Totals |
| Grade School | 45 | 22 | 6 | 73 |
| High School | 94 | 115 | 30 | 239 |
| College | 49 | 52 | 13 | 114 |
| Totals | 188 | 189 | 49 | 426 |

**4.11-9** A random sample of $n = 1362$ persons were classified according to the respondent's education level and whether the respondent was Protestant, Catholic, or Jewish. Use these data to test at an $\alpha = 0.05$ significance level the hypothesis that these attributes of classification are independent.

| Education Level | Protestant | Catholic | Jewish |
|---|---|---|---|
| Less than high school | 359 | 140 | 5 |
| High school or junior college | 462 | 200 | 17 |
| Bachelor's degree | 88 | 39 | 2 |
| Graduate degree | 37 | 10 | 3 |

# CHAPTER
# FOUR COMMENTS

Modern testing of statistical hypotheses is often said to have begun with Karl Pearson's chi-square tests in 1900. It is always difficult to know exactly when certain topics started, but 1900 is a nice round date and Pearson's proposed chi-square is an important statistic. It should be noted that shortly after that date a "then young" Ronald A. Fisher pointed out to the more elderly Pearson that a degree of freedom should be subtracted for each parameter estimated. Pearson never believed Fisher (Fisher was right), and this difference of opinion created a life-long battle between these two giants of the field. Since Pearson was the editor of the journal *Biometrika*, Fisher was not allowed to publish there after their disagreement.

Also in the early part of the twentieth century, W. S. Gosset discovered the *t*-distribution. However, he was working for a brewery and for a certain reason (he did not want the brewery to know that he was spending time on statistics, or the brewery did not want other breweries to know that they were using statistical methods) he published under the pseudonym "A Student." Hence that statistic is usually known as Student's *t*. In a certain sense Gosset was lucky in finding this *t*-distribution and found it by using the first four moments. Later Fisher proved that *T* really had that distribution.

Most of the other confidence intervals and tests of the hypotheses in this chapter resulted from the theories of Fisher, Neyman, and Pearson (actually Egon, Karl's son), which existed in the first half of the twentieth century. Distribution-free techniques (not only those of Section 4.9) were discovered by a collection of statisticians called nonparametricians, mainly in the twentieth century before 1975. Nonparametrics has recently become much broader than only distribution-free methods.

# CHAPTER

# 5

# COMPUTER-ORIENTED TECHNIQUES

## 5.1 COMPUTATION OF STATISTICS

The reader is encouraged to use whatever statistical package is available for finding characteristics of a set of data. Use different graphical techniques to obtain a better understanding of the "shape" of your data.

Many statistical packages can also be used to find probabilities for the standard distributions. The following examples use Minitab to find some probabilities for discrete distributions.

 EXAMPLE 5.1-1   In Example 2.3-1 some probabilities were found for the binomial distribution, $b(10, 0.5)$, using Table II. Here are two of those probabilities found using Minitab. To find these probabilities you use

$$Calc \rightarrow Probability\ Distributions \rightarrow Binomial$$

and input the necessary information.

```
Cumulative Distribution Function
Binomial with n = 10 and p = 0.5
x  P( X <= x )
5     0.623047

Probability Density Function
Binomial with n = 10 and p = 0.5
x  P( X = x )
6     0.205078
```

EXAMPLE 5.1-2    In Example 2.3-4 some Poisson probabilities for the Poisson distribution with $\lambda = 5$ were found using Table III in the Appendix. Here we find those probabilities using Minitab. To find these probabilities you use

$$\text{Calc} \rightarrow \text{Probability Distributions} \rightarrow \text{Poisson}$$

and input the necessary information.

```
Cumulative Distribution Function
Poisson with mean = 5
x   P( X <= x )
6      0.762183


Probability Density Function
Poisson with mean = 5
x   P( X = x )
6      0.146223
```

EXAMPLE 5.1-3    In Example 2.3-8 a hypergeometric probability was calculated. To find hypergeometric probabilities using Minitab, input

$$\text{Calc} \rightarrow \text{Probability Distributions} \rightarrow \text{Hypergeometric}$$

and then the necessary information. (Minitab lets $M = N_1$.)

```
Probability Density Function
Hypergeometric with N = 50, M = 5, and n = 4
x   P( X = x )
0      0.646960


Cumulative Distribution Function
Hypergeometric with N = 50, M = 5, and n = 4
x   P( X <= x )
1      0.955037
```

The next example illustrates how Minitab can be used to group a set of discrete data.

EXAMPLE 5.1-4    Let $X$ equal the number of integers in their natural position in a random permutation of the first 20 positive integers. For example, in the following random permutation,

$$2 \ 15 \ 12 \ 14 \ 5 \ 4 \ 3 \ 10 \ 8 \ 19 \ 16 \ 7 \ 6 \ 18 \ 17 \ 11 \ 1 \ 13 \ 20 \ 9$$

the integer 5 is the only integer in its natural position so $X = 1$. We simulated 100 observations of $X$ yielding the following outcomes.

```
2  2  1  0  0  1  2  1  2  2  1  0  1  3  2  1  2  1  0  1
1  1  3  1  3  1  0  2  1  0  1  2  0  0  1  0  0  0  1  2
0  1  2  0  1  1  5  0  0  1  0  1  1  0  2  1  0  0  1  0
0  0  1  0  0  1  1  3  1  0  2  0  0  1  4  2  2  1  1  1
0  0  2  0  0  1  0  2  1  0  2  2  1  2  2  0  1  0  1  1
```

In Minitab we can find characteristics of these data by putting these data in a column with a column header $x$. Then input

Stat → Basic Statistics → Display Descriptive Statistics

and input $x$. The following information will be calculated for you. Included in the Descriptive Statistics are TrMean and SE Mean. TrMean stands for trimmed mean. A 5% trimmed mean is calculated. Minitab removes the smallest 5% and the largest 5% of the values (rounded to the nearest integer), and then averages the remaining values. The SE Mean is the standard error of the mean and is the standard deviation divided by the square root of $N$, StDev/$\sqrt{N}$ using Minitab notation.

```
Descriptive Statistics
```

| Variable | N | Mean | Median | TrMean | StDev | SE Mean |
|---|---|---|---|---|---|---|
| x | 100 | 1.0100 | 1.0000 | 0.9222 | 0.9898 | 0.0990 |

| Variable | Minimum | Maximum | Q1 | Q3 |
|---|---|---|---|---|
| x | 0.0000 | 5.0000 | 0.0000 | 2.0000 |

You can also have Minitab group the data by selecting

Stat → Tables → Tally

and input $x$. The following tally is given.

```
Tally for Discrete Variables: x
```

| x | Count |
|---|---|
| 0 | 35 |
| 1 | 38 |
| 2 | 21 |
| 3 | 4 |
| 4 | 1 |
| 5 | 1 |
| N= | 100 |

Since the mean and the variance are both close to one, do you think that $X$ has an approximate Poisson distribution with $\lambda = 1$? To help you answer this question, compare the relative frequencies of the outcomes with their respective probabilities if $X$ has a Poisson distribution with $\lambda = 1$.

Minitab can also be used for continuous type data and continuous probability distributions. The next example uses the data in Example 3.1-4 to find a stem-and-leaf diagram and also numerical characteristics of a set of ACT scores.

EXAMPLE 5.1-5    For the ACT scores in Example 3.1-4 we shall use Minitab to find a stem-and-leaf diagram and also numerical characteristics of these data. Place the ACT scores in a column headed by ACT. To find a stem-and-leaf diagram select

Graph → Stem-and-Leaf

and input ACT. Input a 2 for the increment. You will then obtain the following output in which the first column gives the depths.

```
Stem-and-Leaf Display: ACT

Stem-and-leaf of ACT        N  = 60
Leaf Unit = 1.0

     4      1 8999
    11      2 0000111
    20      2 222333333
    30      2 4444455555
    30      2 66666667777
    19      2 8888999999
     9      3 00011
     4      3 223
     1      3 5
```

You can also find the numerical characteristics of these data by selecting

Stat → Basic Statistics → Display Descriptive Statistics

and then selecting ACT. The following characteristics are then calculated.

```
Descriptive Statistics: ACT

Variable     N      Mean    Median    TrMean    StDev    SE Mean
ACT         60    25.450    25.500    25.389    3.933      0.508


Variable        Minimum    Maximum          Q1          Q3
ACT              18.000     35.000      23.000      28.750
```

The following example illustrates how Minitab can be used to find percentiles for a special distribution.

EXAMPLE 5.1-6    The p.d.f. of a Weibull random variable $X$ is

$$f(x) = \frac{\alpha x^{\alpha-1}}{\beta^\alpha}\, e^{-(x/\beta)^\alpha}, \qquad 0 < x < \infty,$$

where the parameters $\alpha$ and $\beta$ are positive. The p.d.f. defined in Example 3.2-5,

$$f(x) = \frac{3x^2}{4^3}\, e^{-(x/4)^3}, \qquad 0 < x < \infty,$$

is a Weibull p.d.f. with parameters $\alpha = 3$ and $\beta = 4$. In that example the 30th and 90th percentiles were calculated. To find these percentiles using Minitab, select

Calc → Probability Distributions → Weibull

and then select "Inverse cumulative probability" and input the two parameters, 3 and 4 and input the constant 0.3. Repeat this inputting the constant 0.9. Minitab will then give you the following:

```
Inverse Cumulative Distribution Function

Weibull with first shape parameter=3.00000 and second=4.00000

P( X <= x )          x
     0.3000       2.8367


P( X <= x )          x
     0.9000       5.2820
```

## EXERCISES 5.1

Remark    For the exercises in this section, use Minitab or whatever statistical software is available.

**5.1-1** Let the distribution of $X$ be $b(23, 0.69)$. Find
  (a) $P(X \leq 17)$;
  (b) $P(X = 16)$.

**5.1-2** Let $X$ have a Poisson distribution with mean, $\lambda = 17.6$. Find
  (a) $P(X \leq 18)$;
  (b) $P(X = 17)$.

**5.1-3** In LOTTO 49 a player selects 6 integers out of the first 49 positive integers. The state then randomly selects 6 integers out of the same set. A cash prize is given to the player who matches 4, 5, or 6 integers. Let $X$ equal the number of integers that match. Find
  (a) $P(X \leq 3)$;
  (b) $P(X = 0)$;
  (c) $P(X = 1)$.

**5.1-4** For this exercise use the data in Exercise 2.1-5.
  (a) Compute the sample means and the sample standard deviations.
  (b) Tally the frequencies of $0, 1, \ldots, 9$.
  (c) Construct histograms of the data.

**5.1-5** Fifty test scores on a statistics examination were used to construct the stem-and-leaf diagrams in Tables 3.1-4 and 3.1-5.
  (a) Use Minitab to construct similar stem-and-leaf diagrams.
  (b) Calculate the sample mean and sample standard deviation of these test scores.
  (c) Find the five-number summary of these scores.
  (d) Construct a box-and-whisker diagram of these data.

**5.1-6** Find the first, second, and third quartiles for the Weibull distribution defined in Examples 3.2-5 and 5.1-6.

**5.1-7** Let $Z$ have a standard normal distribution, $N(0, 1)$. Find
  (a) $z_{0.01}$;
  (b) $z_{0.05}$;
  (c) $z_{0.975}$.

**5.1-8** Let the distribution of $X$ be $\chi^2(23)$. Find

    **(a)** $P(X \leq 23)$;

    **(b)** $a$ so that $P(X \leq a) = 0.07$;

    **(c)** $b$ so $P(X \leq b) = 0.92$.

## 5.2 COMPUTER ALGEBRA SYSTEMS

In this section we give *Maple* solutions for several of the examples in the first four chapters. All of these will be cross referenced. Although *Maple* is being used, it is also possible to use another computer algebra system. For some of these examples, a set of special statistical procedures will be used. These are stored as `stat.m`.

> **Remark**   To read the supplementary procedures, you must specify where they are. Copy the "Maple Examples" folder from the web site into a new folder, named "Tanis-Hogg" on your computer's C drive. The examples are set up assuming that this has been done. *Maple* should first be loaded. Then open the file named "Menu" in "Maple Examples." From this file, you can access the supplementary procedures and all of the examples in this section. For individual examples, use `read 'C:\\Tanis-Hogg\\Maple Examples\\stat.m':` to load `stat.m`.

**EXAMPLE 5.2-1**   In Example 2.1-5 the following p.m.f. was defined:

$$f(x) = P(X = x) = \left(\frac{2}{3}\right)^{x-1}\left(\frac{1}{3}\right), \qquad x = 1, 2, 3, 4, \ldots.$$

Here is a *Maple* solution for finding the mean and the variance of $X$.

**Maple:**

```
>  mu := sum(x*(2/3)^(x-1)*(1/3), x = 1 .. infinity);
```

$$\mu := 3$$

Again using *Maple*, we can find the variance as follows:

```
>  sum((x - mu)^2*(2/3)^(x-1)*(1/3), x = 1 .. infinity);
```

$$6$$

★

That is, $\sigma^2 = 6$. It follows that the standard deviation is $\sigma = \sqrt{6} = 2.449$.

**EXAMPLE 5.2-2**   Let the distribution of $X$ be $b(n, p)$. (See Section 2.2.) Here is a *Maple* solution for finding the mean and the variance of $X$.

**Maple:**
```
>  read 'C:\\Tanis-Hogg\\Maple Examples\\stat.m';
>  f := BinomialPDF(n, p, x);
```

$$f := \text{binomial}(n, x)\, p^x\, (1 - p)^{(n-x)}$$

```
>  assume(p <1 ): additionally(p > 0):
   simplify(sum(f, x = 0..n));
```

1

```
> mu := simplify(sum(x*f, x = 0 .. n));
```

$$\mu := np$$

```
> sigma := simplify(sqrt(sum((x - mu)^2*f, x = 0 .. n)));
```

$$\sigma := \sqrt{np\,(1-p)} \qquad \bigstar$$

We found the value of $\sigma$ so $\sigma^2 = np(1-p)$.

**EXAMPLE 5.2-3**    Let $X$ have a Poisson distribution with $\lambda > 0$. (See Section 2.3.) We shall use *Maple* to first show that the sum of all of the probabilities of the Poisson p.m.f. is equal to one and then find the mean and variance.

Maple:
```
> read 'C:\\Tanis-Hogg\\Maple Examples\\stat.m';
> f := PoissonPDF(lambda, x);
```

$$f := \frac{\lambda^x\, e^{(-\lambda)}}{x!}$$

```
> sum(f, x = 0 .. infinity);
```

$$1$$

```
> mu := sum(x*f, x = 0 .. infinity);
```

$$\mu := \lambda$$

```
> var := sum((x - mu)^2*f, x = 0 .. infinity);
```

$$var := \lambda \qquad \bigstar$$

Note that for the Poisson distribution,

$$\mu = \sigma^2 = \lambda,$$

a model in which the mean and the variance are equal.

**EXAMPLE 5.2-4**    Let $X$ have a negative binomial distribution with p.m.f.

$$g(x) = \binom{x-1}{r-1}p^r(1-p)^{x-r} = \binom{x-1}{r-1}p^r q^{x-r}, \qquad x = r, r+1, \ldots,$$

where $0 < p < 1$ and $q = 1 - p$. (See Section 2.3.) Recall that if $r = 1$, $X$ has a geometric distribution since the p.m.f. consists of terms of a geometric series, namely

$$g(x) = p(1-p)^{x-1}, \qquad x = 1, 2, 3, \cdots.$$

Using *Maple*, we show that the sum of the terms of the negative binomial p.m.f. is equal to one and then find the mean and variance as follows:

Maple:

```
>   read 'C:\\Tanis-Hogg\\Maple Examples\\stat.m';
>   g := NegBinomialPDF(r, p, x);
```

$$g := \text{binomial}(x - 1, r - 1)\, p^r\, (1 - p)^{(x-r)}$$

```
>   sum(g, x = r .. infinity);
```

$$1$$

```
>   mu := sum(x*g, x = r .. infinity);
```

$$\mu := \frac{r}{p}$$

```
>   var := sum((x - mu)^2*g, x = r .. infinity));
```

$$var := -\frac{r(-1+p)}{p^2}$$   ★

That is,

$$\mu = \frac{r}{p} \quad \text{and} \quad \sigma^2 = \frac{r(1-p)}{p^2},$$

which in the special case of $r = 1$, (the geometric distribution) become

$$\mu = \frac{1}{p} \quad \text{and} \quad \sigma^2 = \frac{1-p}{p^2}.$$

**EXAMPLE 5.2-5**   Let $X$ have a hypergeometric distribution with p.m.f. given by

$$f(x) = P(X = x) = \frac{\dbinom{N_1}{x}\dbinom{N_2}{n-x}}{\dbinom{N}{n}},$$

where the space $S$ is the collection of non-negative integers $x$ that satisfies the inequalities $x \le n$, $x \le N_1$, and $n - x \le N_2$. Furthermore, $N_1$ and $N_2$ are positive integers and $N = N_1 + N_2$. (See Section 2.3.)

Here is the *Maple* code to show that the sum of the probabilities is one and to find the mean and the variance of $X$.

Maple:

```
>   read 'C:\\Tanis-Hogg\\Maple Examples\\stat.m';
>   f := HypergeometricPDF(N[1], N[2], n, x);
```

$$f := \frac{\text{binomial}(N_1, x)\,\text{binomial}(N_2, n - x)}{\text{binomial}(N_1 + N_2, n)}$$

```
>   simplify(sum(f, x = 0 .. N[1]));
```

$$1$$

```
>   mu := simplify(sum(x*f, x = 0 .. N[1]));
```

$$\mu := \frac{N_1\, n}{N_1 + N_2}$$

```
>   var := simplify(sum(x^2*f, x = 0 .. N[1]) - mu^2);
```

$$var := \frac{N_2\, N_1\, (N_1 + N_2 - n)\, n}{(N_1 + N_2 - 1)\,(N_1 + N_2)^2}$$   ★

*Maple* does not always write the output in the standard way. So we note that usually we write

$$\mu = n\left(\frac{N_1}{N}\right) = np$$

and

$$\sigma^2 = n\left(\frac{N_1}{N}\right)\left(\frac{N_2}{N}\right)\left(\frac{N-n}{N-1}\right) = np(1-p)\left(\frac{N-n}{N-1}\right),$$

where $p = N_1/N$ is the fraction of orange chips and $1 - p = N_2/N$ is the fraction of blue chips in a bowl, assuming there are $N_1$ orange chips and $N_2$ blue chips in a bowl.

**EXAMPLE 5.2-6**   In Example 3.2-5 the random variable $X$ has a Weibull-type distribution with distribution function

$$F(x) = 1 - e^{-(x/4)^3}, \qquad 0 \le x < \infty.$$

*Maple* can be used to find the 30th and 90th percentiles, the p.d.f. of $X$, and the probability that $X$ falls between these two calculated percentiles as follows:

**Maple:**

```
>   interface(showassumed=0): assume(x>0):
>   F := 1 - exp(-(x/4)^3);
```

$$F := 1 - e^{-x^3/64}$$

```
>   pi[0.3] := fsolve(F = 0.30, x);
```

$$\pi_{0.3} := 2.836726888$$

```
>   pi[0.9] := fsolve(F = 0.90, x);
```

$$\pi_{0.9} := 5.282001914$$

```
>   f := diff(F, x);
```

$$f := \frac{3}{64}\, x^2\, e^{-x^3/64}$$

We now integrate the p.d.f. of $X$ between the 30th and 90th percentiles. Note that there is a little roundoff error.

```
>  int(f, x = pi[0.3] .. pi[0.9]);
```

$$0.6000000001$$

★

See Figure 3.2-2.

EXAMPLE 5.2-7

This example finds the characteristics of the exponential distribution with $\theta > 0$. (See Section 3.3).

**Maple:**
```
>  interface(showassumed=0): assume(theta>0):
>  f := piecewise(x < 0, 0, 1/theta*exp(-x/theta));
```

$$f(x) = \begin{cases} 0, & x < 0, \\ \dfrac{e^{-x/\theta}}{\theta}, & 0 \le x < \infty. \end{cases}$$

```
>  mu := int(x*f, x = -infinity .. infinity);
```

$$\mu := \theta$$

```
>  var := int((x - mu)^2*f, x = -infinity .. infinity);
```

$$var := \theta^2$$

```
>  sigma := sqrt(var);
```

$$\sigma := \theta$$

Here is the distribution function as a function of $u$.
```
>  F := int(f, x = -infinity .. u);
```

$$F(u) = \begin{cases} 0, & u < 0, \\ 1 - e^{-u/\theta}, & 0 \le u < \infty. \end{cases}$$

```
>  m := solve(F = 1/2, u);
```

$$m := \ln(2)\,\theta$$

★

Note that the distribution function was used to find the median.

EXAMPLE 5.2-8

This example uses *Maple* to find the characteristics of the gamma distribution with $\alpha > 0$ and $\theta > 0$. (See Section 3.3). The MGF is defined in Section 6.1.

**Maple:**
```
>  interface(showassumed=0): assume(theta>0):
>  additionally(alpha>0): additionally(t < 1/theta):
>  f := 1/GAMMA(alpha)/theta^alpha*x^(alpha-1)*exp(-x/theta);
```

$$f := \frac{x^{\alpha-1}\,e^{-x/\theta}}{\Gamma(\alpha)\,\theta^\alpha}$$

```
>   MGF := expand(int(exp(t*x)*f, x = 0 .. infinity));
```

$$MGF := \frac{1}{(1 - t\theta)^{\alpha}}$$

```
>   mu := expand(int(x*f, x = 0 .. infinity));
```

$$\mu := \theta\,\alpha$$

```
>   var := expand(int(x^2*f, x = 0 .. infinity) - mu^2);
```

$$var := \theta^2\,\alpha \qquad\qquad ★$$

Note that the assumptions are important for these calculations.

The following example gives a *Maple* solution for Example 2.2-3.

**EXAMPLE 5.2-9**   Find the mean and variance of the random variable $X$ with p.m.f.

$$f(x) = \left(\frac{1}{2}\right)^x, \qquad x = 1, 2, 3, \dots.$$

Maple:
```
>   f := (1/2)^x;
>   mu := sum(x*f, x = 1 .. infinity);
```

$$\mu := 2$$

```
>   var := sum((x - mu)^2*f, x = 1 .. infinity);
```

$$var := 2$$

```
>   sum(x*(x + 1)*f, x = 1 .. infinity);
```

$$8 \qquad\qquad ★$$

Note that the latter sum gives the value of $E[X(X + 1)]$.

**EXAMPLE 5.2-10**   This example uses *Maple* to find the mean and the variance of a normally distributed random variable, $N(a, b^2)$, where $-\infty < a < \infty$ and $b > 0$. It also shows that the p.d.f. of $X$ is indeed a p.d.f.

Maple:
```
>   assume(b > 0):
>   f := 1/b/sqrt(2*Pi)*exp(-(x - a)^2/2/b^2);
```

$$f := \frac{1}{2}\,\frac{\sqrt{2}\,e^{-(x-a)^2/(2b^2)}}{b\,\sqrt{\pi}}$$

```
>   int(f, x = -infinity .. infinity);
```

$$1$$

```
>   mu := int(x*f, x = -infinity .. infinity);
```

$$\mu := a$$

```
>  sigma := sqrt(int((x - mu)^2*f, x = -infinity .. infinity));
```

$$\sigma := b$$ ★

Thus we usually write the normal p.d.f. as

$$f(x) = \frac{1}{\sigma \sqrt{2\pi}} \exp\left[-\frac{(x-\mu)^2}{2\sigma^2}\right], \qquad -\infty < x < \infty,$$

and say that the random variable $X$ with this p.d.f. is $N(\mu, \sigma^2)$.

There are times when we must solve two equations simultaneously for two unknowns. Usually that is not too difficult. However, *Maple* can make it even easier. We illustrate this in the next example by solving Equations 4.3-1 for $c$ and $n$.

**EXAMPLE 5.2-11**

Maple:
```
>  eqn1 := (c - 75)/(10/sqrt(n)) = 1.645;
```

$$eqn1 := \frac{1}{10}(c - 75)\sqrt{n} = 1.645$$

```
>  eqn2 := (c - 76.5)/(10/sqrt(n)) = -1.282;
```

$$eqn2 := \frac{1}{10}(c - 76.5)\sqrt{n} = -1.282$$

```
>  solve({eqn1, eqn2},{n,c});
```

$$\{c = 75.84301332, n = 380.7701778\}$$ ★

Of course, we know that $n$ must be an integer so $n = 381$.

The next three examples use *Maple* to find the p.d.f.s, means, and variances for three random variables whose distributions are defined in terms of other random variables. These distributions are important in applications and they are the $t$, $F$, and beta distributions.

**EXAMPLE 5.2-12**

(**Student's $t$ distribution**)   We let

$$T = \frac{Z}{\sqrt{U/r}},$$

where $Z$ is a random variable that is $N(0, 1)$, $U$ is a random variable that is $\chi^2(r)$, and $Z$ and $U$ are independent. (See Exercise 5.2-2.) In the *Maple* output, we define the joint p.d.f. of $Z$ and $U$. The distribution function of $T$ is given by

$$F(t) = P(T \leq t) = P(Z \leq \sqrt{U/r}\, t).$$

The p.d.f. of $T$ is then found by taking the derivative of $F(t)$.

Maple:
```
>  read 'C:\\Tanis-Hogg\\Maple Examples\\stat.m':
>  interface(showassumed=0): assume(t, real):
```

```
>  g := NormalPDF(0,1,z)*ChisquarePDF(r, u);
```

$$g := \frac{1}{2} \frac{\sqrt{2}\,e^{-z^2/2}\,u^{r/2-1}\,e^{-u/2}}{\sqrt{\pi}\,\Gamma(\frac{r}{2})\,2^{r/2}}$$

```
>  F := int(Int(g, z = -infinity..sqrt(u/r)*t),u = 0..infinity);
```

$$F := \int_0^\infty \int_{-\infty}^{\sqrt{\frac{u}{r}}\,t} \frac{1}{2} \frac{\sqrt{2}\,e^{-z^2/2}\,u^{r/2-1}\,e^{-u/2}}{\sqrt{\pi}\,\Gamma(\frac{r}{2})\,2^{r/2}}\,dz\,du$$

```
>  f := simplify(diff(F, t));
```

$$f := \frac{\sqrt{\frac{1}{r}}\,(\frac{t^2+r}{r})^{(-1/2-r/2)}\,\Gamma(\frac{1}{2}+\frac{r}{2})}{\sqrt{\pi}\,\Gamma(\frac{r}{2})}$$

```
>  mu := int(t*f, t = -infinity .. infinity);
```

$$\mu := 0$$

```
>  var := simplify(int((t - mu)^2*f, t = -infinity .. infinity));
```

$$var := \frac{r}{r-2} \qquad\qquad ★$$

We generally write the p.d.f. as follows:

$$f(t) = \frac{\Gamma((r+1)/2)}{\sqrt{\pi r}\,\Gamma(r/2)} \frac{1}{(1+t^2/r)^{(r+1)/2}}, \qquad -\infty < t < \infty.$$

Note that the mean is defined when $r > 1$ and the variance is defined when $r > 2$.

Graphs of the p.d.f. of $T$ when $r = 1, 3$, and 7 along with the $N(0,1)$ p.d.f. are given in Figure 5.2-1. In this figure we see that the tails of the $t$ distribution are

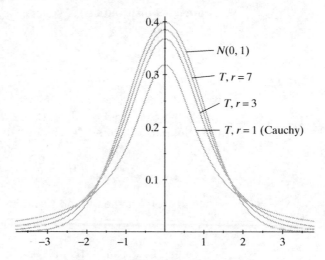

**Figure 5.2-1** $t$ Distribution p.d.f.s

heavier than those of a normal one; that is, there is more extreme probability in the $t$ distribution than in the standardized normal one.

**EXAMPLE 5.2-13**    (**Fisher's or Snedecor's $F$ distribution**) We let

$$F = \frac{U/r_1}{V/r_2},$$

where $U$ and $V$ are independent chi-square variables with $r_1$ and $r_2$ degrees of freedom, respectively. In the *Maple* output, we define the joint p.d.f. of $U$ and $V$. In this derivation we let $W = F$ to avoid using $f$ as a symbol for a variable. The distribution function of $W$ is given by

$$G(w) = P(W \le w) = P(U \le [r_1/r_2]wV).$$

The p.d.f. of $W$ is then found by taking the derivative of $G(w)$.

Maple:
```
> read 'C:\\Tanis-Hogg\\Maple Examples\\stat.m':
> interface(showassumed=0): assume(r[1]>0): additionally(r[2]>4):
> g := ChisquarePDF(r[1], u)*ChisquarePDF(r[2], v);
```

$$g := \frac{u^{(r_1/2-1)} e^{(-u/2)} v^{(r_2/2-1)} e^{(-v/2)}}{\Gamma(\frac{r_1}{2}) 2^{(r_1/2)} \Gamma(\frac{r_2}{2}) 2^{(r_2/2)}}$$

```
> G := int(Int(g, u = 0 .. (r[1]/r[2])*w*v), v = 0 .. infinity);
```

$$G := \int_0^\infty \int_0^{\frac{r_1 w v}{r_2}} \frac{u^{(r_1/2-1)} e^{(-u/2)} v^{(r_2/2-1)} e^{(-v/2)}}{\Gamma(\frac{r_1}{2}) 2^{(r_1/2)} \Gamma(\frac{r_2}{2}) 2^{(r_2/2)}} \, du \, dv$$

```
> h := diff(G, w): h := simplify(h, symbolic);
```

$$h := \frac{r_1^{(r_1/2)} r_2^{(r_2/2)} w^{(r_1/2-1)} (r_1 w + r_2)^{(-r_1/2-r_2/2)} \Gamma(\frac{1}{2} r_1 + \frac{1}{2} r_2)}{\Gamma(\frac{1}{2} r_1) \Gamma(\frac{1}{2} r_2)}$$

```
> mu := simplify(int(w*h, w = 0 .. infinity));
```

$$\mu := \frac{r_2}{r_2 - 2}$$

```
> var := simplify(int((w - mu)^2*h, w = 0 .. infinity));
```

$$var := \frac{2 r_2^2 (r_1 + r_2 - 2)}{r_1 (r_2 - 4)(r_2 - 2)^2} \qquad \bigstar$$

We usually write the $F$ p.d.f. as follows:

$$f(w) = \frac{(r_1/r_2)^{r_1/2} \Gamma[(r_1 + r_2)/2] w^{r_1/2-1}}{\Gamma(r_1/2)\Gamma(r_2/2)[1 + (r_1 w/r_2)]^{(r_1+r_2)/2}}.$$

Graphs of the p.d.f.s for the $F$ distribution are given in Figure 5.2-2.

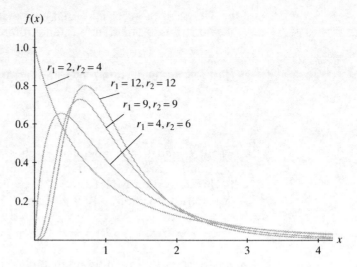

**Figure 5.2-2** Graphs of $F$ p.d.f.s

---

EXAMPLE 5.2-14    (**Beta distribution**) Let $X$ and $Y$ be independent gamma random variables with parameters $\alpha$, $\theta$ and $\beta$, $\theta$, respectively. To find the p.d.f. of $W = X/(X + Y)$, we will use *Maple* to find the distribution function of $W$ and then find the p.d.f. of $W$ by taking the derivative of this distribution function. Note that the distribution function of $W$ is defined by

$$F(w) = P(W \le w) = P\left(\frac{[1 - w]X}{w} \le Y\right).$$

We define the distribution function using the joint p.d.f. of $X$ and $Y$.

Maple:
```
>   read 'C:\\Tanis-Hogg\\Maple Examples\\stat.m':
>   interface(showassumed=0): assume(alpha > 0):additionally(w<1):
>   additionally(beta>0): additionally(theta>0):additionally(w>0):
>   g := GammaPDF(alpha, theta, x)*GammaPDF(beta, theta, y);
```

$$g := \frac{x^{(\alpha-1)} e^{(-x/\theta)} y^{(\beta-1)} e^{(-y/\theta)}}{\Gamma(\alpha) \theta^{\alpha} \Gamma(\beta) \theta^{\beta}}$$

```
>   F := int(Int(g, y=((1 - w)/w)*x..infinity), x=0..infinity);
```

$$F := \int_0^\infty \int_{\frac{(1-w)}{w}x}^\infty \frac{x^{(\alpha-1)} e^{(-x/\theta)} y^{(\beta-1)} e^{(-y/\theta)}}{\Gamma(\alpha) \theta^{\alpha} \Gamma(\beta) \theta^{\beta}} \, dy \, dx$$

```
>   f := simplify(diff(F, w));
```

$$f := \frac{w^{(\alpha-1)} (1 - w)^{\beta-1} \Gamma(\alpha + \beta)}{\Gamma(\alpha) \Gamma(\beta)}$$

```
>   mu := int(w*f, w = 0 .. 1);
```

$$\mu := \frac{\alpha}{\alpha + \beta}$$

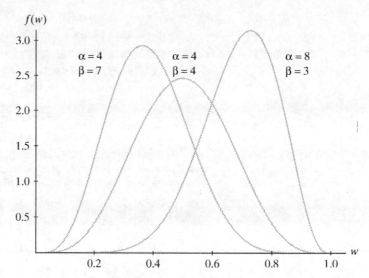

**Figure 5.2-3** Beta distribution p.d.f.s

```
>  var := simplify(int((w - mu)^2*f, w = 0 .. 1));
```

$$var := \frac{\alpha\,\beta}{(\alpha+\beta)^2\,(\alpha+\beta+1)}$$                ★

We have found the p.d.f. of $W$ as well as its mean and variance. Some graphs of beta p.d.f.s are given in Figure 5.2-3.

---

**EXERCISES 5.2**

**5.2-1** In Exercise 4.2-6 we considered a random sample of size $n$, $X_1, X_2, \ldots, X_n$, from a normal distribution, $N(\mu, \sigma^2)$. We were asked to find values of $a$ and $b$ that minimize the length of a confidence interval for $\sigma$. That is, we minimized

$$k = s\sqrt{n-1}\left(\frac{1}{\sqrt{a}} - \frac{1}{\sqrt{b}}\right)$$

under the restriction

$$G(b) - G(a) = \int_a^b g(u)\,du = 1 - \alpha,$$

where $G(u)$ and $g(u)$ are the distribution function and p.d.f. of a $\chi^2(n-1)$ distribution, respectively. This lead to the result that $a$ and $b$ must satisfy

$$a^{n/2}e^{-a/2} - b^{n/2}e^{-b/2} = 0.$$

**(a)** Use *Maple* to show that, when $n = 13$, $a = 5.9397$ and $b = 24.2017$ for a 90% confidence interval for $\sigma$. HINT: In *Maple* you can solve two equations for two unknowns as follows:

Maple:
```
>  restart: read 'C:\\Tanis-Hogg\\Maple Examples\\stat.m':
```

```
>  g := ChisquarePDF(n-1, x):
>  n := 13: alpha := 0.10:
>  eq1 := int(g, x = a .. b) = 1 - alpha:
>  eq2 := a^(n/2)*exp(-a/2) = b^(n/2)*exp(-b/2):
>  fsolve({eq2,eq1},{a,b}, {a=0..n-1, b=n-1..infinity});
```

$$\{a = 5.939654248, \; b = 24.20167765\} \qquad \bigstar$$

   **(b)** Using the data in Example 4.2-1 find a 90% confidence interval for $\sigma$ of shortest length.

   **(c)** Compare the lengths of the shortest confidence interval for $\sigma$ with that of the confidence that uses equal tail probabilities.

**5.2-2** Let $X, Y, Z$ be a random sample of size 3 from a standard normal distribution, $N(0, 1)$.

   **(a)** Explain why $T = X/\sqrt{(Y^2 + Z^2)/2}$ has a $t$ distribution with two degrees of freedom. (See Example 5.2-12.)

   **(b)** Let $U = X/\sqrt{(X^2 + Y^2)/2}$. Use a technique similar to that used in Example 5.2-12 to show that the p.d.f. of $U$ is $f(u) = 1/(\pi\sqrt{2 - u^2})$, $-\sqrt{2} < u < \sqrt{2}$. Explain why the p.d.f. of $U$ is so different from that of the p.d.f. of $T$. Also find the mean and the variance of $U$.

   **(c)** Let $V = U^2/2$ where $U$ is defined in part (b). Find the distribution function of $V$ and take its derivative to show that $V$ has a beta distribution with $\alpha = 1/2$ and $\beta = 1/2$.

**5.2-3** Let $X, Y$ be a random sample of size 2 from a standard normal distribution, $N(0, 1)$. Let $U = X/Y$. Define the distribution function of $U$ and then take its derivative to show that $U$ has a Cauchy distribution.

**5.2-4** Let $X, Y$ be a random sample of size 2 from a Cauchy distribution. (See Equation 5.3-1.) Let $W = (X + Y)/2$ be the sample mean.

   **(a)** Define the joint p.d.f. of $X$ and $Y$.

   **(b)** Define the distribution function of $W$ using this joint p.d.f.

   **(c)** Take the derivative of this distribution function to find the p.d.f. of $W$. How is $W$ distributed?

   **(d)** What is the distribution of the sample mean for any size sample from the Cauchy distribution?

**5.2-5** Let $X$ have the p.d.f. $f(x) = 4x^3, 0 < x < 1$. Find the p.d.f. of $Y = X^2$.

**5.2-6** Let $X$ have a **logistic distribution** with p.d.f.

$$f(x) = \frac{e^{-x}}{(1 + e^{-x})^2}, \qquad -\infty < x < \infty.$$

Show that

$$Y = \frac{1}{1 + e^{-X}}$$

has a $U(0, 1)$ distribution.

**5.2-7** Let $X$ have the p.d.f. $f(x) = xe^{-x^2/2}, 0 < x < \infty$. Find the p.d.f. of $Y = X^2$.

**5.2-8** Let $X$ have the Cauchy p.d.f. $f(x) = [\pi(1 + x^2)]^{-1}, -\infty < x < \infty$. Find the p.d.f. of $Y = X^2$.

## 5.3 SIMULATION

We now prove an important theorem that allows us to simulate observations from a given distribution.

---

**Theorem 5.3-1**

Let $Y$ have a distribution that is $U(0,1)$. Let $F(x)$ have the properties of a distribution function of the continuous type with $F(a) = 0$, $F(b) = 1$, and suppose that $F(x)$ is strictly increasing on the support $a < x < b$, where $a$ and $b$ could be $-\infty$ and $\infty$, respectively. Then the random variable $X$ defined by $X = F^{-1}(Y)$ is a continuous-type random variable with distribution function $F(x)$.

**Proof**

The distribution function of $X$ is

$$P(X \le x) = P[F^{-1}(Y) \le x], \qquad a < x < b.$$

However, $F(x)$ is strictly increasing so $\{F^{-1}(Y) \le x\}$ is equivalent to $\{Y \le F(x)\}$ and hence

$$P(X \le x) = P[Y \le F(x)], \qquad a < x < b.$$

But $Y$ is $U(0,1)$; so $P(Y \le y) = y$ for $0 < y < 1$, and accordingly

$$P(X \le x) = P[Y \le F(x)] = F(x), \qquad 0 < F(x) < 1.$$

That is, the distribution function of $X$ is $F(x)$.

---

We illustrate how Theorem 5.3-1 can be used to simulate observations from a given distribution.

**EXAMPLE 5.3-1**

To see how we can simulate observations from an exponential distribution with a mean of $\theta = 10$, note that the distribution function of $X$ is $F(x) = 1 - e^{-x/10}$ when $0 \le x < \infty$. Solving $y = F(x)$ for $x$ yields $x = F^{-1}(y) = -10\ln(1 - y)$. Observe independent random numbers $y_1, y_2, \dots, y_n$ from the $U(0,1)$ distribution. Then $x_i = -10\ln(1 - y_i)$, $i = 1, 2, \dots, n$, represent $n$ independent observations of an exponential random variable $X$ with mean $\theta = 10$. Table 5.3-1 gives the values of 15 random numbers, $y_i$, along with the values of $x_i = -10\ln(1 - y_i)$.

Figure 5.3-1 shows the **quantile-quantile** ($q$-$q$) plot for these data in which the theoretical quantiles of that exponential distribution are plotted against the corresponding sample quantiles (the ordered $x$ values). The linearity of this $q$-$q$ plot emphasizes that the exponential distribution is the correct model. The box plot, Figure 5.3-1, shows the skewness of the data as it should be for an exponential distribution.

The sample mean is $\bar{x} = 10.4270$ and the sample standard deviation is $s = 10.4311$, both of which are close to $\theta = 10$. We again caution the reader that in similar simulations the fit is sometimes better and sometimes not as good.

You are encouraged to replicate this simulation. Either use the random numbers from Table VIII in the Appendix or use the computer, in which case you can simulate a larger sample.

## Table 5.3-1 Random Exponential Observations and Exponential Quantiles

| $y$ | $x = -10\ln(1-y)$ | Ordered $x$'s | Quantiles |
|---|---|---|---|
| 0.1514 | 1.6417 | 0.0431 | 0.6454 |
| 0.6697 | 11.0775 | 0.5414 | 1.3353 |
| 0.0527 | 0.5414 | 1.0569 | 2.0764 |
| 0.4749 | 6.4417 | 1.6417 | 2.8768 |
| 0.2900 | 3.4249 | 2.6840 | 3.7469 |
| 0.2354 | 2.6840 | 3.4249 | 4.7000 |
| 0.9662 | 33.8729 | 6.4417 | 5.7536 |
| 0.0043 | 0.0431 | 6.8736 | 6.9315 |
| 0.1003 | 1.0569 | 8.9526 | 8.2668 |
| 0.9192 | 25.1578 | 11.0775 | 9.8083 |
| 0.4971 | 6.8736 | 13.0674 | 11.6315 |
| 0.7293 | 13.0674 | 17.2878 | 13.8629 |
| 0.9118 | 24.2815 | 24.2815 | 16.7398 |
| 0.8225 | 17.2878 | 25.1578 | 20.7944 |
| 0.5915 | 8.9526 | 33.8729 | 27.7259 |

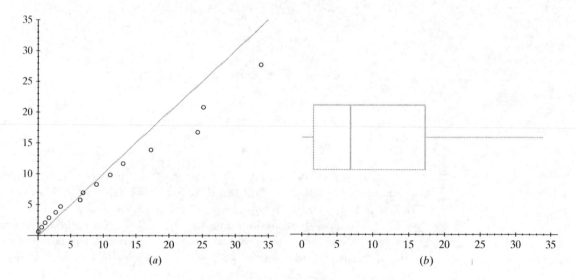

Figure 5.3-1 Exponential $q$-$q$ plot and box plot

Suppose a spinner is mounted at the point $(0, 1)$. Let $w$ be the smallest angle between the $y$ axis and the spinner (see Figure 5.3-2). Assume that $w$ is the value of a random variable $W$ that has a uniform distribution on the interval $(-\pi/2, \pi/2)$. That is, $W$ is $U(-\pi/2, \pi/2)$, and the distribution function of $W$ is

$$P(W \leq w) = F(w) = \begin{cases} 0, & -\infty < w < -\dfrac{\pi}{2}, \\ \left(w + \dfrac{\pi}{2}\right)\left(\dfrac{1}{\pi}\right), & -\dfrac{\pi}{2} \leq w < \dfrac{\pi}{2}, \\ 1, & \dfrac{\pi}{2} \leq w < \infty. \end{cases}$$

The relationship between $x$ and $w$ is given by $x = \tan w$; that is, $x$ is the point on the $x$-axis which is the intersection of that axis and the linear extension of the spinner. To find the distribution of the random variable $X = \tan W$, we see that the distribution function of $X$ is given by

$$G(x) = P(X \le x) = P(\tan W \le x) = P(W \le \arctan x)$$

$$= F(\arctan x) = \left(\arctan x + \frac{\pi}{2}\right)\left(\frac{1}{\pi}\right), \qquad -\infty < x < \infty.$$

The last equality follows because $-\pi/2 < w = \arctan x < \pi/2$. The p.d.f. of $X$ is given by

$$g(x) = G'(x) = \frac{1}{\pi(1 + x^2)}, \qquad -\infty < x < \infty. \tag{5.3-1}$$

In Figure 5.3-2 the graph of this **Cauchy p.d.f.** is given.

EXAMPLE 5.3-2
To help appreciate the large probability in the tails of the Cauchy distribution, it is useful to simulate some observations of a Cauchy random variable, $X$. We can first begin with a random number, $Y$, that is an observation from the $U(0,1)$ distribution. From the distribution function of $X$, namely $G(x)$, we have

$$y = G(x) = \left(\arctan x + \frac{\pi}{2}\right)\left(\frac{1}{\pi}\right), \qquad -\infty < x < \infty,$$

or, equivalently,

$$x = \tan\left(\pi y - \frac{\pi}{2}\right). \tag{5.3-2}$$

This latter expression provides an observations of $X$.

In Table 5.3-2 the values of $y$ are the first 10 random numbers in the last column of Table VIII in the Appendix. The corresponding values of $x$ are given by Equation 5.3-2. Although most of these observations from the Cauchy distribution are relatively small in magnitude, we see that a very large value (in magnitude) occurs occasionally. Another way of looking at this situation is by considering sightings (or firing of a gun) from an observation tower, here with coordinates $(0, 1)$,

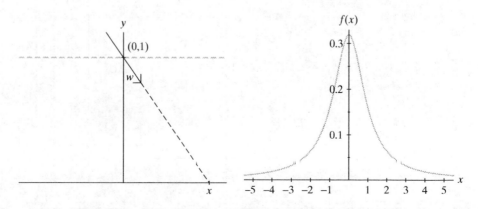

**Figure 5.3-2** Spinner and Cauchy p.d.f.

| Table 5.3-2 Cauchy Observations | |
| --- | --- |
| y | x |
| 0.1514 | −1.9415 |
| 0.6697 | 0.5901 |
| 0.0527 | −5.9847 |
| 0.4749 | −0.0790 |
| 0.2900 | −0.7757 |
| 0.2354 | −1.0962 |
| 0.9662 | 9.3820 |
| 0.0043 | −74.0211 |
| 0.1003 | −3.0678 |
| 0.9192 | 3.8545 |

at independent random angles, each with the uniform distribution $U(-\pi/2, \pi/2)$; the target points would then be at Cauchy observations.

**EXAMPLE 5.3-3**  Take a random sample $X_1, X_2, \ldots, X_6$ of size $n = 6$ from the uniform distribution $U(0, 1)$. Repeat this simulation and, for each sample, compute the sample average, say $\bar{x}$. Now construct a histogram of the $\bar{x}$s. We repeated this 10,000 times to obtain Figure 5.3-3. Note how the histogram takes on the shape of a normal p.d.f. which we studied in detail in Section 3.6.

There are times when we want to simulate observations of a normally distributed random variable. One simple way to do this is by using the Central Limit Theorem. If $X_1, X_2, \ldots, X_{12}$ is a random sample of size 12 from a $U(0, 1)$ distribution (i.e., 12 random numbers), then

$$Z = \frac{\sum_{i=1}^{12}(X_i - 1/2)}{\sqrt{12}\sqrt{1/12}} = \sum_{i=1}^{12} X_i - 6$$

has an approximate $N(0, 1)$ distribution.

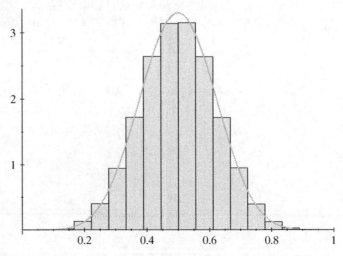

**Figure 5.3-3** Simulation of 10,000 $\bar{x}$s for samples of size 6 from $U(0, 1)$

There is also an interesting transformation that can be used to simulate observations of normally distributed random variables. It is called the Box-Muller transformation. (See the seventh edition of the text by Hogg and Tanis, pp. 286–287.) Let $U_1, U_2$ be independent $U(0, 1)$ random variables. Then

$$Z_1 = \sqrt{-2 \ln U_1} \cos(2\pi U_2) \quad \text{and} \quad Z_2 = \sqrt{-2 \ln U_1} \sin(2\pi U_2) \quad (5.3\text{-}3)$$

have distributions that are $N(0, 1)$, and furthermore $Z_1$ and $Z_2$ are independent. (See Exercise 5.3-15.)

Assume that $Z_1$ and $Z_2$ are independent $N(0, 1)$ random variables. Let

$$X_1 = \sqrt{1 - \rho^2} Z_1 + \rho Z_2 \quad \text{and} \quad X_2 = Z_2, \quad 0 \leq \rho \leq 1. \quad (5.3\text{-}4)$$

Clearly $E(X_1) = E(X_2) = 0$ and $\text{Var}(X_1) = \text{Var}(X_2) = 1$. Also $E(X_1 X_2) = \rho E(Z_2^2) = \rho$ since $E(Z_2^2) = \text{Var}(Z_2) = 1$. Thus we know that the correlation coefficient of $X_1$ and $X_2$ is $\rho$. We say that $X_1$ and $X_2$ have what is called a **bivariate normal distribution** with means equal to zero and variances equal to one and correlation coefficient $\rho$. Suppose, however, we wanted $Y_1$ and $Y_2$ to have a bivariate normal distribution with means $\mu_1$ and $\mu_2$ and variances $\sigma_1^2$ and $\sigma_2^2$ and correlation coefficient $\rho$. Let

$$X_1 = \frac{Y_1 - \mu_1}{\sigma_1}, \quad X_2 = \frac{Y_2 - \mu_2}{\sigma_2}$$

or

$$Y_1 = \sigma_1 X_1 + \mu_1, \quad Y_2 = \sigma_2 X_2 + \mu_2. \quad (5.3\text{-}5)$$

Then $Y_1$ and $Y_2$ would have such a distribution. For illustration, if $\mu_1 = 5.8$ feet, $\mu_2 = 5.3$ feet, $\sigma_1 = \sigma_2 = 0.2$ foot, and $\rho = 0.6$, then $Y_1$ and $Y_2$ might be the respective male and female heights of a married couple with this bivariate distribution.

## EXERCISES 5.3

**5.3-1** Using the last 20 4-digit numbers in the random numbers in Table VIII,

    **(a)** Find the corresponding 20 observations of a random sample from a distribution with p.d.f. $f(x) = 4x^3$, $0 < x < 1$;

    **(b)** Compare $E(X)$ with the sample mean;

    **(c)** Compare $\text{Var}(X)$ with the sample variance.

**5.3-2** Let $f(x) = 1/[\pi(1 + x^2)]$, $-\infty < x < \infty$, be the p.d.f. of the Cauchy random variable $X$. Show that $E(X)$ does not exist. HINT: For $\int_{-\infty}^{\infty} \frac{|x|}{\pi(1 + x^2)} \, dx$ to be finite, the integral $\int_0^{\infty} \frac{x}{\pi(1 + x^2)} \, dx$ must be finite. But

$$\int_0^{\infty} \frac{x}{\pi(1 + x^2)} \, dx = \frac{1}{2\pi} \lim_{b \to \infty} \left[ \ln(1 + x^2) \right]_0^b = \frac{1}{2\pi} \lim_{b \to \infty} \left[ \ln(1 + b^2) - \ln(1) \right] = \infty.$$

**5.3-3** The p.d.f. of $X$ is $f(x) = 2x$, $0 < x < 1$.

    **(a)** Find the distribution function of $X$.

    **(b)** Describe how an observation of $X$ can be simulated.

    **(c)** Simulate 10 observations of $X$.

**5.3-4** Using a random number generator, find $n = 18$ observations from a distribution with p.d.f. $f(x) = 2x$, $0 < x < 1$. Compute $\bar{x}$, the sample mean. Do this 1000 times, obtaining 1000 $\bar{x}$s. Make a histogram for the 1000 values of the sample mean and note how a bell-shaped curve fits this histogram.

**5.3-5** Let $X$ have a Cauchy distribution. Find

   **(a)** $P(X > 1)$.

   **(b)** $P(X > 5)$.

   **(c)** $P(X > 10)$.

**5.3-6** Appendix Table VIII can be used for part (a).

   **(a)** Simulate nine observations of a Cauchy random variable.

   **(b)** Find the sample mean and the sample median for your nine observations. Which of these statistics seems to give the better estimate of the center (here zero) of the distribution? Compare your results with those of other students.

   **(c)** Simulate 1000 observations of a Cauchy random variable. Compare the probabilities in Exercise 5.3-5 with the relative frequencies of these events.

**5.3-7** Note that the "birthday problem" (see Exercises 2.1-10 and 2.1-11) is similar to rolling an $n$-sided die until one of the faces is observed twice. Suppose we use a die with $n = 20$ faces. Roll this die $r$ times where $r \le 21$.

   **(a)** What is the probability that all of the faces are different?

   **(b)** Define the probability as a function of $r$, say $P(r)$, that at least two rolls are equal.

   **(c)** Graph the set of points $\{[r, P(r)], r = 1, 2, \ldots, 20\}$.

   **(d)** Find the value of $r$ for which **(i)** $P(r) \approx 0.42$, **(ii)** $P(r) \approx 0.56$.

   **(e)** If the die is rolled only five times, use simulation to see if there is a match about 0.42 of the time.

   **(f)** If the die is rolled only six times, use simulation to see if there is there a match about 0.56 of the time.

   **(g)** Let the random variable $Y$ equal the number of rolls needed to observe the first match. Define the p.d.f. of $Y$.

   **(h)** Find $\mu = E(Y)$, $\sigma^2 = \text{Var}(Y)$, and $\sigma$.

   **(i)** Illustrate these results empirically. Compare the relative frequency histogram of 200 repetitions of this experiment with the probability histogram, $\bar{x}$ with $\mu$, $s^2$ with $\sigma^2$, $s$ with $\sigma$.

**5.3-8** In craps (see Exercise 1.3-13) it is possible to win or lose with just one roll of the dice. At times it takes several rolls of the dice to determine whether a player wins or loses. Let $N$ equal the number of times the dice are rolled to determine whether a player wins or loses.

   **(a)** Use 1000 simulations to estimate $P(N = k)$, $k = 1, 2, 3, 4, \ldots$. Show that the average of the observations of $N$ is close to $\mu = 557/165 \approx 3.376$. Show that the sample variance of the observations of $N$ is close to $\sigma^2 = 245{,}672/27{,}225 \approx 9.0238$. HINT: There is a procedure in the *Maple* supplement called Craps () that simulates one play of craps. The first argument is 0 if the gambler loses and the first argument is 1 if the gambler wins. The succeeding arguments give the outcomes on the dice.

   **(b)** Verify theoretically that the values for $\mu$ and $\sigma^2$ are correct. First show that

$$P(N = 1) = \frac{12}{36}.$$

For $n = 2, 3, \ldots, P(N = n)$ is equal to

$$2\left[\left(\frac{27}{36}\right)^{n-2}\left(\frac{3}{36}\right)\left(\frac{9}{36}\right) + \left(\frac{26}{36}\right)^{n-2}\left(\frac{4}{36}\right)\left(\frac{10}{36}\right) + \left(\frac{25}{36}\right)^{n-2}\left(\frac{5}{36}\right)\left(\frac{11}{36}\right)\right].$$

Now show that

$$\mu = \sum_{n=1}^{\infty} nP(N=n) = \frac{557}{165},$$

and

$$\sigma^2 = \sum_{n=1}^{\infty} (n - E(N))^2 \, P(N=n) = \frac{245,672}{27,225}.$$

**(c)** Construct a probability histogram and a relative frequency histogram like Figures 2.1-1 and 2.1-2.

**5.3-9** Roll an $N$-sided die where $N$ is both unknown and is very large. You will be allowed to roll this die 10 times observing $x_1, x_2, \ldots, x_{10}$. Using the 10 outcomes you are to guess the value of $N$. There are several possible ways to estimate $N$. Five possibilities are given and we shall use simulation to compare them when $N = 750$. Give a reason why each of the following is a possibility:

- $y_{10} = \max\{x_1, x_2, \ldots, x_{10}\}$. (See Example 2.4-3.)
- $\lceil (11/10) * y_{10} \rceil$. NOTE: Note that $\lceil M \rceil$ is the "ceiling" or greatest integer function.
- $\lceil 2 * \bar{x} - 1 \rceil$.
- $\lceil \sqrt{12}\, s + 1 \rceil$.
- $2M$, where $M$ is the median of the set of observations.

**(a)** Simulate 500 samples of 10 integers selected randomly from 1 to 750, inclusive, sampling with replacement.

**(b)** For each sample calculate the values of your estimates. Then put each set of 500 estimates into a list.

**(c)** Find the means (averages) of each set of 500 estimates.

**(d)** Find the standard deviations of each set of 500 estimates.

**(e)** On the same figure graph box-and-whisker diagrams of each of the sets of 500 estimates.

**(f)** Based on your simulation and the means and standard deviations of each set of estimates, and the box-and-whisker plots, which estimate would you recommend? Why?

**5.3-10** To estimate a population size of mobile animals or fish, some "individuals" are caught, marked, and released. After these are well mixed within the population, a new sample is caught and the number of marked individuals in this second sample is used to estimate the population size.

- $N_1$ is the number caught in the first sample, marked, and released into the population.
- $n$ is the size of the second catch.
- $x$ is the number recaptured (marked and caught again) in the second catch.
- $N$ is the size of the total population.

Recall that the maximum likelihood estimate of $N$ is given by Equation 2.4-3 as

$$\widehat{N} = \left\lfloor \frac{N_1 n}{x} \right\rfloor .$$

**(a)** Instead of fish, we use candies. Given a large bag of m&m's, "capture" 1/4 cup of them, count the captured m&m's, and replace them with an equal number of fruit Skittles[®].

**(b)** From the mixed candies, capture 1/4 cup and estimate the population size of M & M's using the proportion of Skittles in the sample. Return all the candies to their original bags.

(c) Repeat the above process until you have obtained 10 estimates of the population size.

(d) Use the average of your 10 estimates to estimate the population size.

(e) Do the 10 estimates have a large or small standard deviation?

(f) Use a computer simulation to replicate this experiment. Let $N_1 = 50, 100, 150, 200, 250, 300$, and 350. Let $n = 400 - N1$. Compare these different combinations and recommend the combination you think is best.

**5.3-11** Use simulation to compare 90% confidence intervals for the mean when the underlying distribution is not normal. In particular, for each of the following distributions, simulate 50 random samples of size $n = 5$. For each sample, calculate the endpoints for a 90% confidence interval for $\mu$, first using the value of $\sigma$ and $z_{0.05} = 1.645$ and then using $s$ and $t_{0.05}(4) = 2.132$. Do about 90% of these confidence intervals contain the mean? Why? What are the average lengths of the confidence intervals?

(a) $U(0, 2)$.

(b) Exponential distribution with $\theta = 1$.

**5.3-12** Let $X_1, X_2, \ldots, X_7$ be a random sample of size $n = 7$ from a normal distribution, $N(1, 1)$. Then $T = (\overline{X} - 1)/(S/\sqrt{7})$ has a $t$ distribution with 6 degrees of freedom.

(a) Simulate 500 samples of size $n = 7$ from a normal distribution, $N(1, 1)$. For each sample, calculate the value of $T$. Illustrate empirically that these observations come from a $t$ distribution with 6 degrees of freedom by constructing a histogram with the $t$ p.d.f. superimposed and also constructing a $q$-$q$ plot of the percentiles of the $t$ distribution versus the order statistics of the sample.

(b) Repeat part (a) but this time simulate random samples of size $n = 7$ from an exponential distribution with mean 1. How good is the fit in this case? Why?

(c) Simulate 500 samples of size $n = 7$ from a Cauchy distribution. For each sample, calculate the value of $T = \overline{X}/(S/\sqrt{7})$. Again construct a histogram of these data with the $t$ p.d.f. superimposed and also constructing a $q$-$q$ plot of the percentiles of the $t$ distribution with 6 degrees of freedom versus the order statistics of the sample observations of $T$. Describe the fit.

**5.3-13** Let $X_1, X_2, \ldots, X_{11}$ be a random sample of size 11 from a normal distribution, $N(0, 1)$. Let $Y_1, Y_2, \ldots, Y_{11}$ be a random sample of size 11 from a normal distribution, $N(0, 1)$. Then $F = S_X^2/S_Y^2$, the ratio of the sample variances, has an $F$ distribution with $r_1 = 10$ and $r_2 = 10$ degrees of freedom.

(a) Simulate 100 samples of size 11 from each of the normal distributions and for each pair of samples, calculate the value of $F$. Construct a histogram of these data with the $F(10, 10)$ p.d.f. superimposed and construct a $q$-$q$ plot of the percentiles of the $F$ distribution with $r_1 = 10$ and $r_2 = 10$ degrees of freedom versus the order statistics of the sample observations of $F$. Is there a good fit?

(b) Repeat part (a) but this time take the samples from two exponential distributions, each having a mean of 1. Is this fit as good as that in part (a)? Why?

(c) Repeat part (a) but this time take the samples from two $U(-1, 1)$ distributions.

**5.3-14** In this exercise we compare the sampling distributions of $(n-1)S^2/\sigma^2$ when sampling from a normal distribution and from an exponential distribution.

(a) Simulate 200 random samples of size $n = 16$ from a normal distribution, $N(10, 1)$. For each sample, calculate the value of $S^2$ yielding the values $s_1^2, s_2^2, \ldots, s_{200}^2$. Since $(n-1)S^2/\sigma^2$ is $\chi^2(n-1)$, plot a histogram of the values $15s_1^2/1, 15s_2^2/1, \ldots, 15s_{200}^2/1$ with a $\chi^2(15)$ p.d.f. superimposed.

(b) Repeat part (a) but this time sample from a shifted exponential distribution with mean 10 and variance 1; that is, one with p.d.f. $f(x) = e^{-(x-9)}, 9 < x$.

(c) For the 200 values of $s^2$ in each of parts (a) and (b), find a 90 percent confidence interval for $\sigma^2$ by finding the 5th and 95th percentiles of the ordered values of $s^2$. Since we know $\sigma^2 = 1$ in each case, we are 90 percent confident that each of these intervals covers $\sigma^2 = 1$.

**5.3-15** Simulate 200 pairs of observations from a $U(0,1)$ distribution, say $U_1, U_2$. For each pair of random numbers calculate the values of

$$Z_1 = \sqrt{-2\ln U_1}\cos(2\pi U_2) \quad \text{and} \quad Z_2 = \sqrt{-2\ln U_1}\sin(2\pi U_2).$$

(See Equations 5.3-3.)

**(a)** Show that the sample means and the sample variances of the observations of $Z_1$ and $Z_2$ are close to 0 and 1, respectively.

**(b)** Construct relative frequency histograms of each of the sets of observations of $Z_1$ and $Z_2$ and superimpose the $N(0,1)$ p.d.f. on each histogram. Do you have good fits?

**(c)** Construct $q$-$q$ plots of the quantiles of a $N(0,1)$ distribution versus the order statistics of each set of observations of $Z_1$ and $Z_2$. Is this a good fit?

**(d)** Construct a scatterplot of the observations of $(Z_1, Z_2)$ and also calculate the correlation coefficient of these observations. What is your conclusion?

## 5.4   RESAMPLING

Sampling and resampling methods have become more useful in recent years due to the power of computers. These methods are even used in very beginning courses to convince students that statistics have distributions; that is, statistics are random variables with distributions. At this stage in the book, the reader should be convinced that this is true, although we did use some sampling in this chapter to help sell the idea that the sample mean has an approximate normal distribution.

Resampling methods, however, are used for more than showing that statistics have certain distributions. Rather, they are needed in finding approximate distributions of certain statistics which are used to make statistical inferences. We already know a great deal about the distribution of $\overline{X}$, and resampling methods are not needed for $\overline{X}$. In particular, $\overline{X}$ has an approximate normal distribution with mean $\mu$ and standard deviation $\sigma/\sqrt{n}$. Of course, if the latter is unknown, we can estimate it by $s/\sqrt{n}$ and note that $(\overline{X} - \mu)/(s/\sqrt{n})$ has an approximate $N(0,1)$ distribution, provided the sample size is large enough and the underlying distribution is not too badly skewed with a long heavy tail.

We know very little about the distribution of $S^2$. However, the statistic $S^2$ is not very robust in that its distribution changes a great deal as the underlying distribution changes. It is not like $\overline{X}$, which always has an approximate normal distribution provided the mean $\mu$ and variance $\sigma^2$ of the underlying distribution exist. So what do we do about distributions of statistics like the sample variance $S^2$ whose distribution depends so much on having a given underlying distribution? We use resampling methods that essentially substitute computation for theory. We need to have some idea about the distributions of these various estimators to find confidence intervals for the corresponding parameters.

Let us now explain resampling. Suppose that we need to find the distribution of some statistic, like $S^2$, but we do not believe that we are sampling from a normal distribution. We observe the values of $X_1, X_2, \ldots, X_n$ to be $x_1, x_2, \ldots, x_n$. Actually if we know nothing about the underlying distribution, then the empirical distribution found by placing the weight $1/n$ on each $x_i$ is the best estimate of that distribution. Therefore to get some idea about the distribution of $S^2$, let us take a random sample of size $n$ from this empirical distribution. We compute $S^2$ for that sample, say it is $s_1^2$. We then do it again, getting $s_2^2$. And again computing $s_3^2$. We continue to do this a large number of times, say $N$; and $N$ might be 1000, 2000, or even 10,000. Once we have these $N$ values of $S^2$, we can construct a histogram or a stem-and-leaf display or a $q$-$q$ plot, anything to help us get some information about the distribution of $S^2$ when the sample arises from this empirical distribution, which is an estimate of the real underlying distribution. Clearly we must use the computer for all of this sampling. We illustrate this resampling procedure using not $S^2$, but a statistic called the trimmed mean.

While we usually do not know the underlying distribution, we will state that, in this illustration, it is of the Cauchy type because there are certain basic ideas we want to review or introduce for the first time. The p.d.f. of the Cauchy is

$$f(x) = \frac{1}{\pi(1+x^2)}, \qquad -\infty < x < \infty.$$

The distribution function is

$$F(x) = \int_{-\infty}^{x} \frac{1}{\pi(1+w^2)}\, dw = \frac{1}{\pi}\arctan x + \frac{1}{2}, \qquad -\infty < x < \infty.$$

If we want to generate some $X$ values that have this distribution, we let $Y$ have the uniform distribution $U(0,1)$ and define $X$ by

$$Y = F(X) = \frac{1}{\pi}\arctan X + \frac{1}{2}$$

or, equivalently,

$$X = \tan\left[\pi\left(Y - \frac{1}{2}\right)\right].$$

We can generate 40 values of $Y$ on the computer and then calculate the 40 values of $X$. Let us now add $\theta = 5$ to each $X$ value to create a sample from a Cauchy distribution with a median of 5. That is, we have a random sample of 40 $W$ values, where $W = X + 5$. We will consider some statistics used to estimate the median, $\theta$, of this distribution. Of course, usually the value of the median is unknown, but here we know that it is equal to $\theta = 5$ and our statistics are estimates of this known number. These 40 values of $W$ are the following, after ordering:

| | | | | | | | | | |
|---|---|---|---|---|---|---|---|---|---|
| −7.34 | −5.92 | −2.98 | 0.19 | 0.77 | 0.95 | 2.86 | 3.17 | 3.76 | 4.20 |
| 4.20 | 4.27 | 4.31 | 4.42 | 4.60 | 4.73 | 4.84 | 4.87 | 4.90 | 4.96 |
| 4.98 | 5.00 | 5.09 | 5.09 | 5.14 | 5.22 | 5.23 | 5.42 | 5.50 | 5.83 |
| 5.94 | 5.95 | 6.00 | 6.01 | 6.24 | 6.82 | 9.62 | 10.03 | 18.27 | 93.62 |

It is interesting to observe that many of these 40 values are close to the $\theta = 5$, being between 3 and 7; they are almost as if they had arisen from a normal distribution with mean $\mu = 5$ and $\sigma^2 = 1$. But then we note the outliers; these very large or small values occur because of the heavy and long tails of the Cauchy distribution. This suggests that the sample mean $\overline{X}$ is not a very good estimator of the middle. And it is not as $\overline{x} = 6.67$ in this sample. In a more theoretical course, it can be shown that due to the fact that the mean $\mu$ and the variance $\sigma^2$ do not exist for a Cauchy distribution, $\overline{X}$ is not any better than a single observation $X_i$ in estimating the median $\theta$. The sample median $\widetilde{m}$ is a much better estimate of $\theta$ as it is not influenced by the outliers. Here the median equals 4.97, which is fairly close to 5. Actually, the maximum likelihood estimator found by maximizing

$$L(\theta) = \prod_{i=1}^{40} \frac{1}{\pi[1+(x_i-\theta)^2]}$$

is extremely good but requires difficult numerical methods to compute. Then advanced theory shows that, in the case of a Cauchy distribution, a **trimmed mean** found by ordering the sample and discarding the smallest and largest $3/8 = 37.5\%$ of the sample and averaging the middle 25% is almost as good as the maximum likelihood estimator, but it is much easier to compute. This trimmed mean is usually denoted by $\overline{X}_{0.375}$; but we use $\overline{X}_t$ for brevity and here $\overline{x}_t = 4.96$. For this sample it is not quite as good as the median; but, for most samples, it is better. Trimmed means are often very useful and many times are used with a smaller trimming percentage. For example, in sporting events like skating and diving, often the smallest and largest of the judges' scores are discarded.

For this Cauchy example, let us resample from the empirical distribution created by placing the "probability" 1/40 on each of our 40 observations. With each of these samples, we find our trimmed mean $\overline{X}_t$. That is, we order the observations of each resample and average the middle 25% of the order statistics, namely the middle 10 order statistics. We do this $N = 1000$ times, thus obtaining $N = 1000$ values of $\overline{X}_t$. These are summarized with the histogram in Figure 5.4-1(a).

From this resampling procedure, which is called **bootstrapping**, we have some idea about the distribution if the sample arises from the empirical distribution and, hopefully, from the underlying distribution, which is approximated by the empirical distribution. While the distribution of the sample mean, $\overline{X}$, is not normal if the sample arises from a Cauchy-type distribution, the approximate distribution of $\overline{X}_t$ is normal. From the histogram in Figure 5.4-1(a) of trimmed mean values, this looks to be the case. This observation is supported by the q-q plot in Figure 5.4-1(b) of the quantities of a standard normal distribution versus those of the 1000 $\overline{x}_t$ values, which is very close to being a straight line.

How do we find a confidence interval for $\theta$? Recall that the middle of the distribution of $\overline{X}_t - \theta$ is zero. So a guess at $\theta$ would be the amount needed to move the histogram of $\overline{X}_t$ values over so that zero is more or less in the middle of this translated histogram. We recognize that this histogram was just generated using the original sample $X_1, X_2, \ldots, X_{40}$ and thus is really only an estimate of the distribution of $\overline{X}_t$.

We could get a point estimate of $\theta$ by moving it over until its median (or mean) is at zero. Clearly, however, there is some error in doing so; and we really want some bounds for $\theta$ as given by a confidence interval.

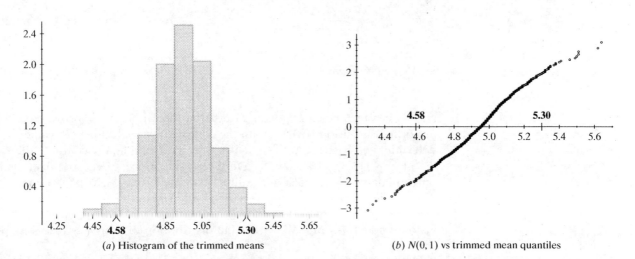

(a) Histogram of the trimmed means    (b) $N(0, 1)$ vs trimmed mean quantiles

Figure 5.4-1  $N = 1000$ observations of trimmed means

To find that confidence interval, let us proceed as follows: In the $N = 1000$ resampled values of $\overline{X}_t$, we find two points, say $c$ and $d$, such that about 25 values are less than $c$ and about 25 are greater than $d$. That is, $c$ and $d$ are about on the respective 2.5 and 97.5 percentiles of the empirical distribution of these $N = 1000$ resampled $\overline{X}_t$ values. Thus, $\theta$ should be big enough so that over 2.5 percent of the $\overline{X}_t$ values are less than $c$ and small enough so that over 2.5 percent of the $\overline{X}_t$ values are greater than $d$. This requires that $c < \theta$ and $\theta < d$; thus $[c, d]$ serves as an approximate 95% confidence interval for $\theta$ as found by the **percentile method**. With our bootstrapped distribution of $N = 1000$ $\overline{X}_t$ values, this 95% confidence interval for $\theta$ is 4.58 to 5.30 and these two points are marked on the histogram and the $q$-$q$ plot. Clearly, we could change this percentage to other values, such as 90%.

This **percentile method**, associated with bootstrap method, is a nonparametric procedure as we make no assumptions about the underlying distribution. It is interesting to compare the answer to that obtained by using the order statistics $Y_1 < Y_2 < \cdots < Y_{40}$. If the sample arises from a continuous-type distribution, we have, when $\theta$ is the median, that (using a calculator or computer)

$$P(Y_{14} < \theta < Y_{27}) = \sum_{k=14}^{26} \binom{40}{k} \left(\frac{1}{2}\right)^{40} = 0.9615.$$

(See Section 4.9.) Since in our illustration $Y_{14} = 4.42$ and $Y_{27} = 5.23$, then the interval [4.42, 5.23] is an approximate 96% confidence interval for $\theta$. Of course, $\theta = 5$ is included in each of the two confidence intervals. In this case, the bootstrap confidence interval is a little more symmetric about $\theta = 5$ and somewhat shorter, but required much more work.

We have now illustrated bootstrapping, which allows us to substitute computation for theory to make statistical inferences about characteristics of the underlying distribution. This method is becoming more important as we encounter complicated data sets which clearly do not satisfy certain underlying assumptions. For illustration, consider the distribution of $T = (\overline{X} - \mu)/(S/\sqrt{n})$ when the random sample arises from an exponential distribution that has p.d.f. $f(x) = e^{-x}$, $0 < x < \infty$, with mean $\mu = 1$. First we will *not* use resampling, but we will simulate the distribution of $T$ when the sample size $n = 16$ by taking $N = 1000$ random samples from this known exponential distribution. Here

$$F(x) = \int_0^x e^{-w}\, dw = 1 - e^{-x}, \qquad 0 < x < \infty.$$

So $Y = F(X)$ means
$$X = -\ln(1 - Y)$$

and $X$ has that given exponential distribution with $\mu = 1$ provided $Y$ has the uniform distribution $U(0, 1)$. With the computer, we select $n = 16$ values of $Y$ and determine the corresponding $n = 16$ values of $X$ and finally compute the value of $T = (\overline{X} - 1)/(S/\sqrt{16})$, say $T_1$. We repeat this process over and over again, not only obtaining $T_1$, but also the values of $T_2, T_3, \ldots, T_{1000}$. We have done this and display the histogram of the 1000 $T$ values in Figure 5.4-2(a). Moreover the $q$-$q$ plot with quantiles of $N(0, 1)$ on the $y$-axis is displayed in Figure 5.4-2(b). Both the histogram and $q$-$q$ plot show that the distribution of $T$ in this case is skewed to the left.

In the preceding illustration we knew the underlying distribution. Let us now sample from the exponential distribution with mean $\mu = 1$ but add a value $\theta$ to

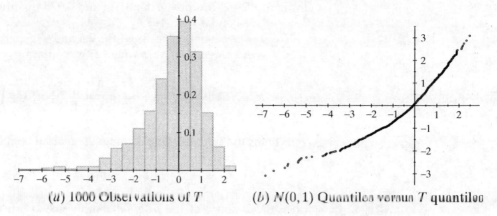

(*a*) 1000 Observations of *T*          (*b*) *N*(0, 1) Quantiles versus *T* quantiles

Figure 5.4-2  *T* Observations from an exponential distribution

each $X$. Thus we will try to estimate the new mean $\theta + 1$. The authors know the value of $\theta$, but the readers do not at this time. The observed 16 values of this random sample are:

|         |        |         |         |         |         |         |         |
|---------|--------|---------|---------|---------|---------|---------|---------|
| 11.9776 | 9.3889 | 9.9798  | 13.4676 | 9.2895  | 10.1242 | 9.5798  | 9.3148  |
| 9.0605  | 9.1680 | 11.0394 | 9.1083  | 10.3720 | 9.0523  | 13.2969 | 10.5852 |

At this point we are trying to find a confidence interval for $\mu = \theta + 1$, and we pretend that we do not know that the underlying distribution is exponential. This is the case in practice: We do not know the underlying distribution. So we use the empirical distribution as the best guess of the underlying distribution; it is found by placing the weight 1/16 on each of the observations. The mean of this empirical distribution is $\bar{x} = 10.3003$. Therefore we obtain some idea about the distribution of $T$ by now simulating

$$T = \frac{\overline{X} - 10.3003}{S/\sqrt{16}}$$

with $N = 1000$ random samples from the empirical distribution.

We obtain $t_1, t_2, \ldots, t_{1000}$ and these are used to construct a histogram, Figure 5.4-3(*a*), and *q-q* plot, Figure 5.4-3(*b*). These two figures look somewhat like those

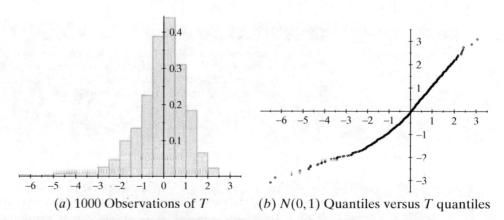

(*a*) 1000 Observations of *T*          (*b*) *N*(0, 1) Quantiles versus *T* quantiles

Figure 5.4-3  *T* Observations from an empirical distribution

in Figure 5.4-2. Moreover the 0.025 and 0.975 quantiles of these 1000 $t$-values are $c = -3.1384$ and $d = 1.8167$.

Now we have some idea about the 2.5 and 97.5 percentiles of the $T$-distribution. Hence we can write as a very rough approximation

$$P\left(-3.1384 \leq \frac{\overline{X} - \mu}{S/\sqrt{16}} \leq 1.8167\right) \approx 0.95.$$

This leads to the rough approximate 95 percent confidence interval

$$[\overline{x} - 1.8167s/\sqrt{16}, \ \overline{x} - (-3.1384)s/\sqrt{16}]$$

once the $\overline{x}$ and $s$ of the *original* sample are substituted. With $\overline{x} = 10.3003$ and $s = 1.4544$ we have

$$[10.3003 - 1.8167(1.4544)/4, \ 10.3003 + 3.1384(1.4544)/4] = [9.6397, 11.4414]$$

as a 95% approximate confidence interval for $\mu = \theta + 1$. As we added $\theta = 9$ to each $x$-value, we see that the interval does cover $\theta + 1 = 10$.

It is easy to see how this procedure gets its name, because it is like "pulling yourself up by your own bootstraps," in which the empirical distribution acts as the bootstraps.

**Remark**    The expression "to pull oneself up by his or her own bootstraps" seems to come from *The Surprising Adventures of Baron Munchausen* by Rudolph Erich Raspe. The Baron had fallen from the sky and found himself in a hole nine fathoms deep and had no idea how to get out. He comments as follows: "Looking down I observed that I had on a pair of boots with exceptionally sturdy straps. Grasping them firmly, I pulled with all my might. Soon I had hoisted myself to the top and stepped out on terra firma without further ado."

Of course, here in *bootstrapping*, statisticians pull themselves up by their bootstraps (the empirical distributions) by recognizing the empirical distribution is the best estimate of the underlying distribution without a lot of other assumptions. So they use the empirical distribution as if it is the underlying distribution to find approximately the distribution of statistics of interest.

## EXERCISES 5.4

**5.4-1** If time and computing facilities are available, consider the following 40 losses, due to wind-related catastrophes, that were recorded to the nearest one million dollars. These data include only those losses of $2 million or more; and, for convenience, they have been ordered and recorded in millions.

| | | | | | | | | | |
|---|---|---|---|---|---|---|---|---|---|
| 2 | 2 | 2 | 2 | 2 | 2 | 2 | 2 | 2 | 2 |
| 2 | 2 | 3 | 3 | 3 | 3 | 4 | 4 | 4 | 5 |
| 5 | 5 | 5 | 6 | 6 | 6 | 6 | 8 | 8 | 9 |
| 15 | 17 | 22 | 23 | 24 | 24 | 25 | 27 | 32 | 43 |

To illustrate bootstrapping, take resamples of size $n = 40$ as many as $N = 100$ times computing $T = (\overline{X} - 5)/(S/\sqrt{40})$ each time. Here the value 5 is the median of the original sample. Construct a histogram of the bootstrapped values of $T$.

**5.4-2  (a)** Consider the 16 observed values, rounded to the nearest tenth, from the exponential distribution that was given in this section, namely

$$
\begin{array}{cccccccc}
12.0 & 9.4 & 10.0 & 13.5 & 9.3 & 10.1 & 9.6 & 9.3 \\
9.1 & 9.2 & 11.0 & 9.1 & 10.4 & 9.1 & 13.3 & 10.6
\end{array}
$$

Take resamples of size $n = 16$ from these about $N = 200$ times and compute $s^2$ each time. Construct a histogram of these 200 bootstrapped values of $S^2$.

**(b)** Simulate $N = 200$ random samples of size $n = 16$ from an exponential distribution with $\theta$ equal to the mean of the data in part (a) minus 9. For each sample calculate the value of $s^2$. Construct a histogram of these 200 values of $S^2$.

**(c)** Construct a $q$-$q$ plot of the two sets of sample variances to compare these two empirical distributions of $S^2$.

**5.4-3** Let $X_1, X_2, \ldots, X_{21}$ and $Y_1, Y_2, \ldots, X_{21}$ be random samples of sizes $n = 21$ and $m = 21$ from independent $N(0, 1)$ distributions. Then $F = S_X^2/S_Y^2$ has an $F$ distribution with 20 and 20 degrees of freedom.

**(a)** Illustrate this empirically by simulating 100 observations of $F$.

(i) Plot a relative frequency histogram with the $F(20, 20)$ p.d.f. superimposed.

(ii) Construct a $q$-$q$ plot of the quantiles of $F(20, 20)$ versus the order statistics of your simulated data. Is the plot linear?

**(b)** Consider the following 21 observations of the $N(0, 1)$ random variable $X$:

$$
\begin{array}{ccccccc}
0.1616 & -0.8593 & 0.3105 & 0.3932 & -0.2357 & 0.9697 & 1.3633 \\
-0.4166 & 0.7540 & -1.0570 & -0.1287 & -0.6172 & 0.3208 & 0.9637 \\
0.2494 & -1.1907 & -2.4699 & -0.1931 & 1.2274 & -1.2826 & -1.1532
\end{array}
$$

and the following 21 observations of the $N(0, 1)$ random variable $Y$:

$$
\begin{array}{ccccccc}
0.4419 & -0.2313 & 0.9233 & -0.1203 & 1.7659 & -0.2022 & 0.9036 \\
-0.4996 & -0.8778 & -0.8574 & 2.7574 & 1.1033 & 0.7066 & 1.3595 \\
-0.0056 & -0.5545 & -0.1491 & -0.9774 & -0.0868 & 1.7462 & -0.2636
\end{array}
$$

Resample with a sample of size 21 from each of these sets of observations, sampling with replacement. Calculate the value of $w = s_x^2/s_y^2$. Repeat this to simulate 100 observations of $W$ from these two empirical distributions. Use the same graphical comparisons that you used in part (a) to see if these represent observations from an approximate $F(20, 20)$ distribution.

**(c)** Consider the following 21 observations of the exponential random variable $X$ with mean 1:

$$
\begin{array}{ccccccc}
0.6958 & 1.6394 & 0.2464 & 1.5827 & 0.0201 & 0.4544 & 0.8427 \\
0.6385 & 0.1307 & 1.0223 & 1.3423 & 1.6653 & 0.0081 & 5.2150 \\
0.5453 & 0.0844 & 1.2346 & 0.5721 & 1.5167 & 0.4843 & 0.9145
\end{array}
$$

and the following 21 observations of the exponential random variable $Y$ with mean 1:

$$
\begin{array}{ccccccc}
1.1921 & 0.3708 & 0.0874 & 0.5696 & 0.1192 & 0.0164 & 1.6482 \\
0.2453 & 0.4522 & 3.2312 & 1.4745 & 0.8870 & 2.8097 & 0.8533 \\
0.1466 & 0.9494 & 0.0485 & 4.4379 & 1.1244 & 0.2624 & 1.3655
\end{array}
$$

Resample with a sample of size 21 from each of these sets of observations, sampling with replacement. Calculate the value of $w = s_x^2/s_y^2$. Repeat this to simulate 100 observations of $W$ from these two empirical distributions. Use the same graphical comparisons that you used in part (a) to see if these represent observations from an approximate $F(20, 20)$ distribution.

**5.4-4** The following 54 pairs of data are from Old Faithful geyser:

(duration in minutes of an eruption, time in minutes until the next eruption).

$$
\begin{array}{cccccc}
(2.500, 72) & (4.467, 88) & (2.333, 62) & (5.000, 87) & (1.683, 57) & (4.500, 94) \\
(4.500, 91) & (2.083, 51) & (4.367, 98) & (1.583, 59) & (4.500, 93) & (4.550, 86) \\
(1.733, 70) & (2.150, 63) & (4.400, 91) & (3.983, 82) & (1.767, 58) & (4.317, 97) \\
(1.917, 59) & (4.583, 90) & (1.833, 58) & (4.767, 98) & (1.917, 55) & (4.433, 107) \\
(1.750, 61) & (4.583, 82) & (3.767, 91) & (1.833, 63) & (4.817, 91) & (1.900, 52) \\
(4.517, 94) & (2.000, 60) & (4.650, 84) & (1.817, 63) & (4.917, 91) & (4.000, 83) \\
(4.317, 84) & (2.133, 71) & (4.783, 83) & (4.217, 70) & (4.733, 81) & (2.000, 60) \\
(4.717, 91) & (1.917, 51) & (4.233, 85) & (1.567, 55) & (4.567, 98) & (2.133, 49) \\
(4.500, 83) & (1.717, 63) & (4.783, 102) & (1.850, 56) & (4.583, 86) & (1.733, 62)
\end{array}
$$

(a) Calculate the correlation coefficient of these data and construct a scatterplot of the data.

(b) To estimate the distribution of the correlation coefficient, $R$, resample 500 samples of size 54 from the empirical distribution and for each sample, calculate the value of $R$.

(c) Construct a histogram of these 500 observations of $R$.

(d) Simulate 500 samples of size 54 from a bivariate normal distribution with correlation coefficient equal to the correlation coefficient of the geyser data. For each sample of 54, calculate the correlation coefficient.

(e) Construct a histogram of the 500 observations of the correlation coefficient.

(f) Construct a $q$-$q$ plot of the 500 observations of $R$ from the bivariate normal distribution of part (d) versus the 500 observations in part (b). Do the two distributions of $R$ appear to be about equal?

5.4-5 Consider the following 12 pairs of data. (They are values of a random sample of size 12 from a bivariate normal distribution with correlation coefficient $\rho = -0.8$.)

| | | | |
|---|---|---|---|
| $(-0.8118, 0.5205)$ | $(-1.7527, 1.2374)$ | $(-0.0639, 1.1181)$ | $(-0.5526, 0.7047)$ |
| $(-0.6387, -0.3772)$ | $(0.1743, 0.2298)$ | $(1.7126, -2.2842)$ | $(0.3114, 0.0846)$ |
| $(-0.6270, 1.0964)$ | $(0.5996, -1.1564)$ | $(-0.9370, 0.6896)$ | $(0.6739, -0.0470)$ |

(a) To estimate the distribution of the correlation coefficient, $R$, resample 500 samples of size 12 from the empirical distribution and for each sample, calculate the value of $R$.

(b) Construct a histogram of these 500 observations of $R$.

(c) Simulate 500 samples of size 12 from a bivariate normal distribution with correlation coefficient equal to the correlation coefficient of the given data. For each sample of 12, calculate the correlation coefficient.

(d) Construct a histogram of the 500 observations of the correlation coefficient.

(e) Construct a $q$-$q$ plot of the 500 observations of $R$ from the bivariate normal distribution of part (c) versus the 500 observations in part (a). Do the two distributions of $R$ appear to be about equal?

5.4-6 Consider the following 20 pairs of data. (They are paired data from two exponential distributions, each having a mean of 5.)

| | | | |
|---|---|---|---|
| $(5.4341, 8.4902)$ | $(33.2097, 4.7063)$ | $(0.4034, 1.8961)$ | $(1.4137, 0.2996)$ |
| $(17.9365, 3.1350)$ | $(4.4867, 6.2089)$ | $(11.5107, 10.9784)$ | $(8.2473, 19.6554)$ |
| $(1.9995, 3.6339)$ | $(1.8965, 1.7850)$ | $(1.7116, 1.1545)$ | $(4.4594, 1.2344)$ |
| $(0.4036, 0.7260)$ | $(3.0578, 19.0489)$ | $(21.4049, 4.6495)$ | $(3.8845, 13.7945)$ |
| $(5.9536, 9.2438)$ | $(11.3942, 1.7863)$ | $(5.4813, 4.3356)$ | $(7.0590, 1.15834)$ |

(a) Calculate the correlation coefficient of these data and also construct a scatterplot of the data.

(b) To estimate the distribution of the correlation coefficient, $R$, resample 500 samples of size 20 from the empirical distribution and for each sample, calculate the value of $R$.

(c) Construct a histogram of these 500 observations of $R$.

(d) Simulate 500 samples of size 20 from a bivariate normal distribution with correlation coefficient 0. For each sample of 20, calculate the correlation coefficient.

(e) Construct a histogram of the 500 observations of the correlation coefficient.

(f) Construct a $q$-$q$ plot of the 500 observations of $R$ from the bivariate normal distribution of part (c) versus the 500 observations in part (a). Do the two distributions of $R$ appear to be about equal?

5.4-7 Consider the following 20 pairs of data. (They are paired data from two Cauchy distributions.)

| | | | |
|---|---|---|---|
| $(-6.0141, 3.1634)$ | $(-1.0420, 16.6134)$ | $(0.0894, 0.1775)$ | $(-4.0680, -0.4082)$ |
| $(2.4561, -3.0045)$ | $(-3.7596, -1.0007)$ | $(0.1678, -1.1852)$ | $(-0.5504, -0.9796)$ |
| $(1.4399, -0.1211)$ | $(-0.2510, -0.7397)$ | $(-5.3430, 0.4839)$ | $(-0.8826, -2.9297$ |
| $(1.4219, 1.7424)$ | $(-5.7973, 0.1788)$ | $(1.0064, -0.2806)$ | $(-2.1313, -0.5802)$ |
| $(5.8900, 0.8383)$ | $(-2.2183, -0.4968)$ | $(1.1598, 1.2758)$ | $(-1.0269, -15.7753)$ |

**(a)** To estimate the distribution of the correlation coefficient, $R$, resample 500 samples of size 20 from the empirical distribution and for each sample, calculate the value of $R$.

**(b)** Construct a histogram of these 500 observations of $R$.

**(c)** Simulate 500 samples of size 20 from a bivariate normal distribution with correlation coefficient 0. For each sample of 20, calculate the correlation coefficient.

**(d)** Construct a histogram of the 500 observations of the correlation coefficient.

**(e)** Construct a $q$-$q$ plot of the 500 observations of $R$ from the bivariate normal distribution of part (c) versus the 500 observations in part (a). Do the two distributions of $R$ appear to be about equal?

# CHAPTER
# FIVE COMMENTS

The first two sections consider the computations of statistics or of certain mathematical solutions. Hopefully you will use the computer in future statistical applications. The last two sections are the type of things in which more modern statisticians like to be involved, namely simulation or resampling. It should be observed that Brad Efron of Stanford University first proposed the concept of bootstrapping and contributed much in its early development.

The computer has helped many modern areas of statistics. While Bayes's Theorem is an old one, the Neo-Bayesian movement is highly dependent upon the computer. Bayesians treat the parameters involved as random variables. For illustration, the p.d.f. (or p.m.f.) $f(x; \theta)$ would be thought of as one of a conditional distribution $f(x \mid \theta)$. Therefore with a random sample $X_1, X_2, \ldots, X_n$ and a prior p.d.f. $g(\theta)$ of the parameter, the joint p.d.f. of the observations and parameter is

$$g(\theta)f(x_1 \mid \theta)f(x_2 \mid \theta) \cdots f(x_n \mid \theta).$$

By summing or integrating out $\theta$, we obtain the joint marginal of $X_1, X_2, \ldots, X_n$, say $h(x_1, x_2, \ldots, x_n)$. Thus the conditional distribution of the parameter, given the observations, is

$$\frac{g(\theta)f(x_1 \mid \theta)f(x_2 \mid \theta) \cdots f(x_n \mid \theta)}{h(x_1, x_2, \ldots, x_n)} = k(\theta \mid x_1, x_2, \ldots, x_n)$$

which in a sense is Bayes's Theorem. The Bayesians use this conditional p.d.f. $k(\theta \mid x_1, x_2, \ldots, x_n)$, called a posterior p.d.f., to find point and interval estimates of $\theta$. For example, the conditional mean $E(\theta \mid x_1, x_2, \ldots, x_n) = u(x_1, x_2, \ldots, x_n)$ is a possible point estimate of $\theta$ and $v(x_1, x_2, \ldots, x_n)$ to $w(x_1, x_2, \ldots, x_n)$ is an interval estimate for $\theta$ where

$$\int_{v(x_1, x_2, \ldots, x_n)}^{w(x_1, x_2, \ldots, x_n)} k(\theta \mid x_1, x_2, \ldots, x_n) \, d\theta = 1 - \alpha.$$

In many simple cases, it is easy to determine $u(x_1, x_2, \ldots, x_n)$, $v(x_1, x_2, \ldots, x_n)$, and $w(x_1, x_2, \ldots, x_n)$; but if the situation becomes at all complicated, the computer must be used to find these values. Two such procedures used by the Bayesians are the *Gibbs Sampler* and *Markov Chain Monte Carlo* (*MCMC*). For further discussion, see Hogg, McKean, and Craig (2005), but the Bayesians have found that the computer has "saved them."

# CHAPTER

# 6

# SOME SAMPLING DISTRIBUTION THEORY

## MOMENT-GENERATING FUNCTION TECHNIQUE

The mean, variance, and standard deviation are important characteristics of a distribution. For some distributions, like the binomial, it is fairly difficult to compute directly $E(X)$ and $E(X^2)$ to find the mean and the variance (unless you use *Maple* or another Computer Algebra System--CAS). In this section, we define a function of $t$ that will help us generate the moments of a distribution, and thus it is called the moment-generating function. While this generating characteristic is extremely important, there is another uniqueness property that is even more important. We first define this new function of $t$ and then explain this uniqueness property before showing how it can be used to compute the moments.

---

**Definition 6.1-1**

Let $X$ be a random variable of the discrete type with p.m.f. $f(x)$ and space $S$. If there is a positive number $h$ such that

$$E(e^{tX}) = \sum_{x \varepsilon S} e^{tx} f(x)$$

exists and is finite for $-h < t < h$, then the function of $t$ defined by

$$M(t) = E(e^{tX})$$

is called the *moment-generating function of* $X$ (or of the distribution of $X$). This is often abbreviated as m.g.f.

First, it is evident that if we set $t = 0$, we have $M(0) = 1$. Moreover, if the space of $S$ is $\{b_1, b_2, b_3, \ldots\}$, the moment-generating function is given by the expansion

$$M(t) = e^{tb_1}f(b_1) + e^{tb_2}f(b_2) + e^{tb_3}f(b_3) + \cdots.$$

Thus we see that the coefficient of $e^{tb_i}$ is the probability

$$f(b_i) = P(X = b_i).$$

Accordingly if two random variables (or two distributions of probability) have the same moment-generating function, they must have the same distribution of probability. That is, if the two had the two probability mass functions $f(x)$ and $g(y)$, and the same space $S = \{b_1, b_2, b_3, \ldots\}$ and

$$e^{tb_1}f(b_1) + e^{tb_2}f(b_2) + \cdots = e^{tb_1}g(b_1) + e^{tb_2}g(b_2) + \cdots \qquad (6.1\text{-}1)$$

for all $t$, $-h < t < h$, mathematical theory requires that

$$f(b_i) = g(b_i), \qquad i = 1, 2, 3, \cdots.$$

So we see that the moment-generating function uniquely determines the distribution of a random variable. That is, if the m.g.f. exists, there is one and only one distribution of probability associated with that m.g.f.

> **Remark**   From elementary algebra, we can get some understanding of why Equation 6.1-1 requires $f(b_i) = g(b_i)$. In that equation, let $e^t = w$ and say the points in the support, namely $b_1, b_2, \ldots, b_k$, are positive integers, the largest of which is $m$. Thus Equation 6.1-1 provides the equality of two $m$th degree polynomials in $w$ for an uncountable number of values of $w$. A fundamental theorem of algebra requires that the corresponding coefficients of the two polynomials be equal; that is, $f(b_i) = g(b_i)$, $i = 1, 2, \ldots, k$.

**EXAMPLE 6.1-1**   If $X$ has the m.g.f.

$$M(t) = e^t\left(\frac{3}{6}\right) + e^{2t}\left(\frac{2}{6}\right) + e^{3t}\left(\frac{1}{6}\right),$$

then the probabilities are

$$P(X = 1) = \frac{3}{6}, \qquad P(X = 2) = \frac{2}{6}, \qquad P(X = 3) = \frac{1}{6}.$$

We could write this, if we choose to do so, by saying $X$ has the p.m.f.

$$f(x) = \frac{4 - x}{6}, \qquad x = 1, 2, 3.$$

**EXAMPLE 6.1-2**   Suppose we are given that the m.g.f. of $X$ is

$$M(t) = \frac{e^t/2}{1 - e^t/2}, \qquad t < \ln 2.$$

Until we expand $M(t)$, we can not detect the coefficients of $e^{b_i t}$. Recalling

$$(1 - z)^{-1} = 1 + z + z^2 + z^3 + \cdots, \qquad -1 < z < 1,$$

we have that

$$\frac{e^t}{2}\left(1 - \frac{e^t}{2}\right)^{-1} = \frac{e^t}{2}\left(1 + \frac{e^t}{2} + \frac{e^{2t}}{2^2} + \frac{e^{3t}}{2^3} + \cdots\right)$$

$$= \left(e^t\right)\left(\frac{1}{2}\right) + \left(e^{2t}\right)\left(\frac{1}{2}\right)^2 + \left(e^{3t}\right)\left(\frac{1}{2}\right)^3 + \cdots,$$

when $e^t/2 < 1$ and thus $t < \ln 2$. That is

$$P(X = x) = \left(\frac{1}{2}\right)^x,$$

when $x$ is a positive integer, or, equivalently, the p.m.f. of $X$ is,

$$f(x) = \left(\frac{1}{2}\right)^x, \qquad x = 1, 2, 3, \cdots.$$

From the theory of mathematical analysis, it can be shown that the existence of $M(t)$, for $-h < t < h$, implies that derivatives of $M(t)$ of all orders exist at $t = 0$; moreover, it is permissible to interchange differentiation and summation. Thus

$$M'(t) = \sum_{x \varepsilon S} x e^{tx} f(x),$$

$$M''(t) = \sum_{x \varepsilon S} x^2 e^{tx} f(x),$$

and, for each positive integer r,

$$M^{(r)}(t) = \sum_{x \varepsilon S} x^r e^{tx} f(x).$$

Setting $t = 0$, we see that

$$M'(0) = \sum_{x \varepsilon S} x f(x) = E(X),$$

$$M''(0) = \sum_{x \varepsilon S} x^2 f(x) = E(X^2),$$

and, in general,

$$M^{(r)}(0) = \sum_{x \varepsilon S} x^r f(x) = E(X^r).$$

In particular, if the moment-generating function exists,

$$\mu = M'(0) \qquad \text{and} \qquad \sigma^2 = M''(0) - [M'(0)]^2.$$

The above argument shows that we can find the moments of $X$ by differentiating $M(t)$. In using this technique, it must be emphasized that in use, we first evaluate the summation representing $M(t)$ to obtain a closed form solution, and then we differentiate that solution to obtain the moments of $X$. Example 6.1-3 illustrates the use of the moment-generating function for finding the first and second moments and then the mean and variance of the important binomial distribution.

EXAMPLE 6.1-3 The p.m.f. of the binomial distribution is

$$f(x) = \binom{n}{x} p^x (1-p)^{n-x}, \qquad x = 0, 1, 2, \ldots, n.$$

Thus the m.g.f. is

$$
\begin{aligned}
M(t) = E(e^{tX}) &= \sum_{x=0}^{n} e^{tx} \binom{n}{x} p^x (1-p)^{n-x} \\
&= \sum_{x=0}^{n} \binom{n}{x} (pe^t)^x (1-p)^{n-x} \\
&= [(1-p) + pe^t]^n, \qquad -\infty < t < \infty,
\end{aligned}
$$

from the expansion of $(a+b)^n$ with $a = 1-p$ and $b = pe^t$. It is interesting to note that here and elsewhere that the m.g.f. is usually rather easy to compute if the p.m.f. has a factor involving an exponential, like $p^x$ in the binomial p.m.f.

The first two derivatives of $M(t)$ are

$$M'(t) = n[(1-p) + pe^t]^{n-1}(pe^t)$$

and

$$M''(t) = n(n-1)[(1-p) + pe^t]^{n-2}(pe^t)^2 + n[(1-p) + pe^t]^{n-1}(pe^t).$$

Thus

$$\mu = E(X) = M'(0) = np$$

and

$$
\begin{aligned}
\sigma^2 = E(X^2) - [E(X)]^2 &= M''(0) - [M'(0)]^2 \\
&= n(n-1)p^2 + np - (np)^2 = np(1-p),
\end{aligned}
$$

as was claimed in Section 2.2.

A CAS could be used to solve Example 6.1-3. (See Exercise 6.1-7.)

For the continuous-type random variables, the definition of the moment-generating function is similar to the discrete case with integrals replacing summations.

EXAMPLE 6.1-4 Let $X$ have the p.d.f.

$$f(x) = \begin{cases} xe^{-x}, & 0 \le x < \infty, \\ 0, & \text{elsewhere.} \end{cases}$$

Then

$$M(t) = \int_0^\infty e^{tx} x e^{-x}\, dx = \lim_{b \to \infty} \int_0^b x e^{-(1-t)x}\, dx$$

$$= \lim_{b \to \infty} \left[ -\frac{x e^{-(1-t)x}}{1-t} - \frac{e^{-(1-t)x}}{(1-t)^2} \right]_0^b$$

$$= \lim_{b \to \infty} \left[ -\frac{b e^{-(1-t)b}}{1-t} - \frac{e^{-(1-t)b}}{(1-t)^2} \right] + \frac{1}{(1-t)^2}$$

$$= \frac{1}{(1-t)^2},$$

provided that $t < 1$. Note that $M(0) = 1$, which is true for every moment-generating function. Now

$$M'(t) = \frac{2}{(1-t)^3} \quad \text{and} \quad M''(t) = \frac{6}{(1-t)^4}.$$

Thus

$$\mu = M'(0) = 2$$

and

$$\sigma^2 = M''(0) - [M'(0)]^2 = 6 - 2^2 = 2.$$

A CAS could be used to solve Example 6.1-4. (See Exercise 6.1-8.)

**EXAMPLE 6.1-5** Let $X$ have a normal distribution, $N(\mu, \sigma^2)$. The moment-generating function of $X$ is

$$M(t) = \int_{-\infty}^{\infty} \frac{e^{tx}}{\sigma \sqrt{2\pi}} \exp\left[ -\frac{(x-\mu)^2}{2\sigma^2} \right] dx$$

$$= \int_{-\infty}^{\infty} \frac{1}{\sigma \sqrt{2\pi}} \exp\left\{ -\frac{1}{2\sigma^2}[x^2 - 2(\mu + \sigma^2 t)x + \mu^2] \right\} dx.$$

To evaluate this integral, we complete the square in the exponent

$$x^2 - 2(\mu + \sigma^2 t)x + \mu^2 = [x - (\mu + \sigma^2 t)]^2 - 2\mu\sigma^2 t - \sigma^4 t^2.$$

Thus

$$M(t) = \exp\left( \frac{2\mu\sigma^2 t + \sigma^4 t^2}{2\sigma^2} \right) \int_{-\infty}^{\infty} \frac{1}{\sigma \sqrt{2\pi}} \exp\left\{ -\frac{1}{2\sigma^2}[x - (\mu + \sigma^2 t)]^2 \right\} dx.$$

Note that the integrand in the last integral is like the p.d.f. of a normal distribution with $\mu$ replaced by $\mu + \sigma^2 t$. However, the normal p.d.f. integrates to one for all real $\mu$, in particular when $\mu$ is replaced by $\mu + \sigma^2 t$. Thus

$$M(t) = \exp\left( \frac{2\mu\sigma^2 t + \sigma^4 t^2}{2\sigma^2} \right) = \exp\left( \mu t + \frac{\sigma^2 t^2}{2} \right).$$

Now

$$M'(t) = (\mu + \sigma^2 t) \exp\left(\mu t + \frac{\sigma^2 t^2}{2}\right)$$

and

$$M''(t) = [(\mu + \sigma^2 t)^2 + \sigma^2] \exp\left(\mu t + \frac{\sigma^2 t^2}{2}\right).$$

Thus

$$E(X) = M'(0) = \mu,$$

$$\text{Var}(X) = M''(0) - [M'(0)]^2 = \mu^2 + \sigma^2 - \mu^2 = \sigma^2.$$

That is, the parameters $\mu$ and $\sigma^2$ in the p.d.f. of $X$ are the mean and the variance of $X$.

A summary of some important discrete and continuous distributions is given in the end pages at the front of this book. Included are the means, variances, and moment-generating functions.

**EXERCISES 6.1**

**6.1-1** **(i)** Give the name of the distribution of $X$ (if it has a name); **(ii)** find the values of $\mu$ and $\sigma^2$; and **(iii)** calculate $P(1 \leq X \leq 2)$ when the moment-generating function of $X$ is given by

**(a)** $M(t) = (0.3 + 0.7e^t)^5$.

**(b)** $M(t) = \dfrac{0.3e^t}{1 - 0.7e^t}, \qquad t < -\ln(0.7)$.

**(c)** $M(t) = 0.45 + 0.55e^t$.

**(d)** $M(t) = 0.3e^t + 0.4e^{2t} + 0.2e^{3t} + 0.1e^{4t}$.

**(e)** $M(t) = (0.6e^t)^2(1 - 0.4e^t)^{-2}, \qquad t < -\ln(0.4)$.

**(f)** $M(t) = \sum_{x=1}^{10} (0.1)e^{tx}$.

**6.1-2** Given a random permutation of the integers in the set $\{1, 2, 3, 4, 5\}$, let $X$ equal the number of integers that are in their natural position. The moment-generating function of $X$ is

$$M(t) = \frac{44}{120} + \frac{45}{120}e^t + \frac{20}{120}e^{2t} + \frac{10}{120}e^{3t} + \frac{1}{120}e^{5t}.$$

**(a)** Find the mean and variance of $X$.

**(b)** Find the probability that at least one integer is in its natural position.

**(c)** Draw a graph of the probability histogram of the p.m.f. of $X$.

**6.1-3** Let the moment-generating function $M(t)$ of $X$ exist for $-h < t < h$. Consider the function $R(t) = \ln M(t)$. The first two derivatives of $R(t)$ are, respectively,

$$R'(t) = \frac{M'(t)}{M(t)} \quad \text{and} \quad R''(t) = \frac{M(t)M''(t) - [M'(t)]^2}{[M(t)]^2}.$$

Setting $t = 0$, show that

**(a)** $\mu = R'(0)$.

**(b)** $\sigma^2 = R''(0)$.

**6.1-4** Find the m.g.f. of the Poisson distribution and use the results of Exercise 6.1-3 to find its mean and variance. HINT: In the summation representing the m.g.f. write $e^{tx}\lambda^x$ as $(\lambda e^t)^x$ to help you recognize the series representation of $e^{\lambda e^t}$.

**6.1-5** Find the m.g.f. of the gamma distribution and use the results of Exercise 6.1-3 to find its mean and variance. HINT: In the integral representing the m.g.f., change variables by letting $y = (1 - \theta t)x/\theta$, where $1 - \theta t > 0$.

**6.1-6** What are the distributions associated with the following moment-generating functions?

    **(a)** $M(t) = \dfrac{1}{1 - 4t}, \qquad t < \dfrac{1}{4}$.

    **(b)** $M(t) = \dfrac{2}{2 - t}, \qquad t < 2$.

    **(c)** $M(t) = (1 - 5t)^{-10}, \qquad t < 1/5$.

    **(d)** $M(t) = (1 - 2t)^{-12}, \qquad t < 1/2$.

    **(e)** $M(t) = e^{-7t+8t^2}, \qquad -\infty < t < \infty$.

**6.1-7** Use a CAS to find the moment-generating function, mean, and variance for the binomial distribution, $b(n, p)$. (See Example 6.1-3.)

**6.1-8** Use a CAS to find the moment-generating function, mean, and variance for the distribution given in Example 6.1-4.

## 6.2   M.G.F. OF LINEAR FUNCTIONS

We are able to find rather easily the m.g.f. of a linear combination of several independent random variables. We give this in detail for two independent discrete-type random variables, $X_1$ and $X_2$, with p.m.f.s $f_1(x_1)$, $x_1 \in S_1$, and $f_2(x_2)$, $x_2 \in S_2$, and m.g.f.s $M_1(t)$ and $M_2(t)$, respectively. However, it is easy to see how this not only extends to $n$ discrete-type independent random variables, but to continuous-type independent random variables with integrals replacing summations. In that continuous case, $X_1, X_2, \ldots, X_n$ are said to be **independent** if their **joint p.d.f.** is equal to

$$f_1(x_1)f_2(x_2) \cdots f_n(x_n),$$

where $f_i(x_i)$ is the p.d.f. of $X_i, i = 1, 2, \ldots, n$.

With $Y = a_1 X_1 + a_2 X_2$, we have

$$M_Y(t) = E(e^{tY}) = E[e^{t(a_1 X_1 + a_2 X_2)}]$$

$$= \sum_{x_1 \varepsilon S_1} \sum_{x_2 \varepsilon S_2} e^{a_1 t x_1} f_1(x_1) e^{a_2 t x_2} f_2(x_2)$$

$$= \left[ \sum_{x_1 \varepsilon S_1} e^{(a_1 t)x_1} f_1(x_1) \right] \left[ \sum_{x_2 \varepsilon S_2} e^{(a_2 t)x_2} f_2(x_2) \right].$$

However, since

$$M_i(t) = \sum_{x_i \varepsilon S_i} e^{tx_i} f_i(x_i),$$

then

$$M_i(a_i t) = \sum_{x_i \varepsilon S_i} e^{(a_i t)x_i} f_i(x_i), \qquad i = 1, 2.$$

Thus

$$M_Y(t) = M_1(a_1 t) M_2(a_2 t).$$

**Remark** In this proof, we see that

$$E[e^{t(a_1 X_1 + a_2 X_2)}] = E[e^{ta_1 X_1} e^{ta_2 X_2}] = E[e^{ta_1 X_1}] E[e^{ta_2 X_2}].$$

More generally, if $X_1, X_2, \ldots, X_n$ are independent random variables,

$$E[u_1(X_1) u_2(X_2) \cdots u_n(X_n)] = E[u_1(X_1)] E[u_2(X_2)] \cdots E[u_n(X_n)].$$

This was also noted in a Remark in Section 2.5.

With $M_i(t)$ being the m.g.f. associated with the p.m.f. (or p.d.f.) of $X_i$, $i = 1, 2, \ldots, n$, the obvious extension is

$$M_Y(t) = \prod_{i=1}^{n} M_i(a_i t),$$

where $Y = a_1 X_1 + a_2 X_2 + \cdots + a_n X_n$ is a linear combination of the independent random variables $X_1, X_2, \ldots, X_n$. In particular, if $a_1 = a_2 = \cdots = a_n = 1$, then

$$M_Y(t) = M_1(t) M_2(t) \cdots M_n(t).$$

**EXAMPLE 6.2-1** Let $X_1, X_2, X_3$ be three independent Poisson random variables with respective means $\lambda_1 = 3, \lambda_2 = 5, \lambda_3 = 2$. Then, if $Y = X_1 + X_2 + X_3$, we have

$$M_Y(t) = e^{3(e^t - 1)} e^{5(e^t - 1)} e^{2(e^t - 1)} = e^{10(e^t - 1)}.$$

This is the m.g.f. of a Poisson random variable with $\lambda = 10$. So, from the uniqueness property of the m.g.f., $Y$ must have a Poisson distribution with $\lambda = 10$.

The preceding example can be generalized.

**Theorem 6.2-1**    If $X_1, X_2, \ldots, X_n$ are $n$ independent Poisson random variables with respective parameters $\lambda_1, \lambda_2, \ldots, \lambda_n$, then $Y = X_1 + X_2 + \cdots + X_n$ has a Poisson distribution with mean $\lambda_1 + \lambda_2 + \cdots + \lambda_n$.

The proof is left as an exercise (Exercise 6.2-3).

**EXAMPLE 6.2-2** Let $X_1, X_2, X_3, X_4$ be four independent chi-square random variables with respective degrees of freedom of $r_1 = 4, r_2 = 2, r_3 = 5, r_4 = 4$. Then the sum $Y = X_1 + X_2 + X_3 + X_4$ has m.g.f.

$$M_Y(t) = (1 - 2t)^{-4/2} (1 - 2t)^{-2/2} (1 - 2t)^{-5/2} (1 - 2t)^{-4/2} = (1 - 2t)^{-15/2},$$

which is the m.g.f. of a chi-square distribution with 15 degrees of freedom. Hence, from the uniqueness of the m.g.f., $Y$ must have that distribution, namely $\chi^2(15)$.

The generalization of Example 6.2-2 is Theorem 4.1-1, which the reader is asked to prove in Exercise 6.2-8.

If $X_1, X_2, \ldots, X_n$ are observations of a random sample from a distribution with m.g.f. $M(t)$ so that each $X_i$ has that m.g.f., then, with $a_1 = a_2 = \cdots = a_n = 1$, the sum $Y = X_1 + X_2 + \cdots + X_n$ has the m.g.f.

$$M_Y(t) = [M(t)]^n.$$

Moreover, if $\overline{X} = (X_1 + X_2 + \cdots + X_n)/n$ so that $a_1 = a_2 = \cdots = a_n = 1/n$, then the m.g.f. of the sample mean is

$$M_{\overline{X}}(t) = \left[ M\left(\frac{t}{n}\right) \right]^n.$$

**EXAMPLE 6.2-3** Let $X_1, X_2, \ldots, X_n$ denote a random sample from

$$f(x) = p^x(1-p)^{1-x}, \qquad x = 0, 1,$$

with m.g.f.

$$M(t) = E(e^{tX}) = (1-p) + pe^t.$$

Hence the sum $Y = X_1 + X_2 + \cdots + X_n$ has m.g.f.

$$M_Y(t) = [1 - p + pe^t]^n,$$

which is the m.g.f. of $b(n, p)$. So again we see $Y$ has that binomial distribution. Note that we can think of each $X_i$ as being a success ($X_i = 1$) or failure ($X_i = 0$) with respective probabilities $p$ and $1 - p$. Then $Y$ is equal to the number of successes in these $n$ independent trials.

**EXAMPLE 6.2-4** Let $\overline{X}$ be the mean of a random sample of size $n$ from $N(\mu, \sigma^2)$. Then since

$$M(t) = e^{\mu t + \sigma^2 t^2/2}$$

we have

$$M_{\overline{X}}(t) = \left[ e^{\mu(t/n) + \sigma^2(t/n)^2/2} \right]^n = e^{\mu t + (\sigma^2/n)t^2/2},$$

which is the m.g.f. of $N(\mu, \sigma^2/n)$. Hence $\overline{X}$ has this normal distribution.

This last example is very important because we now know that if the random sample arises from a normal distribution, then

$$Z = \frac{\overline{X} - \mu}{\sigma/\sqrt{n}} \quad \text{is} \quad N(0, 1).$$

This result is so important, we state it as a theorem.

**Theorem 6.2-2**

If $\overline{X}$ is the mean of a random sample taken from a normal distribution, $N(\mu, \sigma^2)$, then

$$Z = \frac{\overline{X} - \mu}{\sigma/\sqrt{n}} \quad \text{is} \quad N(0, 1).$$

The Central Limit Theorem, which is proved in the next section, states that if the random sample arises from any distribution with mean $\mu$ and variance $\sigma^2$, then

$$Z = \frac{\overline{X} - \mu}{\sigma/\sqrt{n}} \quad \text{is approximately} \quad N(0, 1),$$

provided the sample size $n$ is large enough. Now we find that if the underlying distribution is normal, then $Z$ is exactly normal for every possible sample size $n = 1, 2, 3, \cdots$. Incidently, the underlying normal distribution is the only distribution that achieves that result. So while this latter statement is more difficult to prove, we accept it and say that the exact normality of $Z$ *characterizes* the normal distribution.

We close this section by making an important observation about the distribution of a linear function of independent normally distributed random variables. Let $X_i$ be $N(\mu_i, \sigma_i^2)$, $i = 1, 2, \ldots, n$, and let $X_1, X_2, \ldots, X_n$ be independent. If $Y = \sum_{i=1}^{n} a_i X_i$, then

$$M_Y(t) = \prod_{i=1}^{n} M_i(a_i t) = \prod_{i=1}^{n} e^{\mu_i(a_i t) + \sigma_i^2(a_i t)^2/2}$$

$$= e^{(\Sigma a_i \mu_i)t + (\Sigma a_i^2 \sigma_i^2)t^2/2}.$$

However, this is the m.g.f. of a $N\left(\sum_{i=1}^{n} a_i \mu_i, \sum_{i=1}^{n} a_i^2 \sigma_i^2\right)$ distribution, and hence the linear function $Y$ of the $n$ independent normally distributed variables has this normal distribution. This, of course, is Theorem 4.1-2.

**EXAMPLE 6.2-5**

Let $X_1$ be $N(70, 100)$ and $X_2$ be $N(85, 200)$. If $X_1$ and $X_2$ are independent and $Y = X_1 + 2X_2$, then $\mu_Y = 70 + 2(85) = 240$ and $\sigma_Y^2 = 100 + 2^2(200) = 900$ and so $Y$ is $N(240, 900)$. For example,

$$P(210 < X_1 + 2X_2 < 300) = P\left(\frac{210 - 240}{30} < \frac{Y - 240}{30} < \frac{300 - 240}{30}\right)$$

$$= \Phi(2) - \Phi(-1) = 0.9772 - 0.1587 = 0.8185.$$

Also note that the distribution of $W = X_1 - 2X_2$ is $N(-100, 900)$ since the mean is $\mu_W = 70 - 2(85) = -100$ and the variance is $\sigma_W^2 = 100 + (-2)^2(200) = 900$.

## EXERCISES 6.2

**6.2-1** Let $X_1$ and $X_2$ be independent and have binomial distributions $b(n_1, p)$ and $b(n_2, p)$. Find the moment-generating function of $Y = X_1 + X_2$. How is $Y$ distributed?

**6.2-2** Let $X_1, X_2, X_3$ be mutually independent random variables with Poisson distributions having means 2, 1, 4, respectively.

**(a)** Find the moment-generating function of the sum $Y = X_1 + X_2 + X_3$.

**(b)** How is $Y$ distributed?

**(c)** Compute $P(3 \le Y \le 9)$.

**6.2-3** Prove Theorem 6.2-1, showing that the sum of $n$ independent Poisson random variables with respective means $\lambda_1, \lambda_2, \ldots, \lambda_n$ is Poisson with mean

$$\lambda_1 + \lambda_2 + \cdots + \lambda_n.$$

**6.2-4** Let $X_1, X_2, X_3, X_4, X_5$ be a random sample of size 5 from a geometric distribution with $p = 1/3$.

(a) Find the moment-generating function of $Y = X_1 + X_2 + X_3 + X_4 + X_5$.

(b) How is $Y$ distributed?

**6.2-5** Let $W = X_1 + X_2 + \cdots + X_h$, a sum of $h$ mutually independent and identically distributed exponential random variables with mean $\theta$. Show that $W$ has a gamma distribution with mean $h\theta$.

**6.2-6** The moment-generating function of $X$ is

$$M_X(t) = \left(\frac{1}{4}\right)(e^t + e^{2t} + e^{3t} + e^{4t});$$

the moment-generating function of $Y$ is

$$M_Y(t) = \left(\frac{1}{3}\right)(e^t + e^{2t} + e^{3t});$$

$X$ and $Y$ are independent random variables. Let $W = X + Y$.

(a) Find the moment-generating function of $W$.

(b) Give the p.m.f. of $W$; that is, determine $P(W = w)$, $w = 2, 3, \ldots, 7$, from the moment-generating function of $W$.

**6.2-7** Let $X_1, X_2, \ldots, X_{16}$ be a random sample from $N(77, 25)$, a normal distribution with $\mu = 77$ and $\sigma^2 = 25$. Compute

(a) $P(77 < \overline{X} < 79.5)$.    (b) $P(74.2 < \overline{X} < 78.4)$.

**6.2-8** If $X_1, X_2, \ldots, X_n$ are independent chi-square random variables with degrees of freedom $r_1, r_2, \ldots, r_n$, show that

$$Y = X_1 + X_2 + \cdots + X_n \quad \text{is} \quad \chi^2\left(\sum_{i=1}^{n} r_i\right).$$

**6.2-9** Let $Z_1, Z_2, \ldots, Z_7$ be a random sample from the standard normal distribution $N(0, 1)$. Let $W = Z_1^2 + Z_2^2 + \cdots + Z_7^2$. Find $P(1.69 < W < 14.07)$.

**6.2-10** If $X_1, X_2, \ldots, X_{16}$ is a random sample of size $n = 16$ from the normal distribution $N(50, 100)$, determine

$$P\left(796.2 \le \sum_{i=1}^{16}(X_i - 50)^2 \le 2630\right).$$

**6.2-11** Let $X_1$ and $X_2$ be independent with normal distributions $N(4, 1)$ and $N(5, 4)$, respectively. Find $P(X_1 < X_2)$. HINT: First note that $P(X_1 < X_2) = P(X_1 - X_2 < 0)$ and then find the distribution of $X_1 - X_2$.

**6.2-12** Let the independent random variables $X_1$ and $X_2$ be $b(n_1, p = 1/2)$ and $b(n_2, p = 1/2)$, respectively. Find the m.g.f. of $Y = X_1 + X_2 + n_2$ and thus show that $Y$ is $b(n_1 + n_2, p = 1/2)$.

**6.2-13** Let $X_1, X_2, X_3$ be a random sample from $N(2, 9)$. Compute

$$P(2X_1 + 4X_2 - 4X_3 > 8).$$

**6.2-14** Roll a fair six-sided die until each face has been observed at least once. Call the first new face 1, the second 2, etc. Let $X_i$ equal the number of additional rolls needed to observe the $i$th new face after $i - 1$ faces have been observed, $i = 2, 3, 4, 5, 6$. (The number of rolls needed to observe the first new face is $X_1 = 1$.) Note that the distribution of $X_i$ is geometric with $p_i = (7 - i)/6$, $i = 1, 2, \ldots, 6$. Let $Y = X_1 + X_2 + \cdots + X_6$.

**(a)** Define the moment-generating function of $Y$.

**(b)** Find the mean of $Y$. Either use the fact that

$$\mu_Y = E(Y) = E(X_1) + E(X_2) + \cdots + E(X_6)$$

or use the m.f.g. of $Y$ and a CAS.

**(c)** Find the variance of $Y$. Either use the fact that

$$\sigma_Y^2 = \text{Var}(Y) = \text{Var}(X_1) + \text{Var}(X_2) + \cdots + \text{Var}(X_6)$$

or use a CAS.

**(d)** Simulate this experiment to support your theoretical answers.

**6.2-15** Let $X_1$ and $X_2$ be independent random variables such that $Y = X_1 + X_2$ is Poisson with mean $\lambda$ and $X_1$ is Poisson with mean $\lambda_1$, where $\lambda > \lambda_1$. Show that $X_2$ is Poisson with mean $\lambda - \lambda_1$.

**6.2-16** Let $X_1$ and $X_2$ be independent random variables such that $Y = X_1 + X_2$ is $\chi^2(r)$ and $X_1$ is $\chi^2(r_1)$, where $r > r_1$. show that $X_2$ is $\chi^2(r - r_1)$.

## 6.3 LIMITING MOMENT-GENERATING FUNCTIONS

We would like to prove that the binomial distribution can be approximated by the Poisson distribution when $n$ is sufficiently large and $p$ fairly small by showing that the limit, under certain conditions, of the binomial m.g.f. is that of the Poisson.

Consider the moment-generating function of $Y$, which is $b(n, p)$. We shall take the limit of this as $n \to \infty$ such that $np = \lambda$ is a constant; thus $p \to 0$. The moment-generating function of $Y$ is

$$M(t) = (1 - p + pe^t)^n.$$

Because $p = \lambda/n$, we have that

$$M(t) = \left[1 - \frac{\lambda}{n} + \frac{\lambda}{n}e^t\right]^n = \left[1 + \frac{\lambda(e^t - 1)}{n}\right]^n.$$

Since

$$\lim_{n \to \infty} \left(1 + \frac{b}{n}\right)^n = e^b,$$

we have

$$\lim_{n \to \infty} M(t) = e^{\lambda(e^t - 1)},$$

which exists for all real $t$. But this is the moment-generating function of a Poisson random variable with mean $\lambda$. Hence this Poisson distribution seems like a reasonable approximation to the binomial distribution when $n$ is large and $p$ is small. This approximation is usually found to be fairly successful if $n \geq 20$ and $p \leq 0.05$ and very successful if $n \geq 100$ and $p \leq 0.10$, but it is not bad if these bounds are violated somewhat. That is, it could be used in other situations too; we only want to stress that the approximation becomes better with larger $n$ and smaller $p$.

The preceding result illustrates the theorem we now state without proof.

**Theorem 6.3-1**   If a sequence of moment-generating functions approaches a certain one, say $M(t)$, then the limit of the corresponding distributions must be the distribution corresponding to $M(t)$.

**Remark**   This theorem certainly appeals to one's intuition! In a more advanced course, the proof of this theorem is given and there the existence of the moment-generating function is not even needed, for we would use the characteristic function $\phi(t) = E(e^{itX})$ instead.

The following example illustrates graphically the convergence of the binomial moment-generating functions to that of a Poisson distribution.

**EXAMPLE 6.3-1**   Consider the moment-generating function for the Poisson distribution with $\lambda = 5$ and those for three binomial distributions for which $np = 5$, namely, $b(50, 1/10)$, $b(100, 0.05)$, and $b(200, 0.025)$. These four moment-generating functions are, respectively:

$$M(t) = e^{5(e^t - 1)}$$

$$M(t) = (0.9 + 0.1e^t)^{50}$$

$$M(t) = (0.95 + 0.05e^t)^{100}$$

$$M(t) = (0.975 + 0.025e^t)^{200}$$

The graphs of these moment-generating functions are shown in Figure 6.3-1. Although the proof and the figure show the convergence of the binomial moment-generating functions to that of the Poisson distribution, the other graphs in Figure 6.3-1 show more clearly how the Poisson distribution can be used to approximate binomial probabilities with large $n$ and small $p$.

The next example gives a numerical approximation.

**EXAMPLE 6.3-2**   Let $Y$ be $b(50, 1/25)$. Then

$$P(Y \le 1) = \left(\frac{24}{25}\right)^{50} + 50\left(\frac{1}{25}\right)\left(\frac{24}{25}\right)^{49} = 0.400.$$

Since $\lambda = np = 2$, the Poisson approximation is

$$P(Y \le 1) \approx 0.406,$$

from Table III in the Appendix.

Theorem 6.3-1 is used to prove the Central Limit Theorem. To help in the understanding of this proof, let us first consider a different problem, that of the limiting distribution of the mean $\overline{X}$ of a random sample $X_1, X_2, \ldots, X_n$, from a

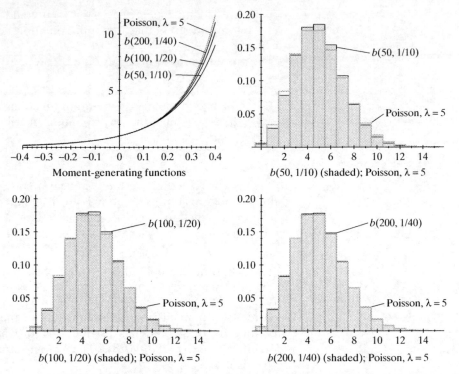

**Figure 6.3-1** Poisson approximation of the binomial distribution

distribution with mean $\mu$. If the distribution has moment-generating function $M(t)$, the moment-generating function of $\overline{X}$ is $[M(t/n)]^n$. But, by Taylor's expansion, there exists a number $t_1$ between 0 and $t/n$ such that

$$M\left(\frac{t}{n}\right) = M(0) + M'(t_1)\frac{t}{n}$$

$$= 1 + \frac{\mu t}{n} + \frac{[M'(t_1) - M'(0)]t}{n}$$

because $M(0) = 1$ and $M'(0) = \mu$. Since $M'(t)$ is continuous at $t = 0$ and since $t_1 \to 0$ as $n \to \infty$, we know that

$$\lim_{n \to \infty}[M'(t_1) - M'(0)] = 0.$$

Thus, using a result from advanced calculus, we obtain

$$\lim_{n \to \infty}\left[M\left(\frac{t}{n}\right)\right]^n = \lim_{n \to \infty}\left\{1 + \frac{\mu t}{n} + \frac{[M'(t_1) - M'(0)]t}{n}\right\}^n$$

$$= e^{\mu t}$$

for all real $t$. But this limit is the moment-generating function of a degenerate distribution with all of the probability on $\mu$. Accordingly, $\overline{X}$ has this limiting distribution, indicating that $\overline{X}$ converges to $\mu$ in a certain sense. This is one form of the law of large numbers.

We have seen that, in some probability sense, $\overline{X}$ converges to $\mu$ in the limit, or, equivalently, $\overline{X} - \mu$ converges to zero. Let us multiply the difference $\overline{X} - \mu$ by some function of $n$ so that the result will not converge to zero. In our search for such a function, it is natural to consider

$$W = \frac{\overline{X} - \mu}{\sigma/\sqrt{n}} = \frac{\sqrt{n}(\overline{X} - \mu)}{\sigma} = \frac{Y - n\mu}{\sqrt{n}\,\sigma},$$

where $Y$ is the sum of the observations of the random sample. The reason for this is that $W$ is a standardized random variable and has mean 0 and variance 1 for each positive integer $n$. We are now ready to prove the Central Limit Theorem.

**Theorem 6.3-2**

**(Central Limit Theorem)** If $\overline{X}$ is the mean of a random sample $X_1, X_2, \ldots, X_n$ of size $n$ from a distribution with a finite mean $\mu$ and a finite positive variance $\sigma^2$, then the distribution of

$$W = \frac{\overline{X} - \mu}{\sigma/\sqrt{n}} = \frac{\displaystyle\sum_{i=1}^{n} X_i - n\mu}{\sqrt{n}\,\sigma}$$

is $N(0, 1)$ in the limit as $n \to \infty$.

**Proof**

We first consider

$$E[\exp(tW)] = E\left\{\exp\left[\left(\frac{t}{\sqrt{n}\sigma}\right)\left(\sum_{i=1}^{n} X_i - n\mu\right)\right]\right\}$$

$$= E\left\{\exp\left[\left(\frac{t}{\sqrt{n}}\right)\left(\frac{X_1 - \mu}{\sigma}\right)\right] \cdots \exp\left[\left(\frac{t}{\sqrt{n}}\right)\left(\frac{X_n - \mu}{\sigma}\right)\right]\right\}$$

$$= E\left\{\exp\left[\left(\frac{t}{\sqrt{n}}\right)\left(\frac{X_1 - \mu}{\sigma}\right)\right]\right\} \cdots E\left\{\exp\left[\left(\frac{t}{\sqrt{n}}\right)\left(\frac{X_n - \mu}{\sigma}\right)\right]\right\},$$

which follows from the independence of $X_1, X_2, \ldots, X_n$. Then

$$E[\exp(tW)] = \left[m\left(\frac{t}{\sqrt{n}}\right)\right]^n, \qquad -h < \frac{t}{\sqrt{n}} < h,$$

where

$$m(t) = E\left\{\exp\left[t\left(\frac{X_i - \mu}{\sigma}\right)\right]\right\}, \qquad -h < t < h,$$

is the common moment-generating function of each

$$Y_i = \frac{X_i - \mu}{\sigma}, \qquad i = 1, 2, \ldots, n.$$

Since $E(Y_i) = 0$ and $E(Y_i^2) = 1$, it must be that

$$m(0) = 1, \qquad m'(0) = E\left(\frac{X_i - \mu}{\sigma}\right) = 0, \qquad m''(0) = E\left[\left(\frac{X_i - \mu}{\sigma}\right)^2\right] = 1.$$

Hence, using Taylor's formula with a remainder, there exists a number $t_1$ between 0 and $t$ such that

$$m(t) = m(0) + m'(0)t + \frac{m''(t_1)t^2}{2} = 1 + \frac{m''(t_1)t^2}{2}.$$

By adding and subtracting $t^2/2$, we have that

$$m(t) = 1 + \frac{t^2}{2} + \frac{[m''(t_1) - 1]t^2}{2}.$$

Using this expression of $m(t)$ in $E[\exp(tW)]$, we can represent the moment-generating function of $W$ by

$$E[\exp(tW)] = \left\{ 1 + \frac{1}{2}\left(\frac{t}{\sqrt{n}}\right)^2 + \frac{1}{2}[m''(t_1) - 1]\left(\frac{t}{\sqrt{n}}\right)^2 \right\}^n$$

$$= \left\{ 1 + \frac{t^2}{2n} + \frac{[m''(t_1) - 1]t^2}{2n} \right\}^n, \qquad -\sqrt{n}h < t < \sqrt{n}h,$$

where now $t_1$ is between 0 and $t/\sqrt{n}$. Since $m''(t)$ is continuous at $t = 0$ and $t_1 \to 0$ as $n \to \infty$, we have that

$$\lim_{n \to \infty} [m''(t_1) - 1] = 1 - 1 = 0.$$

Thus, using a result from advanced calculus, we have that

$$\lim_{n \to \infty} E[\exp(tW)] = \lim_{n \to \infty} \left\{ 1 + \frac{t^2}{2n} + \frac{[m''(t_1) - 1]t^2}{2n} \right\}^n$$

$$= \lim_{n \to \infty} \left\{ 1 + \frac{t^2}{2n} \right\}^n = e^{t^2/2}$$

for all real $t$, which is the m.g.f. of the standard normal distribution, $N(0, 1)$. This means that the limiting distribution of

$$W = \frac{\overline{X} - \mu}{\sigma/\sqrt{n}} = \frac{\sum\limits_{i=1}^{n} X_i - n\mu}{\sqrt{n}\,\sigma}$$

is $N(0, 1)$. This completes the proof of the Central Limit Theorem.

**Remark**  In this proof, we have assumed the existence of the m.g.f. and hence all of the moments. In a more advanced course, we would use the characteristic function $\phi(t) = E(e^{itX})$ and only need the existence of the first two moments as stated in the theorem. With this change, the proof we give here is exactly the one given in that advanced course.

To help appreciate the proof of the Central Limit Theorem, the following example graphically illustrates the convergence of the moment-generating functions for two distributions.

**EXAMPLE 6.3-3**  Let $X_1, X_2, \ldots, X_n$ be a random sample of size $n$ from a $\chi^2(1)$ distribution. The moment-generating function of $(\overline{X} - 1)/(\sqrt{2}/\sqrt{n})$ is

$$M_n(t) = \frac{e^{-t\sqrt{n}/\sqrt{2}}}{(1 - \sqrt{2}\,t/\sqrt{n}\,)^{n/2}}.$$

The Central Limit Theorem says that, as $n$ increases, this moment-generating function approaches that of the standard normal distribution, namely,

$$M(t) = e^{t^2/2}.$$

The moment-generating functions for $M(t)$ and $M_n(t), n = 20, 100, 250$, are shown in Figure 6.3-2(a). Shown in Figure 6.3-2(b) are the p.d.f.s of $(\overline{X} - 1)/(\sqrt{2}/\sqrt{n}\,)$ along with the $N(0, 1)$ p.d.f. Note that because the underlying distribution is skewed (sketch a graph of the $\chi^2(1)$ p.d.f.), the convergence is quite slow in this case.

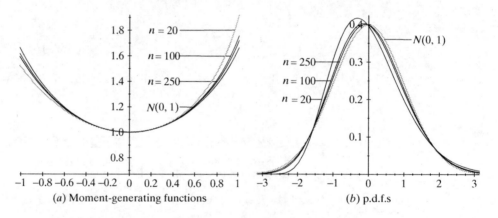

**Figure 6.3-2** Sampling from a $\chi^2(1)$ distribution

**EXAMPLE 6.3-4**  Let the random variable $X$ have a $U$-shaped distribution with p.d.f. $f(x) = (3/2)x^2$, $-1 < x < 1$. For this distribution, $\mu = 0$ and $\sigma^2 = 3/5$. Its moment-generating function, for $t \neq 0$, is

$$M(t) = \frac{3}{2} \frac{e^t t^2 - 2e^t t + 2e^t - e^{-t}t^2 - 2e^{-t}t - 2e^{-t}}{t^3}.$$

Of course, $M(0) = 1$. The moment-generating function of

$$W_n = \frac{\overline{X} - 0}{\sqrt{(3/5)/n}}$$

is

$$E[e^{tW_n}] = \left\{ E\left[ \exp\left( \sqrt{\frac{5}{3n}}\, t \right) \right] \right\}^n = \left[ M\left( \sqrt{\frac{5}{3n}}\, t \right) \right]^n$$

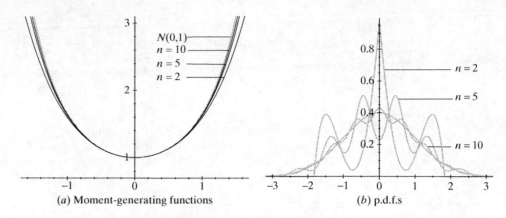

(a) Moment-generating functions          (b) p.d.f.s

**Figure 6.3-3** Sampling from a $U$-shaped distribution

The graphs of these moment-generating functions when $n = 2, 5, 10$ and the graph of the moment-generating function for the standard normal distribution are shown in Figure 6.3-3($a$). A comparison of the p.d.f.s of

$$W_n = \frac{\overline{X} - 0}{\sqrt{(3/5)/n}}$$

with that of the $N(0, 1)$ distribution is shown in Figure 6.3-3($b$). These p.d.f.s were found using *Maple*.

## EXERCISES 6.3

**6.3-1** With probability 0.001, a prize of \$499 is won in the Michigan Daily Lottery when a \$1 straight bet is placed. Let $Y$ equal the number of \$499 prizes won by a gambler after placing $n$ straight bets. Note that $Y$ is $b(n, 0.001)$. After placing $n = 2000$ \$1 bets, a gambler is behind if $\{Y \leq 4\}$. Use the Poisson distribution to approximate $P(Y \leq 4)$ when $n = 2000$.

**6.3-2** Suppose that the probability of suffering a side effect from a certain flu vaccine is 0.005. If 1000 persons are inoculated, find approximately the probability that

**(a)** At most 1 person suffers.

**(b)** 4, 5, or 6 persons suffer.

**6.3-3** Approximate $P(39.75 \leq \overline{X} \leq 41.25)$, where $\overline{X}$ is the mean of a random sample of size 32 from a distribution with mean $\mu = 40$ and variance $\sigma^2 = 8$.

**6.3-4** Let $X_1, X_2, \ldots, X_{18}$ be a random sample of size 18 from a chi-square distribution with $r = 1$. Recall that $\mu = 1, \sigma^2 = 2$.

**(a)** How is $Y = \sum_{i=1}^{18} X_i$ distributed?

**(b)** Using the result of part (a), we see from Table IV in the Appendix that

$$P(Y \leq 9.390) = 0.05 \qquad \text{and} \qquad P(Y \leq 34.80) = 0.99.$$

Compare these two probabilities with the approximations found using the Central Limit Theorem.

**6.3-5** A random sample of size $n = 18$ is taken from the distribution with p.d.f $f(x) = 1 - x/2$, $0 \leq x \leq 2$.

**(a)** Find $\mu$ and $\sigma^2$.

**(b)** Find, approximately, $P(2/3 \leq \overline{X} \leq 5/6)$.

**6.3-6** Let $Y$ be $b(72, 1/3)$. Approximate $P(22 \leq Y \leq 28)$.

**6.3-7** Let $Y$ be the sum of a random sample of size $n = 12$ found by rolling a fair die 12 independent times.

(a) Approximate $P(36 \le Y \le 48)$. HINT: Since the event of interest is $\{Y = 36, 37, \ldots, 48\}$, rewrite the probability as $P(35.5 < Y < 48.5)$.

(b) If possible (perhaps using a CAS) show that

$$P(36 \le Y \le 48) = \frac{49{,}300{,}925}{68{,}024{,}448} = 0.72475.$$

**6.3-8** Let $X$ have a Poisson distribution with mean $\mu = n$. Show that the limit of the m.g.f. of $Z = (X - n)/\sqrt{n}$ is $e^{t^2/2}$. This is the reason we can approximate the Poisson distribution with a large mean by the normal distribution.

## 6.4  USE OF ORDER STATISTICS IN NON-REGULAR CASES

Let us now consider by example how we can use the order statistics to find **confidence intervals for non-regular parameters**. In the case that the random sample $X_1, X_2, \ldots, X_n$ arises from a distribution with p.d.f.

$$f(x; \theta) = \frac{1}{\theta}, \qquad 0 \le x \le \theta,$$

we know that the maximum likelihood estimator of $\theta$ is the largest order statistic, $Y_n = \max(X_i)$. The distribution function of $Y_n$ is, for $0 \le y \le \theta$,

$$G(y) = P(Y_n \le y) = P(\text{all } X_i \le y) = \left( \int_0^y \frac{1}{\theta}\, dx \right)^n = \left( \frac{y}{\theta} \right)^n,$$

from the independence of $X_1, X_2, \ldots, X_n$. Thus the p.d.f. of $Y_n$ is

$$g(y) = G'(y) = \frac{n y^{n-1}}{\theta^n}, \qquad 0 \le y \le \theta.$$

Now consider

$$P(c \le Y_n \le \theta) = \int_c^\theta \frac{n y^{n-1}}{\theta^n}\, dy = 1 - \alpha,$$

where $\alpha$ is a small number like 0.05. This becomes

$$1 - \left( \frac{c}{\theta} \right)^n = 1 - \alpha \qquad \text{or} \qquad c = (\alpha^{1/n})\theta.$$

Thus

$$1 - \alpha = P(\alpha^{1/n}\theta \le Y_n \le \theta)$$

$$= P\left( \frac{\alpha^{1/n}}{Y_n} \le \frac{1}{\theta} \le \frac{1}{Y_n} \right)$$

$$= P\left( Y_n \le \theta \le \frac{Y_n}{\alpha^{1/n}} \right).$$

That is, $[y_n, y_n/\alpha^{1/n}]$ is a $100(1 - \alpha)\%$ confidence interval for $\theta$. Say $n = 10$ and $\alpha = 0.05$, then $\alpha^{1/n} = 0.05^{1/10} = 0.74$. Suppose $Y_{10}$ was observed to be $y_{10} = 2.10$. Then

$$2.10 \le \theta \le \frac{2.10}{0.74} \qquad \text{or} \qquad 2.10 \le \theta \le 2.84 \qquad \text{or} \qquad [2.10, 2.84]$$

is a 95% confidence interval for $\theta$.

The reason that these confidence intervals tend to be shorter than those based on $\overline{X}$ for $\mu$ is explained by considering the variances of $Y_n$ and $\overline{X}$. Here

$$E(Y_n) = \int_0^\theta y \, \frac{ny^{n-1}}{\theta^n} \, dy = \frac{n}{n+1}\theta$$

and

$$\mathrm{Var}(Y_n) = \int_0^\theta y^2 \, \frac{ny^{n-1}}{\theta^n} \, dy - \left(\frac{n}{n+1}\right)^2 \theta^2$$

$$= \frac{n}{n+2}\theta^2 - \frac{n^2}{(n+1)^2}\theta^2 = \frac{n}{(n+1)^2(n+2)}\theta^2.$$

That is, $\mathrm{Var}(Y_n)$ essentially has an $n^2$ in the denominator and $\mathrm{Var}(\overline{X}) = \sigma^2/n = (\theta^2/12)/n$, for this underlying p.d.f. with $\mathrm{Var}(X) = \theta^2/12$. So if $n$ is at all large, $\mathrm{Var}(Y_n) \le \mathrm{Var}(\overline{X})$. Here with $n = 10$, we have

$$\mathrm{Var}(Y_n) = (0.0069)\theta^2 < (0.0083)\theta^2 = \mathrm{Var}(\overline{X}).$$

However, with $n = 100$, this inequality is

$$\mathrm{Var}(Y_n) = (0.000096)\theta^2 < (0.000833)\theta^2 = \mathrm{Var}(\overline{X}),$$

which is much more dramatic with this larger $n$.

**EXAMPLE 6.4-1** Let $X_1, X_2, \ldots, X_n$ be a random sample from the distribution with p.d.f.

$$f(x; \theta) = e^{-(x-\theta)}, \qquad \theta \le x < \infty.$$

To find the maximum likelihood estimator of $\theta$, consider

$$L(\theta) = \prod_{i=1}^n e^{-(x_i-\theta)} = e^{-\Sigma(x_i-\theta)}, \qquad \theta \le x_i < \infty.$$

So we wish to maximize $L(\theta)$ or minimize $\sum_{i=1}^n (x_i - \theta)$. We can accomplish the latter by making $\theta$ as large as possible, namely $\widehat{\theta} = \min(X_i) = Y_1$, the smallest order statistic. We can not make $\theta >$ smallest $x_i$, for then $L(\theta)$ would equal zero, which is certainly not the maximum of $L(\theta)$. Thus $\widehat{\theta}$ is really $Y_1 = \min(X_i)$, the largest possible value for $\theta$. The p.d.f. of $Y_1$ is found by considering, for $\theta \le y < \infty$, the distribution function of $Y_1$, namely

$$G(y) = P(Y_1 \le y) = 1 - P(y < \text{all } X_i)$$

$$= 1 - \left(\int_y^\infty e^{-(x-\theta)}dx\right)^n = 1 - e^{-n(y-\theta)}.$$

Hence the p.d.f. of $Y_1$ is

$$g(y) = G'(y) = ne^{-n(y-\theta)}, \qquad \theta \le y < \infty.$$

To find a confidence interval for $\theta$, we have

$$P(\theta \leq Y_1 \leq c) = 1 - \alpha$$

or, equivalently,

$$G(c) - G(\theta) = 1 - e^{-n(c-\theta)} = 1 - \alpha.$$

Thus

$$e^{-n(c-\theta)} = \alpha \qquad \text{and} \qquad c = \theta - \frac{1}{n} \ln \alpha.$$

It follows that

$$1 - \alpha = P\left( \theta \leq Y_1 \leq \theta - \frac{1}{n} \ln \alpha \right)$$

$$= P\left( -Y_1 \leq -\theta \leq -Y_1 - \frac{1}{n} \ln \alpha \right)$$

$$= P\left( Y_1 + \frac{1}{n} \ln \alpha \leq \theta \leq Y_1 \right).$$

That is, $[y_1 + \ln \alpha / n, y_1]$ is a $100(1 - \alpha)\%$ confidence interval for $\theta$. If $\alpha = 0.05$, $n = 10$, and we observe $y_1 = 2.70$, then

$$[2.70 + (-2.996)/10, 2.70] = [2.40, 2.70]$$

is a 95 percent confidence interval for $\theta$.

The reader is asked to show (see Exercise 6.4-1) that $E(Y_1) = \theta + 1/n$ and $\text{Var}(Y_1) = 1/n^2$ which is smaller than $\text{Var}(\overline{X}) = 1/n$, since $\text{Var}(X) = 1$ with this underlying distribution.

**EXAMPLE 6.4-2**   Let $Y_1$ be the smallest order statistic of a random sample of size $n$ from the shifted exponential distribution with p.d.f.

$$f(x; \theta) = e^{-(x-\theta)}, \qquad \theta \leq x < \infty.$$

In Example 6.4-1 we showed that $[y_1 + \ln(\alpha)/n, y_1]$ is a $100(1 - \alpha)\%$ confidence interval for $\theta$. Since $E(\overline{X}) = \theta + 1$, $[\overline{x} - 1 - 2s/\sqrt{n}, \overline{x} - 1 + 2s/\sqrt{n}]$ is an approximate 95% confidence interval for $\theta$. In order to compare these confidence intervals for $\theta$, we simulated 50 random samples of size $n = 30$ from this shifted exponential distribution with $\theta = 5$. Figure 6.4-1(a) shows these 50 confidence intervals $[y_1 + \ln(0.05)/30, y_1]$ and Figure 6.4-1(b) shows, for the same data, the 50 approximate 95% confidence intervals found using $\overline{x}$. (Note the difference in the scales in these two figures.) The lengths of the confidence intervals in Figure 6.4-1(a) are 0.09986 while the average of the lengths of the approximate confidence intervals in Figure 6.4-1(b) is 0.72068. This is a huge difference in average lengths. Also for this particular simulation, 47 of the confidence intervals using $y_1$ cover $\theta = 5$ while 46 of the approximate confidence intervals using $\overline{x}$ cover $\theta = 5$. Those numbers change from simulation to simulation.

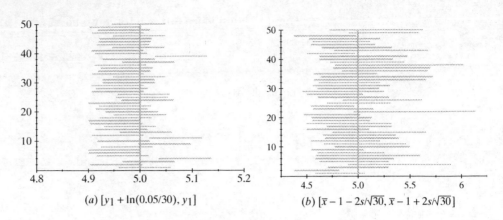

(a) $[y_1 + \ln(0.05/30), y_1]$  (b) $[\bar{x} - 1 - 2s/\sqrt{30}, \bar{x} - 1 + 2s/\sqrt{30}]$

**Figure 6.4-1**  Confidence intervals for $\theta$

## EXERCISES 6.4

**6.4-1** Let $Y_1$ be the first order statistic of a random sample of size $n$ from a distribution with p.d.f. $f(x; \theta) = e^{-(x-\theta)}$, $\theta \le x \le \infty$.

(a) Define the p.d.f. of $Y_1$.

(b) Show that $E(Y_1) = \theta + 1/n$.

(c) Show that $\text{Var}(Y_1) = 1/n^2$.

**6.4-2** Example 6.4-2 shows the results of a simulation for comparing confidence intervals for $\theta$ when sampling from a shifted exponential distribution. Write a computer program to do this simulation and interpret your output.

**6.4-3** Let $Y_1 < Y_2$ be the $n = 2$ order statistics of a random sample $X_1, X_2$ from a distribution with p.d.f. $f(x; \theta) = 1/2\theta$, $-\theta < x < \theta$.

(a) Show that $Z = \max(-Y_1, Y_2)$ is the maximum likelihood estimator of $\theta$.

(b) Show that the distribution function of $Z$ is

$$P(Z \le z) = P(-z \le Y_1 \text{ and } Y_2 \le z) = (z/\theta)^2, \qquad 0 \le z < \theta.$$

(c) If we reject $H_0 : \theta = 1$ if $Z \le c$ or $Z \ge 1$, find $c$ so that $\alpha = 0.05$.

# CHAPTER
# SIX COMMENTS

As pointed out in Section 6.4, often in non-regular cases of estimation, estimators based upon order statistics are better than those based upon $\overline{X}$. That is, they have smaller variances and shorter confidence intervals than those estimators associated with $\overline{X}$. The other sections introduce the moment-generating function and its use in studying the distribution of linear statistics, particularly that of $\overline{X}$. The proof of the Central Limit Theorem (CLT) is given by using the m.f.g.; but, in a more advanced course, the same proof would be given using the characteristic function $\phi(t) = E(e^{itX})$.

We would like to comment on the works of three early probabilists, who worked on aspects of the CLT without using the m.f.g. The first was Abraham de Moivre, a Frenchman who lived most of his gloomy life in England. After publishing *The Doctrine of Chance*, he turned to a project that Nicolaus Bernoulli had suggested to him. Using the fact that $Y$ is $b(n, p)$, he discovered that the relative frequency of successes, namely $Y/n$, had an interesting approximating distribution itself. De Moivre had discovered that well known bell-shaped curve, called the normal distribution. This distribution allowed de Moivre to determine a measure of spread, which we now call a standard deviation. Also he could determine approximately the probabilities of $Y/n$ falling in given intervals containing $p$.

The final two persons who we would like to mention are Carl Friedrich Gauss and Marquis Pierre Simon de Laplace. Gauss was 29 years Laplace's junior, and was so secretive about his work it is difficult to tell who discovered the Central Limit Theorem first. This was a generalization of de Moivre's result. In de Moivre's case he was sampling from a Bernoulli distribution where $X_i = 1$ or $0$ on the $i$th trial, then

$$\frac{Y}{n} = \sum_{i=1}^{n} \frac{X_i}{n} = \overline{X},$$

the relative frequency of success, has that approximate normal distribution of de Moivre's. Laplace and Gauss were sampling from any distribution, provided the second moment existed, and they found that the sample mean $\overline{X}$ had an approximate normal distribution. Seemingly this Central Limit Theorem was published in 1809 by Laplace just before Gauss's *Theoria Motus* in 1810. For some reason, the normal distribution is often referred to as the Gaussian distribution; people seem to forget about Laplace's contribution and, worse than that, de Moivre's original work 83 years earlier. Since then, there have been many more generalizations of the Central Limit Theorem; in particular, most estimators of parameters in regular cases have approximate normal distributions.

# APPENDIX A

# REFERENCES

Anscombe, F. J., "Graphs in Statistical Analysis," *The American Statistician,* *27,* (1973), pp. 17–21.

Aspin, A. A., "Tables for Use in Comparisons Whose Accuracy Involves Two Variances, Separately Estimated," *Biometrika*, **36** (1949), pp. 290–296.

Box, G. E. P., W. G. Hunter, and J. S. Hunter, *Statistics for Experimenters*, John Wiley & Sons, Inc., New York, 1978.

Box, G. E. P., and M. E. Muller, "A Note on the Generation of Random Normal Deviates," *Ann. Math. Statist.*," **29** (1958), p. 610.

Crisman, R., "Shortest Confidence Interval for the Standard Deviation of a Normal Distribution," *J. Undergrad. Math.*, **7**, 2 (1975), p. 57.

Guenther, William C., "Shortest Confidence Intervals," *Amer. Statist.*, **23**, 1(1969), p. 22.

Hogg, R. V., and J. Ledolter, *Applied Statistics for Engineers and Physical Scientists*, 2nd ed., Macmillan Publishing Company, New York, 1992.

Hogg, R. V., J. W. McKean, and A. T. Craig, *Introduction to Mathematical Statistics*, 6th ed., Prentice Hall, NJ, 2005.

Hogg, R. V., and A. T. Craig, "On the Decomposition of Certain Chi-Square Variables," *Ann. Math. Statist.*, **29** (1958), p. 608.

Hogg, R. V., and E. A. Tanis, *Probability and Statistical Inference*, 7th ed., Prentice Hall, NJ, 2006.

Iman, Ronald L., *A Data-Based Approach to Statistics*, Wadsworth Publishing Co., Belmont, CA, 1994.

Johnson, V. E. and J. H. Albert, *Ordinal Data Modeling*, Springer Verlag, New York, 1999.

Kaigh, W. D., "The Case of the Missing Lottery Number," *The College Mathematics Journal*, January, 2001, pp. 15–19.

Karian, Z. A., and E. A. Tanis, *Probability & Statistics: Explorations with MAPLE*, 2nd ed., Prentice Hall, NJ, 1999.

Keating, Jerome P., and Scott, David W., "Ask Dr. STATS," *Stats, The Magazine for Students of Statistics*, **25**, Spring, 1999, pp. 16–22.

Kinney, J., "Mathematica As an Aid in Teaching Probability and Statistics," *Proceedings of the Statistical Computing Section*, (1998), American Statistical Association, pp. 25–32.

Montgomery, D. C., *Design and Analysis of Experiments*, 2nd ed., John Wiley & Sons, Inc., New York, 1984.

Pearson, K., "On the Criterion That a Given System of Deviations from the Probable in the Case of a Correlated System of Variables Is Such That It Can Be Reasonably Supposed to Have Arisen from Random Sampling," *Phil. Mag.*, Series 5, **50** (1900), p. 157.

Putz, John, "The Golden Section and the Piano Sonatas of Mozart," *Mathematics Magazine*, Vol. 68, No. 4, Oct., 1995, pp. 275–282.

Quain, J. R., "Going Mainstream," *PC Magazine*, February, 1994.

Rafter, J. A., M. L. Abell, J. P. Braselton, *Statistics with Maple*, Academic Press, An imprint of Elsevier Science (USA), 2003.

Snee, R. D., L. B. Hare, and J. R. Trout, *Experiments in Industry*, American Society of Quality Conrol, Milwaukee, Wis., 1985.

Tanis, E. A., "Maple Integrated Into the Instruction of Probability and Statistics," *Proceedings of the Statistical Computing Section*, (1998), American Statistical Association, pp. 19–24.

Tate, R. F., and G. W. Klett, "Optimum Confidence Intervals for the Variance of a Normal Distribution," *J. Am. Statist. Assoc.*, **54** (1959), p. 674.

Tukey, John W., *Exploratory Data Analysis*, Addison-Wesley Publishing Company, Reading, Mass., 1977.

Velleman, P. F., and D. C. Hoaglin, *Applications, Basics, and Computing of Exploratory Data Analysis*, Duxbury Press, Boston, 1981.

Yobs, A .R., Swanson, R. A., and Lamotte, L. C. "Laboratory Reliability of the Papanicolaou Smear." *Obstetrics and Gynecology*, Volume 65, February 1985, pp. 235–244.

# APPENDIX B

# TABLES

## Table I Binomial Coefficients

$$\binom{n}{r} = \frac{n!}{r!(n-r)!} = \binom{n}{n-r}$$

| $n$ | $\binom{n}{0}$ | $\binom{n}{1}$ | $\binom{n}{2}$ | $\binom{n}{3}$ | $\binom{n}{4}$ | $\binom{n}{5}$ | $\binom{n}{6}$ | $\binom{n}{7}$ | $\binom{n}{8}$ | $\binom{n}{9}$ | $\binom{n}{10}$ | $\binom{n}{11}$ | $\binom{n}{12}$ | $\binom{n}{13}$ |
|---|---|---|---|---|---|---|---|---|---|---|---|---|---|---|
| 0 | 1 | | | | | | | | | | | | | |
| 1 | 1 | 1 | | | | | | | | | | | | |
| 2 | 1 | 2 | 1 | | | | | | | | | | | |
| 3 | 1 | 3 | 3 | 1 | | | | | | | | | | |
| 4 | 1 | 4 | 6 | 4 | 1 | | | | | | | | | |
| 5 | 1 | 5 | 10 | 10 | 5 | 1 | | | | | | | | |
| 6 | 1 | 6 | 15 | 20 | 15 | 6 | 1 | | | | | | | |
| 7 | 1 | 7 | 21 | 35 | 35 | 21 | 7 | 1 | | | | | | |
| 8 | 1 | 8 | 28 | 56 | 70 | 56 | 28 | 8 | 1 | | | | | |
| 9 | 1 | 9 | 36 | 84 | 126 | 126 | 84 | 36 | 9 | 1 | | | | |
| 10 | 1 | 10 | 45 | 120 | 210 | 252 | 210 | 120 | 45 | 10 | 1 | | | |
| 11 | 1 | 11 | 55 | 165 | 330 | 462 | 462 | 330 | 165 | 55 | 11 | 1 | | |
| 12 | 1 | 12 | 66 | 220 | 495 | 792 | 924 | 792 | 495 | 220 | 66 | 12 | 1 | |
| 13 | 1 | 13 | 78 | 286 | 715 | 1,287 | 1,716 | 1,716 | 1,287 | 715 | 286 | 78 | 13 | 1 |
| 14 | 1 | 14 | 91 | 364 | 1,001 | 2,002 | 3,003 | 3,432 | 3,003 | 2,002 | 1,001 | 364 | 91 | 14 |
| 15 | 1 | 15 | 105 | 455 | 1,365 | 3,003 | 5,005 | 6,435 | 6,435 | 5,005 | 3,003 | 1,365 | 455 | 105 |
| 16 | 1 | 16 | 120 | 560 | 1,820 | 4,368 | 8,008 | 11,440 | 12,870 | 11,440 | 8,008 | 4,368 | 1,820 | 560 |
| 17 | 1 | 17 | 136 | 680 | 2,380 | 6,188 | 12,376 | 19,448 | 24,310 | 24,310 | 19,448 | 12,376 | 6,188 | 2,380 |
| 18 | 1 | 18 | 153 | 816 | 3,060 | 8,568 | 18,564 | 31,824 | 43,758 | 48,620 | 43,758 | 31,824 | 18,564 | 8,568 |
| 19 | 1 | 19 | 171 | 969 | 3,876 | 11,628 | 27,132 | 50,388 | 75,582 | 92,378 | 92,378 | 75,582 | 50,388 | 27,132 |
| 20 | 1 | 20 | 190 | 1,140 | 4,845 | 15,504 | 38,760 | 77,520 | 125,970 | 167,960 | 184,756 | 167,960 | 125,970 | 77,520 |
| 21 | 1 | 21 | 210 | 1,330 | 5,985 | 20,349 | 54,264 | 116,280 | 203,490 | 293,930 | 352,716 | 352,716 | 293,930 | 203,490 |
| 22 | 1 | 22 | 231 | 1,540 | 7,315 | 26,334 | 74,613 | 170,544 | 319,770 | 497,420 | 646,646 | 705,432 | 646,646 | 497,420 |
| 23 | 1 | 23 | 253 | 1,771 | 8,855 | 33,649 | 100,947 | 245,157 | 490,314 | 817,190 | 1,144,066 | 1,352,078 | 1,352,078 | 1,144,066 |
| 24 | 1 | 24 | 276 | 2,024 | 10,626 | 42,504 | 134,596 | 346,104 | 735,471 | 1,307,504 | 1,961,256 | 2,496,144 | 2,704,156 | 2,496,144 |
| 25 | 1 | 25 | 300 | 2,300 | 12,650 | 53,130 | 177,100 | 480,700 | 1,081,575 | 2,042,975 | 3,268,760 | 4,457,400 | 5,200,300 | 5,200,300 |
| 26 | 1 | 26 | 325 | 2,600 | 14,950 | 65,780 | 230,230 | 657,800 | 1,562,275 | 3,124,550 | 5,311,735 | 7,726,160 | 9,657,700 | 10,400,600 |

For $r > 13$ you may use the identity $\binom{n}{r} = \binom{n}{n-r}$.

## Table II  The Binomial Distribution

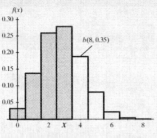

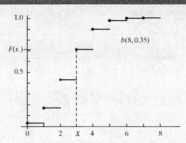

$$F(x) = P(X \le x) = \sum_{k=0}^{x} \frac{n!}{k!(n-k)!} p^k (1-p)^{n-k}$$

| | | | | | | | $p$ | | | | |
|---|---|---|---|---|---|---|---|---|---|---|---|
| $n$ | $x$ | 0.05 | 0.10 | 0.15 | 0.20 | 0.25 | 0.30 | 0.35 | 0.40 | 0.45 | 0.50 |
| 2 | 0 | 0.9025 | 0.8100 | 0.7225 | 0.6400 | 0.5625 | 0.4900 | 0.4225 | 0.3600 | 0.3025 | 0.2500 |
|   | 1 | 0.9975 | 0.9900 | 0.9775 | 0.9600 | 0.9375 | 0.9100 | 0.8775 | 0.8400 | 0.7975 | 0.7500 |
|   | 2 | 1.0000 | 1.0000 | 1.0000 | 1.0000 | 1.0000 | 1.0000 | 1.0000 | 1.0000 | 1.0000 | 1.0000 |
| 3 | 0 | 0.8574 | 0.7290 | 0.6141 | 0.5120 | 0.4219 | 0.3430 | 0.2746 | 0.2160 | 0.1664 | 0.1250 |
|   | 1 | 0.9928 | 0.9720 | 0.9392 | 0.8960 | 0.8438 | 0.7840 | 0.7182 | 0.6480 | 0.5748 | 0.5000 |
|   | 2 | 0.9999 | 0.9990 | 0.9966 | 0.9920 | 0.9844 | 0.9730 | 0.9571 | 0.9360 | 0.9089 | 0.8750 |
|   | 3 | 1.0000 | 1.0000 | 1.0000 | 1.0000 | 1.0000 | 1.0000 | 1.0000 | 1.0000 | 1.0000 | 1.0000 |
| 4 | 0 | 0.8145 | 0.6561 | 0.5220 | 0.4096 | 0.3164 | 0.2401 | 0.1785 | 0.1296 | 0.0915 | 0.0625 |
|   | 1 | 0.9860 | 0.9477 | 0.8905 | 0.8192 | 0.7383 | 0.6517 | 0.5630 | 0.4752 | 0.3910 | 0.3125 |
|   | 2 | 0.9995 | 0.9963 | 0.9880 | 0.9728 | 0.9492 | 0.9163 | 0.8735 | 0.8208 | 0.7585 | 0.6875 |
|   | 3 | 1.0000 | 0.9999 | 0.9995 | 0.9984 | 0.9961 | 0.9919 | 0.9850 | 0.9744 | 0.9590 | 0.9375 |
|   | 4 | 1.0000 | 1.0000 | 1.0000 | 1.0000 | 1.0000 | 1.0000 | 1.0000 | 1.0000 | 1.0000 | 1.0000 |
| 5 | 0 | 0.7738 | 0.5905 | 0.4437 | 0.3277 | 0.2373 | 0.1681 | 0.1160 | 0.0778 | 0.0503 | 0.0312 |
|   | 1 | 0.9774 | 0.9185 | 0.8352 | 0.7373 | 0.6328 | 0.5282 | 0.4284 | 0.3370 | 0.2562 | 0.1875 |
|   | 2 | 0.9988 | 0.9914 | 0.9734 | 0.9421 | 0.8965 | 0.8369 | 0.7648 | 0.6826 | 0.5931 | 0.5000 |
|   | 3 | 1.0000 | 0.9995 | 0.9978 | 0.9933 | 0.9844 | 0.9692 | 0.9460 | 0.9130 | 0.8688 | 0.8125 |
|   | 4 | 1.0000 | 1.0000 | 0.9999 | 0.9997 | 0.9990 | 0.9976 | 0.9947 | 0.9898 | 0.9815 | 0.9688 |
|   | 5 | 1.0000 | 1.0000 | 1.0000 | 1.0000 | 1.0000 | 1.0000 | 1.0000 | 1.0000 | 1.0000 | 1.0000 |
| 6 | 0 | 0.7351 | 0.5314 | 0.3771 | 0.2621 | 0.1780 | 0.1176 | 0.0754 | 0.0467 | 0.0277 | 0.0156 |
|   | 1 | 0.9672 | 0.8857 | 0.7765 | 0.6553 | 0.5339 | 0.4202 | 0.3191 | 0.2333 | 0.1636 | 0.1094 |
|   | 2 | 0.9978 | 0.9842 | 0.9527 | 0.9011 | 0.8306 | 0.7443 | 0.6471 | 0.5443 | 0.4415 | 0.3438 |
|   | 3 | 0.9999 | 0.9987 | 0.9941 | 0.9830 | 0.9624 | 0.9295 | 0.8826 | 0.8208 | 0.7447 | 0.6562 |
|   | 4 | 1.0000 | 0.9999 | 0.9996 | 0.9984 | 0.9954 | 0.9891 | 0.9777 | 0.9590 | 0.9308 | 0.8906 |
|   | 5 | 1.0000 | 1.0000 | 1.0000 | 0.9999 | 0.9998 | 0.9993 | 0.9982 | 0.9959 | 0.9917 | 0.9844 |
|   | 6 | 1.0000 | 1.0000 | 1.0000 | 1.0000 | 1.0000 | 1.0000 | 1.0000 | 1.0000 | 1.0000 | 1.0000 |
| 7 | 0 | 0.6983 | 0.4783 | 0.3206 | 0.2097 | 0.1335 | 0.0824 | 0.0490 | 0.0280 | 0.0152 | 0.0078 |
|   | 1 | 0.9556 | 0.8503 | 0.7166 | 0.5767 | 0.4449 | 0.3294 | 0.2338 | 0.1586 | 0.1024 | 0.0625 |
|   | 2 | 0.9962 | 0.9743 | 0.9262 | 0.8520 | 0.7564 | 0.6471 | 0.5323 | 0.4199 | 0.3164 | 0.2266 |
|   | 3 | 0.9998 | 0.9973 | 0.9879 | 0.9667 | 0.9294 | 0.8740 | 0.8002 | 0.7102 | 0.6083 | 0.5000 |
|   | 4 | 1.0000 | 0.9998 | 0.9988 | 0.9953 | 0.9871 | 0.9712 | 0.9444 | 0.9037 | 0.8471 | 0.7734 |
|   | 5 | 1.0000 | 1.0000 | 0.9999 | 0.9996 | 0.9987 | 0.9962 | 0.9910 | 0.9812 | 0.9643 | 0.9375 |

| Table II *continued* | | | | | | | | | | |
|---|---|---|---|---|---|---|---|---|---|---|
| | | | | | | *p* | | | | | |
| *n* | *x* | 0.05 | 0.10 | 0.15 | 0.20 | 0.25 | 0.30 | 0.35 | 0.40 | 0.45 | 0.50 |
| | 6 | 1.0000 | 1.0000 | 1.0000 | 1.0000 | 0.9999 | 0.9998 | 0.9994 | 0.9984 | 0.9963 | 0.9922 |
| | 7 | 1.0000 | 1.0000 | 1.0000 | 1.0000 | 1.0000 | 1.0000 | 1.0000 | 1.0000 | 1.0000 | 1.0000 |
| 8 | 0 | 0.6634 | 0.4305 | 0.2725 | 0.1678 | 0.1001 | 0.0576 | 0.0319 | 0.0168 | 0.0084 | 0.0039 |
| | 1 | 0.9428 | 0.8131 | 0.6572 | 0.5033 | 0.3671 | 0.2553 | 0.1691 | 0.1064 | 0.0632 | 0.0352 |
| | 2 | 0.9942 | 0.9619 | 0.8948 | 0.7969 | 0.6785 | 0.5518 | 0.4278 | 0.3154 | 0.2201 | 0.1445 |
| | 3 | 0.9996 | 0.9950 | 0.9786 | 0.9437 | 0.8862 | 0.8059 | 0.7064 | 0.5941 | 0.4770 | 0.3633 |
| | 4 | 1.0000 | 0.9996 | 0.9971 | 0.9896 | 0.9727 | 0.9420 | 0.8939 | 0.8263 | 0.7396 | 0.6367 |
| | 5 | 1.0000 | 1.0000 | 0.9998 | 0.9988 | 0.9958 | 0.9887 | 0.9747 | 0.9502 | 0.9115 | 0.8555 |
| | 6 | 1.0000 | 1.0000 | 1.0000 | 0.9999 | 0.9996 | 0.9987 | 0.9964 | 0.9915 | 0.9819 | 0.9648 |
| | 7 | 1.0000 | 1.0000 | 1.0000 | 1.0000 | 1.0000 | 0.9999 | 0.9998 | 0.9993 | 0.9983 | 0.9961 |
| | 8 | 1.0000 | 1.0000 | 1.0000 | 1.0000 | 1.0000 | 1.0000 | 1.0000 | 1.0000 | 1.0000 | 1.0000 |
| 9 | 0 | 0.6302 | 0.3874 | 0.2316 | 0.1342 | 0.0751 | 0.0404 | 0.0207 | 0.0101 | 0.0046 | 0.0020 |
| | 1 | 0.9288 | 0.7748 | 0.5995 | 0.4362 | 0.3003 | 0.1960 | 0.1211 | 0.0705 | 0.0385 | 0.0195 |
| | 2 | 0.9916 | 0.9470 | 0.8591 | 0.7382 | 0.6007 | 0.4628 | 0.3373 | 0.2318 | 0.1495 | 0.0898 |
| | 3 | 0.9994 | 0.9917 | 0.9661 | 0.9144 | 0.8343 | 0.7297 | 0.6089 | 0.4826 | 0.3614 | 0.2539 |
| | 4 | 1.0000 | 0.9991 | 0.9944 | 0.9804 | 0.9511 | 0.9012 | 0.8283 | 0.7334 | 0.6214 | 0.5000 |
| | 5 | 1.0000 | 0.9999 | 0.9994 | 0.9969 | 0.9900 | 0.9747 | 0.9464 | 0.9006 | 0.8342 | 0.7461 |
| | 6 | 1.0000 | 1.0000 | 1.0000 | 0.9997 | 0.9987 | 0.9957 | 0.9888 | 0.9750 | 0.9502 | 0.9102 |
| | 7 | 1.0000 | 1.0000 | 1.0000 | 1.0000 | 0.9999 | 0.9996 | 0.9986 | 0.9962 | 0.9909 | 0.9805 |
| | 8 | 1.0000 | 1.0000 | 1.0000 | 1.0000 | 1.0000 | 1.0000 | 0.9999 | 0.9997 | 0.9992 | 0.9980 |
| | 9 | 1.0000 | 1.0000 | 1.0000 | 1.0000 | 1.0000 | 1.0000 | 1.0000 | 1.0000 | 1.0000 | 1.0000 |
| 10 | 0 | 0.5987 | 0.3487 | 0.1969 | 0.1074 | 0.0563 | 0.0282 | 0.0135 | 0.0060 | 0.0025 | 0.0010 |
| | 1 | 0.9139 | 0.7361 | 0.5443 | 0.3758 | 0.2440 | 0.1493 | 0.0860 | 0.0464 | 0.0233 | 0.0107 |
| | 2 | 0.9885 | 0.9298 | 0.8202 | 0.6778 | 0.5256 | 0.3828 | 0.2616 | 0.1673 | 0.0996 | 0.0547 |
| | 3 | 0.9990 | 0.9872 | 0.9500 | 0.8791 | 0.7759 | 0.6496 | 0.5138 | 0.3823 | 0.2660 | 0.1719 |
| | 4 | 0.9999 | 0.9984 | 0.9901 | 0.9672 | 0.9219 | 0.8497 | 0.7515 | 0.6331 | 0.5044 | 0.3770 |
| | 5 | 1.0000 | 0.9999 | 0.9986 | 0.9936 | 0.9803 | 0.9527 | 0.9051 | 0.8338 | 0.7384 | 0.6230 |
| | 6 | 1.0000 | 1.0000 | 0.9999 | 0.9991 | 0.9965 | 0.9894 | 0.9740 | 0.9452 | 0.8980 | 0.8281 |
| | 7 | 1.0000 | 1.0000 | 1.0000 | 0.9999 | 0.9996 | 0.9984 | 0.9952 | 0.9877 | 0.9726 | 0.9453 |
| | 8 | 1.0000 | 1.0000 | 1.0000 | 1.0000 | 1.0000 | 0.9999 | 0.9995 | 0.9983 | 0.9955 | 0.9893 |
| | 9 | 1.0000 | 1.0000 | 1.0000 | 1.0000 | 1.0000 | 1.0000 | 1.0000 | 0.9999 | 0.9997 | 0.9990 |
| | 10 | 1.0000 | 1.0000 | 1.0000 | 1.0000 | 1.0000 | 1.0000 | 1.0000 | 1.0000 | 1.0000 | 1.0000 |
| 11 | 0 | 0.5688 | 0.3138 | 0.1673 | 0.0859 | 0.0422 | 0.0198 | 0.0088 | 0.0036 | 0.0014 | 0.0005 |
| | 1 | 0.8981 | 0.6974 | 0.4922 | 0.3221 | 0.1971 | 0.1130 | 0.0606 | 0.0302 | 0.0139 | 0.0059 |
| | 2 | 0.9848 | 0.9104 | 0.7788 | 0.6174 | 0.4552 | 0.3127 | 0.2001 | 0.1189 | 0.0652 | 0.0327 |
| | 3 | 0.9984 | 0.9815 | 0.9306 | 0.8389 | 0.7133 | 0.5696 | 0,4256 | 0.2963 | 0.1911 | 0.1133 |
| | 4 | 0.9999 | 0.9972 | 0.9841 | 0.9496 | 0.8854 | 0.7897 | 0.6683 | 0.5328 | 0.3971 | 0.2744 |
| | 5 | 1.0000 | 0.9997 | 0.9973 | 0.9883 | 0.9657 | 0.9218 | 0.8513 | 0.7535 | 0.6331 | 0.5000 |
| | 6 | 1.0000 | 1.0000 | 0.9997 | 0.9980 | 0.9924 | 0.9784 | 0.9499 | 0.9006 | 0.8262 | 0.7256 |
| | 7 | 1.0000 | 1.0000 | 1.0000 | 0.9998 | 0.9988 | 0.9957 | 0.9878 | 0.9707 | 0.9390 | 0.8867 |
| | 8 | 1.0000 | 1.0000 | 1.0000 | 1.0000 | 0.9999 | 0.9994 | 0.9980 | 0.9941 | 0.9852 | 0.9673 |
| | 9 | 1.0000 | 1.0000 | 1.0000 | 1.0000 | 1.0000 | 1.0000 | 0.9998 | 0.9993 | 0.9978 | 0.9941 |

| | | | | | | | | | | |
|---|---|---|---|---|---|---|---|---|---|---|
| | | | | | | Table II *continued* | | | | |

| | | | | | | *p* | | | | |
|---|---|---|---|---|---|---|---|---|---|---|
| *n* | *x* | 0.05 | 0.10 | 0.15 | 0.20 | 0.25 | 0.30 | 0.35 | 0.40 | 0.45 | 0.50 |
| | 10 | 1.0000 | 1.0000 | 1.0000 | 1.0000 | 1.0000 | 1.0000 | 1.0000 | 1.0000 | 0.9998 | 0.9995 |
| | 11 | 1.0000 | 1.0000 | 1.0000 | 1.0000 | 1.0000 | 1.0000 | 1.0000 | 1.0000 | 1.0000 | 1.0000 |
| 12 | 0 | 0.5404 | 0.2824 | 0.1422 | 0.0687 | 0.0317 | 0.0138 | 0.0057 | 0.0022 | 0.0008 | 0.0002 |
| | 1 | 0.8816 | 0.6590 | 0.4435 | 0.2749 | 0.1584 | 0.0850 | 0.0424 | 0.0196 | 0.0083 | 0.0032 |
| | 2 | 0.9804 | 0.8891 | 0.7358 | 0.5583 | 0.3907 | 0.2528 | 0.1513 | 0.0834 | 0.0421 | 0.0193 |
| | 3 | 0.9978 | 0.9744 | 0.9078 | 0.7946 | 0.6488 | 0.4925 | 0.3467 | 0.2253 | 0.1345 | 0.0730 |
| | 4 | 0.9998 | 0.9957 | 0.9761 | 0.9274 | 0.8424 | 0.7237 | 0.5833 | 0.4382 | 0.3044 | 0.1938 |
| | 5 | 1.0000 | 0.9995 | 0.9954 | 0.9806 | 0.9456 | 0.8822 | 0.7873 | 0.6652 | 0.5269 | 0.3872 |
| | 6 | 1.0000 | 0.9999 | 0.9993 | 0.9961 | 0.9857 | 0.9614 | 0.9154 | 0.8418 | 0.7393 | 0.6128 |
| | 7 | 1.0000 | 1.0000 | 0.9999 | 0.9994 | 0.9972 | 0.9905 | 0.9745 | 0.9427 | 0.8883 | 0.8062 |
| | 8 | 1.0000 | 1.0000 | 1.0000 | 0.9999 | 0.9996 | 0.9983 | 0.9944 | 0.9847 | 0.9644 | 0.9270 |
| | 9 | 1.0000 | 1.0000 | 1.0000 | 1.0000 | 1.0000 | 0.9998 | 0.9992 | 0.9972 | 0.9921 | 0.9807 |
| | 10 | 1.0000 | 1.0000 | 1.0000 | 1.0000 | 1.0000 | 1.0000 | 0.9999 | 0.9997 | 0.9989 | 0.9968 |
| | 11 | 1.0000 | 1.0000 | 1.0000 | 1.0000 | 1.0000 | 1.0000 | 1.0000 | 1.0000 | 0.9999 | 0.9998 |
| | 12 | 1.0000 | 1.0000 | 1.0000 | 1.0000 | 1.0000 | 1.0000 | 1.0000 | 1.0000 | 1.0000 | 1.0000 |
| 13 | 0 | 0.5133 | 0.2542 | 0.1209 | 0.0550 | 0.0238 | 0.0097 | 0.0037 | 0.0013 | 0.0004 | 0.0001 |
| | 1 | 0.8646 | 0.6213 | 0.3983 | 0.2336 | 0.1267 | 0.0637 | 0.0296 | 0.0126 | 0.0049 | 0.0017 |
| | 2 | 0.9755 | 0.8661 | 0.6920 | 0.5017 | 0.3326 | 0.2025 | 0.1132 | 0.0579 | 0.0269 | 0.0112 |
| | 3 | 0.9969 | 0.9658 | 0.8820 | 0.7473 | 0.5843 | 0.4206 | 0.2783 | 0.1686 | 0.0929 | 0.0461 |
| | 4 | 0.9997 | 0.9935 | 0.9658 | 0.9009 | 0.7940 | 0.6543 | 0.5005 | 0.3530 | 0.2279 | 0.1334 |
| | 5 | 1.0000 | 0.9991 | 0.9924 | 0.9700 | 0.9198 | 0.8346 | 0.7159 | 0.5744 | 0.4268 | 0.2905 |
| | 6 | 1.0000 | 0.9999 | 0.9987 | 0.9930 | 0.9757 | 0.9376 | 0.8705 | 0.7712 | 0.6437 | 0.5000 |
| | 7 | 1.0000 | 1.0000 | 0.9998 | 0.9988 | 0.9944 | 0.9818 | 0.9538 | 0.9023 | 0.8212 | 0.7095 |
| | 8 | 1.0000 | 1.0000 | 1.0000 | 0.9998 | 0.9990 | 0.9960 | 0.9874 | 0.9679 | 0.9302 | 0.8666 |
| | 9 | 1.0000 | 1.0000 | 1.0000 | 1.0000 | 0.9999 | 0.9993 | 0.9975 | 0.9922 | 0.9797 | 0.9539 |
| | 10 | 1.0000 | 1.0000 | 1.0000 | 1.0000 | 1.0000 | 0.9999 | 0.9997 | 0.9987 | 0.9959 | 0.9888 |
| | 11 | 1.0000 | 1.0000 | 1.0000 | 1.0000 | 1.0000 | 1.0000 | 1.0000 | 0.9999 | 0.9995 | 0.9983 |
| | 12 | 1.0000 | 1.0000 | 1.0000 | 1.0000 | 1.0000 | 1.0000 | 1.0000 | 1.0000 | 1.0000 | 0.9999 |
| | 13 | 1.0000 | 1.0000 | 1.0000 | 1.0000 | 1.0000 | 1.0000 | 1.0000 | 1.0000 | 1.0000 | 1.0000 |
| 14 | 0 | 0.4877 | 0.2288 | 0.1028 | 0.0440 | 0.0178 | 0.0068 | 0.0024 | 0.0008 | 0.0002 | 0.0001 |
| | 1 | 0.8470 | 0.5846 | 0.3567 | 0.1979 | 0.1010 | 0.0475 | 0.0205 | 0.0081 | 0.0029 | 0.0009 |
| | 2 | 0.9699 | 0.8416 | 0.6479 | 0.4481 | 0.2811 | 0.1608 | 0.0839 | 0.0398 | 0.0170 | 0.0065 |
| | 3 | 0.9958 | 0.9559 | 0.8535 | 0.6982 | 0.5213 | 0.3552 | 0.2205 | 0.1243 | 0.0632 | 0.0287 |
| | 4 | 0.9996 | 0.9908 | 0.9533 | 0.8702 | 0.7415 | 0.5842 | 0.4227 | 0.2793 | 0.1672 | 0.0898 |
| | 5 | 1.0000 | 0.9985 | 0.9885 | 0.9561 | 0.8883 | 0.7805 | 0.6405 | 0.4859 | 0.3373 | 0.2120 |
| | 6 | 1.0000 | 0.9998 | 0.9978 | 0.9884 | 0.9617 | 0.9067 | 0.8164 | 0.6925 | 0.5461 | 0.3953 |
| | 7 | 1.0000 | 1.0000 | 0.9997 | 0.9976 | 0.9897 | 0.9685 | 0.9247 | 0.8499 | 0.7414 | 0.6047 |
| | 8 | 1.0000 | 1.0000 | 1.0000 | 0.9996 | 0.9978 | 0.9917 | 0.9757 | 0.9417 | 0.8811 | 0.7880 |
| | 9 | 1.0000 | 1.0000 | 1.0000 | 1.0000 | 0.9997 | 0.9983 | 0.9940 | 0.9825 | 0.9574 | 0.9102 |
| | 10 | 1.0000 | 1.0000 | 1.0000 | 1.0000 | 1.0000 | 0.9998 | 0.9989 | 0.9961 | 0.9886 | 0.9713 |

## Table II continued

| n | x | 0.05 | 0.10 | 0.15 | 0.20 | 0.25 | 0.30 | 0.35 | 0.40 | 0.45 | 0.50 |
|---|---|------|------|------|------|------|------|------|------|------|------|
| | | | | | | | $p$ | | | | |
| | 11 | 1.0000 | 1.0000 | 1.0000 | 1.0000 | 1.0000 | 1.0000 | 0.9999 | 0.9994 | 0.9978 | 0.9935 |
| | 12 | 1.0000 | 1.0000 | 1.0000 | 1.0000 | 1.0000 | 1.0000 | 1.0000 | 0.9999 | 0.9997 | 0.9991 |
| | 13 | 1.0000 | 1.0000 | 1.0000 | 1.0000 | 1.0000 | 1.0000 | 1.0000 | 1.0000 | 1.0000 | 0.9999 |
| | 14 | 1.0000 | 1.0000 | 1.0000 | 1.0000 | 1.0000 | 1.0000 | 1.0000 | 1.0000 | 1.0000 | 1.0000 |
| 15 | 0 | 0.4633 | 0.2059 | 0.0874 | 0.0352 | 0.0134 | 0.0047 | 0.0016 | 0.0005 | 0.0001 | 0.0000 |
| | 1 | 0.8290 | 0.5490 | 0.3186 | 0.1671 | 0.0802 | 0.0353 | 0.0142 | 0.0052 | 0.0017 | 0.0005 |
| | 2 | 0.9638 | 0.8159 | 0.6042 | 0.3980 | 0.2361 | 0.1268 | 0.0617 | 0.0271 | 0.0107 | 0.0037 |
| | 3 | 0.9945 | 0.9444 | 0.8227 | 0.6482 | 0.4613 | 0.2969 | 0.1727 | 0.0905 | 0.0424 | 0.0176 |
| | 4 | 0.9994 | 0.9873 | 0.9383 | 0.8358 | 0.6865 | 0.5155 | 0.3519 | 0.2173 | 0.1204 | 0.0592 |
| | 5 | 0.9999 | 0.9978 | 0.9832 | 0.9389 | 0.8516 | 0.7216 | 0.5643 | 0.4032 | 0.2608 | 0.1509 |
| | 6 | 1.0000 | 0.9997 | 0.9964 | 0.9819 | 0.9434 | 0.8689 | 0.7548 | 0.6098 | 0.4522 | 0.3036 |
| | 7 | 1.0000 | 1.0000 | 0.9994 | 0.9958 | 0.9827 | 0.9500 | 0.8868 | 0.7869 | 0.6535 | 0.5000 |
| | 8 | 1.0000 | 1.0000 | 0.9999 | 0.9992 | 0.9958 | 0.9848 | 0.9578 | 0.9050 | 0.8182 | 0.6964 |
| | 9 | 1.0000 | 1.0000 | 1.0000 | 0.9999 | 0.9992 | 0.9963 | 0.9876 | 0.9662 | 0.9231 | 0.8491 |
| | 10 | 1.0000 | 1.0000 | 1.0000 | 1.0000 | 0.9999 | 0.9993 | 0.9972 | 0.9907 | 0.9745 | 0.9408 |
| | 11 | 1.0000 | 1.0000 | 1.0000 | 1.0000 | 1.0000 | 0.9999 | 0.9995 | 0.9981 | 0.9937 | 0.9824 |
| | 12 | 1.0000 | 1.0000 | 1.0000 | 1.0000 | 1.0000 | 1.0000 | 0.9999 | 0.9987 | 0.9989 | 0.9963 |
| | 13 | 1.0000 | 1.0000 | 1.0000 | 1.0000 | 1.0000 | 1.0000 | 1.0000 | 1.0000 | 0.9999 | 0.9995 |
| | 14 | 1.0000 | 1.0000 | 1.0000 | 1.0000 | 1.0000 | 1.0000 | 1.0000 | 1.0000 | 1.0000 | 1.0000 |
| | 15 | 1.0000 | 1.0000 | 1.0000 | 1.0000 | 1.0000 | 1.0000 | 1.0000 | 1.0000 | 1.0000 | 1.0000 |
| 16 | 0 | 0.4401 | 0.1853 | 0.0743 | 0.0281 | 0.0100 | 0.0033 | 0.0010 | 0.0003 | 0.0001 | 0.0000 |
| | 1 | 0.8108 | 0.5147 | 0.2839 | 0.1407 | 0.0635 | 0.0261 | 0.0098 | 0.0033 | 0.0010 | 0.0003 |
| | 2 | 0.9571 | 0.7892 | 0.5614 | 0.3518 | 0.1971 | 0.0994 | 0.0451 | 0.0183 | 0.0066 | 0.0021 |
| | 3 | 0.9930 | 0.9316 | 0.7899 | 0.5981 | 0.4050 | 0.2459 | 0.1339 | 0,0651 | 0.0281 | 0.0106 |
| | 4 | 0.9991 | 0.9830 | 0.9209 | 0.7982 | 0.6302 | 0.4499 | 0.2892 | 0.1666 | 0.0853 | 0.0384 |
| | 5 | 0.9999 | 0.9967 | 0.9765 | 0.9183 | 0.8103 | 0.6598 | 0.4900 | 0.3288 | 0.1976 | 0.1051 |
| | 6 | 1.0000 | 0.9995 | 0.9944 | 0.9733 | 0.9204 | 0.8247 | 0.6881 | 0.5272 | 0.3660 | 0.2272 |
| | 7 | 1.0000 | 0.9999 | 0.9989 | 0.9930 | 0.9729 | 0.9256 | 0.8406 | 0.7161 | 0.5629 | 0.4018 |
| | 8 | 1.0000 | 1.0000 | 0.9998 | 0.9985 | 0.9925 | 0.9743 | 0.9329 | 0.8577 | 0.7441 | 0.5982 |
| | 9 | 1.0000 | 1.0000 | 1.0000 | 0.9998 | 0.9984 | 0.9929 | 0.9771 | 0.9417 | 0.8759 | 0.7728 |
| | 10 | 1.0000 | 1.0000 | 1.0000 | 1.0000 | 0.9997 | 0.9984 | 0.9938 | 0.9809 | 0.9514 | 0.8949 |
| | 11 | 1.0000 | 1.0000 | 1.0000 | 1.0000 | 1.0000 | 0.9997 | 0.9987 | 0.9951 | 0.9851 | 0.9616 |
| | 12 | 1.0000 | 1.0000 | 1.0000 | 1.0000 | 1.0000 | 1.0000 | 0.9998 | 0.9991 | 0.9965 | 0.9894 |
| | 13 | 1.0000 | 1.0000 | 1.0000 | 1.0000 | 1.0000 | 1.0000 | 1.0000 | 0.9999 | 0.9994 | 0.9979 |
| | 14 | 1.0000 | 1.0000 | 1.0000 | 1.0000 | 1.0000 | 1.0000 | 1.0000 | 1.0000 | 0.9999 | 0.9997 |
| | 15 | 1.0000 | 1.0000 | 1.0000 | 1.0000 | 1.0000 | 1.0000 | 1.0000 | 1.0000 | 1.0000 | 1.0000 |
| | 16 | 1.0000 | 1.0000 | 1.0000 | 1.0000 | 1.0000 | 1.0000 | 1.0000 | 1.0000 | 1.0000 | 1.0000 |
| 20 | 0 | 0.3585 | 0.1216 | 0.0388 | 0.0115 | 0.0032 | 0.0008 | 0.0002 | 0.0000 | 0.0000 | 0.0000 |
| | 1 | 0.7358 | 0.3917 | 0.1756 | 0.0692 | 0.0243 | 0.0076 | 0.0021 | 0.0005 | 0.0001 | 0.0000 |
| | 2 | 0.9245 | 0.6769 | 0.4049 | 0.2061 | 0.0913 | 0.0355 | 0.0121 | 0.0036 | 0.0009 | 0.0002 |
| | 3 | 0.9841 | 0.8670 | 0.6477 | 0.4114 | 0.2252 | 0.1071 | 0.0444 | 0.0160 | 0.0049 | 0.0013 |
| | 4 | 0.9974 | 0.9568 | 0.8298 | 0.6296 | 0.4148 | 0.2375 | 0.1182 | 0.0510 | 0.0189 | 0.0059 |

| | | | | | | p | | | | | |
|---|---|---|---|---|---|---|---|---|---|---|---|
| n | x | 0.05 | 0.10 | 0.15 | 0.20 | 0.25 | 0.30 | 0.35 | 0.40 | 0.45 | 0.50 |
| | 5 | 0.9997 | 0.9887 | 0.9327 | 0.8042 | 0.6172 | 0.4164 | 0.2454 | 0.1256 | 0.0553 | 0.0207 |
| | 6 | 1.0000 | 0.9976 | 0.9781 | 0.9133 | 0.7858 | 0.6080 | 0.4166 | 0.2500 | 0.1299 | 0.0577 |
| | 7 | 1.0000 | 0.9996 | 0.9941 | 0.9679 | 0.8982 | 0.7723 | 0.6010 | 0.4159 | 0.2520 | 0.1316 |
| | 8 | 1.0000 | 0.9999 | 0.9987 | 0.9900 | 0.9591 | 0.8867 | 0.7624 | 0.5956 | 0.4143 | 0.2517 |
| | 9 | 1.0000 | 1.0000 | 0.9998 | 0.9974 | 0.9861 | 0.9520 | 0.8782 | 0.7553 | 0.5914 | 0.4119 |
| | 10 | 1.0000 | 1.0000 | 1.0000 | 0.9994 | 0.9961 | 0.9829 | 0.9468 | 0.8725 | 0.7507 | 0.5881 |
| | 11 | 1.0000 | 1.0000 | 1.0000 | 0.9999 | 0.9991 | 0.9949 | 0.9804 | 0.9435 | 0.8692 | 0.7483 |
| | 12 | 1.0000 | 1.0000 | 1.0000 | 1.0000 | 0.9998 | 0.9987 | 0.9940 | 0.9790 | 0.9420 | 0.8684 |
| | 13 | 1.0000 | 1.0000 | 1.0000 | 1.0000 | 1.0000 | 0.9997 | 0.9985 | 0.9935 | 0.9786 | 0.9423 |
| | 14 | 1.0000 | 1.0000 | 1.0000 | 1.0000 | 1.0000 | 1.0000 | 0.9997 | 0.9984 | 0.9936 | 0.9793 |
| | 15 | 1.0000 | 1.0000 | 1.0000 | 1.0000 | 1.0000 | 1.0000 | 1.0000 | 0.9997 | 0.9985 | 0.9941 |
| | 16 | 1.0000 | 1.0000 | 1.0000 | 1.0000 | 1.0000 | 1.0000 | 1.0000 | 1.0000 | 0.9997 | 0.9987 |
| | 17 | 1.0000 | 1.0000 | 1.0000 | 1.0000 | 1.0000 | 1.0000 | 1.0000 | 1.0000 | 1.0000 | 0.9998 |
| | 18 | 1.0000 | 1.0000 | 1.0000 | 1.0000 | 1.0000 | 1.0000 | 1.0000 | 1.0000 | 1.0000 | 1.0000 |
| | 19 | 1.0000 | 1.0000 | 1.0000 | 1.0000 | 1.0000 | 1.0000 | 1.0000 | 1.0000 | 1.0000 | 1.0000 |
| | 20 | 1.0000 | 1.0000 | 1.0000 | 1.0000 | 1.0000 | 1.0000 | 1.0000 | 1.0000 | 1.0000 | 1.0000 |
| 25 | 0 | 0.2774 | 0.0718 | 0.0172 | 0.0038 | 0.0008 | 0.0001 | 0.0000 | 0.0000 | 0.0000 | 0.0000 |
| | 1 | 0.6424 | 0.2712 | 0.0931 | 0.0274 | 0.0070 | 0.0016 | 0.0003 | 0.0001 | 0.0000 | 0.0000 |
| | 2 | 0.8729 | 0.5371 | 0.2537 | 0.0982 | 0.0321 | 0.0090 | 0.0021 | 0.0004 | 0.0001 | 0.0000 |
| | 3 | 0.9659 | 0.7636 | 0.4711 | 0.2340 | 0.0962 | 0.0332 | 0.0097 | 0.0024 | 0.0005 | 0.0001 |
| | 4 | 0.9928 | 0.9020 | 0.6821 | 0.4207 | 0.2137 | 0.0905 | 0.0320 | 0.0095 | 0.0023 | 0.0005 |
| | 5 | 0.9988 | 0.9666 | 0.8385 | 0.6167 | 0.3783 | 0.1935 | 0.0826 | 0.0294 | 0.0086 | 0.0020 |
| | 6 | 0.9998 | 0.9905 | 0.9305 | 0.7800 | 0.5611 | 0.3407 | 0.1734 | 0.0736 | 0.0258 | 0.0073 |
| | 7 | 1.0000 | 0.9977 | 0.9745 | 0.8909 | 0.7265 | 0.5118 | 0.3061 | 0.1536 | 0.0639 | 0.0216 |
| | 8 | 1.0000 | 0.9995 | 0.9920 | 0.9532 | 0.8506 | 0.6769 | 0.4668 | 0.2735 | 0.1340 | 0.0539 |
| | 9 | 1.0000 | 0.9999 | 0.9979 | 0.9827 | 0.9287 | 0.8106 | 0.6303 | 0.4246 | 0.2424 | 0.1148 |
| | 10 | 1.0000 | 1.0000 | 0.9995 | 0.9944 | 0.9703 | 0.9022 | 0.7712 | 0.5858 | 0.3843 | 0.2122 |
| | 11 | 1.0000 | 1.0000 | 0.9999 | 0.9985 | 0.9893 | 0.9558 | 0.8746 | 0.7323 | 0.5426 | 0.3450 |
| | 12 | 1.0000 | 1.0000 | 1.0000 | 0.9996 | 0.9966 | 0.9825 | 0.9396 | 0.8462 | 0.6937 | 0.5000 |
| | 13 | 1.0000 | 1.0000 | 1.0000 | 0.9999 | 0.9991 | 0.9940 | 0.9745 | 0.9222 | 0.8173 | 0.6550 |
| | 14 | 1.0000 | 1.0000 | 1.0000 | 1.0000 | 0.9998 | 0.9982 | 0.9907 | 0.9656 | 0.9040 | 0.7878 |
| | 15 | 1.0000 | 1.0000 | 1.0000 | 1.0000 | 1.0000 | 0.9995 | 0.9971 | 0.9868 | 0.9560 | 0.8852 |
| | 16 | 1.0000 | 1.0000 | 1.0000 | 1.0000 | 1.0000 | 0.9999 | 0.9992 | 0.9957 | 0.9826 | 0.9461 |
| | 17 | 1.0000 | 1.0000 | 1.0000 | 1.0000 | 1.0000 | 1.0000 | 0.9998 | 0.9988 | 0.9942 | 0.9784 |
| | 18 | 1.0000 | 1.0000 | 1.0000 | 1.0000 | 1.0000 | 1.0000 | 1.0000 | 0.9997 | 0.9984 | 0.9927 |
| | 19 | 1.0000 | 1.0000 | 1.0000 | 1.0000 | 1.0000 | 1.0000 | 1.0000 | 0.9999 | 0.9996 | 0.9980 |
| | 20 | 1.0000 | 1.0000 | 1.0000 | 1.0000 | 1.0000 | 1.0000 | 1.0000 | 1.0000 | 0.9999 | 0.9995 |
| | 21 | 1.0000 | 1.0000 | 1.0000 | 1.0000 | 1.0000 | 1.0000 | 1.0000 | 1.0000 | 1.0000 | 0.9999 |
| | 22 | 1.0000 | 1.0000 | 1.0000 | 1.0000 | 1.0000 | 1.0000 | 1.0000 | 1.0000 | 1.0000 | 1.0000 |
| | 23 | 1.0000 | 1.0000 | 1.0000 | 1.0000 | 1.0000 | 1.0000 | 1.0000 | 1.0000 | 1.0000 | 1.0000 |
| | 24 | 1.0000 | 1.0000 | 1.0000 | 1.0000 | 1.0000 | 1.0000 | 1.0000 | 1.0000 | 1.0000 | 1.0000 |
| | 25 | 1.0000 | 1.0000 | 1.0000 | 1.0000 | 1.0000 | 1.0000 | 1.0000 | 1.0000 | 1.0000 | 1.0000 |

## Table III  The Poisson Distribution

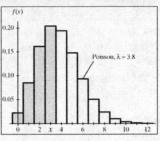
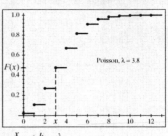

$$F(x) = P(X \le x) = \sum_{k=0}^{x} \frac{\lambda^k e^{-\lambda}}{k!}$$

| | | | | | $\lambda = E(X)$ | | | | | |
|---|---|---|---|---|---|---|---|---|---|---|
| $x$ | 0.1 | 0.2 | 0.3 | 0.4 | 0.5 | 0.6 | 0.7 | 0.8 | 0.9 | 1.0 |
| 0 | 0.905 | 0.819 | 0.741 | 0.670 | 0.607 | 0.549 | 0.497 | 0.449 | 0.407 | 0.368 |
| 1 | 0.995 | 0.982 | 0.963 | 0.938 | 0.910 | 0.878 | 0.844 | 0.809 | 0.772 | 0.736 |
| 2 | 1.000 | 0.999 | 0.996 | 0.992 | 0.986 | 0.977 | 0.966 | 0.953 | 0.937 | 0.920 |
| 3 | 1.000 | 1.000 | 1.000 | 0.999 | 0.998 | 0.997 | 0.994 | 0.991 | 0.987 | 0.981 |
| 4 | 1.000 | 1.000 | 1.000 | 1.000 | 1.000 | 1.000 | 0.999 | 0.999 | 0.998 | 0.996 |
| 5 | 1.000 | 1.000 | 1.000 | 1.000 | 1.000 | 1.000 | 1.000 | 1.000 | 1.000 | 0.999 |
| 6 | 1.000 | 1.000 | 1.000 | 1.000 | 1.000 | 1.000 | 1.000 | 1.000 | 1.000 | 1.000 |

| $x$ | 1.1 | 1.2 | 1.3 | 1.4 | 1.5 | 1.6 | 1.7 | 1.8 | 1.9 | 2.0 |
|---|---|---|---|---|---|---|---|---|---|---|
| 0 | 0.333 | 0.301 | 0.273 | 0.247 | 0.223 | 0.202 | 0.183 | 0.165 | 0.150 | 0.135 |
| 1 | 0.699 | 0.663 | 0.627 | 0.592 | 0.558 | 0.525 | 0.493 | 0.463 | 0.434 | 0.406 |
| 2 | 0.900 | 0.879 | 0.857 | 0.833 | 0.809 | 0.783 | 0.757 | 0.731 | 0.704 | 0.677 |
| 3 | 0.974 | 0.966 | 0.957 | 0.946 | 0.934 | 0.921 | 0.907 | 0.891 | 0.875 | 0.857 |
| 4 | 0.995 | 0.992 | 0.989 | 0.986 | 0.981 | 0.976 | 0.970 | 0.964 | 0.956 | 0.947 |
| 5 | 0.999 | 0.998 | 0.998 | 0.997 | 0.996 | 0.994 | 0.992 | 0.990 | 0.987 | 0.983 |
| 6 | 1.000 | 1.000 | 1.000 | 0.999 | 0.999 | 0.999 | 0.998 | 0.997 | 0.997 | 0.995 |
| 7 | 1.000 | 1.000 | 1.000 | 1.000 | 1.000 | 1.000 | 1.000 | 0.999 | 0.999 | 0.999 |
| 8 | 1.000 | 1.000 | 1.000 | 1.000 | 1.000 | 1.000 | 1.000 | 1.000 | 1.000 | 1.000 |

| $x$ | 2.2 | 2.4 | 2.6 | 2.8 | 3.0 | 3.2 | 3.4 | 3.6 | 3.8 | 4.0 |
|---|---|---|---|---|---|---|---|---|---|---|
| 0 | 0.111 | 0.091 | 0.074 | 0.061 | 0.050 | 0.041 | 0.033 | 0.027 | 0.022 | 0.018 |
| 1 | 0.355 | 0.308 | 0.267 | 0.231 | 0.199 | 0.171 | 0.147 | 0.126 | 0.107 | 0.092 |
| 2 | 0.623 | 0.570 | 0.518 | 0.469 | 0.423 | 0.380 | 0.340 | 0.303 | 0.269 | 0.238 |
| 3 | 0.819 | 0.779 | 0.736 | 0.692 | 0.647 | 0.603 | 0.558 | 0.515 | 0.473 | 0.433 |
| 4 | 0.928 | 0.904 | 0.877 | 0.848 | 0.815 | 0.781 | 0.744 | 0.706 | 0.668 | 0.629 |
| 5 | 0.975 | 0.964 | 0.951 | 0.935 | 0.916 | 0.895 | 0.871 | 0.844 | 0.816 | 0.785 |
| 6 | 0.993 | 0.988 | 0.983 | 0.976 | 0.966 | 0.955 | 0.942 | 0.927 | 0.909 | 0.889 |
| 7 | 0.998 | 0.997 | 0.995 | 0.992 | 0.988 | 0.983 | 0.977 | 0.969 | 0.960 | 0.949 |
| 8 | 1.000 | 0.999 | 0.999 | 0.998 | 0.996 | 0.994 | 0.992 | 0.988 | 0.984 | 0.979 |
| 9 | 1.000 | 1.000 | 1.000 | 0.999 | 0.999 | 0.998 | 0.997 | 0.996 | 0.994 | 0.992 |
| 10 | 1.000 | 1.000 | 1.000 | 1.000 | 1.000 | 1.000 | 0.999 | 0.999 | 0.998 | 0.997 |
| 11 | 1.000 | 1.000 | 1.000 | 1.000 | 1.000 | 1.000 | 1.000 | 1.000 | 0.999 | 0.999 |
| 12 | 1.000 | 1.000 | 1.000 | 1.000 | 1.000 | 1.000 | 1.000 | 1.000 | 1.000 | 1.000 |

## Table III *continued*

| x | 4.2 | 4.4 | 4.6 | 4.8 | 5.0 | 5.2 | 5.4 | 5.6 | 5.8 | 6.0 |
|---|------|------|------|------|------|------|------|------|------|------|
| 0 | 0.015 | 0.012 | 0.010 | 0.008 | 0.007 | 0.006 | 0.005 | 0.004 | 0.003 | 0.002 |
| 1 | 0.078 | 0.066 | 0.056 | 0.048 | 0.040 | 0.034 | 0.029 | 0.024 | 0.021 | 0.017 |
| 2 | 0.210 | 0.185 | 0.163 | 0.143 | 0.125 | 0.109 | 0.095 | 0.082 | 0.072 | 0.062 |
| 3 | 0.395 | 0.359 | 0.326 | 0.294 | 0.265 | 0.238 | 0.213 | 0.191 | 0.170 | 0.151 |
| 4 | 0.590 | 0.551 | 0.513 | 0.476 | 0.440 | 0.406 | 0.373 | 0.342 | 0.313 | 0.285 |
| 5 | 0.753 | 0.720 | 0.686 | 0.651 | 0.616 | 0.581 | 0.546 | 0.512 | 0.478 | 0.446 |
| 6 | 0.867 | 0.844 | 0.818 | 0.791 | 0.762 | 0.732 | 0.702 | 0.670 | 0.638 | 0.606 |
| 7 | 0.936 | 0.921 | 0.905 | 0.887 | 0.867 | 0.845 | 0.822 | 0.797 | 0.771 | 0.744 |
| 8 | 0.972 | 0.964 | 0.955 | 0.944 | 0.932 | 0.918 | 0.903 | 0.886 | 0.867 | 0.847 |
| 9 | 0.989 | 0.985 | 0.980 | 0.975 | 0.968 | 0.960 | 0.951 | 0.941 | 0.929 | 0.916 |
| 10 | 0.996 | 0.994 | 0.992 | 0.990 | 0.986 | 0.982 | 0.977 | 0.972 | 0.965 | 0.957 |
| 11 | 0.999 | 0.998 | 0.997 | 0.996 | 0.995 | 0.993 | 0.990 | 0.988 | 0.984 | 0.980 |
| 12 | 1.000 | 0.999 | 0.999 | 0.999 | 0.998 | 0.997 | 0.996 | 0.995 | 0.993 | 0.991 |
| 13 | 1.000 | 1.000 | 1.000 | 1.000 | 0.999 | 0.999 | 0.999 | 0.998 | 0.997 | 0.996 |
| 14 | 1.000 | 1.000 | 1.000 | 1.000 | 1.000 | 1.000 | 0.999 | 0.999 | 0.999 | 0.999 |
| 15 | 1.000 | 1.000 | 1.000 | 1.000 | 1.000 | 1.000 | 1.000 | 1.000 | 1.000 | 0.999 |
| 16 | 1.000 | 1.000 | 1.000 | 1.000 | 1.000 | 1.000 | 1.000 | 1.000 | 1.000 | 1.000 |

| x | 6.5 | 7.0 | 7.5 | 8.0 | 8.5 | 9.0 | 9.5 | 10.0 | 10.5 | 11.0 |
|---|------|------|------|------|------|------|------|------|------|------|
| 0 | 0.002 | 0.001 | 0.001 | 0.000 | 0.000 | 0.000 | 0.000 | 0.000 | 0.000 | 0.000 |
| 1 | 0.011 | 0.007 | 0.005 | 0.003 | 0.002 | 0.001 | 0.001 | 0.000 | 0.000 | 0.000 |
| 2 | 0.043 | 0.030 | 0.020 | 0.014 | 0.009 | 0.006 | 0.004 | 0.003 | 0.002 | 0.001 |
| 3 | 0.112 | 0.082 | 0.059 | 0.042 | 0.030 | 0.021 | 0.015 | 0.010 | 0.007 | 0.005 |
| 4 | 0.224 | 0.173 | 0.132 | 0.100 | 0.074 | 0.055 | 0.040 | 0.029 | 0.021 | 0.015 |
| 5 | 0.369 | 0.301 | 0.241 | 0.191 | 0.150 | 0.116 | 0.089 | 0.067 | 0.050 | 0.038 |
| 6 | 0.527 | 0.450 | 0.378 | 0.313 | 0.256 | 0.207 | 0.165 | 0.130 | 0.102 | 0.079 |
| 7 | 0.673 | 0.599 | 0.525 | 0.453 | 0.386 | 0.324 | 0.269 | 0.220 | 0.179 | 0.143 |
| 8 | 0.792 | 0.729 | 0.662 | 0.593 | 0.523 | 0.456 | 0.392 | 0.333 | 0.279 | 0.232 |
| 9 | 0.877 | 0.830 | 0.776 | 0.717 | 0.653 | 0.587 | 0.522 | 0.458 | 0.397 | 0.341 |
| 10 | 0.933 | 0.901 | 0.862 | 0.816 | 0.763 | 0.706 | 0.645 | 0.583 | 0.521 | 0.460 |
| 11 | 0.966 | 0.947 | 0.921 | 0.888 | 0.849 | 0.803 | 0.752 | 0.697 | 0.639 | 0.579 |
| 12 | 0.984 | 0.973 | 0.957 | 0.936 | 0.909 | 0.876 | 0.836 | 0.792 | 0.742 | 0.689 |
| 13 | 0.993 | 0.987 | 0.978 | 0.966 | 0.949 | 0.926 | 0.898 | 0.864 | 0.825 | 0.781 |
| 14 | 0.997 | 0.994 | 0.990 | 0.983 | 0.973 | 0.959 | 0.940 | 0.917 | 0.888 | 0.854 |
| 15 | 0.999 | 0.998 | 0.995 | 0.992 | 0.986 | 0.978 | 0.967 | 0.951 | 0.932 | 0.907 |
| 16 | 1.000 | 0.999 | 0.998 | 0.996 | 0.993 | 0.989 | 0.982 | 0.973 | 0.960 | 0.944 |
| 17 | 1.000 | 1.000 | 0.999 | 0.998 | 0.997 | 0.995 | 0.991 | 0.986 | 0.978 | 0.968 |
| 18 | 1.000 | 1.000 | 1.000 | 0.999 | 0.999 | 0.998 | 0.096 | 0.993 | 0.988 | 0.982 |
| 19 | 1.000 | 1.000 | 1.000 | 1.000 | 0.999 | 0.999 | 0.998 | 0.997 | 0.994 | 0.991 |
| 20 | 1.000 | 1.000 | 1.000 | 1.000 | 1.000 | 1.000 | 0.999 | 0.998 | 0.997 | 0.995 |
| 21 | 1.000 | 1.000 | 1.000 | 1.000 | 1.000 | 1.000 | 1.000 | 0.999 | 0.999 | 0.998 |
| 22 | 1.000 | 1.000 | 1.000 | 1.000 | 1.000 | 1.000 | 1.000 | 1.000 | 0.999 | 0.999 |
| 23 | 1.000 | 1.000 | 1.000 | 1.000 | 1.000 | 1.000 | 1.000 | 1.000 | 1.000 | 1.000 |

| | Table III *continued* | | | | | | | | | |
|---|---|---|---|---|---|---|---|---|---|---|
| x | 11.5 | 12.0 | 12.5 | 13.0 | 13.5 | 14.0 | 14.5 | 15.0 | 15.5 | 16.0 |
| 0 | 0.000 | 0.000 | 0.000 | 0.000 | 0.000 | 0.000 | 0.000 | 0.000 | 0.000 | 0.000 |
| 1 | 0.000 | 0.000 | 0.000 | 0.000 | 0.000 | 0.000 | 0.000 | 0.000 | 0.000 | 0.000 |
| 2 | 0.001 | 0.001 | 0.000 | 0.000 | 0.000 | 0.000 | 0.000 | 0.000 | 0.000 | 0.000 |
| 3 | 0.003 | 0.002 | 0.002 | 0.001 | 0.001 | 0.000 | 0.000 | 0.000 | 0.000 | 0.000 |
| 4 | 0.011 | 0.008 | 0.005 | 0.004 | 0.003 | 0.002 | 0.001 | 0.001 | 0.001 | 0.000 |
| 5 | 0.028 | 0.020 | 0.015 | 0.011 | 0.008 | 0.006 | 0.004 | 0.003 | 0.002 | 0.001 |
| 6 | 0.060 | 0.046 | 0.035 | 0.026 | 0.019 | 0.014 | 0.010 | 0.008 | 0.006 | 0.004 |
| 7 | 0.114 | 0.090 | 0.070 | 0.054 | 0.041 | 0.032 | 0.024 | 0.018 | 0.013 | 0.010 |
| 8 | 0.191 | 0.155 | 0.125 | 0.100 | 0.079 | 0.062 | 0.048 | 0.037 | 0.029 | 0.022 |
| 9 | 0.289 | 0.242 | 0.201 | 0.166 | 0.135 | 0.109 | 0.088 | 0.070 | 0.055 | 0.043 |
| 10 | 0.402 | 0.347 | 0.297 | 0.252 | 0.211 | 0.176 | 0.145 | 0.118 | 0.096 | 0.077 |
| 11 | 0.520 | 0.462 | 0.406 | 0.353 | 0.304 | 0.260 | 0.220 | 0.185 | 0.154 | 0.127 |
| 12 | 0.633 | 0.576 | 0.519 | 0.463 | 0.409 | 0.358 | 0.311 | 0.268 | 0.228 | 0.193 |
| 13 | 0.733 | 0.682 | 0.629 | 0.573 | 0.518 | 0.464 | 0.413 | 0.363 | 0.317 | 0.275 |
| 14 | 0.815 | 0.772 | 0.725 | 0.675 | 0.623 | 0.570 | 0.518 | 0.466 | 0.415 | 0.368 |
| 15 | 0.878 | 0.844 | 0.806 | 0.764 | 0.718 | 0.669 | 0.619 | 0.568 | 0.517 | 0.467 |
| 16 | 0.924 | 0.899 | 0.869 | 0.835 | 0.798 | 0.756 | 0.711 | 0.664 | 0.615 | 0.566 |
| 17 | 0.954 | 0.937 | 0.916 | 0.890 | 0.861 | 0.827 | 0.790 | 0.749 | 0.705 | 0.659 |
| 18 | 0.974 | 0.963 | 0.948 | 0.930 | 0.908 | 0.883 | 0.853 | 0.819 | 0.782 | 0.742 |
| 19 | 0.986 | 0.979 | 0.969 | 0.957 | 0.942 | 0.923 | 0.901 | 0.875 | 0.846 | 0.812 |
| 20 | 0.992 | 0.988 | 0.983 | 0.975 | 0.965 | 0.952 | 0.936 | 0.917 | 0.894 | 0.868 |
| 21 | 0.996 | 0.994 | 0.991 | 0.986 | 0.980 | 0.971 | 0.960 | 0.947 | 0.930 | 0.911 |
| 22 | 0.999 | 0.997 | 0.995 | 0.992 | 0.989 | 0.983 | 0.976 | 0.967 | 0.956 | 0.942 |
| 23 | 0.999 | 0.999 | 0.998 | 0.996 | 0.994 | 0.991 | 0.986 | 0.981 | 0.973 | 0.963 |
| 24 | 1.000 | 0.999 | 0.999 | 0.998 | 0.997 | 0.995 | 0.992 | 0.989 | 0.984 | 0.978 |
| 25 | 1.000 | 1.000 | 0.999 | 0.999 | 0.998 | 0.997 | 0.996 | 0.994 | 0.991 | 0.987 |
| 26 | 1.000 | 1.000 | 1.000 | 1.000 | 0.999 | 0.999 | 0.998 | 0.997 | 0.995 | 0.993 |
| 27 | 1.000 | 1.000 | 1.000 | 1.000 | 1.000 | 0.999 | 0.999 | 0.998 | 0.997 | 0.996 |
| 28 | 1.000 | 1.000 | 1.000 | 1.000 | 1.000 | 1.000 | 0.999 | 0.999 | 0.999 | 0.998 |
| 29 | 1.000 | 1.000 | 1.000 | 1.000 | 1.000 | 1.000 | 1.000 | 1.000 | 0.999 | 0.999 |
| 30 | 1.000 | 1.000 | 1.000 | 1.000 | 1.000 | 1.000 | 1.000 | 1.000 | 1.000 | 0.999 |
| 31 | 1.000 | 1.000 | 1.000 | 1.000 | 1.000 | 1.000 | 1.000 | 1.000 | 1.000 | 1.000 |
| 32 | 1.000 | 1.000 | 1.000 | 1.000 | 1.000 | 1.000 | 1.000 | 1.000 | 1.000 | 1.000 |
| 33 | 1.000 | 1.000 | 1.000 | 1.000 | 1.000 | 1.000 | 1.000 | 1.000 | 1.000 | 1.000 |
| 34 | 1.000 | 1.000 | 1.000 | 1.000 | 1.000 | 1.000 | 1.000 | 1.000 | 1.000 | 1.000 |
| 35 | 1.000 | 1.000 | 1.000 | 1.000 | 1.000 | 1.000 | 1.000 | 1.000 | 1.000 | 1.000 |

## Table IV  The Chi-Square Distribution

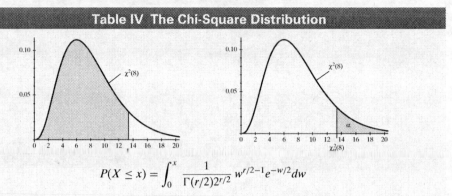

$$P(X \leq x) = \int_0^x \frac{1}{\Gamma(r/2)2^{r/2}} w^{r/2-1}e^{-w/2}dw$$

| | $P(X \leq x)$ | | | | | | | |
|---|---|---|---|---|---|---|---|---|
| | 0.010 | 0.025 | 0.050 | 0.100 | 0.900 | 0.950 | 0.975 | 0.990 |
| $r$ | $\chi^2_{0.99}(r)$ | $\chi^2_{0.975}(r)$ | $\chi^2_{0.95}(r)$ | $\chi^2_{0.90}(r)$ | $\chi^2_{0.10}(r)$ | $\chi^2_{0.05}(r)$ | $\chi^2_{0.025}(r)$ | $\chi^2_{0.01}(r)$ |
| 1 | 0.000 | 0.001 | 0.004 | 0.016 | 2.706 | 3.841 | 5.024 | 6.635 |
| 2 | 0.020 | 0.051 | 0.103 | 0.211 | 4.605 | 5.991 | 7.378 | 9.210 |
| 3 | 0.115 | 0.216 | 0.352 | 0.584 | 6.251 | 7.815 | 9.348 | 11.34 |
| 4 | 0.297 | 0.484 | 0.711 | 1.064 | 7.779 | 9.488 | 11.14 | 13.28 |
| 5 | 0.554 | 0.831 | 1.145 | 1.610 | 9.236 | 11.07 | 12.83 | 15.09 |
| 6 | 0.872 | 1.237 | 1.635 | 2.204 | 10.64 | 12.59 | 14.45 | 16.81 |
| 7 | 1.239 | 1.690 | 2.167 | 2.833 | 12.02 | 14.07 | 16.01 | 18.48 |
| 8 | 1.646 | 2.180 | 2.733 | 3.490 | 13.36 | 15.51 | 17.54 | 20.09 |
| 9 | 2.088 | 2.700 | 3.325 | 4.168 | 14.68 | 16.92 | 19.02 | 21.67 |
| 10 | 2.558 | 3.247 | 3.940 | 4.865 | 15.99 | 18.31 | 20.48 | 23.21 |
| 11 | 3.053 | 3.816 | 4.575 | 5.578 | 17.28 | 19.68 | 21.92 | 24.72 |
| 12 | 3.571 | 4.404 | 5.226 | 6.304 | 18.55 | 21.03 | 23.34 | 26.22 |
| 13 | 4.107 | 5.009 | 5.892 | 7.042 | 19.81 | 22.36 | 24.74 | 27.69 |
| 14 | 4.660 | 5.629 | 6.571 | 7.790 | 21.06 | 23.68 | 26.12 | 29.14 |
| 15 | 5.229 | 6.262 | 7.261 | 8.547 | 22.31 | 25.00 | 27.49 | 30.58 |
| 16 | 5.812 | 6.908 | 7.962 | 9.312 | 23.54 | 26.30 | 28.84 | 32.00 |
| 17 | 6.408 | 7.564 | 8.672 | 10.08 | 24.77 | 27.59 | 30.19 | 33.41 |
| 18 | 7.015 | 8.231 | 9.390 | 10.86 | 25.99 | 28.87 | 31.53 | 34.80 |
| 19 | 7.633 | 8.907 | 10.12 | 11.65 | 27.20 | 30.14 | 32.85 | 36.19 |
| 20 | 8.260 | 9.591 | 10.85 | 12.44 | 28.41 | 31.41 | 34.17 | 37.57 |
| 21 | 8.897 | 10.28 | 11.59 | 13.24 | 29.62 | 32.67 | 35.48 | 38.93 |
| 22 | 9.542 | 10.98 | 12.34 | 14.04 | 30.81 | 33.92 | 36.78 | 40.29 |
| 23 | 10.20 | 11.69 | 13.09 | 14.85 | 32.01 | 35.17 | 38.08 | 41.64 |
| 24 | 10.86 | 12.40 | 13.85 | 15.66 | 33.20 | 36.42 | 39.36 | 42.98 |
| 25 | 11.52 | 13.12 | 14.61 | 16.47 | 34.38 | 37.65 | 40.65 | 44.31 |
| 26 | 12.20 | 13.84 | 15.38 | 17.29 | 35.56 | 38.88 | 41.92 | 45.64 |
| 27 | 12.88 | 14.57 | 16.15 | 18.11 | 36.74 | 40.11 | 43.19 | 46.96 |
| 28 | 13.56 | 15.31 | 16.93 | 18.94 | 37.92 | 41.34 | 44.46 | 48.28 |
| 29 | 14.26 | 16.05 | 17.71 | 19.77 | 39.09 | 42.56 | 45.72 | 49.59 |
| 30 | 14.95 | 16.79 | 18.49 | 20.60 | 40.26 | 43.77 | 46.98 | 50.89 |
| 40 | 22.16 | 24.43 | 26.51 | 29.05 | 51.80 | 55.76 | 59.34 | 63.69 |
| 50 | 29.71 | 32.36 | 34.76 | 37.69 | 63.17 | 67.50 | 71.42 | 76.15 |
| 60 | 37.48 | 40.48 | 43.19 | 46.46 | 74.40 | 79.08 | 83.30 | 88.38 |
| 70 | 45.44 | 48.76 | 51.74 | 55.33 | 85.53 | 90.53 | 95.02 | 100.4 |
| 80 | 53.34 | 57.15 | 60.39 | 64.28 | 96.58 | 101.9 | 106.6 | 112.3 |

## Table Va: The Standard Normal Distribution Function

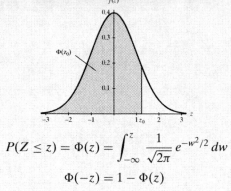

$$P(Z \leq z) = \Phi(z) = \int_{-\infty}^{z} \frac{1}{\sqrt{2\pi}} e^{-w^2/2} \, dw$$

$$\Phi(-z) = 1 - \Phi(z)$$

| z | 0.00 | 0.01 | 0.02 | 0.03 | 0.04 | 0.05 | 0.06 | 0.07 | 0.08 | 0.09 |
|---|------|------|------|------|------|------|------|------|------|------|
| 0.0 | 0.5000 | 0.5040 | 0.5080 | 0.5120 | 0.5160 | 0.5199 | 0.5239 | 0.5279 | 0.5319 | 0.5359 |
| 0.1 | 0.5398 | 0.5438 | 0.5478 | 0.5517 | 0.5557 | 0.5596 | 0.5636 | 0.5675 | 0.5714 | 0.5753 |
| 0.2 | 0.5793 | 0.5832 | 0.5871 | 0.5910 | 0.5948 | 0.5987 | 0.6026 | 0.6064 | 0.6103 | 0.6141 |
| 0.3 | 0.6179 | 0.6217 | 0.6255 | 0.6293 | 0.6331 | 0.6368 | 0.6406 | 0.6443 | 0.6480 | 0.6517 |
| 0.4 | 0.6554 | 0.6591 | 0.6628 | 0.6664 | 0.6700 | 0.6736 | 0.6772 | 0.6808 | 0.6844 | 0.6879 |
| 0.5 | 0.6915 | 0.6950 | 0.6985 | 0.7019 | 0.7054 | 0.7088 | 0.7123 | 0.7157 | 0.7190 | 0.7224 |
| 0.6 | 0.7257 | 0.7291 | 0.7324 | 0.7357 | 0.7389 | 0.7422 | 0.7454 | 0.7486 | 0.7517 | 0.7549 |
| 0.7 | 0.7580 | 0.7611 | 0.7642 | 0.7673 | 0.7703 | 0.7734 | 0.7764 | 0.7794 | 0.7823 | 0.7852 |
| 0.8 | 0.7881 | 0.7910 | 0.7939 | 0.7967 | 0.7995 | 0.8023 | 0.8051 | 0.8078 | 0.8106 | 0.8133 |
| 0.9 | 0.8159 | 0.8186 | 0.8212 | 0.8238 | 0.8264 | 0.8289 | 0.8315 | 0.8340 | 0.8365 | 0.8389 |
| 1.0 | 0.8413 | 0.8438 | 0.8461 | 0.8485 | 0.8508 | 0.8531 | 0.8554 | 0.8577 | 0.8599 | 0.8621 |
| 1.1 | 0.8643 | 0.8665 | 0.8686 | 0.8708 | 0.8729 | 0.8749 | 0.8770 | 0.8790 | 0.8810 | 0.8830 |
| 1.2 | 0.8849 | 0.8869 | 0.8888 | 0.8907 | 0.8925 | 0.8944 | 0.8962 | 0.8980 | 0.8997 | 0.9015 |
| 1.3 | 0.9032 | 0.9049 | 0.9066 | 0.9082 | 0.9099 | 0.9115 | 0.9131 | 0.9147 | 0.9162 | 0.9177 |
| 1.4 | 0.9192 | 0.9207 | 0.9222 | 0.9236 | 0.9251 | 0.9265 | 0.9279 | 0.9292 | 0.9306 | 0.9319 |
| 1.5 | 0.9332 | 0.9345 | 0.9357 | 0.9370 | 0.9382 | 0.9394 | 0.9406 | 0.9418 | 0.9429 | 0.9441 |
| 1.6 | 0.9452 | 0.9463 | 0.9474 | 0.9484 | 0.9495 | 0.9505 | 0.9515 | 0.9525 | 0.9535 | 0.9545 |
| 1.7 | 0.9554 | 0.9564 | 0.9573 | 0.9582 | 0.9591 | 0.9599 | 0.9608 | 0.9616 | 0.9625 | 0.9633 |
| 1.8 | 0.9641 | 0.9649 | 0.9656 | 0.9664 | 0.9671 | 0.9678 | 0.9686 | 0.9693 | 0.9699 | 0.9706 |
| 1.9 | 0.9713 | 0.9719 | 0.9726 | 0.9732 | 0.9738 | 0.9744 | 0.9750 | 0.9756 | 0.9761 | 0.9767 |
| 2.0 | 0.9772 | 0.9778 | 0.9783 | 0.9788 | 0.9793 | 0.9798 | 0.9803 | 0.9808 | 0.9812 | 0.9817 |
| 2.1 | 0.9821 | 0.9826 | 0.9830 | 0.9834 | 0.9838 | 0.9842 | 0.9846 | 0.9850 | 0.9854 | 0.9857 |
| 2.2 | 0.9861 | 0.9864 | 0.9868 | 0.9871 | 0.9875 | 0.9878 | 0.9881 | 0.9884 | 0.9887 | 0.9890 |
| 2.3 | 0.9893 | 0.9896 | 0.9898 | 0.9901 | 0.9904 | 0.9906 | 0.9909 | 0.9911 | 0.9913 | 0.9916 |
| 2.4 | 0.9918 | 0.9920 | 0.9922 | 0.9925 | 0.9927 | 0.9929 | 0.9931 | 0.9932 | 0.9934 | 0.9936 |
| 2.5 | 0.9938 | 0.9940 | 0.9941 | 0.9943 | 0.9945 | 0.9946 | 0.9948 | 0.9949 | 0.9951 | 0.9952 |
| 2.6 | 0.9953 | 0.9955 | 0.9956 | 0.9957 | 0.9959 | 0.9960 | 0.9961 | 0.9962 | 0.9963 | 0.9964 |
| 2.7 | 0.9965 | 0.9966 | 0.9967 | 0.9968 | 0.9969 | 0.9970 | 0.9971 | 0.9972 | 0.9973 | 0.9974 |
| 2.8 | 0.9974 | 0.9975 | 0.9976 | 0.9977 | 0.9977 | 0.9978 | 0.9979 | 0.9979 | 0.9980 | 0.9981 |
| 2.9 | 0.9981 | 0.9982 | 0.9982 | 0.9983 | 0.9984 | 0.9984 | 0.9985 | 0.9985 | 0.9986 | 0.9986 |
| 3.0 | 0.9987 | 0.9987 | 0.9987 | 0.9988 | 0.9988 | 0.9989 | 0.9989 | 0.9989 | 0.9990 | 0.9990 |

| $\alpha$ | 0.400 | 0.300 | 0.200 | 0.100 | 0.050 | 0.025 | 0.020 | 0.010 | 0.005 | 0.001 |
|---|------|------|------|------|------|------|------|------|------|------|
| $z_\alpha$ | 0.253 | 0.524 | 0.842 | 1.282 | 1.645 | 1.960 | 2.054 | 2.326 | 2.576 | 3.090 |
| $z_{\alpha/2}$ | 0.842 | 1.036 | 1.282 | 1.645 | 1.960 | 2.240 | 2.326 | 2.576 | 2.807 | 3.291 |

## Table Vb:  The Standard Normal Right Tail Probabilities

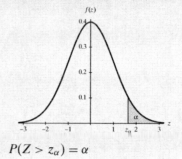

$$P(Z > z_\alpha) = \alpha$$

$$P(Z > z) = 1 - \Phi(z) = \Phi(-z)$$

| $z_\alpha$ | 0.00 | 0.01 | 0.02 | 0.03 | 0.04 | 0.05 | 0.06 | 0.07 | 0.08 | 0.09 |
|---|---|---|---|---|---|---|---|---|---|---|
| 0.0 | 0.5000 | 0.4960 | 0.4920 | 0.4880 | 0.4840 | 0.4801 | 0.4761 | 0.4721 | 0.4681 | 0.4641 |
| 0.1 | 0.4602 | 0.4562 | 0.4522 | 0.4483 | 0.4443 | 0.4404 | 0.4364 | 0.4325 | 0.4286 | 0.4247 |
| 0.2 | 0.4207 | 0.4168 | 0.4129 | 0.4090 | 0.4052 | 0.4013 | 0.3974 | 0.3936 | 0.3897 | 0.3859 |
| 0.3 | 0.3821 | 0.3783 | 0.3745 | 0.3707 | 0.3669 | 0.3632 | 0.3594 | 0.3557 | 0.3520 | 0.3483 |
| 0.4 | 0.3446 | 0.3409 | 0.3372 | 0.3336 | 0.3300 | 0.3264 | 0.3228 | 0.3192 | 0.3156 | 0.3121 |
| 0.5 | 0.3085 | 0.3050 | 0.3015 | 0.2981 | 0.2946 | 0.2912 | 0.2877 | 0.2843 | 0.2810 | 0.2776 |
| 0.6 | 0.2743 | 0.2709 | 0.2676 | 0.2643 | 0.2611 | 0.2578 | 0.2546 | 0.2514 | 0.2483 | 0.2451 |
| 0.7 | 0.2420 | 0.2389 | 0.2358 | 0.2327 | 0.2296 | 0.2266 | 0.2236 | 0.2206 | 0.2177 | 0.2148 |
| 0.8 | 0.2119 | 0.2090 | 0.2061 | 0.2033 | 0.2005 | 0.1977 | 0.1949 | 0.1922 | 0.1894 | 0.1867 |
| 0.9 | 0.1841 | 0.1814 | 0.1788 | 0.1762 | 0.1736 | 0.1711 | 0.1685 | 0.1660 | 0.1635 | 0.1611 |
| 1.0 | 0.1587 | 0.1562 | 0.1539 | 0.1515 | 0.1492 | 0.1469 | 0.1446 | 0.1423 | 0.1401 | 0.1379 |
| 1.1 | 0.1357 | 0.1335 | 0.1314 | 0.1292 | 0.1271 | 0.1251 | 0.1230 | 0.1210 | 0.1190 | 0.1170 |
| 1.2 | 0.1151 | 0.1131 | 0.1112 | 0.1093 | 0.1075 | 0.1056 | 0.1038 | 0.1020 | 0.1003 | 0.0985 |
| 1.3 | 0.0968 | 0.0951 | 0.0934 | 0.0918 | 0.0901 | 0.0885 | 0.0869 | 0.0853 | 0.0838 | 0.0823 |
| 1.4 | 0.0808 | 0.0793 | 0.0778 | 0.0764 | 0.0749 | 0.0735 | 0.0721 | 0.0708 | 0.0694 | 0.0681 |
| 1.5 | 0.0668 | 0.0655 | 0.0643 | 0.0630 | 0.0618 | 0.0606 | 0.0594 | 0.0582 | 0.0571 | 0.0559 |
| 1.6 | 0.0548 | 0.0537 | 0.0526 | 0.0516 | 0.0505 | 0.0495 | 0.0485 | 0.0475 | 0.0465 | 0.0455 |
| 1.7 | 0.0446 | 0.0436 | 0.0427 | 0.0418 | 0.0409 | 0.0401 | 0.0392 | 0.0384 | 0.0375 | 0.0367 |
| 1.8 | 0.0359 | 0.0351 | 0.0344 | 0.0336 | 0.0329 | 0.0322 | 0.0314 | 0.0307 | 0.0301 | 0.0294 |
| 1.9 | 0.0287 | 0.0281 | 0.0274 | 0.0268 | 0.0262 | 0.0256 | 0.0250 | 0.0244 | 0.0239 | 0.0233 |
| 2.0 | 0.0228 | 0.0222 | 0.0217 | 0.0212 | 0.0207 | 0.0202 | 0.0197 | 0.0192 | 0.0188 | 0.0183 |
| 2.1 | 0.0179 | 0.0174 | 0.0170 | 0.0166 | 0.0162 | 0.0158 | 0.0154 | 0.0150 | 0.0146 | 0.0143 |
| 2.2 | 0.0139 | 0.0136 | 0.0132 | 0.0129 | 0.0125 | 0.0122 | 0.0119 | 0.0116 | 0.0113 | 0.0110 |
| 2.3 | 0.0107 | 0.0104 | 0.0102 | 0.0099 | 0.0096 | 0.0094 | 0.0091 | 0.0089 | 0.0087 | 0.0084 |
| 2.4 | 0.0082 | 0.0080 | 0.0078 | 0.0075 | 0.0073 | 0.0071 | 0.0069 | 0.0068 | 0.0066 | 0.0064 |
| 2.5 | 0.0062 | 0.0060 | 0.0059 | 0.0057 | 0.0055 | 0.0054 | 0.0052 | 0.0051 | 0.0049 | 0.0048 |
| 2.6 | 0.0047 | 0.0045 | 0.0044 | 0.0043 | 0.0041 | 0.0040 | 0.0039 | 0.0038 | 0.0037 | 0.0036 |
| 2.7 | 0.0035 | 0.0034 | 0.0033 | 0.0032 | 0.0031 | 0.0030 | 0.0029 | 0.0028 | 0.0027 | 0.0026 |
| 2.8 | 0.0026 | 0.0025 | 0.0024 | 0.0023 | 0.0023 | 0.0022 | 0.0021 | 0.0021 | 0.0020 | 0.0019 |
| 2.9 | 0.0019 | 0.0018 | 0.0018 | 0.0017 | 0.0016 | 0.0016 | 0.0015 | 0.0015 | 0.0014 | 0.0014 |
| 3.0 | 0.0013 | 0.0013 | 0.0013 | 0.0012 | 0.0012 | 0.0011 | 0.0011 | 0.0011 | 0.0010 | 0.0010 |
| 3.1 | 0.0010 | 0.0009 | 0.0009 | 0.0009 | 0.0008 | 0.0008 | 0.0008 | 0.0008 | 0.0007 | 0.0007 |
| 3.2 | 0.0007 | 0.0007 | 0.0006 | 0.0006 | 0.0006 | 0.0006 | 0.0006 | 0.0005 | 0.0005 | 0.0005 |
| 3.3 | 0.0005 | 0.0005 | 0.0005 | 0.0004 | 0.0004 | 0.0004 | 0.0004 | 0.0004 | 0.0004 | 0.0003 |
| 3.4 | 0.0003 | 0.0003 | 0.0003 | 0.0003 | 0.0003 | 0.0003 | 0.0003 | 0.0003 | 0.0003 | 0.0002 |

## Table VI The *t* Distribution

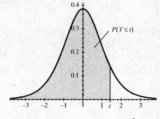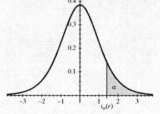

$$P(T \le t) = \int_{-\infty}^{t} \frac{\Gamma[(r+1)/2]}{\sqrt{\pi r}\, \Gamma(r/2)(1+w^2/r)^{(r+1)/2}}\, dw$$

$$P(T \le -t) = 1 - P(T \le t)$$

| | $P(T \le t)$ | | | | | | |
|---|---|---|---|---|---|---|---|
| | 0.60 | 0.75 | 0.90 | 0.95 | 0.975 | 0.99 | 0.995 |
| $r$ | $t_{0.40}(r)$ | $t_{0.25}(r)$ | $t_{0.10}(r)$ | $t_{0.05}(r)$ | $t_{0.025}(r)$ | $t_{0.01}(r)$ | $t_{0.005}(r)$ |
| 1 | 0.325 | 1.000 | 3.078 | 6.314 | 12.706 | 31.821 | 63.657 |
| 2 | 0.289 | 0.816 | 1.886 | 2.920 | 4.303 | 6.965 | 9.925 |
| 3 | 0.277 | 0.765 | 1.638 | 2.353 | 3.182 | 4.541 | 5.841 |
| 4 | 0.271 | 0.741 | 1.533 | 2.132 | 2.776 | 3.747 | 4.604 |
| 5 | 0.267 | 0.727 | 1.476 | 2.015 | 2.571 | 3.365 | 4.032 |
| 6 | 0.265 | 0.718 | 1.440 | 1.943 | 2.447 | 3.143 | 3.707 |
| 7 | 0.263 | 0.711 | 1.415 | 1.895 | 2.365 | 2.998 | 3.499 |
| 8 | 0.262 | 0.706 | 1.397 | 1.860 | 2.306 | 2.896 | 3.355 |
| 9 | 0.261 | 0.703 | 1.383 | 1.833 | 2.262 | 2.821 | 3.250 |
| 10 | 0.260 | 0.700 | 1.372 | 1.812 | 2.228 | 2.764 | 3.169 |
| 11 | 0.260 | 0.697 | 1.363 | 1.796 | 2.201 | 2.718 | 3.106 |
| 12 | 0.259 | 0.695 | 1.356 | 1.782 | 2.179 | 2.681 | 3.055 |
| 13 | 0.259 | 0.694 | 1.350 | 1.771 | 2.160 | 2.650 | 3.012 |
| 14 | 0.258 | 0.692 | 1.345 | 1.761 | 2.145 | 2.624 | 2.997 |
| 15 | 0.258 | 0.691 | 1.341 | 1.753 | 2.131 | 2.602 | 2.947 |
| 16 | 0.258 | 0.690 | 1.337 | 1.746 | 2.120 | 2.583 | 2.921 |
| 17 | 0.257 | 0.689 | 1.333 | 1.740 | 2.110 | 2.567 | 2.898 |
| 18 | 0.257 | 0.688 | 1.330 | 1.734 | 2.101 | 2.552 | 2.878 |
| 19 | 0.257 | 0.688 | 1.328 | 1.729 | 2.093 | 2.539 | 2.861 |
| 20 | 0.257 | 0.687 | 1.325 | 1.725 | 2.086 | 2.528 | 2.845 |
| 21 | 0.257 | 0.686 | 1.323 | 1.721 | 2.080 | 2.518 | 2.831 |
| 22 | 0.256 | 0.686 | 1.321 | 1.717 | 2.074 | 2.508 | 2.819 |
| 23 | 0.256 | 0.685 | 1.319 | 1.714 | 2.069 | 2.500 | 2.807 |
| 24 | 0.256 | 0.685 | 1.318 | 1.711 | 2.064 | 2.492 | 2.797 |
| 25 | 0.256 | 0.684 | 1.316 | 1.708 | 2.060 | 2.485 | 2.787 |
| 26 | 0.256 | 0.684 | 1.315 | 1.706 | 2.056 | 2.479 | 2.779 |
| 27 | 0.256 | 0.684 | 1.314 | 1.703 | 2.052 | 2.473 | 2.771 |
| 28 | 0.256 | 0.683 | 1.313 | 1.701 | 2.048 | 2.467 | 2.763 |
| 29 | 0.256 | 0.683 | 1.311 | 1.699 | 2.045 | 2.462 | 2.756 |
| 30 | 0.256 | 0.683 | 1.310 | 1.697 | 2.042 | 2.457 | 2.750 |
| $\infty$ | 0.253 | 0.674 | 1.282 | 1.645 | 1.960 | 2.326 | 2.576 |

## Table VII    The F Distribution

$$P(F \leq f) = \int_0^f \frac{\Gamma[(r_1 + r_2)/2](r_1/r_2)^{r_1/2} w^{r_1/2-1}}{\Gamma(r_1/2)\Gamma(r_2/2)(1 + r_1 w/r_2)^{(r_1+r_2)/2}} \, dw$$

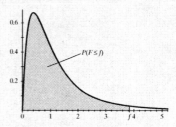

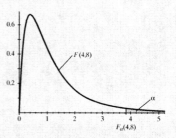

## Table VII continued

$$P(F \le f) = \int_0^f \frac{\Gamma[(r_1+r_2)/2][(r_1/r_2)]^{r_1/2}w^{r_1/2-1}}{\Gamma(r_1/2)\Gamma(r_2/2)(1+r_1w/r_2)^{(r_1+r_2)/2}}\, dw$$

Numerator Degrees of Freedom, $r_1$

| $\alpha$ | $P(F \le f)$ | Den. d.f. $r_2$ | 1 | 2 | 3 | 4 | 5 | 6 | 7 | 8 | 9 | 10 |
|---|---|---|---|---|---|---|---|---|---|---|---|---|
| 0.05 | 0.95 | 1 | 161.4 | 199.5 | 215.7 | 224.6 | 230.2 | 234.0 | 236.8 | 238.9 | 240.5 | 241.9 |
| 0.025 | 0.975 | | 647.79 | 799.50 | 864.16 | 899.58 | 921.85 | 937.11 | 948.22 | 956.66 | 963.28 | 968.63 |
| 0.01 | 0.99 | | 4052 | 4999.5 | 5403 | 5625 | 5764 | 5859 | 5928 | 5981 | 6022 | 6056 |
| 0.05 | 0.95 | 2 | 18.51 | 19.00 | 19.16 | 19.25 | 19.30 | 19.33 | 19.35 | 19.37 | 19.38 | 19.40 |
| 0.025 | 0.975 | | 38.51 | 39.00 | 39.17 | 39.25 | 39.30 | 39.33 | 39.36 | 39.37 | 39.39 | 39.40 |
| 0.01 | 0.99 | | 98.50 | 99.00 | 99.17 | 99.25 | 99.30 | 99.33 | 99.36 | 99.37 | 99.39 | 99.40 |
| 0.05 | 0.95 | 3 | 10.13 | 9.55 | 9.28 | 9.12 | 9.01 | 8.94 | 8.89 | 8.85 | 8.81 | 8.79 |
| 0.025 | 0.975 | | 17.44 | 16.04 | 15.44 | 15.10 | 14.88 | 14.73 | 14.62 | 14.54 | 14.47 | 14.42 |
| 0.01 | 0.99 | | 34.12 | 30.82 | 29.46 | 28.71 | 28.24 | 27.91 | 27.67 | 27.49 | 27.35 | 27.23 |
| 0.05 | 0.95 | 4 | 7.71 | 6.94 | 6.59 | 6.39 | 6.26 | 6.16 | 6.09 | 6.04 | 6.00 | 5.96 |
| 0.025 | 0.975 | | 12.22 | 10.65 | 9.98 | 9.60 | 9.36 | 9.20 | 9.07 | 8.98 | 8.90 | 8.84 |
| 0.01 | 0.99 | | 21.20 | 18.00 | 16.69 | 15.98 | 15.52 | 15.21 | 14.98 | 14.80 | 14.66 | 14.55 |
| 0.05 | 0.95 | 5 | 6.61 | 5.79 | 5.41 | 5.19 | 5.05 | 4.95 | 4.88 | 4.82 | 4.77 | 4.74 |
| 0.025 | 0.975 | | 10.01 | 8.43 | 7.76 | 7.39 | 7.15 | 6.98 | 6.85 | 6.76 | 6.68 | 6.62 |
| 0.01 | 0.99 | | 16.26 | 13.27 | 12.06 | 11.39 | 10.97 | 10.67 | 10.46 | 10.29 | 10.16 | 10.05 |
| 0.05 | 0.95 | 6 | 5.99 | 5.14 | 4.76 | 4.53 | 4.39 | 4.28 | 4.21 | 4.15 | 4.10 | 4.06 |
| 0.025 | 0.975 | | 8.81 | 7.26 | 6.60 | 6.23 | 5.99 | 5.82 | 5.70 | 5.60 | 5.52 | 5.46 |
| 0.01 | 0.99 | | 13.75 | 10.92 | 9.78 | 9.15 | 8.75 | 8.47 | 8.26 | 8.10 | 7.98 | 7.87 |
| 0.05 | 0.95 | 7 | 5.59 | 4.74 | 4.35 | 4.12 | 3.97 | 3.87 | 3.79 | 3.73 | 3.68 | 3.64 |
| 0.025 | 0.975 | | 8.07 | 6.54 | 5.89 | 5.52 | 5.29 | 5.12 | 4.99 | 4.90 | 4.82 | 4.76 |
| 0.01 | 0.99 | | 12.25 | 9.55 | 8.45 | 7.85 | 7.46 | 7.19 | 6.99 | 6.84 | 6.72 | 6.62 |
| 0.05 | 0.95 | 8 | 5.32 | 4.46 | 4.07 | 3.84 | 3.69 | 3.58 | 3.50 | 3.44 | 3.39 | 3.35 |
| 0.025 | 0.975 | | 7.57 | 6.06 | 5.42 | 5.05 | 4.82 | 4.65 | 4.53 | 4.43 | 4.36 | 4.30 |
| 0.01 | 0.99 | | 11.26 | 8.65 | 7.59 | 7.01 | 6.63 | 6.37 | 6.18 | 6.03 | 5.91 | 5.81 |
| 0.05 | 0.95 | 9 | 5.12 | 4.26 | 3.86 | 3.63 | 3.48 | 3.37 | 3.29 | 3.23 | 3.18 | 3.14 |
| 0.025 | 0.975 | | 7.21 | 5.71 | 5.08 | 4.72 | 4.48 | 4.32 | 4.20 | 4.10 | 4.03 | 3.96 |
| 0.01 | 0.99 | | 10.56 | 8.02 | 6.99 | 6.42 | 6.06 | 5.80 | 5.61 | 5.47 | 5.35 | 5.26 |
| 0.05 | 0.95 | 10 | 4.96 | 4.10 | 3.71 | 3.48 | 3.33 | 3.22 | 3.14 | 3.07 | 3.02 | 2.98 |
| 0.025 | 0.975 | | 6.94 | 5.46 | 4.83 | 4.47 | 4.24 | 4.07 | 3.95 | 3.85 | 3.78 | 3.72 |
| 0.01 | 0.99 | | 10.04 | 7.56 | 6.55 | 5.99 | 5.64 | 5.39 | 5.20 | 5.06 | 4.94 | 4.85 |

## Table VII  continued

$$P(F \le f) = \int_0^f \frac{\Gamma[(r_1+r_2)/2](r_1/r_2)^{r_1/2}w^{r_1/2-1}}{\Gamma(r_1/2)\Gamma(r_2/2)(1+r_1w/r_2)^{(r_1+r_2)/2}}\, dw$$

| Den. d.f. $r_2$ | $P(F \le f)$ | $\alpha$ | \(r_1=1\) | 2 | 3 | 4 | 5 | 6 | 7 | 8 | 9 | 10 |
|---|---|---|---|---|---|---|---|---|---|---|---|---|
| 12 | 0.95 | 0.05 | 4.75 | 3.89 | 3.49 | 3.26 | 3.11 | 3.00 | 2.91 | 2.85 | 2.80 | 2.75 |
|  | 0.975 | 0.025 | 6.55 | 5.10 | 4.47 | 4.12 | 3.89 | 3.73 | 3.61 | 3.51 | 3.44 | 3.37 |
|  | 0.99 | 0.01 | 9.33 | 6.93 | 5.95 | 5.41 | 5.06 | 4.82 | 4.64 | 4.50 | 4.39 | 4.30 |
| 15 | 0.95 | 0.05 | 4.54 | 3.68 | 3.29 | 3.06 | 2.90 | 2.79 | 2.71 | 2.64 | 2.59 | 2.54 |
|  | 0.975 | 0.025 | 6.20 | 4.77 | 4.15 | 3.80 | 3.58 | 3.41 | 3.29 | 3.20 | 3.12 | 3.06 |
|  | 0.99 | 0.01 | 8.68 | 6.36 | 5.42 | 4.89 | 4.56 | 4.32 | 4.14 | 4.00 | 3.89 | 3.80 |
| 20 | 0.95 | 0.05 | 4.35 | 3.49 | 3.10 | 2.87 | 2.71 | 2.60 | 2.51 | 2.45 | 2.39 | 2.35 |
|  | 0.975 | 0.025 | 5.87 | 4.46 | 3.86 | 3.51 | 3.29 | 3.13 | 3.01 | 2.91 | 2.84 | 2.77 |
|  | 0.99 | 0.01 | 8.10 | 5.85 | 4.94 | 4.43 | 4.10 | 3.87 | 3.70 | 3.56 | 3.46 | 3.37 |
| 24 | 0.95 | 0.05 | 4.26 | 3.40 | 3.01 | 2.78 | 2.62 | 2.51 | 2.42 | 2.36 | 2.30 | 2.25 |
|  | 0.975 | 0.025 | 5.72 | 4.32 | 3.72 | 3.38 | 3.15 | 2.99 | 2.87 | 2.78 | 2.70 | 2.64 |
|  | 0.99 | 0.01 | 7.82 | 5.61 | 4.72 | 4.22 | 3.90 | 3.67 | 3.50 | 3.36 | 3.26 | 3.17 |
| 30 | 0.95 | 0.05 | 4.17 | 3.32 | 2.92 | 2.69 | 2.53 | 2.42 | 2.33 | 2.27 | 2.21 | 2.16 |
|  | 0.975 | 0.025 | 5.57 | 4.18 | 3.59 | 3.25 | 3.03 | 2.87 | 2.75 | 2.65 | 2.57 | 2.51 |
|  | 0.99 | 0.01 | 7.56 | 5.39 | 4.51 | 4.02 | 3.70 | 3.47 | 3.30 | 3.17 | 3.07 | 2.98 |
| 40 | 0.95 | 0.05 | 4.08 | 3.23 | 2.84 | 2.61 | 2.45 | 2.34 | 2.25 | 2.18 | 2.12 | 2.08 |
|  | 0.975 | 0.025 | 5.42 | 4.05 | 3.46 | 3.13 | 2.90 | 2.74 | 2.62 | 2.53 | 2.45 | 2.39 |
|  | 0.99 | 0.01 | 7.31 | 5.18 | 4.31 | 3.83 | 3.51 | 3.29 | 3.12 | 2.99 | 2.89 | 2.80 |
| 60 | 0.95 | 0.05 | 4.00 | 3.15 | 2.76 | 2.53 | 2.37 | 2.25 | 2.17 | 2.10 | 2.04 | 1.99 |
|  | 0.975 | 0.025 | 5.29 | 3.93 | 3.34 | 3.01 | 2.79 | 2.63 | 2.51 | 2.41 | 2.33 | 2.27 |
|  | 0.99 | 0.01 | 7.08 | 4.98 | 4.13 | 3.65 | 3.34 | 3.12 | 2.95 | 2.82 | 2.72 | 2.63 |
| 120 | 0.95 | 0.05 | 3.92 | 3.07 | 2.68 | 2.45 | 2.29 | 2.17 | 2.09 | 2.02 | 1.96 | 1.91 |
|  | 0.975 | 0.025 | 5.15 | 3.80 | 3.23 | 2.89 | 2.67 | 2.52 | 2.39 | 2.30 | 2.22 | 2.16 |
|  | 0.99 | 0.01 | 6.85 | 4.79 | 3.95 | 3.48 | 3.17 | 2.96 | 2.79 | 2.66 | 2.56 | 2.47 |
| $\infty$ | 0.95 | 0.05 | 3.84 | 3.00 | 2.60 | 2.37 | 2.21 | 2.10 | 2.01 | 1.94 | 1.88 | 1.83 |
|  | 0.975 | 0.025 | 5.02 | 3.69 | 3.12 | 2.79 | 2.57 | 2.41 | 2.29 | 2.19 | 2.11 | 2.05 |
|  | 0.99 | 0.01 | 6.63 | 4.61 | 3.78 | 3.32 | 3.02 | 2.80 | 2.64 | 2.51 | 2.41 | 2.32 |

Numerator Degrees of Freedom, $r_1$

## Table VII continued

$$P(F \le f) = \int_0^f \frac{\Gamma[(r_1+r_2)/2](r_1/r_2)^{r_1/2}w^{r_1/2-1}}{\Gamma(r_1/2)\Gamma(r_2/2)(1+r_1w/r_2)^{(r_1+r_2)/2}}\, dw$$

| α | $P(F \le f)$ | Den. d.f. $r_2$ | \multicolumn Numerator Degrees of Freedom, $r_1$ ||||||||| |
|---|---|---|---|---|---|---|---|---|---|---|---|
| | | | 12 | 15 | 20 | 24 | 30 | 40 | 60 | 120 | ∞ |
| 0.05 | 0.95 | 1 | 243.9 | 245.9 | 248.0 | 249.1 | 250.1 | 251.1 | 252.2 | 253.3 | 254.3 |
| 0.025 | 0.975 | | 976.71 | 984.87 | 993.10 | 997.25 | 1001.4 | 1005.6 | 1009.8 | 1014.0 | 1018.3 |
| 0.01 | 0.99 | | 6106 | 6157 | 6209 | 6235 | 6261 | 6287 | 6313 | 6339 | 6366 |
| 0.05 | 0.95 | 2 | 19.41 | 19.43 | 19.45 | 19.45 | 19.46 | 19.47 | 19.48 | 19.49 | 19.50 |
| 0.025 | 0.975 | | 39.42 | 39.43 | 39.45 | 39.46 | 39.47 | 39.47 | 39.48 | 39.49 | 39.50 |
| 0.01 | 0.99 | | 99.42 | 99.43 | 99.45 | 99.46 | 99.47 | 99.47 | 99.48 | 99.49 | 99.50 |
| 0.05 | 0.95 | 3 | 8.74 | 8.70 | 8.66 | 8.64 | 8.62 | 8.59 | 8.57 | 8.55 | 8.53 |
| 0.025 | 0.975 | | 14.34 | 14.25 | 14.17 | 14.12 | 14.08 | 14.04 | 13.99 | 13.95 | 13.90 |
| 0.01 | 0.99 | | 27.05 | 26.87 | 26.69 | 26.60 | 26.50 | 26.41 | 26.32 | 26.22 | 26.13 |
| 0.05 | 0.95 | 4 | 5.91 | 5.86 | 5.80 | 5.77 | 5.75 | 5.72 | 5.69 | 5.66 | 5.63 |
| 0.025 | 0.975 | | 8.75 | 8.66 | 8.56 | 8.51 | 8.46 | 8.41 | 8.36 | 8.31 | 8.26 |
| 0.01 | 0.99 | | 14.37 | 14.20 | 14.02 | 13.93 | 13.84 | 13.75 | 13.65 | 13.56 | 13.46 |
| 0.05 | 0.95 | 5 | 4.68 | 4.62 | 4.56 | 4.53 | 4.50 | 4.46 | 4.43 | 4.40 | 4.36 |
| 0.025 | 0.975 | | 6.52 | 6.43 | 6.33 | 6.28 | 6.23 | 6.18 | 6.12 | 6.07 | 6.02 |
| 0.01 | 0.99 | | 9.89 | 9.72 | 9.55 | 9.47 | 9.38 | 9.29 | 9.20 | 9.11 | 9.02 |
| 0.05 | 0.95 | 6 | 4.00 | 3.94 | 3.87 | 3.84 | 3.81 | 3.77 | 3.74 | 3.70 | 3.67 |
| 0.025 | 0.975 | | 5.37 | 5.27 | 5.17 | 5.12 | 5.07 | 5.01 | 4.96 | 4.90 | 4.85 |
| 0.01 | 0.99 | | 7.72 | 7.56 | 7.40 | 7.31 | 7.23 | 7.14 | 7.06 | 6.97 | 6.88 |
| 0.05 | 0.95 | 7 | 3.57 | 3.51 | 3.41 | 3.41 | 3.38 | 3.34 | 3.30 | 3.27 | 3.23 |
| 0.025 | 0.975 | | 4.67 | 4.57 | 4.47 | 4.42 | 4.36 | 4.31 | 4.25 | 4.20 | 4.14 |
| 0.01 | 0.99 | | 6.47 | 6.31 | 6.16 | 6.07 | 5.99 | 5.91 | 5.82 | 5.74 | 5.65 |
| 0.05 | 0.95 | 8 | 3.28 | 3.22 | 3.15 | 3.12 | 3.08 | 3.04 | 3.01 | 2.97 | 2.93 |
| 0.025 | 0.975 | | 4.20 | 4.10 | 4.00 | 3.95 | 3.89 | 3.84 | 3.78 | 3.73 | 3.67 |
| 0.01 | 0.99 | | 5.67 | 5.52 | 5.36 | 5.28 | 5.20 | 5.12 | 5.03 | 4.95 | 4.86 |
| 0.05 | 0.95 | 9 | 3.07 | 3.01 | 2.94 | 2.90 | 2.86 | 2.83 | 2.79 | 2.75 | 2.71 |
| 0.025 | 0.975 | | 3.87 | 3.77 | 3.67 | 3.61 | 3.56 | 3.51 | 3.45 | 3.39 | 3.33 |
| 0.01 | 0.99 | | 5.11 | 4.96 | 4.81 | 4.73 | 4.65 | 4.57 | 4.48 | 4.40 | 4.31 |

## Table VII continued

$$P(F \le f) = \int_0^f \frac{\Gamma[(r_1 + r_2)/2](r_1/r_2)^{r_1/2}w^{r_1/2-1}}{\Gamma(r_1/2)\Gamma(r_2/2)(1 + r_1 w/r_2)^{(r_1+r_2)/2}}\, dw$$

| Den. d.f. $r_2$ | $\alpha$ | $P(F \le f)$ | Numerator Degrees of Freedom, $r_1$ ||||||||||
|---|---|---|---|---|---|---|---|---|---|---|---|
| | | | 12 | 15 | 20 | 24 | 30 | 40 | 60 | 120 | $\infty$ |
| 10 | 0.05 | 0.95 | 2.91 | 2.85 | 2.77 | 2.74 | 2.70 | 2.66 | 2.62 | 2.58 | 2.54 |
| | 0.025 | 0.975 | 3.62 | 3.52 | 3.42 | 3.37 | 3.31 | 3.26 | 3.20 | 3.14 | 3.08 |
| | 0.01 | 0.99 | 4.71 | 4.56 | 4.41 | 4.33 | 4.25 | 4.17 | 4.08 | 4.00 | 3.91 |
| 12 | 0.05 | 0.95 | 2.69 | 2.62 | 2.54 | 2.51 | 2.47 | 2.43 | 2.38 | 2.34 | 2.30 |
| | 0.025 | 0.975 | 3.28 | 3.18 | 3.07 | 3.02 | 2.96 | 2.91 | 2.85 | 2.79 | 2.72 |
| | 0.01 | 0.99 | 4.16 | 4.01 | 3.86 | 3.78 | 3.70 | 3.62 | 3.54 | 3.45 | 3.36 |
| 15 | 0.05 | 0.95 | 2.48 | 2.40 | 2.33 | 2.29 | 2.25 | 2.20 | 2.16 | 2.11 | 2.07 |
| | 0.025 | 0.975 | 2.96 | 2.86 | 2.76 | 2.70 | 2.64 | 2.59 | 2.52 | 2.46 | 2.40 |
| | 0.01 | 0.99 | 3.67 | 3.52 | 3.37 | 3.29 | 3.21 | 3.13 | 3.05 | 2.96 | 2.87 |
| 20 | 0.05 | 0.95 | 2.28 | 2.20 | 2.12 | 2.08 | 2.04 | 1.99 | 1.95 | 1.90 | 1.84 |
| | 0.025 | 0.975 | 2.68 | 2.57 | 2.46 | 2.41 | 2.35 | 2.29 | 2.22 | 2.16 | 2.09 |
| | 0.01 | 0.99 | 3.23 | 3.09 | 2.94 | 2.86 | 2.78 | 2.69 | 2.61 | 2.52 | 2.42 |
| 24 | 0.05 | 0.95 | 2.18 | 2.11 | 2.03 | 1.98 | 1.94 | 1.89 | 1.84 | 1.79 | 1.73 |
| | 0.025 | 0.975 | 2.54 | 2.44 | 2.33 | 2.27 | 2.21 | 2.15 | 2.08 | 2.01 | 1.94 |
| | 0.01 | 0.99 | 3.03 | 2.89 | 2.74 | 2.66 | 2.58 | 2.49 | 2.40 | 2.31 | 2.21 |
| 30 | 0.05 | 0.95 | 2.09 | 2.01 | 1.93 | 1.89 | 1.84 | 1.79 | 1.74 | 1.68 | 1.62 |
| | 0.025 | 0.975 | 2.41 | 2.31 | 2.20 | 2.14 | 2.07 | 2.01 | 1.94 | 1.87 | 1.79 |
| | 0.01 | 0.99 | 2.84 | 2.70 | 2.55 | 2.47 | 2.39 | 2.30 | 2.21 | 2.11 | 2.01 |
| 40 | 0.05 | 0.95 | 2.00 | 1.92 | 1.84 | 1.79 | 1.74 | 1.69 | 1.64 | 1.58 | 1.51 |
| | 0.025 | 0.975 | 2.29 | 2.18 | 2.07 | 2.01 | 1.94 | 1.88 | 1.80 | 1.72 | 1.64 |
| | 0.01 | 0.99 | 2.66 | 2.52 | 2.37 | 2.29 | 2.20 | 2.11 | 2.02 | 1.92 | 1.80 |
| 60 | 0.05 | 0.95 | 1.92 | 1.84 | 1.75 | 1.70 | 1.65 | 1.59 | 1.53 | 1.47 | 1.39 |
| | 0.025 | 0.975 | 2.17 | 2.06 | 1.94 | 1.88 | 1.82 | 1.74 | 1.67 | 1.58 | 1.48 |
| | 0.01 | 0.99 | 2.50 | 2.35 | 2.20 | 2.12 | 2.03 | 1.94 | 1.84 | 1.73 | 1.60 |
| 120 | 0.05 | 0.95 | 1.83 | 1.75 | 1.66 | 1.61 | 1.55 | 1.50 | 1.43 | 1.35 | 1.25 |
| | 0.025 | 0.975 | 2.05 | 1.95 | 1.82 | 1.76 | 1.69 | 1.61 | 1.53 | 1.43 | 1.31 |
| | 0.01 | 0.99 | 2.34 | 2.19 | 2.03 | 1.95 | 1.86 | 1.76 | 1.66 | 1.53 | 1.38 |
| $\infty$ | 0.05 | 0.95 | 1.75 | 1.67 | 1.57 | 1.52 | 1.46 | 1.39 | 1.32 | 1.22 | 1.00 |
| | 0.025 | 0.975 | 1.94 | 1.83 | 1.71 | 1.64 | 1.57 | 1.48 | 1.39 | 1.27 | 1.00 |
| | 0.01 | 0.99 | 2.18 | 2.04 | 1.88 | 1.79 | 1.70 | 1.59 | 1.47 | 1.32 | 1.00 |

| Table VIII Random Numbers on the Interval (0, 1) | | | | | | |
|------|------|------|------|------|------|------|
| 3407 | 1440 | 6960 | 8675 | 5649 | 5793 | 1514 |
| 5044 | 9859 | 4658 | 7779 | 7986 | 0520 | 6697 |
| 0045 | 4999 | 4930 | 7408 | 7551 | 3124 | 0527 |
| 7536 | 1448 | 7843 | 4801 | 3147 | 3071 | 4749 |
| 7653 | 4231 | 1233 | 4409 | 0609 | 6448 | 2900 |
| | | | | | | |
| 6157 | 1144 | 4779 | 0951 | 3757 | 9562 | 2354 |
| 6593 | 8668 | 4871 | 0946 | 3155 | 3941 | 9662 |
| 3187 | 7434 | 0315 | 4418 | 1569 | 1101 | 0043 |
| 4780 | 1071 | 6814 | 2733 | 7968 | 8541 | 1003 |
| 9414 | 6170 | 2581 | 1398 | 2429 | 4763 | 9192 |
| | | | | | | |
| 1948 | 2360 | 7244 | 9682 | 5418 | 0596 | 4971 |
| 1843 | 0914 | 9705 | 7861 | 6861 | 7865 | 7293 |
| 4944 | 8903 | 0460 | 0188 | 0530 | 7790 | 9118 |
| 3882 | 3195 | 8287 | 3298 | 9532 | 9066 | 8225 |
| 6596 | 9009 | 2055 | 4081 | 4842 | 7852 | 5915 |
| | | | | | | |
| 4793 | 2503 | 2906 | 6807 | 2028 | 1075 | 7175 |
| 2112 | 0232 | 5334 | 1443 | 7306 | 6418 | 9639 |
| 0743 | 1083 | 8071 | 9779 | 5973 | 1141 | 4393 |
| 8856 | 5352 | 3384 | 8891 | 9189 | 1680 | 3192 |
| 8027 | 4975 | 2346 | 5786 | 0693 | 5615 | 2047 |
| | | | | | | |
| 3134 | 1688 | 4071 | 3766 | 0570 | 2142 | 3492 |
| 0633 | 9002 | 1305 | 2256 | 5956 | 9256 | 8979 |
| 8771 | 6069 | 1598 | 4275 | 6017 | 5946 | 8189 |
| 2672 | 1304 | 2186 | 8279 | 2430 | 4896 | 3698 |
| 3136 | 1916 | 8886 | 8617 | 9312 | 5070 | 2720 |
| | | | | | | |
| 6490 | 7491 | 6562 | 5355 | 3794 | 3555 | 7510 |
| 8628 | 0501 | 4618 | 3364 | 6709 | 1289 | 0543 |
| 9270 | 0504 | 5018 | 7013 | 4423 | 2147 | 4089 |
| 5723 | 3807 | 4997 | 4699 | 2231 | 3193 | 8130 |
| 6228 | 8874 | 7271 | 2621 | 5746 | 6333 | 0345 |
| | | | | | | |
| 7645 | 3379 | 8376 | 3030 | 0351 | 8290 | 3640 |
| 6842 | 5836 | 6203 | 6171 | 2698 | 4086 | 5469 |
| 6126 | 7792 | 9337 | 7773 | 7286 | 4236 | 1788 |
| 4956 | 0215 | 3468 | 8038 | 6144 | 9753 | 3131 |
| 1327 | 4736 | 6229 | 8965 | 7215 | 6458 | 3937 |
| | | | | | | |
| 9188 | 1516 | 5279 | 5433 | 2254 | 5768 | 8718 |
| 0271 | 9627 | 9442 | 9217 | 4656 | 7603 | 8826 |
| 2127 | 1847 | 1331 | 5122 | 8332 | 8195 | 3322 |
| 2102 | 9201 | 2911 | 7318 | 7670 | 6079 | 2676 |
| 1706 | 6011 | 5280 | 5552 | 5180 | 4630 | 4747 |
| | | | | | | |
| 7501 | 7635 | 2301 | 0889 | 6955 | 8113 | 4364 |
| 5705 | 1900 | 7144 | 8707 | 9065 | 8163 | 9846 |
| 3234 | 2599 | 3295 | 9160 | 8441 | 0085 | 9317 |
| 5641 | 4935 | 7971 | 8917 | 1978 | 5649 | 5799 |
| 2127 | 1868 | 3664 | 9376 | 1984 | 6315 | 8396 |

# APPENDIX C

# ANSWERS TO ODD-NUMBERED EXERCISES

## CHAPTER 1

**1.1-1**  **(a)**  $O = \{1, 2, 3, \ldots, 36\}$;

**(b)**  $O = \{w : 19 \leq w \leq 23\}$, where $w$ is the weight in grams;

**(c)**  $O = \{HHH, HHT, HTH, THH, HTT, THT, TTH, TTT\}$.

**1.1-3**  **(a)** 12/52; **(b)** 2/52; **(c)** 16/52; **(d)** 1; **(e)** 0.

**1.1-5**  0.6.

**1.1-7**  **(a)**  $O = \{00, 0, 1, 2, 3, \ldots, 36\}$;

**(b)**  $P(A) = 2/38$;

**(c)**  $P(B) = 4/38$;

**(d)**  $P(D) = 18/38$.

**1.1-9**  2/3.

**1.2-1**  4096.

**1.2-3**  **(a)** 6,760,000; **(b)** 17,576,000.

**1.2-5**  **(a)** 60; **(b)** 125.

**1.2-7**  **(a)** 2; **(b)** 8; **(c)** 20; **(d)** 40.

**1.2-9**  **(a)** 362,880; **(b)** 84; **(c)** 512.

**1.2-11**  **(a)** 0.00024; **(b)** 0.00144; **(c)** 0.02113; **(d)** 0.04754; **(e)** 0.42257.

**1.3-1**  **(a)** 5000/1,000,000; **(b)** 78,515/1,000,000;

**(c)** 73,630/995,000; **(d)** 4,885/78,515.

**1.3-3**  **(a)** 5/35; **(b)** 26/35; **(c)** 5/19; **(d)** 9/23; **(e)** left.

**1.3-5**  **(a)**  $O = \{(R, R), (R, W), (W, R), (W, W)\}$; **(b)** 1/3.

**1.3-7**  1/5.

**1.3-9**  **(a)** $365^r$; **(b)** $_{365}P_r$; **(c)** $1 - {_{365}P_r}/365^r$; **(d)** 23.

1.3-11    **(a)**   49/153; **(b)** 4/7.

1.3-13    **(b)**   8/36; **(c)** 5/11;

    **(e)**   $8/36 + 2[(5/36)(5/11) + (4/36)(4/10) + (3/36)(3/9)] = 0.49293.$

1.4-1    **(a)**   0.14; **(b)** 0.76; **(c)** 0.86.

1.4-3    **(a)**   1/6; **(b)** 1/12; **(c)** 1/4; **(d)** 1/4; **(e)** 1/2.

1.4-5    **(a)**   0.29; **(b)** 0.44.

1.4-7    **(a)**   0.36; **(b)** 0.49; **(c)** 0.01.

1.4-9    $(2/3)^3(1/3)^2; (2/3)^3(1/3)^2.$

1.4-11    **(a)**   $10(1/2)^6;$

    **(b)**   $[(120)(45)/15,504][7/15];$

    **(c)**   neither model is very good.

1.4-13    **(b)**   $1 - 1/e.$

1.5-1    **(a)**   21/32; **(b)** 16/21.

1.5-3    **(a)**   11/24; **(b)** 2/11.

1.5-5    $60/95 = 0.632.$

1.5-7    $50/95 = 0.526.$

1.5-9    **(a)**   $495/30,480 = 0.016;$ **(b)** $29,985/30,480 = 0.984.$


## CHAPTER 2

2.1-1    **(a)**   15, 50; **(b)** 5, 0; **(c)** 5/3, 5/9.

2.1-3    **(a)**   $c = 1/21;$ **(b)** $c = 1/2;$ **(c)** $c = 1/14;$ **(d)** $c = 1/5.$

2.1-5    **(a)**   $f(x) = 1/10, \qquad x = 0, 1, 2, \ldots, 9;$

    **(b)**   The respective frequencies are: 21, 15, 16, 13, 6, 15, 13, 17, 17, 17;

    **(c)**   $\mu = 9/2 = 4.5, \bar{x} = 671/150 = 4.473; \sigma^2 = 33/4 = 8.25, s_x^2 = 9.6067;$

    **(d)**   The respective frequencies are: 8, 13, 15, 10, 12, 10, 4, 9, 15, 0;

    $\bar{y} = 63/16 = 3.9375; s_y^2 = 6.9645.$

2.1-7    **(a)**   $f(x) = \dfrac{6 - |x - 7|}{36}, \qquad x = 2, 3, 4, \ldots, 12;$

    **(c)**   $\mu = 7, \sigma^2 = 35/6.$

2.1-9    $\dfrac{79,137}{190,120} = 0.41625.$

2.1-11    **(a)**   $g(y) = \dfrac{365!(y - 1)}{(366 - y)!365^y}, \qquad y = 2, 3, \ldots, 366;$

    **(b)**   $\mu = 24.6166, \sigma^2 = 148.6403, \sigma = 12.1918.$

2.2-1    $E(X) = 3, E(X^2) = 11, E[(X + 2)^2] = 27.$

2.2-3    $E(X) = 0, E(X^2) = 8/9, E(3X^2 - 2X + 4) = 20/3.$

2.2-5      $E(X) = -\$0.50$.

2.2-7   **(a)**   50;

   **(b)**
   $$f(x) = \begin{cases} 0.4, & x = 25, \\ 0.3, & x = 100, \\ 0.3, & x = 300, \end{cases}$$

   **(c)**   $E(X) = 130$.

2.2-9   **(a)**   **(ii)** 127/64; **(iii)** $\sqrt{7359}/64$; **(iv)** 1.6635;

   **(b)**   **(ii)** 385/64; **(iii)** $\sqrt{7359}/64$; **(iv)** $-1.6635$.

2.2-11      $E(|X - c|) = \dfrac{1}{7} \sum_{x \in S} |x - c|$, where $S = \{1, 2, 3, 5, 15, 25, 50\}$.

   When $c = 5$,

   $$E(|X - 5|) = \frac{1}{7}[(5 - 1) + (5 - 2) + (5 - 3) + (5 - 5) + (15 - 5) + (25 - 5) + (50 - 5)].$$

   If $c$ is either increased or decreased by 1, this expectation is increased by 1/7. Thus $c = 5$, the median, minimizes this expectation while $b = E(X) = \mu$, the mean, minimizes $E[(X - b)^2]$. You could also let $h(c) = E(|X - c|)$ and show that $h'(c) = 0$ when $c = 5$.

2.2-13   **(a)**   $(1/5)^2(4/5)^4 = 0.0164$;

   **(b)**   $\dfrac{6!}{2!4!}(1/5)^2(4/5)^4 = 0.2458$;

   **(c)**   $f(x) = \dbinom{6}{x}\left(\dfrac{1}{5}\right)^x\left(\dfrac{4}{5}\right)^{6-x}, \qquad x = 0, 1, \ldots, 6,$
   $\mu = 6/5, \sigma^2 = 24/25$.

2.3-1   **(a)**   $X$ is $b(7, 0.15)$;

   **(b)**   **(i)** 0.2834, **(ii)** 0.3960, **(iii)** 0.9879.

2.3-3   **(a)**   $P(X \le 5) = 0.5269$;

   **(b)**   $P(X \ge 6) = 0.4731$;

   **(c)**   $P(X = 7) = 0.1490$;

   **(d)**   $\mu = 5.4, \sigma^2 = 2.97, \ \sigma = \sqrt{2.97} = 1.723$.

2.3-5      $n = 5$.

2.3-7      1/6.

2.3-9   **(a)**   0.8197;

   **(b)**   0.8153.

2.3-11      0.540.

2.3-13
   $$\begin{aligned} P(\mu - 2\sigma < X < \mu + 2\sigma) &= P(0 < X < 8) \\ &= P(X \le 7) - P(X = 0) \\ &= 0.931. \end{aligned}$$

2.3-15
$$OC(p) = P(X \leq 3) \approx \sum_{x=0}^{3} \frac{(400p)^x e^{-400p}}{x!};$$
$$OC(0.002) \approx 0.991;$$
$$OC(0.004) \approx 0.921;$$
$$OC(0.006) \approx 0.779;$$
$$OC(0.01) \approx 0.433;$$
$$OC(0.02) \approx 0.042.$$

2.3-17    **(a)**    Negative binomial with $r = 10, p = 0.6$ so
$$\mu = \frac{10}{0.60} = 16.667, \sigma^2 = \frac{10(0.40)}{(0.60)^2} = 11.111, \sigma = 3.333;$$

**(b)**    $P(X = 16) = 0.1240$.

2.3-19
$$f(x) = \frac{\binom{20}{x}\binom{180}{10-x}}{\binom{200}{10}}, \qquad x = 0, 1, \ldots, 10;$$

$$\mu = 1; \sigma^2 = 171/199 = 0.8593;$$

$$0.3398, 0.3487, 0.368.$$

2.3-21    **(a)**    $f(x) = P(X = x) = \dfrac{\binom{6}{x}\binom{43}{6-x}}{\binom{49}{6}}, \qquad x = 0, 1, 2, 3, 4, 5, 6;$

**(b)**    $\mu = 0.7347, \sigma^2 = 0.5776, \sigma = 0.7600;$

**(c)**    $f(0) = \dfrac{435,461}{998,844} > \dfrac{412,542}{998,844} = f(1);$    $X = 0$ is most likely to occur.

**(d)**    The numbers are reasonable because
$$(25,000,000)f(6) = 1.79;$$
$$(25,000,000)f(5) = 461.25;$$
$$(25,000,000)f(4) = 24,215.49.$$

2.4-1    **(a)**    $\bar{x} = 4/3 = 1.333, s^2 = 88/69 = 1.275;$

**(b)**    $\hat{\lambda} = \bar{x}$, yes.

2.4-3    **(a)**    $\bar{x} = 397/50 = 7.940, s^2 = 18,941/2,450 = 7.731;$

**(b)**    both the Poisson, $\lambda = 7.940$, and the $b(302, 0.0263)$ provide good fits.

2.4-5    **(a)**    $\bar{x} = 223/45 = 4.956, s^2 = 4093/990 = 4.134;$

**(b)**    both the Poisson with $\lambda = 4.956$ and the $b(30, 0.1657)$ provide good fits.

2.4-7    $\lceil p(n+1) - 1 \rceil$, where $\lceil y \rceil$ is the ceiling of $y$, the smallest integer greater than or equal to $y$.

2.5-1    $E(Y) = -1, \mathrm{Var}(Y) = 72$.

2.5-3     $E(\overline{X}) = 7/3$, $\mathrm{Var}(\overline{X}) = 5/81$.

2.5-5   **(a)** 0.0035; **(b)** 8; **(c)** $E(Y) = 6$, $\mathrm{Var}(Y) = 4$.

2.5-7     [8.07, 15.83].

2.5-9     [0.646, 0.774].

2.5-11     $n = 2500$.

2.6-1   **(a)** $f_1(1) = f_1(2) = 1/2$; $f_2(1) = f_2(2) = 1/2$;

　　　 **(b)** $\mu_1 = \mu_2 = 3/2$;

　　　 **(c)** $\sigma_1^2 = \sigma_2^2 = 1/4$;

　　　 **(d)** $\mathrm{Cov}(X_1, X_2) = 1/8$;

　　　 **(e)** $\rho = 1/2$;

　　　 **(f)** no.

2.6-3   **(a)** $\mu_1 = \dfrac{7}{3}$, $\mu_2 = \dfrac{5}{3}$, $\sigma_1^2 = \dfrac{5}{9}$, $\sigma_2^2 = \dfrac{2}{9}$;

　　　 **(b)** $\dfrac{1}{3}$;

　　　 **(c)** $\dfrac{2}{9}$;

　　　 **(d)**

$$
g(y) = \begin{cases}
\dfrac{1}{18}, & y = 2, \\[2mm]
\dfrac{4}{18}, & y = 3, \\[2mm]
\dfrac{7}{18}, & y = 4, \\[2mm]
\dfrac{6}{18}, & y = 5,
\end{cases}
$$

　　　 **(e)** $\mu_Y = 4$, $\sigma_Y^2 = \dfrac{7}{9}$;

　　　 **(f)** same.

2.6-5   **(a)** $\mu_Y = 2$, $\sigma_Y^2 = 18(1 - \rho)$; **(b)** $\rho = 1$.

2.6-7   **(d)** **(i)** 9/14, **(ii)** 7/18, **(iii)** 5/9; **(e)** 20/7, 55/49.

2.6-9     $r = 0.383$.

## CHAPTER 3

3.1-1   **(a)** The respective class frequencies are 2, 8, 15, 13, 5, 6, 1;

　　　 **(c)** $\bar{x} = 8.773$, $\bar{u} = 8.785$, $s_x = 0.365$, $s_u = 0.352$;

　　　 **(d)** $800\bar{u} = 7028$, $800(\bar{u} + 2s_u) = 7591.2$. The answer depends on the cost of the nails as well as the time and distance required if too few nails are purchased.

3.1-3  (a)

| Stems | Leaves | Frequency | Depths |
|-------|--------|-----------|--------|
| 34 | 0 0 | 2 | 2 |
| 35 | 0 4 9 | 3 | 5 |
| 36 | 0 0 | 2 | 7 |
| 37 | 0 0 3 4 5 9 9 | 7 | 14 |
| 38 | 0 0 0 0 2 3 5 | 7 | 21 |
| 39 | 0 0 0 0 0 3 5 5 | 8 | (8) |
| 40 | 0 5 7 8 | 4 | 22 |
| 41 | 0 1 5 6 6 8 9 | 7 | 18 |
| 42 | 0 0 0 3 5 7 | 6 | 11 |
| 43 | 2 2 4 | 3 | 5 |
| 44 | 0 | 1 | 2 |
| 45 | 2 | 1 | 1 |

(b)   Frequencies: [2, 3, 2, 7, 7, 8, 4, 7, 6, 3, 1, 1];

(d)   Five-number summary: 340, 379, 390, 418, 452;

(e)   $\bar{x} = 394.82$, $s_x^2 = 709.0282$, $s_x = 26.63$;

(f)   $\bar{u} = 396.46$, $s_u^2 = 688.0784$, $s_u = 26.23$.

3.1-5  (a)

| Stems | Leaves | Frequency | Depths |
|-------|--------|-----------|--------|
| 127 | 8 | 1 | 1 |
| 128 | 8 | 1 | 2 |
| 129 | 5 8 9 | 3 | 5 |
| 130 | 8 | 1 | 6 |
| 131 | 2 3 4 4 5 5 7 | 7 | (7) |
| 132 | 2 7 7 8 | 4 | 7 |
| 133 | 7 9 | 2 | 3 |
| 134 | 8 | 1 | 1 |

(Multiply numbers by $10^{-1}$.)

(b)   7.0, 2.575, 131.45, 131.47, 3.034.

(c)   Five-number summary: 127.80, 130.125, 131.45, 132.70, 134.80.

**3.1-7**  **(a)**

| Stems | Leaves | Frequency | Depths |
|-------|--------|-----------|--------|
| 0• | 612 | 1 | 1 |
| 1∗ | 450 | 1 | 2 |
| 1• | 560 889 961 994 | 4 | 6 |
| 2∗ | 065 142 151 172 195 290 | 6 | 12 |
| 2• | 510 545 788 817 880 921 938 | 7 | (7) |
| 3∗ | 011 041 051 060 062 080 090 | 7 | 7 |

(Multiply numbers by $10^{-2}$.)

**(b)**  Five-number summary: 6.12, 20.4725, 25.275, 30.185, 30.90;

**(c)**  skewed to the left.

**3.1-9**  **(a)**

| Stems | Leaves | Frequency | Depths |
|-------|--------|-----------|--------|
| 101 | 7 | 1 | 1 |
| 102 | 0 0 0 | 3 | 4 |
| 103 |  | 0 | 4 |
| 104 |  | 0 | 4 |
| 105 | 8 9 | 2 | 6 |
| 106 | 1 3 3 6 6 7 7 8 8 | 9 | (9) |
| 107 | 3 7 9 | 3 | 10 |
| 108 | 8 | 1 | 7 |
| 109 | 1 3 9 | 3 | 6 |
| 110 | 0 2 2 | 3 | 3 |

(Multiply numbers by $10^{-1}$.)

**(b)**  Five-number summary: 101.7, 106.0, 106.7, 108.95, 110.2.

**3.2-1**  **(a)**  1/16;

**(b)**  5/16;

**(c)**  81/256.

**3.2-3**  **(a)**  1/2, 3/4;

**(b)**  2/9, 1/3.

**3.2-5**  **(a)**  $1/\sqrt{2}, 1/2^{1/4}, 3^{1/4}/\sqrt{2}$;

**(b)**  −1, 0, 1.

3.2-7  **(a)**  $\mu = 1/2, \sigma^2 = 1/20$;

    **(b)**  $\mu = 2, \sigma^2$ does not exist.

    **(c)**  neither the mean nor the variance exists.

3.2-9  **(a)**  **(i)** $c = 2$,  **(ii)** $F(x) = \begin{cases} 0, & x < 0, \\ x^4/16, & 0 \le x < 2, \\ 1, & 2 \le x < \infty. \end{cases}$

    **(b)**  **(i)** $c = 3/16$,  **(ii)** $F(x) = \begin{cases} 0, & x < 0, \\ (1/8)x^{3/2}, & 0 \le x < 4, \\ 1, & 4 \le x < \infty. \end{cases}$

    **(c)**  **(i)** $c = 1/2$, this p.d.f. is unbounded,  **(ii)**  $F(x) = \begin{cases} 0, & x < 0, \\ \sqrt{x}, & 0 \le x < 1, \\ 1, & 1 \le x < \infty. \end{cases}$

    **(d)**  **(i)** $c = 1$,  **(ii)** $F(x) = \begin{cases} 0, & x < 1, \\ 1 - 1/x, & 1 \le x < \infty. \end{cases}$

3.2-11  $f(x) = e^{-x}/(1 + e^{-x})^2, \qquad -\infty < x < \infty$;

$$f(-x) = \frac{e^x}{(1 + e^{-x})^2} \cdot \frac{(e^{-x})^2}{(e^{-x})^2} = \frac{e^{-x}}{(e^{-x} + 1)^2} = f(x).$$

3.3-3  **(a)**  $G(w) = (w - a)/(b - a), \qquad a \le w \le b$;

    **(b)**  $U(a, b)$.

3.3-5  **(a)**  10.524; 9.320; yes.

3.3-7  **(a)**  $\mu = 240/7 = 34.286, \bar{x} = 32.636$;

    **(b)**  $\sigma^2 = 28,800/49 = 587.755, s^2 = 548.338$;

    **(c)**  $0.605 \approx 0.591 = 13/22$.

3.3-9  $\theta = 2, 0.950$.

3.3-11  $a = 8 - 4\sqrt{3}, b = 8 + 4\sqrt{3}$.

3.4-1  **(a)** 0.3078;  **(b)** 0.4959;  **(c)** 0.2711;  **(d)** 0.1646;

    **(e)** 0.0526;  **(f)** 0.3174 (0.3173);  **(g)** 0.0456 (0.0455);  **(h)** 0.0026 (0.0027).

3.4-3  **(a)**  0.3849; **(b)** 0.5403; **(c)** 0.0603; **(d)** 0.0013;

    **(e)**  0.6826; **(f)** 0.9544; **(g)** 0.9974; **(h)** 0.99.

3.4-7   **(a)**

| Stems | Leaves | Frequencies | Depths |
|-------|--------|-------------|--------|
| 11• | 8 | 1 | 1 |
| 12* | 0 3 | 2 | 3 |
| 12• | 5 6 | 2 | 5 |
| 13* | 1 3 4 | 3 | 8 |
| 13• | 5 5 7 7 7 9 | 6 | 14 |
| 14* | 0 0 2 2 3 4 4 4 | 8 | 22 |
| 14• | 6 6 7 7 7 8 9 9 | 8 | 30 |
| 15* | 0 0 0 0 1 1 1 2 3 3 4 4 | 12 | 30 |
| 15• | 5 5 6 7 8 8 8 9 | 8 | 18 |
| 16* | 0 0 0 2 3 4 | 6 | 10 |
| 16• | 5 5 | 2 | 4 |
| 17* | 1 | 1 | 2 |
| 17• | 5 | 1 | 1 |

3.4-9   0.025.

3.4-11   0.514.

3.4-13   $c = 1/\sqrt{\pi/\ln 3}$.

3.5-1   $\widehat{\theta}_1 = \widehat{\mu} = 33.4267;\ \widehat{\theta}_2 = \widehat{\sigma^2} = 5.0980$.

3.5-3   **(c)**   **(i)** $\widehat{\theta} = 0.5493, \widetilde{\theta} = 0.5975$,

  **(ii)** $\widehat{\theta} = 2.2101, \widetilde{\theta} = 2.4004$,

  **(iii)** $\widehat{\theta} = 0.9588, \widetilde{\theta} = 0.8646$.

3.5-5   $\widehat{\theta} = \min(X_1, X_2, \ldots, X_n)$.

3.5-7   $[\overline{x}(1 - 2/\sqrt{n}),\ \overline{x}(1 + 2/\sqrt{n})]$  or  $\left[\dfrac{\overline{x}}{1 + 2/\sqrt{n}}, \dfrac{\overline{x}}{1 - 2/\sqrt{n}}\right]$.

3.6-1   0.4772.

3.6-3   0.8185.

3.6-5   0.6247.

3.6-7   0.95.

3.6-9   **(a)**   The frequencies are 5, 6, 12, 18, 31, 31, 20, 17, 6, 4; the fifth and sixth classes; **(c)** yes.

3.6-11   0.9522.

3.6-13   $P(Y \le 35) \approx 0.6462$ using normal approximation; $P(Y \le 35) = 0.6725$ using the gamma distribution with $\alpha = 16$ and $\theta = 2$.

3.7-1   **(a)**   0.2878, 0.2881; **(b)** 0.4428, 0.4435; **(c)** 0.1550, 0.1554.

3.7-3   0.9258 using normal approximation, 0.9258 using binomial.

3.7-5　　　0.6915 using normal approximation, 0.7030 using binomial.

3.7-7　　　0.3085.

3.7-9　　　0.6247 using normal approximation, 0.6148 using Poisson.

3.7-11　(a)　0.5548; (b) 0.3823.

3.7-13　　　0.6813 using normal approximation, 0.6788 using binomial.

3.7-15　(a)　0.3802; (b) 0.7571.

3.7-17　　　0.4734 using normal approximation, 0.4749 using Poisson approximation with $\lambda = 50$, 0.4769 using $b(5000, 0.01)$.

3.7-19　　　0.6455 using normal approximation, 0.6449 using Poisson.

3.7-21　(a)　0.8289 using normal approximation, 0.8294 using Poisson.

　　　　(b)　0.0261 using tables in book, 0.0218 using *Maple*.

# CHAPTER 4

4.1-3　(a)　0.890; (b) 0.05.

4.1-3　　　0.8644.

4.1-5　(a)　$d = 2.131$;

　　　(b)　$u(\bar{x}, s) = \bar{x} - 2.131s/4, v(\bar{x}, s) = \bar{x} + 2.131s/4$.

4.1-7　(a)　$d = 1.746$;

　　　(b)　$\overline{X} - \overline{Y} \pm 1.746\sqrt{\dfrac{7S_X^2 + 9S_Y^2}{16}}\sqrt{\dfrac{1}{8} + \dfrac{1}{10}}$;

　　　(c)　$[-8.5517, 10.3517]$;

　　　(d)　$c = 1/4.82, d = 4.20$;

　　　(e)　$P\left(\dfrac{S_Y^2}{4.82S_X^2} < \dfrac{\sigma_Y^2}{\sigma_X^2} < \dfrac{4.20S_Y^2}{S_X^2}\right) = 0.95$;

　　　(f)　$[0.1365, 2.7626]$.

4.2-1　(a)　$s = 6.144$; (b) $[4.406, 10.142]$ or $[4.107, 9.521]$.

4.2-3　　　$\left[\dfrac{\sum_{i=1}^{n}(X_i - \mu)^2}{\chi_{\alpha/2}^2(n)}, \dfrac{\sum_{i=1}^{n}(X_i - \mu)^2}{\chi_{1-\alpha/2}^2(n)}\right]$.

4.2-5　(a)　0.4987; (b) $[0, 1.835]$.

4.2-7　　　$[19.47, 22.33]$.

4.2-9　(a)　$\bar{x} = 3.580$;

　　　(b)　$s = 0.512$;

　　　(c)　$[0, 3.877]$.

4.2-11　(a)　$\bar{x} = 25.475, s = 2.4935$; (b) $[24.059, \infty)$.

4.2-13　(a)　$[-115.480, 129.105]$; (c) no.

4.2-15　　　$\bar{x} - \bar{y} \pm t_{\alpha/2}(n+m-2)\sqrt{\dfrac{(n-1)s_x^2/d + (m-1)s_y^2}{n+m-2}\left(\dfrac{d}{n} + \dfrac{1}{m}\right)}$.

4.2-17     $n = 135$ or $n = 136$.

4.3-1   **(a)**   $1.4 < 1.645$, do not reject $H_0$;

     **(b)**   $1.4 > 1.282$, reject $H_0$.

4.3-3   **(a)**   $\alpha = 0.0478$; **(b)** $\bar{x} = 24.1225 > 22.5$, do not reject $H_0$.

4.3-5   **(a)**   $n = 25$, $c = 1.6$.

4.3-7     $c = 19.5$, $n = 164$ or $165$.

4.4-1   **(a)**   $t = 3.0 > 1.753$, reject $H_0$;

     **(b)**   Since $t_{0.005}(15) = 2.947$, the approximate $p$-value of this test is $0.005$.

4.4-3   **(a)**   $t = (\bar{x} - 47)/(s/\sqrt{20}) \leq -1.729$;

     **(b)**   $-1.789 < -1.729$, reject $H_0$;

     **(c)**   $0.025 < p$-value $< 0.05$, $p$-value $\approx 0.045$.

4.4-5     $1.477 < 1.833$, do not reject $H_0$.

4.4-7   **(a)**   $\chi^2 = 8.895 < 10.12$ so the company was successful.

     **(b)**   Since $\chi^2_{0.975}(19) = 8.907$, $p$-value $\approx 0.025$.

4.4-9   **(a)**   $0.3032$ using $b(100, 0.08)$, $0.313$ using Poisson approximation, $0.2902$ using normal approximation;

     **(b)**   $0.1064$ using $b(100, 0.04)$, $0.111$ using Poisson approximation, $0.1010$ using normal approximation.

4.4-11   **(a)**   $\alpha = 0.1056$, $\alpha = 0.1040$ using $b(192, 0.75)$;

     **(b)**   $\beta = 0.3524$, $\beta = 0.3467$ using $b(192, 0.8)$.

4.4-13     With $n = 130$, $c = 8.5$, $\alpha \approx 0.055$, $\beta \approx 0.094$.

4.5-1   **(a)**
$$t = \frac{\bar{x} - \bar{y}}{\sqrt{\dfrac{15s_x^2 + 12s_y^2}{27}\left(\dfrac{1}{16} + \dfrac{1}{13}\right)}} \geq t_{0.01}(27) = 2.473;$$

     **(b)**   $t = 5.570 > 2.473$, reject $H_0$.

     **(c)**   The critical region is

$$\frac{s_x^2}{s_y^2} \geq F_{0.025}(15, 12) = 3.18 \text{ or } \frac{s_y^2}{s_x^2} \geq F_{0.025}(12, 15) = 2.96.$$

Since $\dfrac{s_x^2}{s_y^2} = \dfrac{1356.75}{692.21} = 1.96 < 3.18$ and $\dfrac{s_y^2}{s_x^2} = 0.51 < 2.96$, accept $H_0$.

4.5-3   **(a)**   $t < -1.706$; **(b)** $-1.714 < -1.706$, reject $H_0$;

     **(c)**   $0.025 < p$-value $< 0.05$; **(e)** $1.836 < 3.28$, do not reject.

4.5-5   **(a)**   $\dfrac{4.88}{5.81} = 0.84 < 2.53 = F_{0.01}(24, 28)$,

$\dfrac{5.81}{4.88} = 1.19 < 2.91 = F_{0.01}(28, 24)$,

do not reject $\sigma_X^2 = \sigma_Y^2$;

     **(b)**   $3.402 > 2.326 = z_{0.01}$, reject $\mu_X = \mu_Y$.

4.5-7     $[0.007, 0.071]$, уез.

4.5-9   **(a)**   $z = \dfrac{\widehat{p}_1 - \widehat{p}_2}{\sqrt{\widehat{p}(1 - \widehat{p})(1/n_1 + 1/n_2)}} \geq 1.645;$

**(b)** $z = 2.341 > 1.645$, reject $H_0$.

**(c)** $z = 2.341 > 2.326$, reject $H_0$.

**(d)** The $p$-value $\approx P(Z \geq 2.341) = 0.0096$.

4.6-1 **(a)** $\widehat{y} = 86.8 + (842/829)(x - 74.5)$;

**(c)** $\widehat{\sigma^2} = 17.9998$.

4.6-3 **(a)** $r = 0.953, \widehat{y} = 0.9810 + 0.0249x$;

**(c)** quadratic regression is more appropriate.

4.6-5
$$P\left[ \chi^2_{1-\gamma/2}(n-2) \leq \frac{n\widehat{\sigma^2}}{\sigma^2} \leq \chi^2_{\gamma/2}(n-2) \right] = 1 - \gamma$$

$$P\left[ \frac{n\widehat{\sigma^2}}{\chi^2_{\gamma/2}(n-2)} \leq \sigma^2 \leq \frac{n\widehat{\sigma^2}}{\chi^2_{1-\gamma/2}(n-2)} \right] = 1 - \gamma.$$

4.6-7 **(a)** $\widehat{y} = 0.506x + 14.657$;

**(c)** $\widehat{\alpha} = 26.33, \widehat{\beta} = 0.506, \widehat{\sigma^2} = 14.126$;

**(d)** $[24.081, 28.585]$ for $\alpha$, $[0.044, 0.968]$ for $\beta$, $[8.566, 42.301]$ for $\sigma^2$.

4.6-9    $2.148 < 2.160$, do not reject $H_0$.

4.6-11    $\widehat{\beta} = (1/n)\sum_{i=1}^{n} y_i/x_i$;    $\widehat{\gamma^2} = (1/n)\sum_{i=1}^{n}(y_i - \widehat{\beta}x_i)^2/x_i^2$.

4.7-1 **(a)** $[75.283, 85.113]$, $[83.838, 90.777]$, $[89.107, 99.728]$;

**(b)** $[68.206, 92.190]$, $[75.833, 98.783]$, $[82.258, 106.577]$.

4.7-3 **(a)** $[62.793, 129.495]$, $[101.854, 135.266]$, $[119.483, 162.470]$;

**(b)** $[28.926, 163.361]$, $[57.856, 179.265]$, $[78.784, 203.169]$.

4.7-5 **(a)** $\widehat{y} = 208.467 + 140.479x$;

**(c)** $1753.733 \pm 160.368$ or $[1593.365, 1914.101]$.

4.7-7 **(b)** $\widehat{y} = 0.8387 + 0.4891x$;

**(c)** $\widehat{\beta} = 0.4891, t = 14.228 > 1.895$, reject $H_0$;

**(d)** $[45.926, 63.355]$.

4.8-1    $7.875 > 4.26$, reject $H_0$.

4.8-3    $F = 15.4 > 3.59$, reject $H_0$.

4.8-5 **(a)** $F \geq F_{0.05}(3, 24) = 3.01$;

**(b)**

| Source | SS | DF | MS | F | p-value |
|--------|-----|-----|-----|-----|---------|
| Treatment | 12,280.86 | 3 | 4,093.62 | 3.455 | 0.0323 |
| Error | 28,434.57 | 24 | 1,184.77 | | |
| Total | 40,715.43 | 27 | | | |

$F = 3.455 > 3.01$, reject $H_0$;

**(c)** $0.025 < p\text{-value} < 0.05$.

4.8-7        $10.224 > 4.26$, reject $H_0$.

4.8-9    **(a)**    $F \geq 5.61$; **(b)** $6.337 > 5.61$, reject $H_0$.

4.9-1    **(a)**    $(6.31, 7.40)$; **(b)** $(6.58, 7.22)$, $0.8204$.

4.9-3    **(a)**    $(y_3 = 5.2, y_{10} = 6.6)$; **(b)** $(y_1 = 4.9, y_7 = 6.2)$, $0.9476$.

4.9-5        $(15.40, 17.05)$.

4.9-7    **(a)**

| Stems | Leaves | Frequencies | Depths |
|-------|--------|-------------|--------|
| 0.8● | 7 8 9 | 3 | 3 |
| 0.9* | 0 0 1 2 2 2 3 4 | 8 | 11 |
| 0.9● | 5 5 5 7 8 8 8 8 9 | 9 | (9) |
| 1.0* | 0 0 1 2 | 4 | 5 |
| 1.0● | 6 | 1 | 1 |

 **(b)**    $\tilde{m} = 0.95$;

 **(c)**    $(y_8 = 0.92, y_{18} = 0.98)$, $0.9568$;

 **(d)**    $\tilde{q}_1 = 0.915$;

 **(e)**    $(y_3 = 0.89, y_{10} = 0.93)$, $0.8966$;

 **(f)**    $\tilde{q}_3 = 0.985$;

 **(g)**    $(y_{15} = 0.97, y_{23} = 1.01)$, $0.9382$.

4.9-9    **(a)**    $\gamma = 0.9992$; **(b)** $\gamma \approx 0.9999$.

4.9-11        $0.8852$.

4.10-1        $6.25 < 7.815$, do not reject if $\alpha = 0.05$; $p$-value $\approx 0.10$.

4.10-3        $202.4 > 21.67$, reject at $\alpha = 0.01$.

4.10-5        $q_3 \geq 7.815$; **(b)** $q_3 = 1.744 < 7.815$; do not reject $H_0$.

4.10-7        $2.010 < 11.07 = \chi^2_{0.05}(5)$, do not reject the null hypothesis.

4.10-9    **(a)**    Frequencies are 2, 4, 21, 33, 20, 14, 4, 1, 1;

 **(b)**    $q = 4.653 < 9.488$, do not reject $H_0$.

4.11-1        $3.23 < 11.07$, do not reject $H_0$.

4.11-3        $q = 10.4 > 9.488$, reject equality of three distributions.

4.11-5        $8.449 < 9.488$, do not reject the null hypothesis.

4.11-7        $23.78 > 21.03$, reject hypothesis of independence.

4.11-9        $9.925 < 12.59$, do not reject independence.

## CHAPTER 5

**5.1-1** **(a)** 0.7642; **(b)** 0.1781.

**5.1-3** **(a)** 0.999013; **(b)** 0.435965; **(c)** 0.413019.

**5.1-5** **(a)**
```
Stem-and-leaf of TestScores   N = 50
Leaf Unit = 1.0
    2      3 48
    6      4 2257
   13      5 1247889
   23      6 0135567899
  (13)     7 0112334556679
   14      8 112334457
    5      9 01337


Stem-and-leaf of TestScores   N = 50
Leaf Unit = 1.0
    1      3 4
    2      3 8
    4      4 22
    6      4 57
    9      5 124
   13      5 7889
   16      6 013
   23      6 5567899
   (7)     7 0112334
   20      7 556679
   14      8 1123344
    7      8 57
    5      9 0133
    1      9 7
```

**(b)** and **(c)**

```
Descriptive Statistics: TestScores
```

| Variable | N | Mean | Median | TrMean | StDev | SE Mean |
|----------|-----|-------|--------|--------|-------|---------|
| TestScor | 50 | 69.32 | 71.00 | 69.75 | 15.25 | 2.16 |

| Variable | Minimum | Maximum | Q1 | Q3 |
|----------|---------|---------|-------|-------|
| TestScor | 34.00 | 97.00 | 58.75 | 81.25 |

**5.1-7** **(a)** 2.32635; **(b)** 1.64485; **(c)** −1.95996.

**5.2-1** **(b)** [2.3034, 4.6496]; **(c)** 2.3462 < 2.49.

**5.2-3** $g(x,y) = (1/2\pi)e^{-(x^2+y^2)/2}, \qquad -\infty < x < \infty, \qquad -\infty < y < \infty;$

$F(u) = \int_0^\infty \int_{-\infty}^{uy} g(x,y)\,dx\,dy + \int_{-\infty}^0 \int_{uy}^\infty g(x,y)\,dx\,dy;$

$f(u) = F'(u) = \dfrac{1}{\pi(1+u^2)}, \qquad -\infty < u < \infty.$

**5.2-5** $g(y) = 2y, \qquad 0 < y < 1.$

**5.2-7** $g(y) = (1/2)e^{-y/2}, \qquad 0 < y < \infty.$

**5.3-1**  **(a)**  0.9309, 0.4827, 0.7997, 0.9496, 0.4310, 0.7767, 0.8600, 0.6503, 0.7480, 0.7921, 0.9663, 0.9693, 0.7592, 0.7192, 0.8301, 0.8128, 0.9961, 0.9825, 0.8726, 0.9572;

**(b)**  $\mu = 4/5 = 0.80, \bar{x} = 0.8143$;

**(c)**  $\sigma^2 = 2/75 = 0.0267, s^2 = 0.0247$.

**5.3-3**  **(a)**  $F(x) = x^2, \qquad 0 < x < 1$;

**(b)**  Let the distribution of $Y$ be $U(0,1)$. Then $x = \sqrt{y}$ is an observation of $X$ when $y$ is an observation of $Y$.

**5.3-5**  **(a)**  $1/4 = 0.25$; **(b)** 0.0628; **(c)** 0.0317.

**5.3-7**  **(a)**  $\dfrac{20!}{(20-r)!\,20^r}, \qquad r = 1, 2, \dots, 20$;

**(h)**  $P(r) = 1 - \dfrac{20!}{(20-r)!\,20^r} \qquad r = 2, 3, \dots, 21$;

**(d)**  **(i)** 5, **(ii)** 6;

**(g)**  $g(y) = \dfrac{20!(y-1)}{(21-y)!\,20^y}, \qquad y = 2, 3, \dots, 21$;

**(h)**  $\mu = 6.2936, \sigma^2 = 6.6844, \sigma = 2.5854$.

## CHAPTER 6

**6.1-1**  **(a)**  **(i)** $b(5, 0.7)$; **(ii)** $\mu = 3.5, \sigma^2 = 1.05$; **(iii)** 0.1607;

**(b)**  **(i)** geometric, $p = 0.3$; **(ii)** $\mu = 10/3, \sigma^2 = 70/9$; **(iii)** 0.51;

**(c)**  **(i)** Bernoulli, $p = 0.55$; **(ii)** $\mu = 0.55, \sigma^2 = 0.2475$; **(iii)** 0.55;

**(d)**  **(ii)** $\mu = 2.1, \sigma^2 = 0.89$; **(iii)** 0.7;

**(e)**  **(i)** negative binomial, $p = 0.6, r = 2$; **(ii)** $\mu = 10/3, \sigma^2 = 20/9$; **(iii)** 0.36;

**(f)**  **(i)** discrete uniform on $1, 2, \dots, 10$; **(ii)** 5.5, 8.25; **(iii)** 0.2.

**6.1-5**  $M(t) = (1 - \theta t)^{-\alpha}, t < 1/\theta; \; R(t) = -\alpha \ln(1 - \theta t); \; \mu = \alpha\theta; \; \sigma^2 = \alpha\theta^2$.

**6.1-7**  $M(t) = (1 - p + pe^t)^n; \; \mu = np; \; \sigma^2 = np(1-p)$.

**6.2-1**  $M(t) = (1 - p + pe^t)^{n_1 + n_2}; b(n_1 + n_2, p)$.

**6.2-7**  **(a)**  0.4772; **(b)** 0.8561.

**6.2-9**  0.925.

**6.2-11**  0.6726.

**6.2-13**  0.4120.

**6.3-1**  0.947.

**6.3-3**  0.6853.

**6.3-5**  **(a)**  $\mu = 2/3, \sigma^2 = 2/9$; **(b)** 0.4332.

**6.3-7**  **(a)**  0.7282.

**6.4-1**  **(a)**  $g_1(y_1) = ne^{-n(y_1 - \theta)}, \qquad \theta \le y_1 < \infty$.

**6.4-3**  **(c)**  $c = 0.224$.

# INDEX

# Discrete Distributions

**Bernoulli**
$0 < p < 1$

$$f(x) = p^x(1-p)^{1-x}, \qquad x = 0, 1$$
$$M(t) = 1 - p + pe^t$$
$$\mu = p, \qquad \sigma^2 = p(1-p)$$

**Binomial**
$b(n, p)$
$0 < p < 1$

$$f(x) = \frac{n!}{x!(n-x)!} p^x(1-p)^{n-x}, \qquad x = 0, 1, 2, \ldots, n$$
$$M(t) = (1 - p + pe^t)^n$$
$$\mu = np, \qquad \sigma^2 = np(1-p)$$

**Geometric**
$0 < p < 1$

$$f(x) = (1-p)^{x-1}p, \qquad x = 1, 2, 3, \ldots$$
$$M(t) = \frac{pe^t}{1 - (1-p)e^t}, \qquad t < -\ln(1-p)$$
$$\mu = \frac{1}{p}, \qquad \sigma^2 = \frac{1-p}{p^2}$$

**Hypergeometric**
$N_1 > 0, \ N_2 > 0$
$N = N_1 + N_2$

$$f(x) = \frac{\binom{N_1}{x}\binom{N_2}{n-x}}{\binom{N}{n}}, \qquad x \le n, x \le N_1, n - x \le N_2$$
$$\mu = n\left(\frac{N_1}{N}\right), \qquad \sigma^2 = n\left(\frac{N_1}{N}\right)\left(\frac{N_2}{N}\right)\left(\frac{N-n}{N-1}\right)$$

**Negative Binomial**
$0 < p < 1$
$r = 1, 2, 3, \ldots$

$$f(x) = \binom{x-1}{r-1} p^r(1-p)^{x-r}, \qquad x = r, r+1, r+2, \ldots$$
$$M(t) = \frac{(pe^t)^r}{[1 - (1-p)e^t]^r}, \qquad t < -\ln(1-p)$$
$$\mu = r\left(\frac{1}{p}\right), \qquad \sigma^2 = \frac{r(1-p)}{p^2}$$

**Poisson**
$0 < \lambda$

$$f(x) = \frac{\lambda^x e^{-\lambda}}{x!}, \qquad x = 0, 1, 2, \ldots$$
$$M(t) = e^{\lambda(e^t - 1)}$$
$$\mu = \lambda, \qquad \sigma^2 = \lambda$$

**Uniform**
$m > 0$

$$f(x) = \frac{1}{m}, \qquad x = 1, 2, \ldots, m$$
$$\mu = \frac{m+1}{2}, \qquad \sigma^2 = \frac{m^2 - 1}{12}$$

# Continuous Distributions

**Beta**
$0 < \alpha$
$0 < \beta$

$$f(x) = \frac{\Gamma(\alpha + \beta)}{\Gamma(\alpha)\Gamma(\beta)} x^{\alpha-1}(1 - x)^{\beta-1}, \qquad 0 < x < 1$$

$$\mu = \frac{\alpha}{\alpha + \beta}, \qquad \sigma^2 = \frac{\alpha\beta}{(\alpha + \beta + 1)(\alpha + \beta)^2}$$

**Chi-square**
$\chi^2(r)$
$r = 1, 2, \ldots$

$$f(x) = \frac{1}{\Gamma(r/2)2^{r/2}} x^{r/2-1}e^{-x/2}, \qquad 0 \le x < \infty$$

$$M(t) = \frac{1}{(1 - 2t)^{r/2}}, \qquad t < \frac{1}{2}$$

$$\mu = r, \qquad \sigma^2 = 2r$$

**Exponential**
$0 < \theta$

$$f(x) = \frac{1}{\theta} e^{-x/\theta}, \qquad 0 \le x < \infty$$

$$M(t) = \frac{1}{1 - \theta t}, \qquad t < \frac{1}{\theta}$$

$$\mu = \theta, \qquad \sigma^2 = \theta^2$$

**Gamma**
$0 < \alpha$
$0 < \theta$

$$f(x) = \frac{1}{\Gamma(\alpha)\theta^\alpha} x^{\alpha-1}e^{-x/\theta}, \qquad 0 \le x < \infty$$

$$M(t) = \frac{1}{(1 - \theta t)^\alpha}, \qquad t < \frac{1}{\theta}$$

$$\mu = \alpha\theta, \qquad \sigma^2 = \alpha\theta^2$$

**Normal**
$N(\mu, \sigma^2)$
$-\infty < \mu < \infty$
$0 < \sigma$

$$f(x) = \frac{1}{\sigma\sqrt{2\pi}} e^{-(x-\mu)^2/2\sigma^2}, \qquad -\infty < x < \infty$$

$$M(t) = e^{\mu t + \sigma^2 t^2/2}$$

$$E(X) = \mu, \qquad \mathrm{Var}(X) = \sigma^2$$

**Uniform**
$U(a, b)$
$a < b$

$$f(x) = \frac{1}{b - a}, \qquad a \le x \le b$$

$$M(t) = \frac{e^{tb} - e^{ta}}{t(b - a)}, \qquad t \ne 0; \qquad M(0) = 1$$

$$\mu = \frac{a + b}{2}, \qquad \sigma^2 = \frac{(b - a)^2}{12}$$

# Confidence Intervals

| Parameter | Assumptions | Endpoints |
|---|---|---|
| $\mu$ | $N(\mu, \sigma^2)$ or $n$ large, $\sigma^2$ known | $\bar{x} \pm z_{\alpha/2} \dfrac{\sigma}{\sqrt{n}}$ |
| $\mu$ | $N(\mu, \sigma^2)$ $\sigma^2$ unknown | $\bar{x} \pm t_{\alpha/2}(n-1)\dfrac{s}{\sqrt{n}}$ |
| $\mu_X - \mu_Y$ | $N(\mu_X, \sigma_X^2)$ $N(\mu_Y, \sigma_Y^2)$ $\sigma_X^2, \sigma_Y^2$ known | $\bar{x} - \bar{y} \pm z_{\alpha/2}\sqrt{\dfrac{\sigma_X^2}{n} + \dfrac{\sigma_Y^2}{m}}$ |
| $\mu_X - \mu_Y$ | Variances unknown, large samples | $\bar{x} - \bar{y} \pm z_{\alpha/2}\sqrt{\dfrac{s_x^2}{n} + \dfrac{s_y^2}{m}}$ |
| $\mu_X - \mu_Y$ | $N(\mu_X, \sigma_X^2)$ $N(\mu_Y, \sigma_Y^2)$ $\sigma_X^2 = \sigma_Y^2$, unknown | $\bar{x} - \bar{y} \pm t_{\alpha/2}(n+m-2)s_p\sqrt{\dfrac{1}{n} + \dfrac{1}{m}},$ $s_p = \sqrt{\dfrac{(n-1)s_x^2 + (m-1)s_y^2}{n+m-2}}$ |
| $\mu_D = \mu_X - \mu_Y$ | $X$ and $Y$ normal, but dependent | $\bar{d} \pm t_{\alpha/2}(n-1)\dfrac{s_d}{\sqrt{n}}$ |
| $\sigma^2$ | $N(\mu, \sigma^2)$ | $\dfrac{(n-1)s^2}{\chi_{\alpha/2}^2(n-1)}, \dfrac{(n-1)s^2}{\chi_{1-\alpha/2}^2(n-1)}$ |
| $\dfrac{\sigma_X^2}{\sigma_Y^2}$ | $N(\mu_X, \sigma_X^2)$ $N(\mu_Y, \sigma_Y^2)$ | $\dfrac{s_x^2/s_y^2}{F_{\alpha/2}(n-1, m-1)}, F_{\alpha/2}(m-1, n-1)\dfrac{s_x^2}{s_y^2}$ |
| $p$ | $b(n, p)$ $n$ is large | $\dfrac{y}{n} \pm z_{\alpha/2}\sqrt{\dfrac{(y/n)(1 - y/n)}{n}}$ |
| $p_1 - p_2$ | $b(n_1, p_1)$ $b(n_2, p_2)$ $n_1, n_2$ large | $\dfrac{y_1}{n_1} - \dfrac{y_2}{n_2} \pm z_{\alpha/2}\sqrt{\dfrac{\hat{p}_1(1 - \hat{p}_1)}{n_1} + \dfrac{\hat{p}_2(1 - \hat{p}_2)}{n_2}},$ $\hat{p}_1 = y_1/n_1, \ \hat{p}_2 = y_2/n_2$ |

# Tests of Hypotheses

| Hypotheses | Assumptions | Critical Region |
|---|---|---|
| $H_0: \mu = \mu_0$ <br> $H_1: \mu > \mu_0$ | $N(\mu, \sigma^2)$ or $n$ large, <br> $\sigma^2$ known | $z = \dfrac{\bar{x} - \mu_0}{\sigma/\sqrt{n}} \geq z_\alpha$ |
| $H_0: \mu = \mu_0$ <br> $H_1: \mu > \mu_0$ | $N(\mu, \sigma^2)$ <br> $\sigma^2$ unknown | $t = \dfrac{\bar{x} - \mu_0}{s/\sqrt{n}} \geq t_\alpha(n-1)$ |
| $H_0: \mu_X - \mu_Y = 0$ <br> $H_1: \mu_X - \mu_Y > 0$ | $N(\mu_X, \sigma_X^2)$ <br> $N(\mu_Y, \sigma_Y^2)$ <br> $\sigma_X^2, \sigma_Y^2$ known | $z = \dfrac{\bar{x} - \bar{y} - 0}{\sqrt{\sigma_X^2/n + \sigma_Y^2/m}} \geq z_\alpha$ |
| $H_0: \mu_X - \mu_Y = 0$ <br> $H_1: \mu_X - \mu_Y > 0$ | Variances unknown, <br> large samples | $z = \dfrac{\bar{x} - \bar{y} - 0}{\sqrt{s_x^2/n + s_y^2/m}} \geq z_\alpha$ |
| $H_0: \mu_X - \mu_Y = 0$ <br> $H_1: \mu_X - \mu_Y > 0$ | $N(\mu_X, \sigma_X^2)$ <br> $N(\mu_Y, \sigma_Y^2)$ <br><br> $\sigma_X^2 = \sigma_Y^2$, unknown | $t = \dfrac{\bar{x} - \bar{y} - 0}{s_p\sqrt{1/n + 1/m}} \geq t_\alpha(n+m-2)$ <br><br> $s_p = \sqrt{\dfrac{(n-1)s_x^2 + (m-1)s_y^2}{n+m-2}}$ |
| $H_0: \mu_D = \mu_X - \mu_Y = 0$ <br> $H_1: \mu_D = \mu_X - \mu_Y > 0$ | $X$ and $Y$ normal, <br> but dependent | $t = \dfrac{\bar{d} - 0}{s_d/\sqrt{n}} \geq t_\alpha(n-1)$ |
| $H_0: \sigma^2 = \sigma_0^2$ <br> $H_1: \sigma^2 > \sigma_0^2$ | $N(\mu, \sigma^2)$ | $\chi^2 = \dfrac{(n-1)s^2}{\sigma_0^2} \geq \chi_\alpha^2(n-1)$ |
| $H_0: \sigma_X^2/\sigma_Y^2 = 1$ <br> $H_1: \sigma_X^2/\sigma_Y^2 > 1$ | $N(\mu_X, \sigma_X^2)$ <br> $N(\mu_Y, \sigma_Y^2)$ | $F = \dfrac{s_x^2}{s_y^2} \geq F_\alpha(n-1, m-1)$ |
| $H_0: p = p_0$ <br> $H_1: p > p_0$ | $b(n, p)$ <br> $n$ is large | $z = \dfrac{y/n - p_0}{\sqrt{p_0(1 - p_0)/n}} \geq z_\alpha$ |
| $H_0: p_1 - p_2 = 0$ <br> $H_1: p_1 - p_2 > 0$ | $b(n_1, p_1)$ <br> $b(n_2, p_2)$ <br> $n_1, n_2$ large | $z = \dfrac{y_1/n_1 - y_2/n_2 - 0}{\sqrt{\left(\dfrac{y_1 + y_2}{n_1 + n_2}\right)\left(1 - \dfrac{y_1 + y_2}{n_1 + n_2}\right)\left(\dfrac{1}{n_1} + \dfrac{1}{n_2}\right)}} \geq z_\alpha$ |